BIOLOGY for TODAY

Complete, single volume, revised edition

Ernest G. Neal
MBE, M.Sc., Ph.D., F.I.Biol.
formerly Second Master and Head of the Science Department
Taunton School

and

Keith R. C. Neal
M.A., F.I.Biol.
Head of the Biology Department
Manchester Grammar School

Blandford Press
Poole Dorset

First published as two volumes in the U.K. 1974 by
Blandford Press, Link House, West Street,
Poole, Dorset, BH15 1LL.

First published as a composite edition in the U.K. in 1983.

British Library Cataloguing in Publication Data

Neal, Ernest G.
Biology for today.—Rev. and updated ed
1. Biology
I. Title II. Neal, Keith R. C.
574 QH308.7

ISBN 0 7137 1248 1

Typeset by Asco Trade Typesetting Ltd., Hong Kong
Printed by South China Printing Co., Hong Kong

Contents

Authors' Preface

This new, extensively revised and re-designed edition of *Biology for Today* aims to meet the requirements of students preparing for first examinations in Biology.

We are convinced that any secondary school course in biology should be educational in the widest sense of the word and should provide future citizens with a better understanding of themselves and the world in which they live. For this reason considerable emphasis has been given to Man and to the application of biological principles to human affairs at all levels—personal, social and world-wide. We have also given emphasis to such subjects as health, behaviour, ecology and conservation, because of their great interest and relevance today.

Special attention has been given to the language used in the text in order to make it easily understood by students within the O. L.—C.S.E. range of abilities. The sequence of topics is designed to reflect the changing interests and powers of comprehension of the student which are associated with increasing age and experience. The reading age has been independently assessed as 14 years for the first half of the book and 15 years for the second.

The major changes in this edition have been made after extensive consultation with teachers. We believe that publication of the text in one volume will allow teachers greater flexibility in choosing the order of subjects taught and will make cross-referencing easier. Significant changes have also been made in the treatment of practical investigations. Suggestions for these are included at the appropriate places within the text but all factual information has now been confined to the main body of the text, thus making the practical instructions more precise. We hope that these changes will make it easier for teachers to decide which practicals they wish to use.

We would like to thank all those teachers who have made valuable suggestions for this new edition; also Messrs John Haller and Stephen Wood of Philip Harris Biological Ltd for providing many of the photographs and for their enthusiastic co-operation over the whole project; also to Barry Jones and Marion Mills for the original line drawings and to Ray Hollidge and Mike O'Malley of Chartwell Illustrators for re-drawing them for this edition.

1

Introduction to biology

What is biology?

Biology is the study of life—the study of all living things. Because we are alive, it also includes the study of ourselves. We share this amazing characteristic of being alive with bacteria and fungi, ferns and worms, buttercups and frogs, fish and spiders, horses and lions.

One striking fact about living organisms is their fantastic variety. They range from the microscopic plants and animals found in a drop of pond water to the giant redwood trees of California and the majestic blue whales of the Antarctic ocean. Some form of life is found practically everywhere on earth, even in unlikely places such as hot springs and deserts. But no matter how deeply we study animals and plants, life itself still retains an element of mystery and grandeur which, perhaps, we will never fully understand.

Why study biology?

Biology brings a new appreciation of life. It helps to open our eyes to the amazing variety and marvellous complexity of living things. It stirs our imagination and increases our sense of wonder. Look down a microscope at a drop of pond water and a new world comes into view; a world of strange, exciting organisms, each a marvel of intricate construction. A swallow flying overhead may appear to some as just another bird, but think of its perfect command of flight in all weathers and the navigational skill which enables it to migrate many thousands of kilometres and return the next year to the exact place where it had previously nested. Watch a bee busily collecting pollen from a flower. Is it just another insect? An insect, yes, but one that is capable of steering by sun-compass to track down a patch of flowers and then pass on to other bees the exact location of that source of nectar.

We often marvel at our own technology, and rightly so, but it is a humbling thought that many of man's achievements are often clumsy imitations of phenomena found in living things. For example, millions of years before man discovered radar, bats were using a very similar technique for locating objects in the dark. Man still has much to learn from nature. Biology is a young science and research prospects for the further benefit of mankind are limitless. There is still much to discover and much to wonder at.

Those of you who hope one day to become doctors or nurses, veterinary surgeons or farmers, horticulturalists or foresters, pharmacists or bacteriologists, naturalists or conservationists, to give just a few examples, will find that your work is concerned with the application of biological principles. But whatever job we do in life, biology is most important because it helps us to understand ourselves. Studying the structure of animals and the way they live helps us to understand how our *own* bodies work; observing the behaviour of animals and how they react with each other, and with their environment, helps us to understand why *we* behave as we do.

Man, like other animals, is basically dependent on plants for food; his survival depends on the success of his crops. Every day we use products made from substances acquired from plants and animals. Many of our clothes are made from cotton or wool and the leather for our shoes originates from the hides of animals; our newspapers are made from wood pulp and the tyres of cars and bicycles are manufactured from the latex of rubber trees. We need a home in which to live and bring up a family. Think of your own home and see how much of it originated from plants—wood is one of our basic raw materials. Although synthetic substances are replacing some of these products it is clear that man cannot do without animals and plants. They are essential for his existence.

Without an understanding of how plants and animals live, man is in great danger of over-exploiting them. Already irreparable damage has been done and man has only just begun to realise that subduing nature is not enough; in order to survive he must live in harmony with it. This is not going to be easy as his rapid rise in numbers and the increase in his standard of living are putting more and

Fig. 1 : 1 Biology includes the study of ourselves.

more strain on available resources. To succeed, biological understanding will be essential. It is an exciting challenge to all of us.

What have all living organisms in common?

Although there is so much diversity among living organisms there are certain characteristic features which they all have in common, and this, of course, applies to us too. They can be summarised as follows:

1. Nutrition. All plants and animals need **food** or the raw materials for making it. The ways in which a grass, a mushroom and a cow obtain it are all different, but the function of the food is the same: to provide energy for all living processes and to provide the raw materials necessary for growth, for the replacement of worn-out parts of the body and for the reproduction of the species. Nutrition is the term used to describe both the taking in of food (feeding) and the subsequent chemical changes that take place before it is used. Green plants feed differently from other living things as they can use light energy to build up sugar from carbon dioxide and water and to build up other complex foods with the help of simple salts. Other organisms have to use food already made by plants or other animals and have to digest it first before it can be absorbed into the body. Some bacteria and blue-green algae have their own unique methods of nutrition.

2. Respiration. This is the process whereby energy is released from food. For this to happen **oxygen** is usually required and **carbon dioxide** and **water** are given off as by-products. The methods of obtaining oxygen vary a great deal; for example when a fish pumps water over its gills the oxygen in the water diffuses into the blood which circulates through them, but in a jellyfish or a tree no special movements are apparent. The important point is that respiration, the release of energy, takes place in every cell of the body—in both plants and animals.

3. Excretion. The chemical processes, such as respiration, taking place within the cells and tissues of an organism are all described by the term **metabolism**. One result of these processes is that waste substances are produced. If they built up to too high a level they would be harmful, so through the process of excretion these substances are removed.

Excretion can therefore be defined as the removal of the waste products of metabolism. It differs from **defaecation**, the removal of waste from the gut, in that most of this waste matter has not originated from processes taking place within the cells of the body. Substances in the urine, however, are truly excretory as they have been formed in the body and removed from the blood by the kidneys.

4. Growth. This is the permanent increase in size of an organism due to the formation of new living matter or **protoplasm**. For this to take place there must be an adequate supply of food. Except for the simplest organisms, growth also leads to an increase in complexity. Thus the germination and growth of an acorn lead eventually to a very complex oak tree. It is incredible, too, to think that each one of us has grown from just one fertilized egg cell about one-fifth of a millimetre in diameter!

5. Reproduction. This is the process whereby life itself is passed on from one generation to another. Living organisms cannot exist indefinitely—the bodies of animals and plants eventually die through old age, disease, accident or the inability to cope with environmental conditions. Reproduction usually takes place when the organism has reached its maximum size, although you will be able to think of exceptions. There are special mechanisms for ensuring that the characteristics of the parents are passed on to the offspring. Thus the eggs of a robin always hatch into robins rather than into any other kind of bird.

6. Irritability. In order for organisms to survive (and thus to reproduce) they must be able to respond to changes in their environment, a characteristic known as irritability (or sensitivity). This involves being able to detect a particular **stimulus**, such as light, and to make an appropriate **response** to it. You may have noticed that when you expose a plant to one-sided illumination its shoot responds very slowly by bending towards the light; this is a growth response. But if you shine a light on to an earthworm half out of its burrow at night, it rapidly disappears; the response in this case is one of movement. Movement is such a characteristic response to a stimulus, especially in animals, that some biologists prefer to consider it separately as a seventh characteristic of all living things, but we consider it more logical to retain it as an example of irritability.

The examples given above of responses to stimuli are advantageous to the organism. If a plant shoot did not grow towards the light it might eventually die, because light is essential for the manufacture of its food. Similarly, if an earthworm did not quickly disappear down its burrow when dawn came it would soon be eaten by a bird. You might think what other stimuli animals and plants respond to and whether these responses help them to survive. Consider also your own responses to stimuli.

Fig. 1:2 Flowers in abundance.

Living and non-living things

Some inanimate objects appear to show some of the characteristics of living things. A computer will 'respond' when certain 'stimuli' are fed into it; a crystal of copper sulphate will 'grow' when kept in a saturated solution of the salt. A

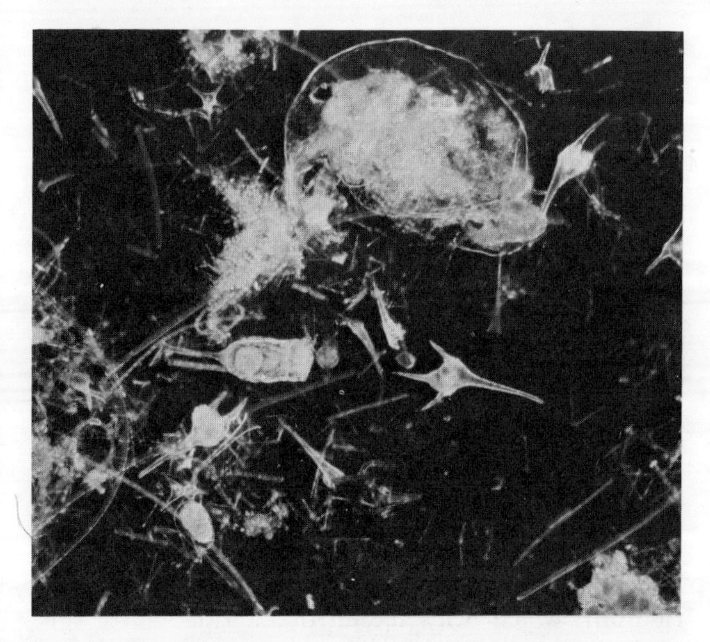

Fig. 1:3 Some of the forms of life found in a drop of lake water.

motor car is 'fed' with petrol, it 'respires' it with the help of oxygen (combustion), the energy released is used to propel the car, and the waste products are 'excreted' through the exhaust system. However, the fundamental difference between living and non-living things is that a living organism controls all the processes itself. For example, in the case of the car, somebody has to fill the petrol tank, turn on the engine and drive it along. Even the incredibly complex and seemingly automatic movements of robots are ultimately controlled by the men who write the computer programs.

Viruses—living or non-living?

It is not easy to say whether viruses are living or non-living. Some can be extracted, crystallised and stored in a bottle indefinitely. They are minute particles, so small that they can only be seen with the electron microscope. They will reproduce, but only when they are inside a cell of a living organism; it seems that they take over the chemical machinery of the cell and use it to make more of themselves. These peculiar structures which only show some of the characteristics of living things will be considered further in Chapter 19.

Branches of biology

Because the science of biology covers such a wide field, various terms are used to describe particular areas of investigation.

Taxonomy is concerned with classification, that is, the sorting out of animals and plants into groups according to their common characteristics. The term **morphology** is often used to describe the general form of an organism, but internal morphology is more commonly known as **anatomy**. Human anatomy, for example, deals with the structure and arrangement of all the parts of the body such as bones, blood vessels, nerves and muscles. In contrast, **physiology** is concerned with how these organs function, e.g. how a muscle contracts. **Biochemistry** is really a branch of physiology but concentrates more on the chemistry of these processes. The study of the early development of a plant or animal is known as **embryology**. In man this would cover the period from the fertilization of the egg to birth. The science of animal behaviour is known as **ethology**, whereas the study of the relationships between animals and plants and their environment comprises **ecology**. **Evolution** is the branch of biology which attempts to explain how new kinds of organisms may have originated. This is closely linked with the study of heredity, or **genetics**, which is concerned with the way the characteristics of organisms are passed on from one generation to another.

2

Cells and organisms

How are plants and animals constructed?

In 1665, **Robert Hooke** described how he cut some very thin slices of cork and examined them under his extremely simple microscope. He found that its structure resembled a honeycomb, so he called the units of which it was composed **cells**. He also looked at thin sections of plant stems and roots and found that they also consisted of cells, though not always of the same shape. Later it was discovered that the bodies of animals were also made of cells. Our bodies are made of billions of them; a single drop of blood contains about 5 million! Cells can be described as the units of living matter, the microscopic 'bricks' from which animals and plants are made.

Although cells vary widely in shape and size they are remarkably similar in basic structure. Being alive, they all contain protoplasm, which is usually present in two forms, a rounded structure, the **nucleus**, surrounded by an almost transparent jelly-like substance, the **cytoplasm**. The nucleus controls the activities of the cell and plays an important part when cells reproduce (Chapter 25), while the cytoplasm carries out all the complex chemical processes going on in the cell.

Examine various kinds of cells for yourself.

1. Onion cells. Cut an onion into four quarters. It consists largely of white scale leaves which easily separate. It is the thin skin or **epidermis** covering the inner surface of one of these scale leaves that you should examine. You will need two pieces about 5 mm square. Prepare the first piece by cutting the epidermis with a sharp scalpel; then peel it off carefully with forceps and mount it in a drop of water on a slide. Prod it gently to remove any air bubbles which may cling to it and place a coverslip on top (Fig. 2:1). Repeat for the second square, but mount it in iodine solution instead of water.

Examine the preparation in water first. Have all the cells the same number of sides? Do the walls appear to be rigid? Can you see under the high power any sign of the living protoplasm? Now examine your iodine preparation. Do the nuclei and cytoplasm show up clearer? Compare your preparation with Fig. 2:2a. Does it show up better? Biologists use many kinds of stains to make things clearer. Draw 2 or 3 cells carefully.

2. Moss cells. Mount the leaf of a moss in a drop of water and examine its cells. Do you notice an important difference from the cells of the onion? The large number of rounded green bodies which make the whole leaf look green are called **chloroplasts** (Fig. 2:2b). They are concerned with the making of food substances and are typical of most plant cells which are exposed to light. Make a drawing of a few of these cells.

3. Cheek cells. As an example of animal cells, examine some of your own. Cells lining the inside of the cheek come off readily when rubbed gently. First sterilise the *handle* of a scalpel. If it is made of metal, pass it through a flame several times and wait until it is cool. If it is made of wood, dip it in alcohol and allow the latter to evaporate. Now use it to rub the inside of your cheek. Put a small drop of saliva on a slide and stir the scalpel handle in it to get off some of the cells. Mount in iodine as before and examine under a microscope. Some cells will be separate, others will still be attached in groups (Fig. 2:2c). Draw some of them under the high power and compare them with the plant cells.

4. Examine specially prepared slides of thin sections of parts of plants and animals.

Although the nucleus and cytoplasm look very simple under the microscope, a photograph taken under an electron microscope (Fig. 2:3) reveals that cytoplasm is extremely complex and is made up of structures called **organelles** which carry out specific functions. These include rod-shaped **mitochondria** concerned with respiration; a network of membranes, the **endoplasmic reticulum**, which helps the distribution of soluble substances within the cell; and **ribosomes** which are attached to the membranes and manufacture proteins. Non-living substances temporarily present in cells include oil globules, starch grains and excretory granules.

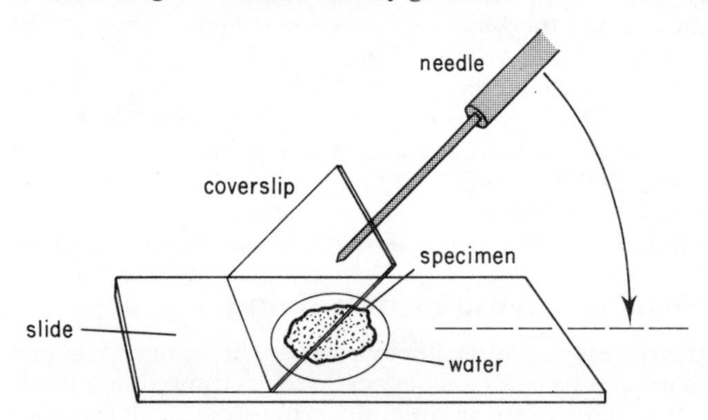

Fig. 2:1 Method of lowering a coverslip on to a preparation.

How do plant and animal cells differ?

Typical plant cells have rigid cell walls made of a non-living organic substance called **cellulose**, while animal cells merely have a very thin, flexible membrane, the **plasma membrane**, which is part of the living cytoplasm. Plant cells usually have spaces or **vacuoles** in them which are filled with a watery fluid called **cell sap**; if they are present in animal cells they are usually so small that they cannot easily be seen under an ordinary microscope. Animal cells contain no chloroplasts while the majority of plant cells do; but those deep inside stems or roots have none as light is necessary for their formation.

Why are cells not all alike?

When a house is built, different materials are used in certain places to serve different purposes; likewise, the cells of the body of an organism differ according to their function. Thus in our own body, **muscle** cells are able to contract and relax and are used in movement; **nerve** cells are able to receive and send out nervous impulses through their long processes; cells from the lining of the stomach secrete digestive juices on to the food, and **sperms** are motile cells

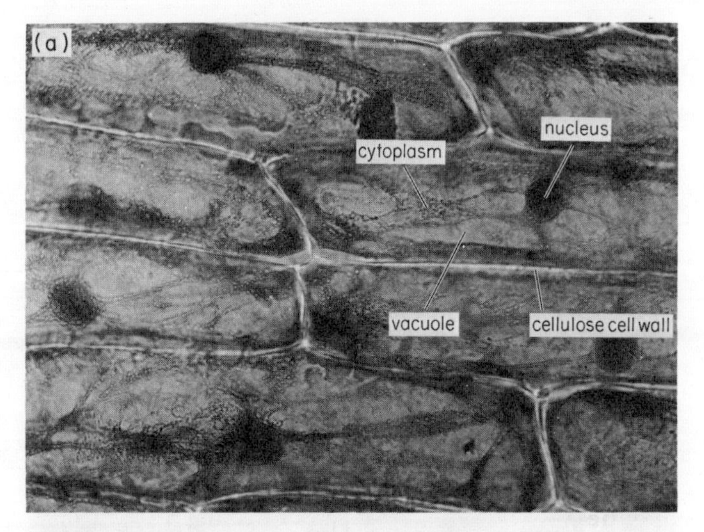

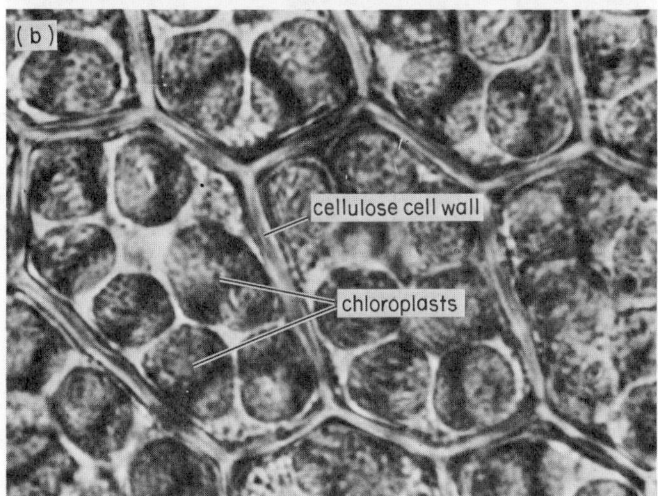

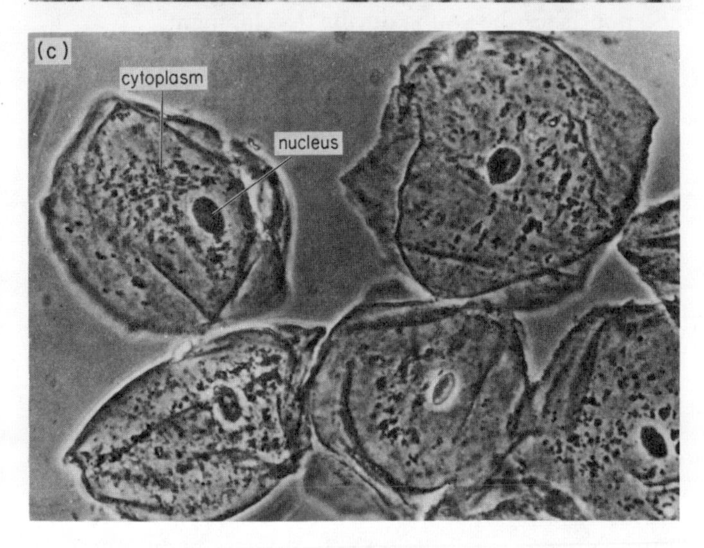

Fig. 2:2 High power photomicrographs of various cells: a) from the epidermis of a scale leaf of an onion bulb b) from a moss leaf c) from the lining of the cheek.

which have a long tail with which they can swim towards an egg and fertilize it (Fig. 2:4). Similarly, in plants there are **mesophyll** cells in the leaf which use their chloroplasts to build up food, **parenchyma** cells which act as packing between the veins and help to keep a stalk rigid, and **root hair** cells which absorb water and nutrients from the soil

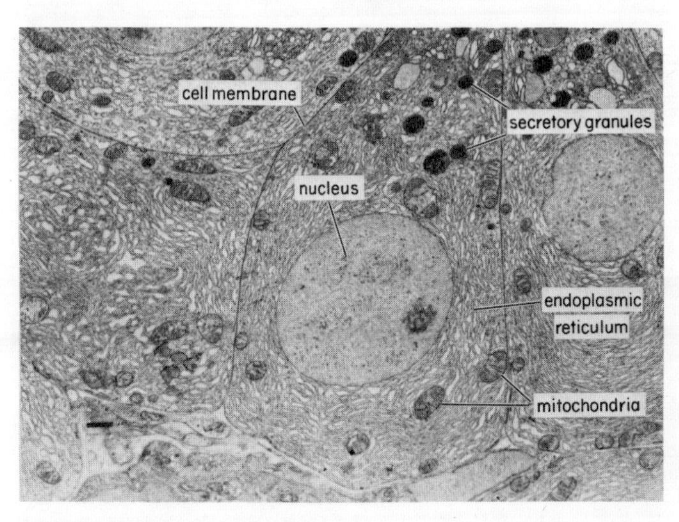

Fig. 2:3 Electronmicrograph of a cell.

(Fig. 2:5). All are adapted in their structure to fulfil a particular function.

How our body is built up

Tissues

Cells are not arranged in a haphazard way, but usually occur in groups according to their type. These aggregates of similar cells which carry out a particular function are called **tissues**.

There are four main groups of tissues making up our bodies:

1. Epithelia (sing. epithelium)

These are sheets of cells which line the inside, or cover the

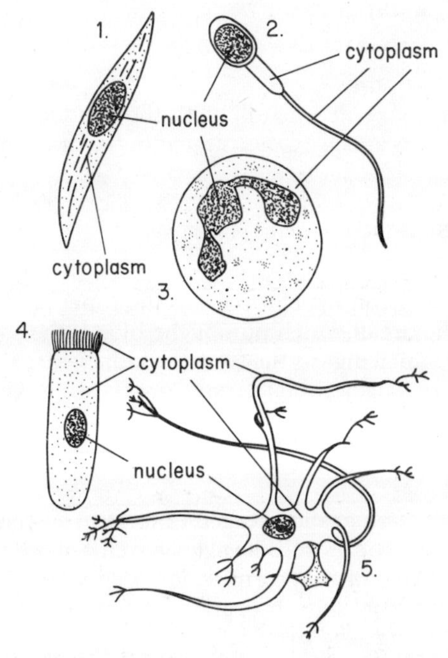

Fig. 2:4 Animal cells: 1. Involuntary muscle. 2. Sperm. 3. White blood cell. 4. Ciliated cell from lining of wind pipe. 5. Nerve cell from brain. (Not drawn to scale.)

outside of the various parts of the body, e.g. the lining cells of the stomach and the surface layer of the skin.

2. Connective tissues
These are very variable. Some help to bind parts of the body together, others like bone and cartilage help to support the body, others such as the blood are fluid and are used for transport.

3. Muscular tissues
These are concerned with movement. Some are used for strong contractions as in limbs, others for rhythmic contractions such as those in the walls of the gut which help squeeze the food along, others in the heart wall keep up regular movements causing the blood to be pumped round the body.

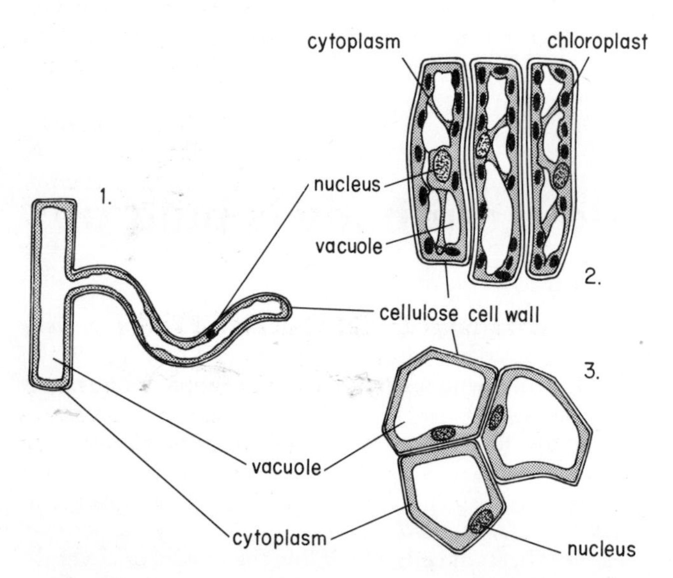

Fig. 2:5 Plant cells: 1. Root hair cell. 2. Mesophyll cells of a leaf. 3. Parenchyma cells.

4. Nervous tissues
These are concerned mainly with the conduction of nervous impulses and are concentrated in the brain and spinal cord.

Organs

These are more complex structures which are built up from various tissues and perform a specific function. Thus the heart is the organ which pumps the blood, the stomach is the organ which digests the food, the kidneys are the organs which excrete urine, and the lungs are the organs which are concerned with breathing.

Systems

Organs are often grouped together to form systems which serve a general purpose. The main systems of our body are:
1. The **alimentary system** which is concerned with the intake, digestion and absorption of food and the elimination of undigested material.
2. The **respiratory system** which brings oxygen into the body and gives out the unwanted carbon dioxide and water.

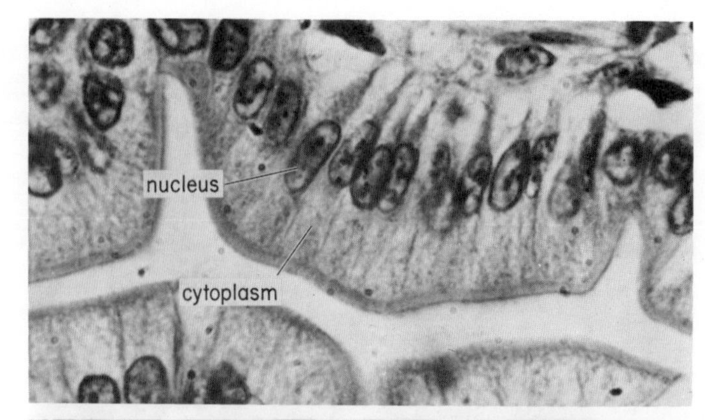

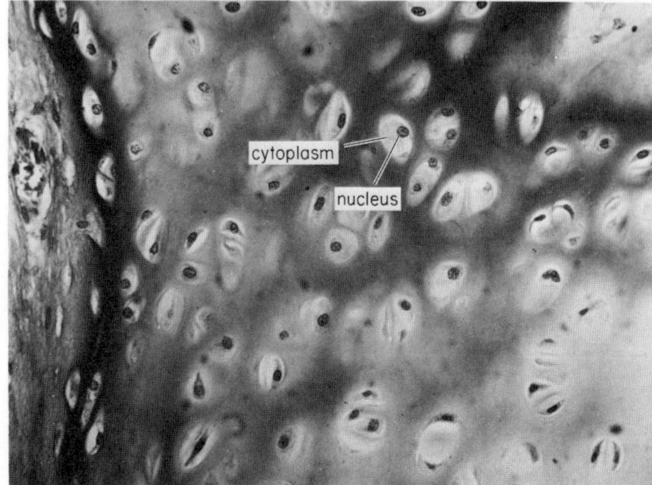

Fig. 2:6 (above) Epithelial cells from the lining of the small intestine; (below) Cartilage cells.

3. The **reproductive system** which produces the sex cells (sperms in male, eggs in female) and which protects and nourishes the young during development.
4. The **blood system** which consists of heart and blood vessels and enables materials to be transported to all the other organs and tissues.
5. The **excretory system** which eliminates useless or harmful products which have been made in the body.
6. The **muscular system** for movement of the whole body or its parts.
7. The **skeletal system** which comprises the bones which give support and protection to other parts of the body.
8. The **sensory system** which receives information from outside and inside the body and consists of sense organs such as the eyes, ears and nose.
9. The **nervous system** which controls and co-ordinates the actions of the body and consists of the brain, spinal cord and nerves.
10. The **endocrine system** which consists of a number of glands in various parts of the body. These pass secretions into the blood which help to co-ordinate body functions and influence growth and reproduction.

These systems do not work independently; they all co-operate with one another so that the body functions as a whole—a complete **organism**. You will see from Figs. 2:7 and 2:8 how some of these organs and systems are arranged; also watch an animal such as a rat being dissected or look at a prepared dissection to see the marvellous way in which all the organs and systems fit together. In Chapter 5 we shall see how plants are also divided into systems.

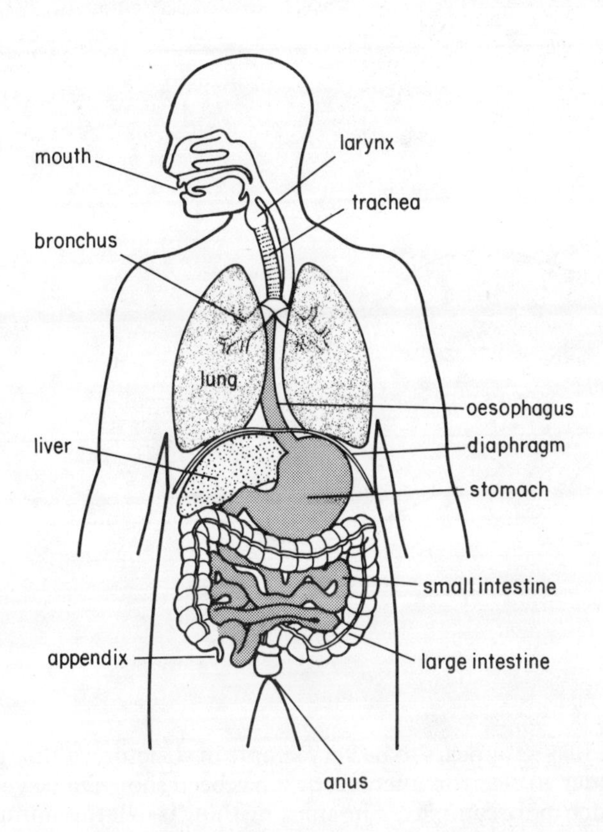

Fig. 2:7 Alimentary and respiratory systems of man.

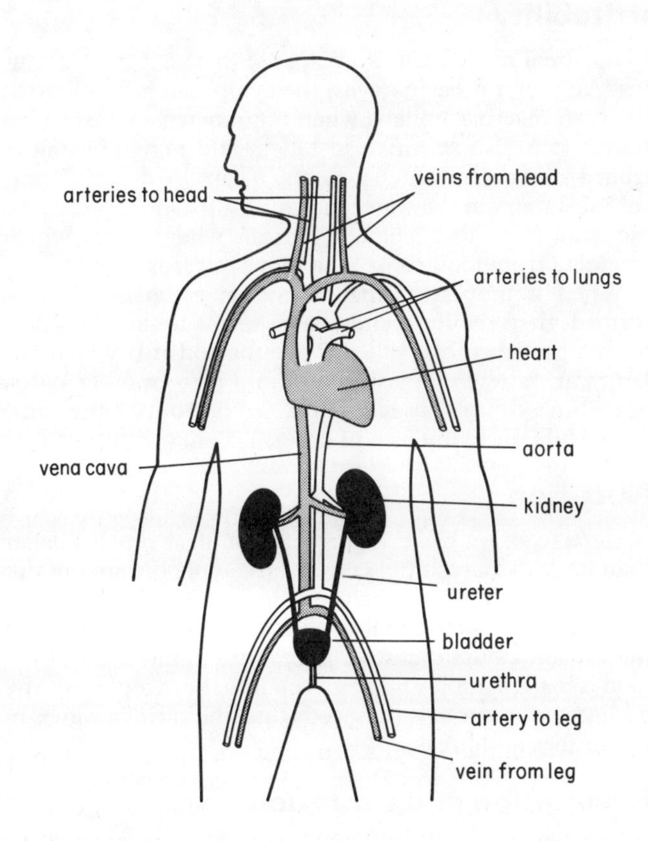

Fig. 2:8 Excretory and blood systems of man (veins lightly shaded, arteries unshaded).

Division of labour

This is an important principle which arises from what we have learnt in this chapter about cells and their functions. Division of labour means the specialisation of different structures for various functions. We have seen that this occurs at many levels:

1. Division of labour within the cell, e.g. the nucleus for cell division and control, the chloroplasts for the manufacture of food.
2. Division of labour between cells or groups of similar cells (tissues), e.g. muscles for movement, bones for support.
3. Division of labour between organs, e.g. the heart for pumping blood, the stomach for digestion.

There is also a fourth level of division of labour between individual organisms living in the same colony or society, e.g. in a bee colony where there are queens, drones and workers which are specialised to serve the society in different ways (p. 55).

We have also seen how the structure of cells and organs is always related to the functions they perform. Later on we shall see an extension of this principle in that the structure of whole organisms is also closely related to the way they live and the place they live in.

Simple organisms

We are now going to study some of the simplest organisms of all, those which have no tissues or organs, but nevertheless can carry out all the properties of living things. The two we shall choose are *Amoeba*, one of the simplest animals alive today, and *Spirogyra* which is a simple plant.

Amoeba

The larger species of amoebae are just visible to the naked eye as tiny white specks. Most species live in freshwater where they move about on the mud at the bottom. Some live in the sea, others in damp soil and in the bodies of animals, causing such diseases as amoebic dysentery. You will probably examine a rather large species, *Amoeba proteus*, which is readily cultured in the laboratory (Fig. 2:9).

Amoebae belong to the group of animals known as

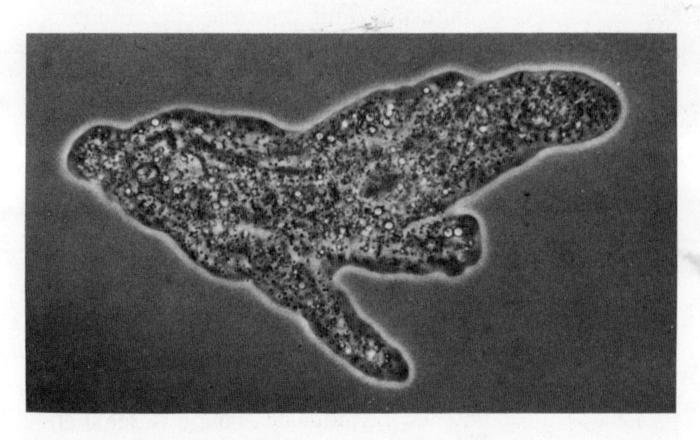

Fig. 2:9 Low power photomicrograph of *Amoeba*.

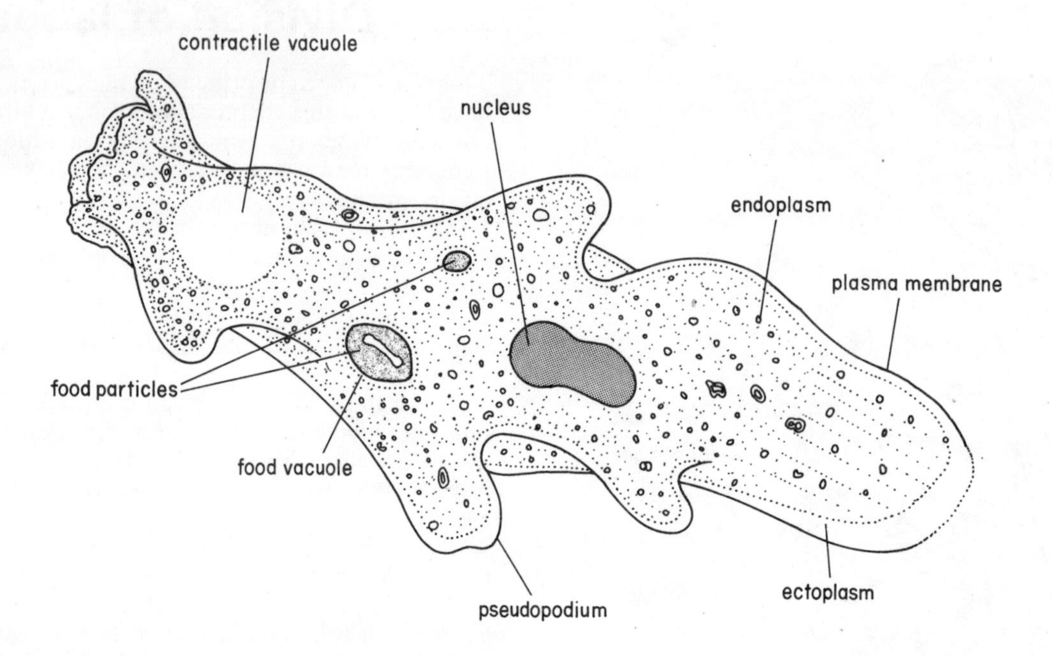

Fig. 2:10 *Amoeba proteus.*

Protozoa (p. 13). These animals are not divided into cells and are best described as **non-cellular**, although in many ways they resemble a single cell and are often called **unicellular** in consequence.

Structure

Amoeba (Fig. 2:10) appears under the high power of a microscope as a grey mass of protoplasm constantly changing its shape as it flows along. The cytoplasm is clear and jelly-like near the surface (**ectoplasm**) and granular and more watery inside (**endoplasm**). Some of the granules in the endoplasm act as food reserves, others are excretory. There is no cell wall, and you may wonder why the contents of the amoeba do not just dissolve in the water and the animal does not disintegrate. It does not do so because of its **plasma membrane**, an extremely thin 'skin' forming the outer layer of the ectoplasm. This holds it together and controls to some extent which substances go in and out.

The nucleus appears oblong and somewhat biconcave in shape.

Within the cytoplasm is a transparent spherical object, the **contractile vacuole**. This controls the amount of water inside the animal and acts rather like a safety valve. An amoeba cannot avoid absorbing water all the time through its surface by a process called osmosis (p. 83); it collects in the contractile vacuole which gradually enlarges until it suddenly disappears, squeezing out its contents into the surrounding water. It then starts enlarging again.

Watch an amoeba under the high power of a microscope or look at a film of one. Notice the streaming of the granules in the endoplasm as it flows along, and the clear ectoplasm. The nucleus cannot always be seen in a living specimen, but you see it best when the amoeba is actively moving and it is carried along in the cytoplasm. Also look out for the contractile vacuole; you may be fortunate enough to see it enlarge and burst.

Amoeba appears to be very simple in structure, although under an electron microscope it has been shown to be very much more complex. The amazing thing is that this minute mass of protoplasm is a complete organism capable of carrying out all the properties of life.

Irritability

Directional movement is a response to an external stimulus. Although it has no eyes, the cytoplasm is sensitive to light; an inactive amoeba when illuminated soon starts to move. It is also sensitive to touch, and may move away from a solid object which it meets. It can also detect chemical substances in the water; harmful substances cause it to move away or to become spherical, while others which originate from food cause it to move towards them.

When it moves, a small bump of ectoplasm is first formed, then the fluid endoplasm seems to flow into it to form a protuberance called a **pseudopodium** which enlarges as more endoplasm flows into it from other parts. Sometimes several pseudopodia are put out at the same time while the cytoplasm of others is being withdrawn.

Nutrition

Amoeba feeds on bacteria, protozoa or algae much smaller than itself. As a result of contact, pseudopodia are put out all round the food to enclose it in a drop of water, now called a **food vacuole** (Fig. 2:11). The cytoplasm round the vacuole pours digestive juices into it and most of the food is made soluble and absorbed into the cytoplasm. Any undigested material is removed when the vacuole comes to the surface and breaks.

Respiration and excretion

Amoeba, like any other organism, needs energy and in the process of respiration this is liberated from the food by means of oxygen. The oxygen diffuses through its surface

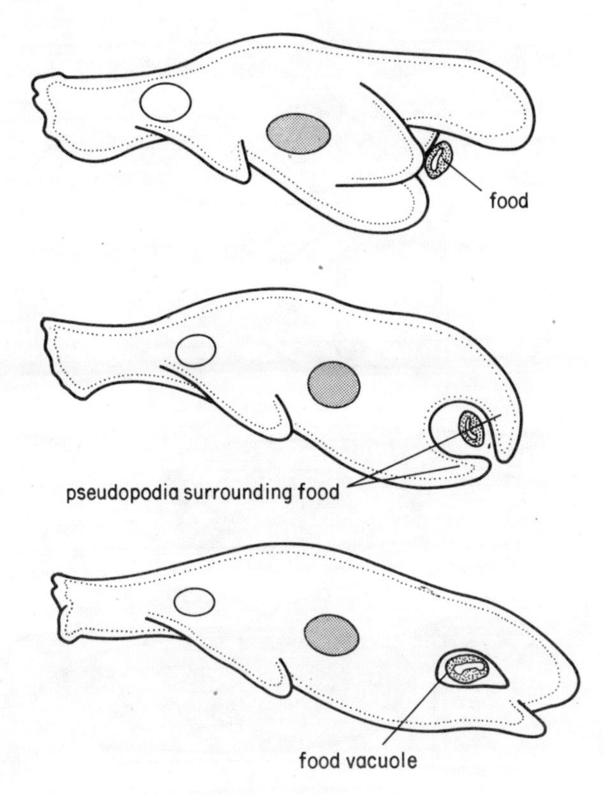

Fig. 2:11 Method of feeding in *Amoeba*.

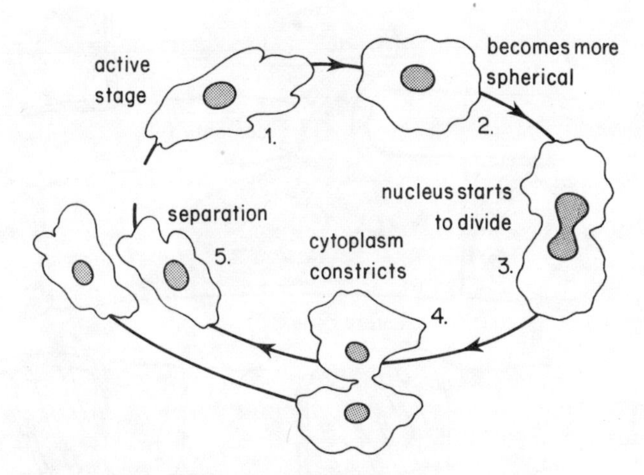

Fig. 2:12 Asexual reproduction in *Amoeba*.

from the surrounding water. Carbon dioxide is formed which diffuses out through the surface with other excretory substances in solution.

Growth and reproduction

With good feeding an amoeba increases in size rapidly, but when it reaches a certain size the nucleus divides into two (Fig. 2:12) and the cytoplasm begins to constrict as the two nuclei move apart; eventually the two parts separate as new individuals. This method of asexual (non-sexual) reproduction is called **binary fission** and is one of the simplest methods. No sexual method (one resulting from the fusion of cells) is known in *Amoeba*.

Some large amoebae are also capable of producing minute spores within themselves, each spore containing some nuclear material and part of the cytoplasm. The amoeba eventually breaks up and liberates these spores which have resistant coats. If the pond dries up these spores can remain dormant for long periods, and with the return of damp conditions they hatch out and grow into amoebae. In the dormant dry state they may be blown to other places and start new colonies. Large amoebae can also form hard protective cysts around themselves if conditions are bad and then they remain dormant for a period.

Spirogyra

In contrast to *Amoeba*, which is a non-cellular animal, *Spirogyra* is a simple green plant. You find this species in ponds and ditches as a mass of hair-like, dark green filaments which are slimy to touch because of a film of mucilage on their surface.

Structure

Each filament (Figs. 2:13 and 14) is a hollow cylinder of many cells arranged end to end. The rigid cell wall is made of cellulose and ribbon-like chloroplasts are arranged spirally along each cell in the cytoplasm lining the cell wall. Different species have a different number of chloroplasts in each cell; the one in Fig. 2:14 has two. Small spherical structures called **pyrenoids** occur at intervals along the chloroplasts. They store starch.

In the centre of each cell is the nucleus which is surrounded by cytoplasm. The cytoplasm lining the cell wall is connected to that around the nucleus by threads of the same material. The spaces contained by the cytoplasm form the vacuole, which is filled with cell sap.

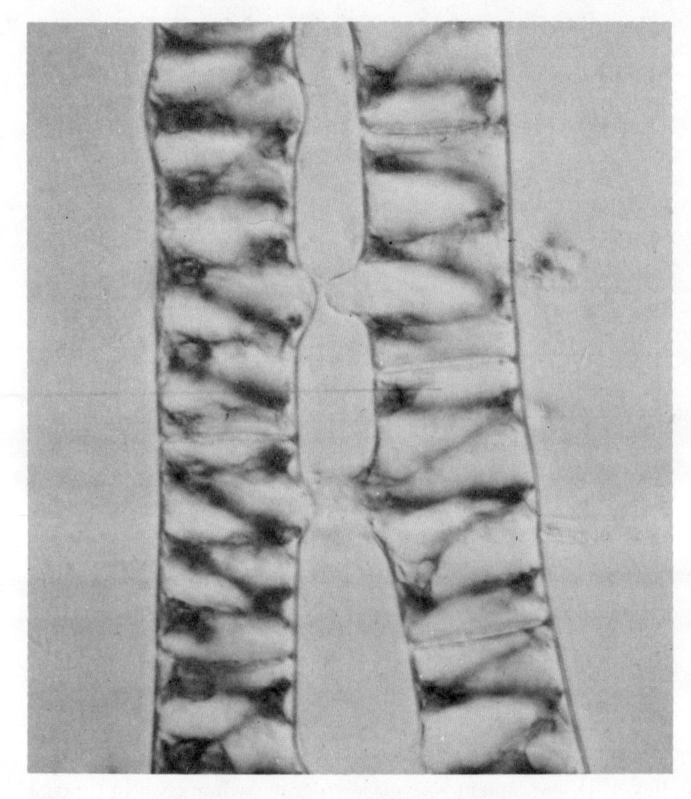
Fig. 2:13 *Spirogyra* filaments under high magnification.

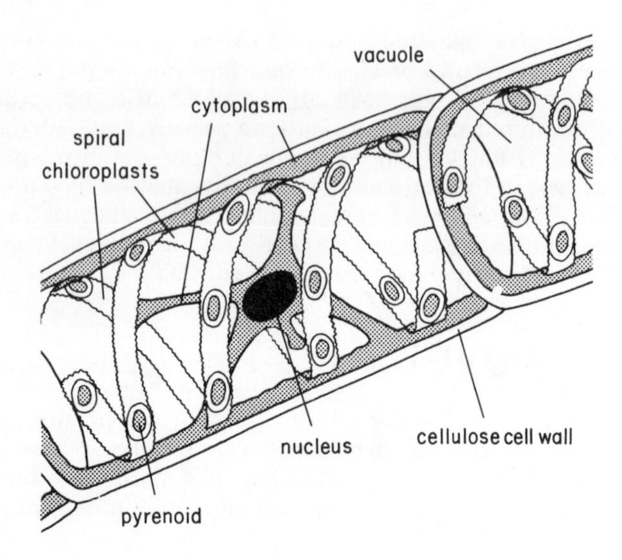

Fig. 2:14 *Spirogyra*: cell enlarged.

Examine some filaments in a drop of water under the low and high powers of the microscope. To make the structures show up better, raise the coverslip, add a drop of iodine solution and replace the coverslip. Iodine stains most things brown, but starch turns blue-black. Is there any starch in the pyrenoids? Note the distribution of the cytoplasm.

Nutrition

Spirogyra, like other green plants, builds up sugar by a process called **photosynthesis** from water and carbon dioxide which it absorbs from the surrounding water. Both light energy and the chlorophyll in the chloroplast are needed for the process and the sugar is then turned to starch for storage in the pyrenoids. *Spirogyra* also obtains salts from the water which are used in the formation of other complex substances such as proteins.

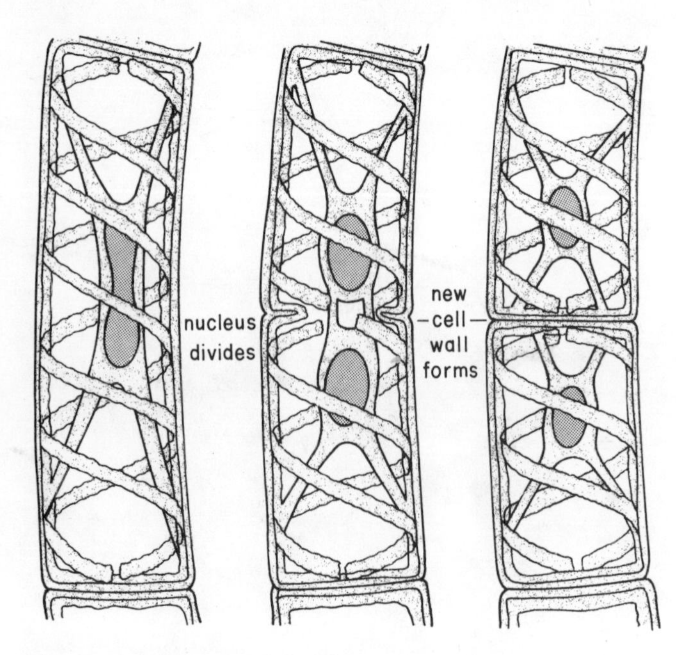

Fig. 2:15 A filament of *Spirogyra* grows when each of its cells divides and the new cells then enlarge.

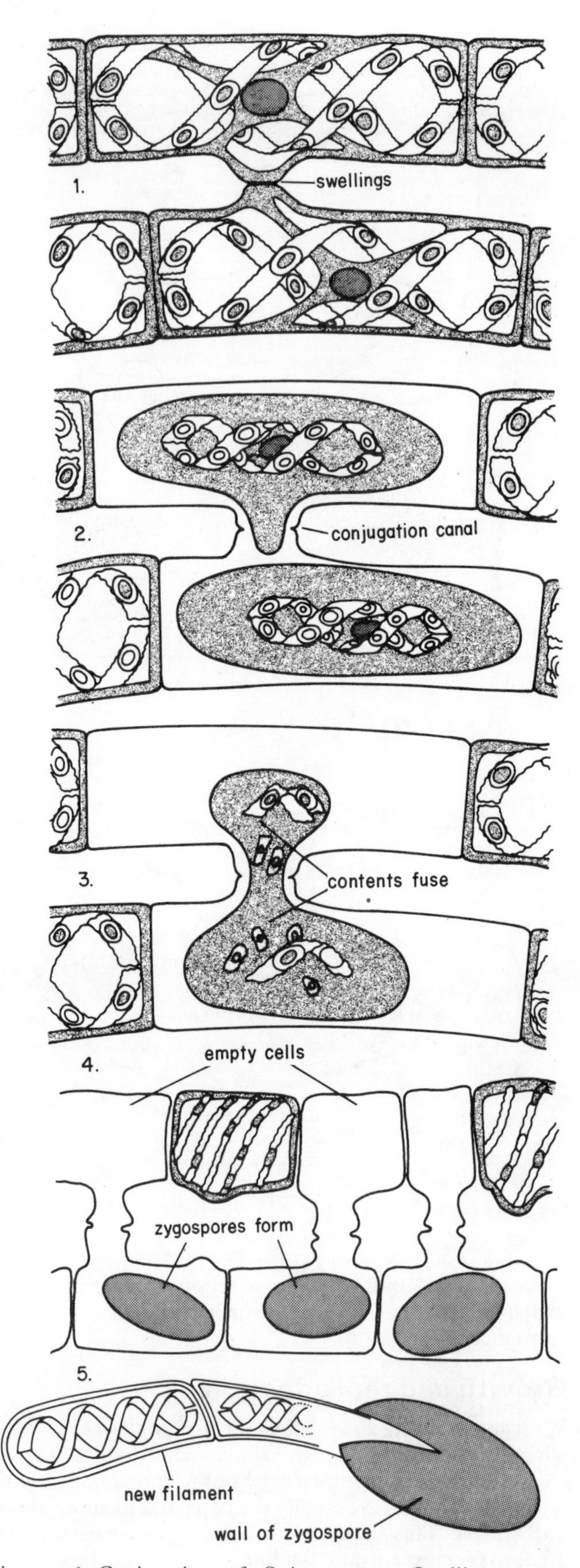

Fig. 2:16 Conjugation of *Spirogyra*: 1. Swellings occur. 2. Cell wall breaks down forming conjugation canal. 3. Contents of one cell passes through conjugation canal. 4. Zygospores are formed following the fusion of the contents of opposite cells. 5. Germination of a zygospore. (4. is on a smaller scale).

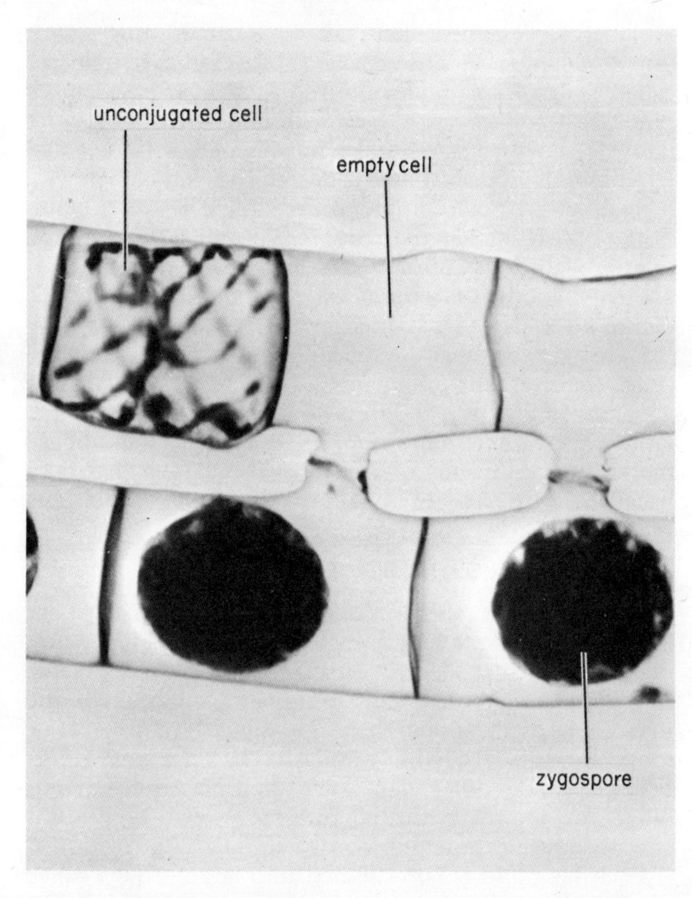

Fig. 2:17 Photomicrograph of *Spirogyra* showing zygospores.

Respiration and excretion

Energy is released from the sugar formed in photosynthesis by respiration with the help of oxygen which is dissolved in the surrounding water and absorbed through the cell wall. Carbon dioxide formed as a by-product is excreted in the reverse direction.

Growth

Enlargement of the filament takes place when the nucleus of each cell divides into two and a new cell wall is formed between them (Fig. 2:15). The cells then gradually increase in size until they get as big as the original cells, when they divide again. Thus the filament grows longer.

Reproduction

This occurs when more filaments are formed. One method is called **fragmentation** when a filament simply breaks up into smaller parts; this is an asexual process.

A second method is more complex and is known as **conjugation** (Fig. 2:16). It is a sexual process involving the fusion of the contents of two cells and results in the formation of **zygospores** with hard resistant walls which protect them if the water dries up. *Spirogyra* has two kinds of filaments which look exactly alike, but behave differently during conjugation. Normally reproduction only occurs between two filaments of different kinds. If two such filaments come into contact lengthwise, swellings form from opposite cells and these push the strands apart as they get larger. (This is an example of irritability—a response to a contact stimulus.) Eventually the separating wall breaks down to form a conjugation canal which connects the two cells. Meanwhile the cytoplasm of each cell becomes a rounded mass, the chloroplasts break up and the contents of one cell pass down the tube and fuse with that of the opposite partner, nucleus fusing with nucleus. This fusion is called **fertilization**.

This process may take place all down the filament to give a ladder-like appearance, the cells of one plant being empty while those of the other contain zygospores (Fig. 2:17). When the filament eventually decays, the zygospores may remain dormant during the winter in the mud, be carried to other parts by the water current, or be dispersed to other watery habitats on the feet of birds.

Thus this simple plant, consisting of cells which are all alike, carries out all the functions characteristic of living things. When we compare *Spirogyra* with *Amoeba*, we notice differences in the ways they both carry out these functions and these illustrate the differences between plants and animals generally.

3

Classifying organisms

Kinds of classification

The classification of organisms is complex because there are over $1\frac{1}{2}$ million different kinds of living organisms which are known. The problem is sorting them out, according to their similarities, into well-defined groups and giving them each a different name.

What similarities are the best ones to use in such a classification? If you use such features as colour or size, method of movement or number of legs, you find you have grouped together a very odd assortment of animals and produced an extremely **artificial classification**. This is not very helpful. The classification we use today is based on structural similarities. This has the great advantage in that it groups together organisms which are nearly related. It is therefore described as a **natural classification**.

The modern theory

This is based on the work of the great Swedish naturalist, **Linnaeus**, who as long ago as 1758 described his method of classification in a book called *Systema Naturae* (The Order of Nature). One of his problems was to think of a different name for every kind of plant and animal. Obviously there were not enough names to go round! This is what he did: just as we have at least two names, a surname which we share with our brothers and sisters, and a first name which distinguishes us from them, so Linnaeus gave each kind of organism two names. This method was therefore called the **binomial classification**. The name corresponding to the surname is the name of the **genus** which it shares with nearly-related forms, the other is the name of the **species**. Linnaeus realised it was very important that the name should be recognised internationally. What confusion would arise if a harmful pest or a disease-producing organism was called by different names in different countries! He therefore named them all in Latin as this was a universal language. The generic name is put first. Thus he called a man *Homo sapiens*, a frog *Rana temporaria* and a pine tree *Pinus sylvestris*. You will notice that the generic name is usually a noun and always begins with a capital letter and the specific name is descriptive and is spelt with a small letter. For example, different species of ladybird beetle have a characteristic number of spots; the common one is called *Coccinella septempunctata* (meaning 7-spotted), another *C. bipunctata* (2-spotted) and a third *C. decempunctata* (10-spotted).

By using two names the specific name may be used many times for different organisms as for each organism the generic name will be different; thus *Pieris brassicae* is the large white butterfly and *Barathra brassicae* is the cabbage moth. *Brassicae* means 'of the cabbage', and is given to both because the caterpillars of both species feed on cabbages.

Levels of classification

Linnaeus used seven main categories in his classification. The largest was the *kingdom* of which he recognised two: plants and animals. However, it is not always easy to separate some simple organisms into plants and animals and in some classifications these are placed in a separate kingdom, the **Protista**. Similarly, bacteria and fungi, being so unlike other plants, are often placed in separate kingdoms. However, for our purpose we will keep to the simpler classification and use two kingdoms only.

Each kingdom is sub-divided into large groups called **phyla** (sing. phylum), which in turn are divided into **classes** and these into **orders** which are composed of **families** which contain various **genera** (sing. genus) which include certain **species**.

Some examples are given in Table 3.1.

Structural characteristics used in classification

If you refer to the Table you will see that man and elephant are both put into the phylum Vertebrata; this is because their general plan of body structure is similar, including a backbone composed of vertebrae. They are both placed in the class Mammalia because they show the characteristic of having mammary glands by which they suckle their young. But they are put in different orders because, for one reason, man has quite different teeth from an elephant and eats very different things; he also has a relatively larger brain. In

TABLE 3:1. CLASSIFICATION OF FIVE REPRESENTATIVE ORGANISMS

	Man	*African elephant*	*Common ladybird*	*Scots pine tree*	*Beech tree*
Kingdom	Animalia	Animalia	Animalia	Plantae	Plantae
Phylum	Vertebrata	Vertebrata	Arthropoda	Spermatophyta	Spermatophyta
Class	Mammalia	Mammalia	Insecta	Gymnospermae	Angiospermae
Order	Primates	Proboscidea	Coleoptera	Coniferales	Fagales
Family	Hominidae	Elephantidae	Coccinellidae	Pinaceae	Fagaceae
Genus	*Homo*	*Loxodonta*	*Coccinella*	*Pinus*	*Fagus*
Species	*H. sapiens*	*L. africana*	*C. septempunctata*	*P. sylvestris*	*F. sylvatica*

TABLE 3:2. LEVELS OF CLASSIFICATION

Level	Group	Examples
Kingdom	Animalia	Otter, Starfish, Snail, *Insect*, *Worm*, *Jellyfish*
Phylum	Vertebrata	Otter, Bird, Reptile, *Amphibian*, *Fish*
Class	Mammalia	Otter, Sheep, Rabbit, *Bat*, *Monkey*
Order	Carnivore	Otter, Dog, Lion, *Bear*, *Seal*
Family	Mustelidae	Otter, Badger, Stoat, *Weasel*
Genus	*Lutra*	Otter
Species		European otter (*Lutra lutra*), Canadian otter (*Lutra canadensis*)

similarities becoming greater ↓

this he is much more like the apes and monkeys, so he is put in the same order as these—the Primates. But there are obvious differences between man and the apes or the monkeys, so he is separated off into the family Hominidae which only includes man-like creatures. Other primitive types of man once belonged to this family, but they are now all extinct and there is only one species left which includes all the races of mankind living today.

Look at Table 3:2. Each line represents a different level of classification. Notice how there are great differences between all the animals in the top line, rather fewer differences between those in the next, fewer still in the next, and so on until with animals belonging to the same genus the differences are extremely small (see also p. 44).

Now let us take a plant example. In the above table you see that the Scots pine and the beech tree are both put in the phylum Spermatophyta; this is because they both produce seeds, but they are put into different classes because the pine has no flowers and forms seeds from ovules which are freely exposed, while the beech has flowers and forms seeds inside a fruit. The pine is put in the order Coniferales because the seeds are produced in special cones, and in the family Pinaceae because it shows, with other species of pine, the characteristic needle leaves which occur in small clusters.

What is a species?

Members of a species are different in structure from all other organisms and they vary little between themselves apart from differences between male and female and variations due to living under different conditions. They can also interbreed and produce offspring which are themselves able to reproduce. Thus the lion and the tiger are separate species although they occasionally breed to produce tigrons, but the latter are normally incapable of having young themselves. The same applies to horses and donkeys when they produce mules. All dogs are considered to be the same species as theoretically they are all able to breed together and the offspring are fertile. The great variation in appearance between different breeds is due to man's selection of the puppies in a litter from which he wishes to breed.

From this definition of a species you will see why the races of mankind are all one species. The variation between the races, for example in skin colour, is due to the fact that they have lived under different conditions for many thousands of years.

Besides structural differences between species there are other factors of importance. Members of the same species often exhibit very similar behaviour; a song thrush in Scotland makes the same kind of nest and lines it with mud as does a song thrush in England; a grey squirrel buries nuts in the ground in a particular way whether it lives in America or in Britain. But although there are many *general* behavioural similarities, it has to be remembered that individual members of a species often behave very differently; consider the members of your class! Members of the same species are also very similar biochemically, i.e. in the sort of chemical reactions going on in the body and in the kind of substances produced; this applies particularly to the proteins which are formed. Thus similarities of structure, biochemistry, behaviour and the ability to produce fertile offspring are all characteristic of a species.

Classification of the animal kingdom

For our purpose, it is only necessary to give a much simplified classification including only the most important phyla and classes; the illustrations will give you an idea of what they look like. The animals in the first eight phyla are often called **invertebrates** because they have no backbone (Figs. 3:1 and 3:2). Those which have backbones are called **vertebrates** (Fig. 3:3).

The main animal phyla

1. PROTOZOA

Microscopic animals which are not made up of cells, e.g. *Amoeba*, *Paramecium*, *Euglena* (Fig. 4:2) and many parasites some of which cause diseases such as malaria and sleeping sickness.

2. COELENTERATA

Animals with a jelly-like body composed of two layers of cells only, and a central mouth surrounded by tentacles. Sometimes there is a hard skeleton on the outside as in corals. Other examples are jellyfish, anemones and *Hydra*.

3. PLATYHELMINTHES (Flatworms)

Worm-like creatures with flattened or ribbon-shaped bodies. Some called planarians, live in freshwater under stones, others are parasites in other animals, e.g. tapeworms and flukes.

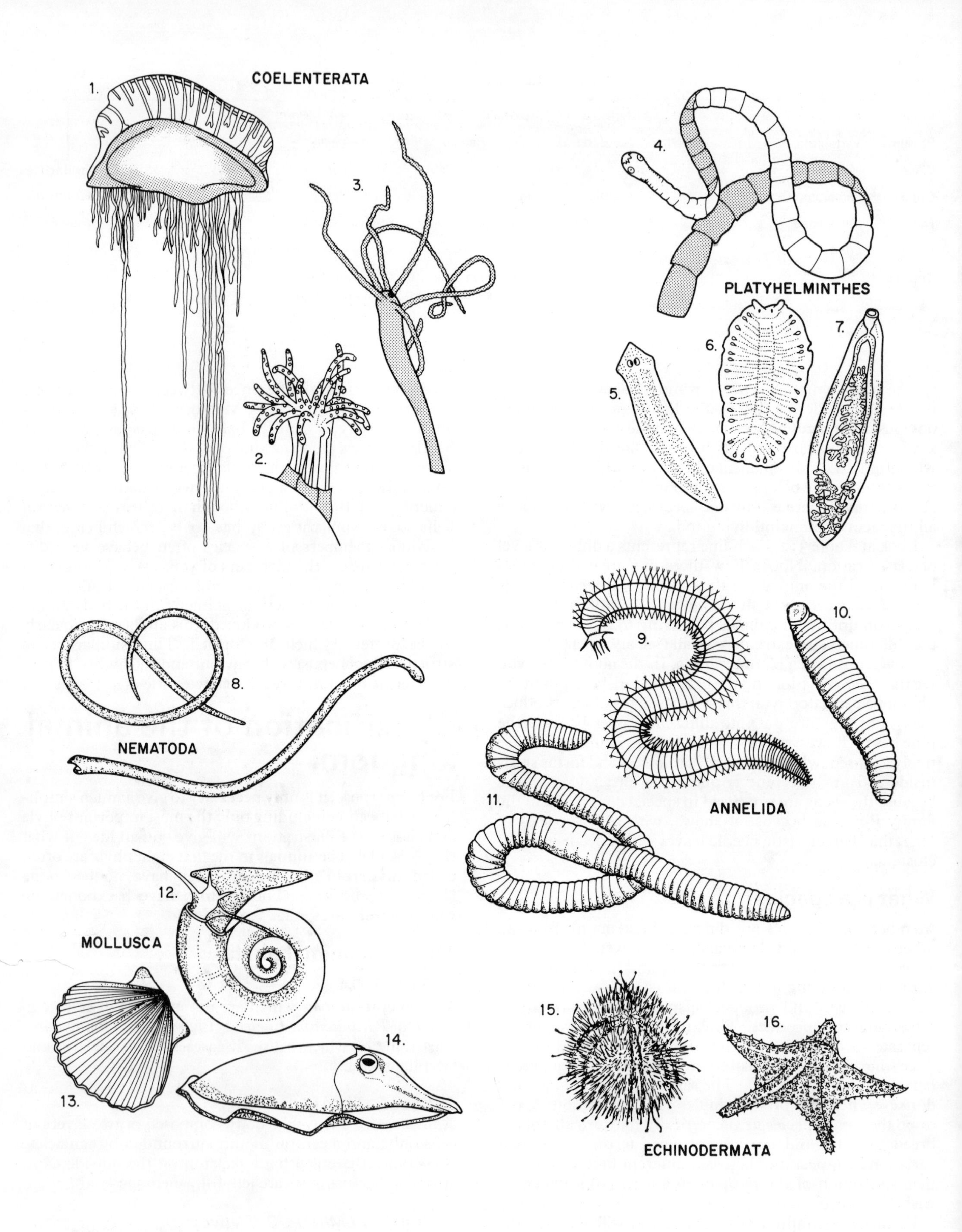

Fig. 3:1 Various invertebrates: Coelenterata: 1. Portuguese man-o-war $\times\frac{1}{6}$. 2. Coral $\times 2$. 3. Hydra $\times 4$. Platyhelminthes: 4. Tapeworm $\times 2$. 5, 6. Planarians. 7. Parasitic flukes $\times 4$. Nematoda: 8. Parasitic worms $\times 20$. Annelida: 9. Ragworm $\times 1$. 10. Leech $\times 1$. 11. Earthworm $\times 1$. Mollusca: 12. Pond snail (coiled shell) $\times 2$. 13. Scallop (bivalve) $\times\frac{1}{2}$. 14. Squid $\times\frac{1}{2}$. Echinodermata: 15. Sea urchin $\times\frac{1}{4}$. 16. Starfish $\times\frac{1}{3}$.

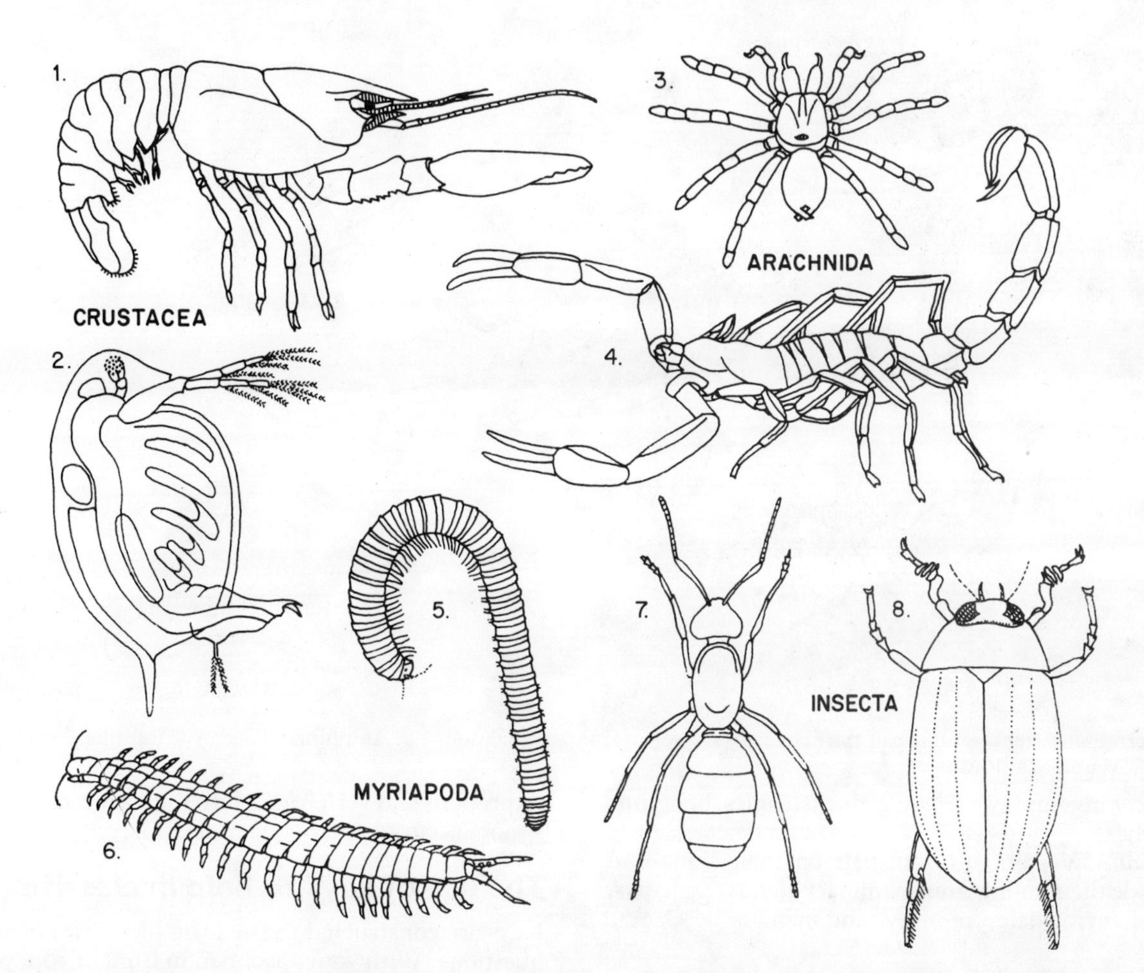

Fig. 3:2 Phylum Arthropoda: representatives of four classes. Crustacea: 1. Crayfish $\times\frac{2}{3}$. 2. Water flea $\times$25. Arachnida: 3. Spider $\times$1. 4. Scorpion $\times$1. Myriapoda: 5. Millipede $\times$2. 6. Centipede $\times$2. Insecta: 7. Ant $\times$8. 8. Beetle.

4. NEMATODA (Roundworms)

Worms with round bodies, pointed at both ends, and no rings or segments. Many are parasites such as the hook-worm, others are free-living in the soil.

5. ANNELIDA (Segmented worms)

Worms with round bodies marked externally into rings or segments. They may have one or more pairs of bristles on each segment. The main groups comprise the earthworms, marine worms and leeches (no bristles).

6. ARTHROPODA

The body is segmented and has a hard outer covering (exoskeleton). Jointed limbs are present. This phylum is the largest of all and is divided into four large classes:

a) *Crustacea*. Arthropods with limbs attached to most segments. Usually aquatic, breathing by means of gills, e.g. crabs, prawns, lobsters, water fleas and wood lice.

b) *Insecta*. Arthropods with the body divided into three parts. They have three pairs of legs and usually two pairs of wings, e.g. butterflies, bees, beetles, flies and locusts.

c) *Arachnida*. Arthropods with the body divided into two parts. They have four pairs of legs and sometimes one pair of leg-like mouth appendages, e.g. spiders, scorpions, mites and ticks.

d) *Myriapoda*. Arthropods with the body divided into a lot of similar segments with one or two pairs of legs to most segments, e.g. centipedes and millipedes.

7. MOLLUSCA

Usually have a shell which may be single and often coiled, e.g. snails; double, e.g. mussels; or internal, e.g. octopuses and squids. Their bodies are soft and they have an organ called a 'foot'.

8. ECHINODERMATA

Marine animals built on a 5-radial plan as in starfish, brittle stars, feather stars, sea cucumbers and sea urchins. They have suckers called tube feet and they circulate water round their bodies in tubes.

9. VERTEBRATA

This single phylum contains all the animals which have backbones. It is divided into five large classes:

a) *Pisces*. Aquatic vertebrates breathing by means of gills, having scales on their skins and possessing fins (no legs), e.g. sticklebacks, trout, eels, and sharks.

b) *Amphibia*. Vertebrates with four legs and scale-less skins, which are usually moist. They usually have a larval stage (tadpole) which is aquatic, and an adult stage which has lungs and lives most of the time on land, e.g. frogs, toads and newts.

c) *Reptilia*. Vertebrates with hard scaly skins. They breathe by means of lungs (even though some live in water), e.g. snakes, lizards, turtles, tortoises and crocodiles.

d) *Aves*. Vertebrates with feathers covering the body and

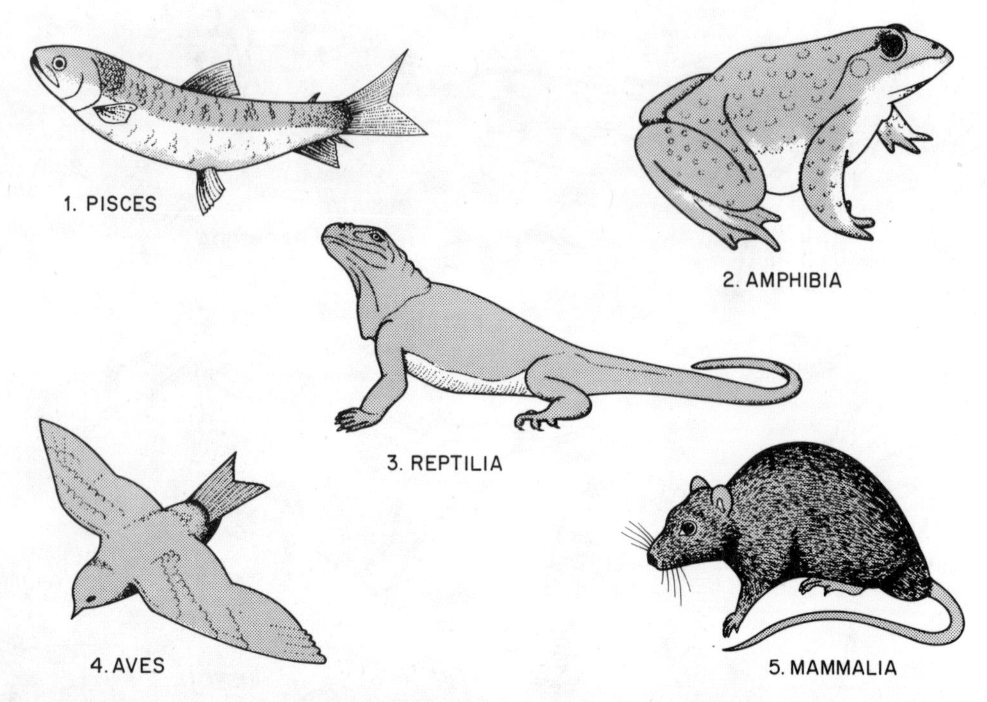

Fig. 3:3 Vertebrates, representative of the five major classes: 1. Pisces: Salmon. 2. Amphibia: Toad. 3. Reptilia: Lizard. 4. Aves: Martin. 5. Mammalia: Rat.

having two wings and two legs, e.g. ducks, eagles, penguins and ostriches.

e) *Mammalia*. Vertebrates with hair on their skins and which suckle their young from mammary glands, e.g. cows, lions, elephants, whales, monkeys and man.

By referring to this classification and the relevant illustrations you should soon be able to put the majority of animals you find into their right phylum, and in the case of arthropods and vertebrates into the correct class. However, a more accurate method is to use a **key**.

The use of keys to help in classification

Keys are constructed in the form of a series of alternative questions. With your specimen in front of you, you decide which of the first two descriptions fits your specimen best. The one you choose then leads you to the next set of questions, and so on until you end up at the correct group.

KEY TO CONSPICUOUS TERRESTRIAL ANIMALS

1.	Body not divided into segments and having no limbs	2
	Body clearly segmented *or* having obvious limbs	3
2.	Soft-bodied and slimy, with a muscular foot	Molluscs (slugs and snails)
	Small, shiny and worm-like, with the body pointed at both ends	Nematodes
	Large and snake-like, with the skin covered with scales	Reptiles (snakes and slow worms)
3.	Jointed legs absent	4
	Jointed legs present	5
4.	Worm-like with more than 15 segments	Annelids (earthworms)
	Less than 15 segments	Certain insect larvae
5.	1 or 2 pairs of walking legs	6
	3 pairs of true walking legs and usually one or two pairs of wings	Insects
	4 pairs of walking legs + 1 pair of leg-like head appendages (pedipalps). Body divided into two main portions	Arachnids
	6 or 7 pairs of walking legs	Crustaceans (woodlice)
	More than 9 pairs of walking legs	Myriapods (centipedes and millipedes)
6.	2 pairs of wings in addition to 2 pairs of walking legs	Insects (a few butterflies)
	2 pairs of walking legs; *or* 1 pair of walking legs + 1 pair of wings as in birds and bats	7
7.	Skin with no scales, feathers or hairs	Amphibians
	Skin covered with hard scales	Reptiles
	Skin with feathers	Birds
	Skin with hair	Mammals

The simple key shown opposite should enable you to put any *conspicuous terrestrial animal* found in Britain into its correct phylum and, in the case of vertebrates and arthropods, into the correct class.

There is no comprehensive key for all the animals and plants you may find. All keys have their limitations because of the number of species involved. However, many useful keys have been constructed which will help you to name many of the animals which belong to the better-known groups.

With the help of this key and the illustrations of the main animal groups classify as many terrestrial animals as you can. Here are some ideas on how to find them; make sure that each is placed in a suitable container.

1. Use a beating tray. This is any object like an inverted umbrella or a sheet, held or placed on the ground under a bush or tree; the branches are then tapped sharply with a stick and some of the animals are dislodged and fall into it. The best time to do this is in May, but summer and autumn months are quite good. Try different kinds of trees and bushes; oak is one of the best. Be careful not to damage the trees (Fig. 3:4).
2. Use a sweep net. This is a tough kind of net which can be swept to and fro amongst grasses and low vegetation. If the contents are turned out on to a black cloth the animals show up better.
3. Scrape together some of the leaf litter which covers the ground in a wood or under a hedge, put it in a large polythene bag and sieve it through a wide-meshed sieve on to a sheet; the animals fall through and can easily be seen (Fig. 3:5).
4. Look in the garden in damp sheltered places such as under large stones or pieces of wood, in compost heaps and under bushy plants and ivy. Many animals seek shelter and moisture during the day. To avoid harming the animals always replace stones and logs exactly as you found them.

You can also collect specimens from streams, ponds, ditches or the sea-shore, according to where you are living, but always remember that small organisms in the daytime usually hide away and so get protection from their enemies; therefore they have to be looked for in the places which give them shelter.

Classification of the plant kingdom

Only a simplified system will be used to give some idea of the main phyla (Figs 3:7 and 8).

The main plant phyla

1. BACTERIA

Simple microscopic plants of various shapes with no well-defined nuclei. The majority have no chlorophyll.

2. ALGAE

Plants without roots, stems or leaves. All contain chlorophyll and the majority are green, but sometimes the green is masked by other colours as in the brown and red seaweeds. *Spirogyra* and microscopic forms such as desmids and diatoms also belong to this phylum.

3. FUNGI

Like algae, they have no roots, stems or leaves, but they lack chlorophyll. Some are saprophytes (p. 29) and cause decay, others are parasites living on other animals and plants. They include moulds, mildews, yeasts, mushrooms and toadstools.

4. BRYOPHYTA

Small green plants which usually have leaves and stems but no proper roots. They are placed in two classes:

a) *Liverworts*. Most are small, branched green plants which lie flat on the ground in damp places, often forming large mats.

Fig. 3:4 Using a beating tray.

Fig. 3:5 Sieving leaf litter.

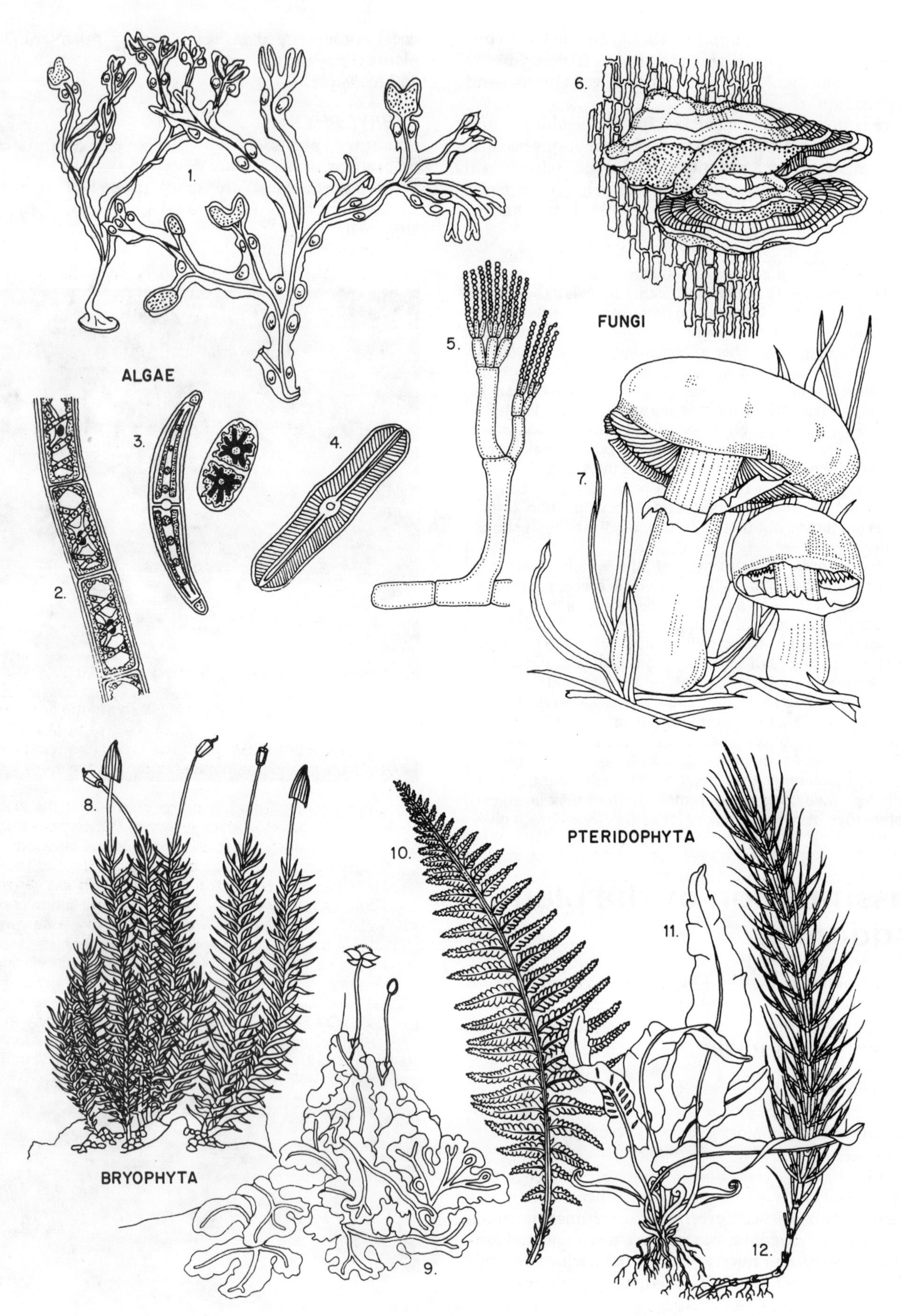

Fig. 3:6 Simpler plants: Algae: 1. Brown seaweed $\times\frac{1}{2}$. 2. Filamentous alga $\times$100. 3. Desmids $\times$200. 4. Diatom $\times$200. Fungi: 5. A mould $\times$1000. 6. Bracket fungus $\times\frac{1}{6}$. 7. Toadstool $\times$1. Bryophyta: 8. Moss $\times$1. 9. Liverwort $\times$1. Pteridophyta: 10, 11. Ferns $\times\frac{1}{8}$. 12. Horsetail $\times\frac{1}{8}$.

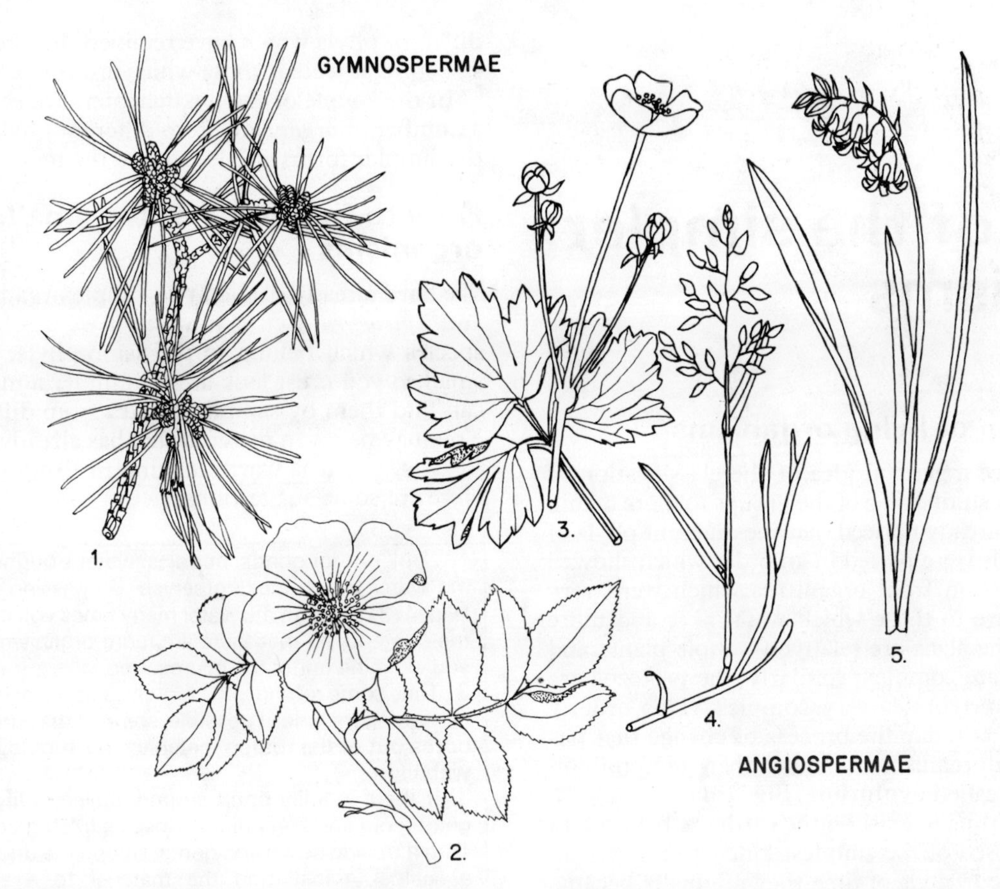

Fig. 3 : 7 Higher plants (Spermatophyta). Gymnospermae: 1. Pine. Angiospermae: 2. Rose. 3. Buttercup. 4. Grass. 5. Bluebell.

b) *Mosses*. These usually grow together in cushions or compact masses. They have distinct stems and the leaves have a mid-rib.

5. *PTERIDOPHYTA*

This phylum includes the ferns, horsetails and club mosses. They have distinct roots, stems and leaves. The ferns produce spores, usually from the underside of the fronds; the horsetails and club mosses produce them from special cone-like structures at the ends of the stems.

6. *SPERMATOPHYTA*

A very large group containing the majority of familiar plants. They all produce pollen and form seeds.

a) *Gymnospermae*. These bear cones and most have needle-shaped leaves. The majority are called conifers, e.g. pine, larch, fir.

b) *Angiospermae*. All have flowers and produce seeds inside a fruit. There are two important groups of these flowering plants:

1. **Monocotyledons**. Usually narrow-leaved plants which include families to which grasses, rushes, lilies, irises and orchids belong. The only common trees in this group are the palms.
2. **Dicotyledons**. Usually broad-leaved plants which include families to which buttercups, roses, peas, dead-nettles and daisies belong. All the trees which are not coniferous belong to this group, e.g. oak, beech, ash, maple.

You will find it quite easy to recognise members of the Spermatophyta, but get some practice with the others and see how many kinds of ferns, mosses, liverworts, algae and fungi you can find. Here are suggestions for finding them.

1. You will find algae mainly in water (although one common species called *Pleurococcus* occurs on tree trunks and turns them green). Look in ponds, canals, water troughs and rivers. If you are near the sea, remember that all the seaweeds are algae.
2. Liverworts are found in damp places—on the banks of ditches, near drains, at water level where a river goes under a bridge, near waterfalls and on damp ground in woods.
3. Mosses are often found in similar places, but they also occur in drier habitats such as on walls, roofs and tree trunks.
4. Ferns also like the damp; bracken is an exception and does well in dry places. Look in woods, damp hedgerows and shady places; also on old walls.
5. Fungi. You will find the larger kinds in greatest variety in the autumn, but some may be found during any month. Different varieties occur near particular trees, so look in oak woods, pine woods, beech woods, etc. Also look in open fields and on old logs. Smaller fungi are all too common when things go mouldy, and some, like the rust fungi, form bright patches of yellow or orange on the leaves of plants, e.g. groundsel plants in the garden.

4

Some of the simpler organisms

The evolution of living organisms

Now that we have a general idea of the classification of organisms we can study some of the groups in more detail.

You will have already noticed that the different phyla of plants and animals were placed in an order which showed a general progression from organisms which were very simple in structure to those which were more and more complex. Thus the algae are relatively simple plants and the angiosperms are complex; similarly, the protozoa are simple and the vertebrates very complex. This orderly progression reflects today the process of change that has been going on in organisms for more than 3400 million years—a process called **evolution** (Fig. 4:1).

The first organisms to exist on the earth are believed to have been composed of the simplest kind of protoplasm, but over immense periods of time they gradually became more complex, giving rise to all the species which have existed in the past and have become extinct, and all which still exist today. However, the rate of change in different organisms has varied enormously; some, having reached a very efficient stage for living in a particular kind of habitat, have hardly changed for many millions of years, while others have gone on changing much more rapidly. As a result, we still have living today plants and animals from different phyla which have retained their relatively simple structure as well as those which are very complex.

In the course of the next chapters we shall be studying a number of organisms from different phyla, starting with the simpler forms and leading to the more complex.

How to find some of the smaller organisms

We have already studied two simple organisms—*Amoeba* and *Spirogyra*. Let us now see if we can find some other species which belong to the same phyla. Most are very small so you must look at them under a microscope. You can find them by sampling water from different habitats. You may be given material that has already been collected for you, but it is worth looking in likely places yourself. Here are some suggestions:

1. Look out for ponds, puddles, water troughs or any standing water which looks greenish. By passing a plankton net (Fig. 4:3) through the water many times you can concentrate the contents so that there are more organisms in each drop you examine under the microscope.
2. Take some rotting lawn mowings and place them in a little rain water. Include, if possible, some of the liquid which often oozes out of the rotting material; you should find it teeming with life.
3. If there is a lily pond around, collect a lily leaf which is getting old and beginning to lose its fresh green appearance. Turn it upside down and gently scrape the undersurface with a scalpel, transferring the material to a specimen tube containing a small amount of water. Examine a drop of this.
4. Collect a large clump of water weed from any pond or slow-moving stream and wash it vigorously, a little at a time, in a dish of water collected from the same habitat. Discard most of the weed, leaving just a few pieces in the water. Allow the sediment to settle overnight and then take samples from the mud at the bottom and examine them. Also look at the sides of the dish to see if any *Hydra* are sticking to it. You will be examining these later (p. 23).

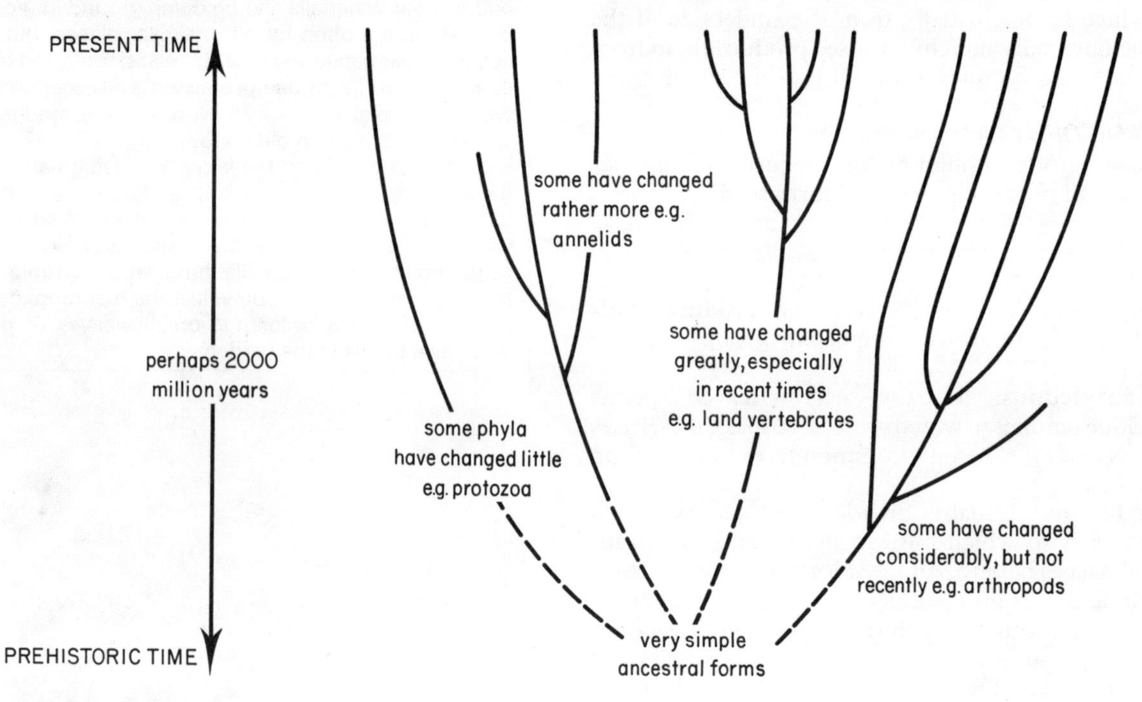

Fig. 4:1 Simplified diagram to show how some phyla may have evolved.

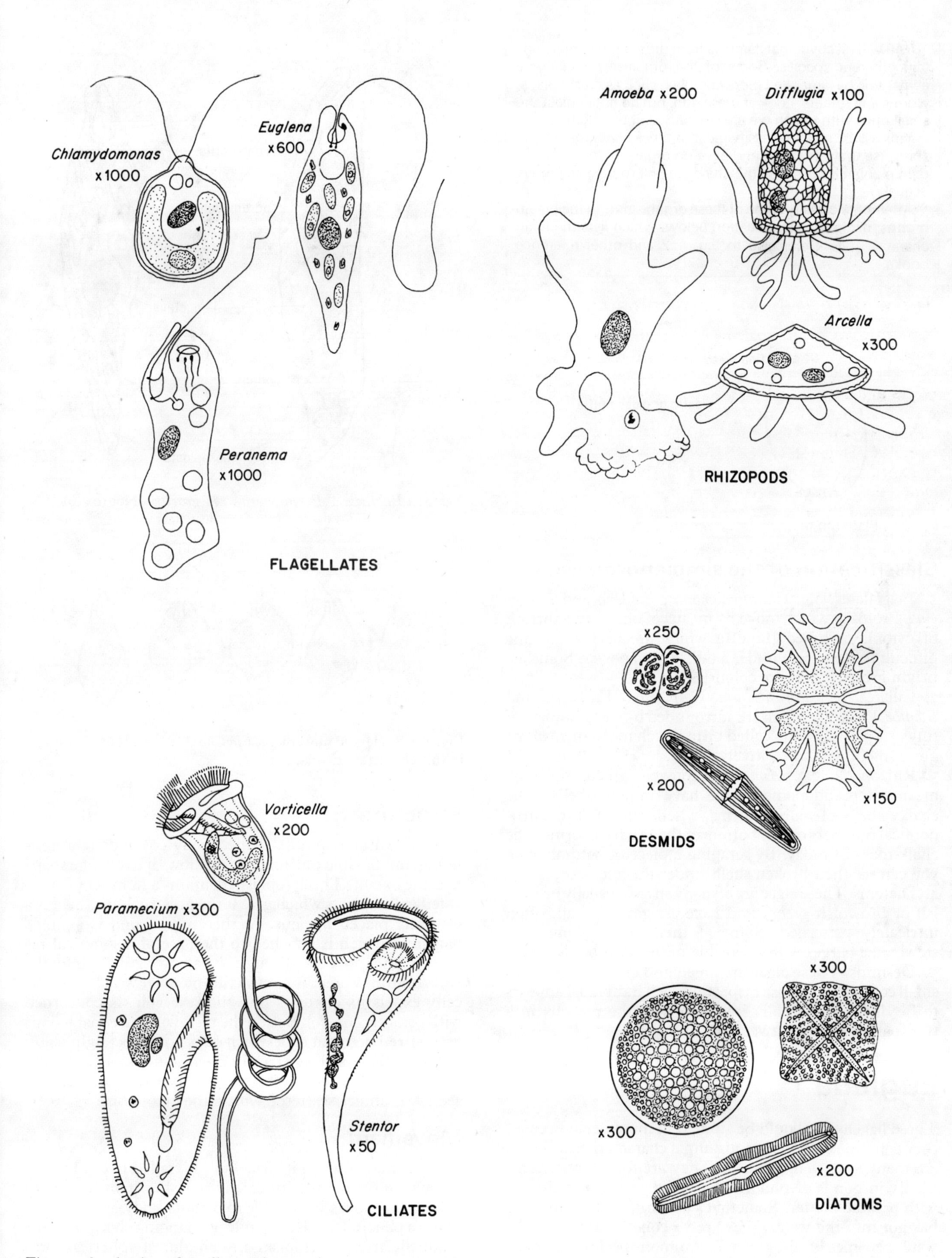

Fig. 4:2 A selection of small aquatic organisms.

You will discover that samples from different habitats contain different species. Some of the organisms you come across will be larger and more complex, such as crustaceans, worms and rotifers. Look at these, too, but do not neglect the small ones with which we are concerned at the moment.

You will be able to classify most of these either into plants (because they are green and have no organs of locomotion) or into animals (because they have no chlorophyll and move actively).

Attempt to classify some of these organisms into their main groups, using the classification below, which lists the main characteristics. Refer also to Fig. 4:2 and other reference books.

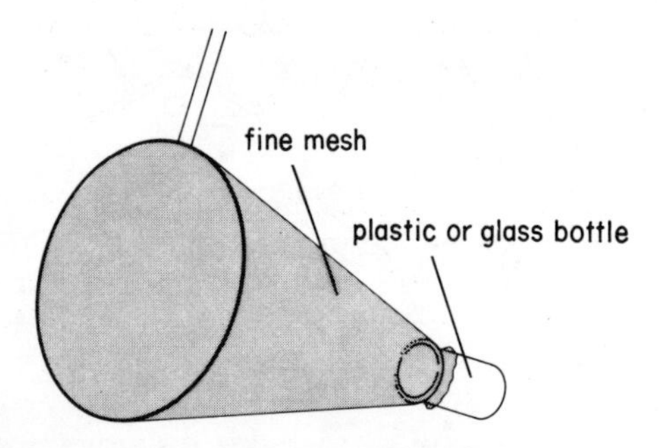

Fig. 4:3 Plankton net.

Classification of the simpler organisms

1. Flagellates, e.g. *Chlamydomonas*, *Euglena* and *Peranema*. Protozoa which move by means of one or two threads of cytoplasm called **flagella** which are very small and difficult to see except under a rather high power. Some are bright green, others are colourless.

2. Ciliates, e.g. *Paramecium*, *Colpidium*, *Vorticella* and *Stentor*. Protozoa which are surrounded by vast numbers of tiny cytoplasmic hairs called **cilia** which move them along quite smoothly and relatively fast.

3. Rhizopods, e.g. *Amoeba*. Protozoa which move by means of pseudopodia. Some have minute shells. The chalky shells of marine forms, when deposited in astronomical numbers on the bottom of the sea, have formed the chalk rocks of today. By scraping a piece of natural chalk you can see their broken shells under the microscope.

4. Diatoms. These algae are of many shapes, usually brownish or brownish-green, and have a rigid cell wall, often intricately patterned. Some of them move along very slowly like barges with no visible means of propulsion.

5. Desmids. These algae are green and either single or in small colonies. On close inspection each individual appears to be divided internally into two with the nucleus in a transparent region between the two parts.

Euglena

These flagellates should be studied in more detail because they combine both plant and animal characteristics. There are many species of *Euglena*. They are found most commonly in ponds or puddles that have been contaminated with organic matter. Sometimes on farms, where manure has got into the water, they are so concentrated that the water becomes bright green. A common species is *Euglena gracilis*, but *E. spirogyra* is larger (Fig. 4:4).

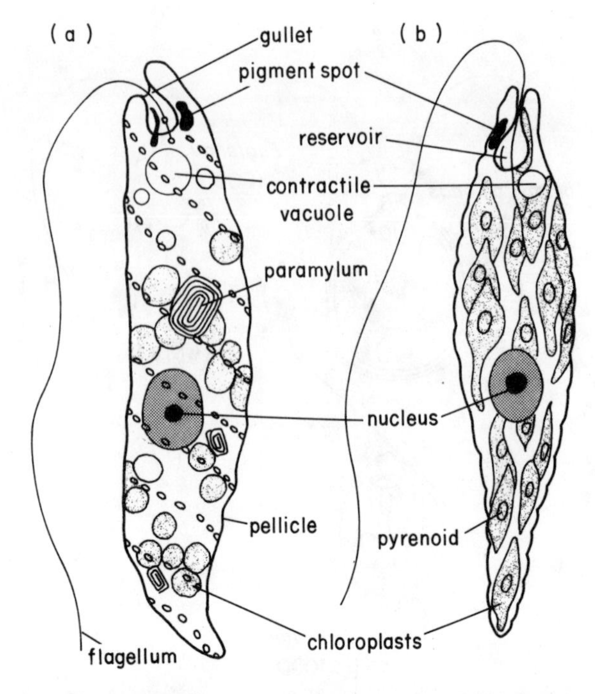

Fig. 4:4 *Euglena*: a) *E. spirogyra* b) *E. gracilis*. (Not to scale.)

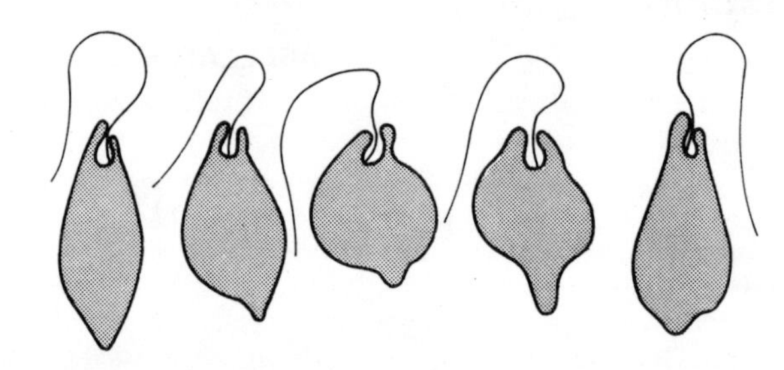

Fig. 4:5 Euglenoid movement. *Euglena* constantly changes shape in this characteristic manner.

Structure

E. gracilis has a spindle-shaped body covered by a non-living, but flexible **pellicle** which allows it to change shape to some extent. The cytoplasm contains a number of elongated chloroplasts which give it the green colour; there is a centrally placed nucleus. At the anterior end is a single flagellum which is attached to the base of a spherical reservoir and projects through a narrow 'gullet' into the surrounding water. A contractile vacuole, which periodically expels its contents into the reservoir, is concerned with osmotic control. Beside the reservoir is a conspicuous orange-red pigment spot which does not detect light itself, but nevertheless plays a part in *Euglena*'s response to light. A carbohydrate food reserve called **paramylum** can be seen as granules scattered within the cytoplasm.

Movement

Euglena moves when its flagellum is trailed behind it and thrown into waves; periodically it lashes it from side to side. As it moves forward it rotates its body as when one twists a pencil. It is also capable of changing shape, passing gradually from an elongated to an almost spherical form (Fig. 4:5).

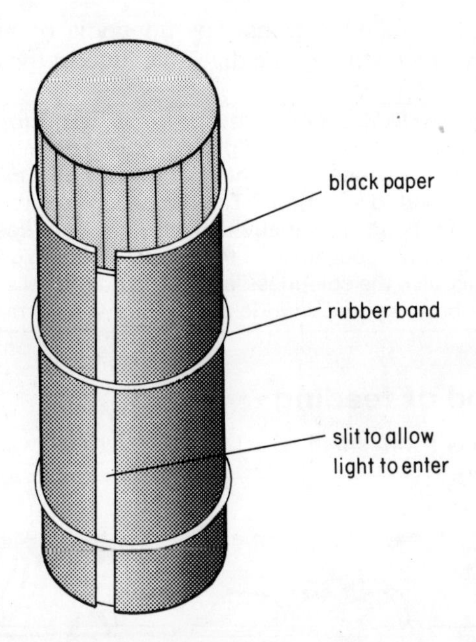

Fig. 4:6 A method of demonstrating the effect of light on the directional movements of *Euglena*.

Irritability and nutrition

Euglena responds to the stimulus of light by swimming towards it. You can demonstrate this example of irritability in the following way:

> Take some 'green' water containing *Euglena* in a specimen tube and fix black paper round it as in Fig. 4:6, leaving a narrow slit down one side. Place it in moderate light for 24 hours and then carefully remove the paper without shaking the tube. Observe the distribution of *Euglena* in relation to light.

This response is useful because, having chlorophyll, *Euglena* is able to feed rather like a plant by using light and forming carbohydrates from carbon dioxide and water. However, if a colony of *Euglena* is put in the dark (you could try this) they lose their colour and can no longer photosynthesise. Nevertheless, they remain healthy as long as organic matter is present in the water. Without their chlorophyll they are extremely like those flagellates which are *always* colourless and, like them, they feed on the organic matter in the water by absorbing it in solution through their surface, thus feeding more like animals. So here is an organism which feeds like a plant in the light and like an animal in the dark. But many biologists think they resemble animals more than plants, because most species are found to need some organic matter to live healthily even when in the light. They are also capable of locomotion and reproduce lengthwise into two like other simple animals.

Some of the simpler animals

Hydra

The animals we have studied so far have all been non-cellular. Now we come to *Hydra*, one of the simplest of the multicellular (many-celled) animals (Fig. 4:7).

The various species of *Hydra*, some green, some brown, belong to the phylum Coelenterata along with the sea anemones, corals and jellyfish. All members of this phylum consist of cells which are arranged in two layers, an outer **ectoderm** and an inner **endoderm**. The body is sac-like having one opening which acts as the mouth, and round it are a number of tentacles.

Hydra may be found in ponds, lakes or streams and may be collected by the method suggested on p. 20. They are easiest to find in summer and autumn, as by then their numbers will have built up after the winter. They attach themselves to water weed, stones or submerged sticks by means of an adhesive disc at the end of the body. When fully expanded they are about 1 cm long.

Structure

Hydra (Fig. 4:8) has a sac-like body, its cavity acting as a gut and its one opening serving as a mouth. The mouth is situated on top of a small cone which is surrounded by 7–10 tentacles. The body wall consists of cells which are arranged in two layers, an outer **ectoderm** and an inner **endoderm** (Fig. 4:9). They are separated by a thin layer of jelly, the **mesogloea** (in jellyfish this is a very thick layer, hence the name). The colour of the different *Hydra* species is caused by microscopic green or brown algae living within the endoderm cells. The cells of *Hydra* vary greatly in both structure and function, thus illustrating once more the principle of division of labour between the cells of an organism. The kinds of cells and their functions are summarised below:

Ectoderm

1. Muscle cells which have contractile tails all lying parallel to the long axis of the body or tentacles. When these contract the body or the tentacles become shorter.

Fig. 4:7 *Hydra* with bud.

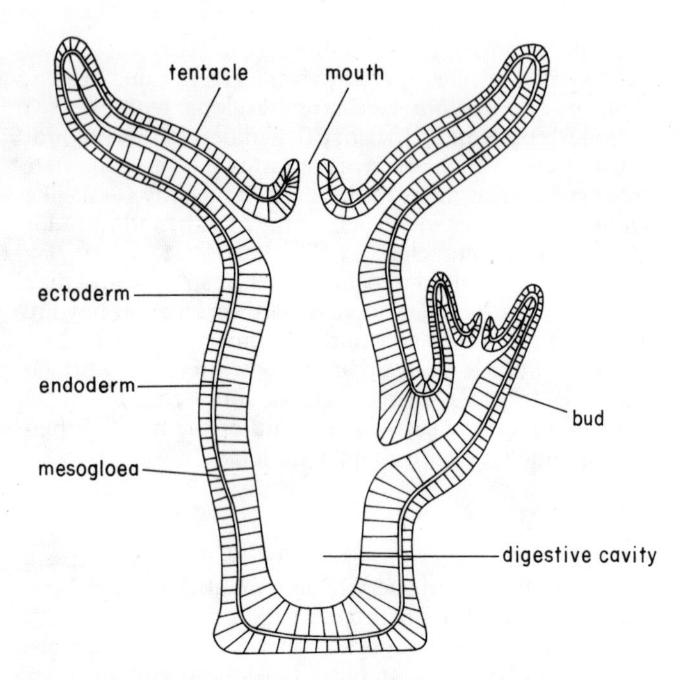

Fig. 4:8 A longitudinal section of *Hydra* with bud.

2. Sensory cells which detect touch or other stimuli.
3. Nerve cells which conduct 'messages' from one part of the body to another.
4. Sting cells which are used for paralysing and catching prey.
5. Reserve cells which are capable of growing into other kinds of cells during growth or into sting cells when they need replacing.
6. Reproductive cells. These are only formed during sexual reproduction.

Endoderm
1. Gland cells which secrete digestive juices into the central cavity.
2. Food-absorbing cells. These also act like muscle cells as they have contractile tails arranged in a circular direction which cause elongation of body or tentacles when they contract. When this happens the muscle cells of the ectoderm are relaxed. The food-absorbing cells ingest partially digested food by means of pseudopodia, others have a flagellum which stirs up the digestive fluid in the cavity.

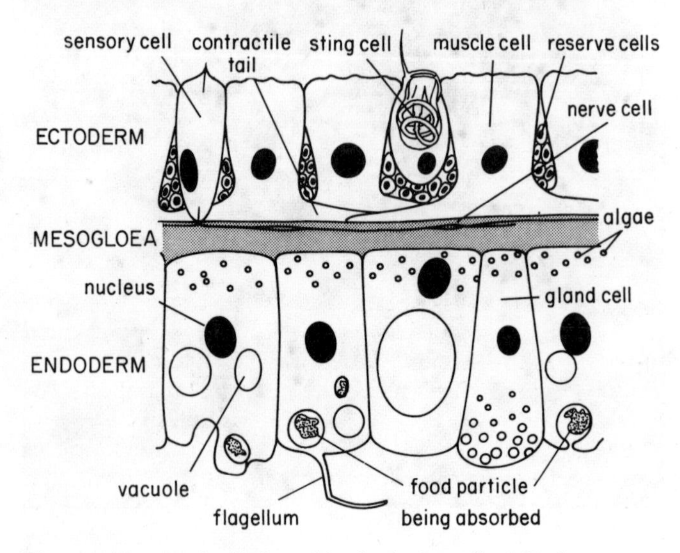

Fig. 4:9 Types of cells found in the body wall of *Hydra*.

Examine a *Hydra* under the low power of the microscope in a drop of water on a cavity slide (why not a normal slide?). Note how it expands and contracts. Does a fully expanded *Hydra* respond to vibrations? Tap the bench near the microscope to find out. How many tentacles has your specimen? Note the warty appearance of the tentacles due to groups of sting cells; also the colourless layer of ectoderm cells and the green or brown of the endoderm cells showing through.

Method of feeding

Hydra feeds on small crustaceans such as water fleas (*Daphnia*).

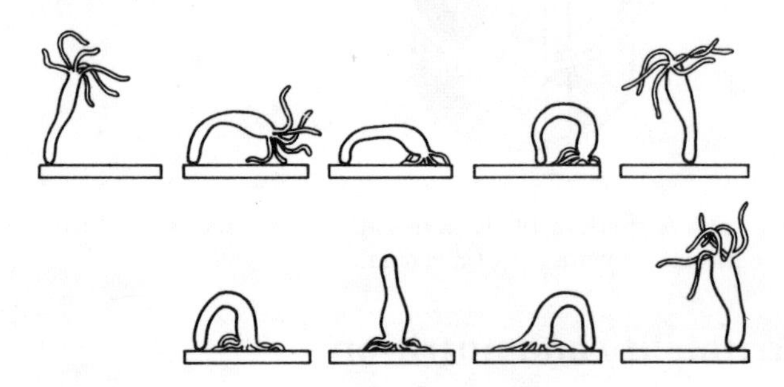

Fig. 4:10 Locomotion in *Hydra* is usually effected by looping (top); occasionally it carries out complete somersaults (bottom).

The tentacles are spread out stiffly like a rigid net thus covering a wider area and making capture more probable. When a water flea touches a tentacle it is quickly paralysed by the sting cells and the tentacles rapidly curl round the prey and help to push it towards the mouth.

When the prey is forced through the mouth, which is greatly enlarged during the process, the gland cells of the endoderm secrete their digestive juices and the soft parts are broken up into semi-digested particles. These are later engulfed by the food-absorbing cells where digestion is completed in food vacuoles as in *Amoeba*. The skeleton of the water flea which is indigestible is then squeezed out through the mouth and removed.

Watch the method of capture by placing some *Hydra* (which have been deprived of food for a few days) in a solid watch glass. When they have become attached and fully expanded, introduce with a pipette a number of water fleas. In the confined space of the tube they will soon make a capture. Use a lens or microprojector to see what happens.

The sting cells (Fig. 4:11) which paralyse the prey are wonderfully specialised cells arranged in groups on the tentacles and less frequently on other parts of the body. Each has a capsule with a hollow thread coiled up inside a bath of poison. When a water flea touches the trigger which projects from the surface of the cell, the capsule contracts and the thread is shot out, turning inside out in the process. Other sting cells have no poison, but their threads curl around the bristles of the prey and grip it. All typical coelenterates have these sting cells. If you stroke the tentacles of a sea anemone with your finger it feels rough; this is because it has shot a number of threads into your skin and under a microscope it is possible to see them. Those

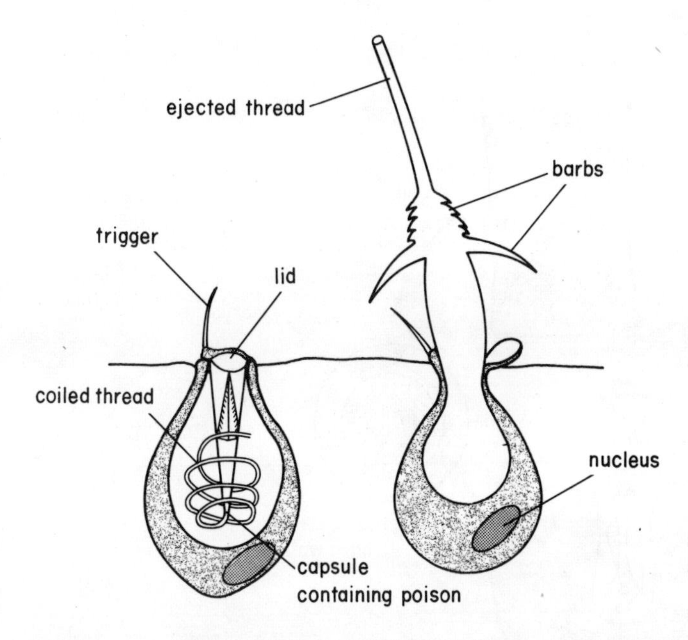

Fig. 4:11 A sting cell before and after discharge.

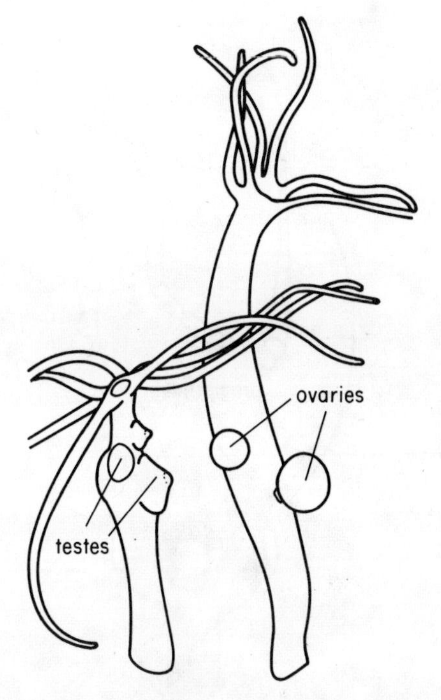

Fig. 4:12 Sexual reproduction of *Hydra*. Although hermaphrodite, the testes usually ripen before the ovaries.

from sea anemones are not powerful enough to sting you, but those of some jellyfish are. In the Portuguese Man o' War, another large coelenterate, the stings can be extremely painful, and people who bathe in tropical seas need to be careful when they are about.

Reproduction

Asexual

Hydra multiplies quickly if kept in a well-lit aquarium and given plenty of water fleas; you can observe the process. First a bump appears on the side and this grows into a bud, which eventually forms tentacles and a mouth at the far end. The cavity of the bud is continuous with that of the parent, so the developing *Hydra* can obtain food from its parent as well as catch its own when its tentacles and mouth have developed (Figs 4:7 and 8). Later the new *Hydra* constricts at the base and pulls itself away from its parent by gripping some water weed with its tentacles. This method of reproduction is asexual and is called **budding**. It occurs constantly throughout the spring and summer when food is plentiful.

Sexual

Under certain circumstances, especially in the autumn, a sexual method of reproduction takes place (Fig. 4:12). When this happens several bumps develop from ectoderm cells in the region below the tentacles; these are the male organs or **testes**. Within them vast numbers of sperms are formed which are shed into the water when the testis wall bursts; they are able to swim with their long cytoplasmic tails. Later on, the same *Hydra* produces one or more swellings further down the body below the testes, and these are the female organs or **ovaries**. Inside the ovary a single egg cell develops, the **ovum**, which gets larger and larger as it stores more food. Eventually the wall of the ovary splits sufficiently to admit sperms, one of which fertilizes the ovum. Now that the egg has been fertilized it can develop into a new individual. By cell division it quickly forms a spherical embryo which secretes round itself a hard horny case which can easily be seen with the naked eye as a dark spot. Later, this drops off into the mud where it remains dormant through the winter months. In the spring the horny case breaks and a new *Hydra* emerges. Thus by forming eggs with protective coats *Hydra* can survive the winter when the adults die.

Hydra is said to be **hermaphrodite** because a single individual forms both male and female organs. However, self-fertilization is usually avoided as the testes become mature before the ova are ripe.

Tapeworms

These are parasitic flatworms belonging to the phylum Platyhelminthes which live in the intestines of vertebrates, especially mammals and birds. A **parasite** can be defined as an organism which obtains its food from another living organism called the **host** by living on or in it; it is completely dependent upon the host which gets no benefit from the parasite in return. Usually a parasite harms the host to some extent.

Structure

A tapeworm's body is ribbon-like and can be very long; one removed from a whale was over 30 m! The body is divided into very many sections called **proglottides** (sing. proglottis) (Fig. 4:13). It has no mouth or gut, but absorbs food digested by its host through the surface of its body.

The head end is very small and bears suckers, and in some species there are also hooks by which it attaches itself to the wall of the intestine. This is essential as otherwise the muscular movements of the wall of the intestine would squeeze it along with the food and it would pass out of the

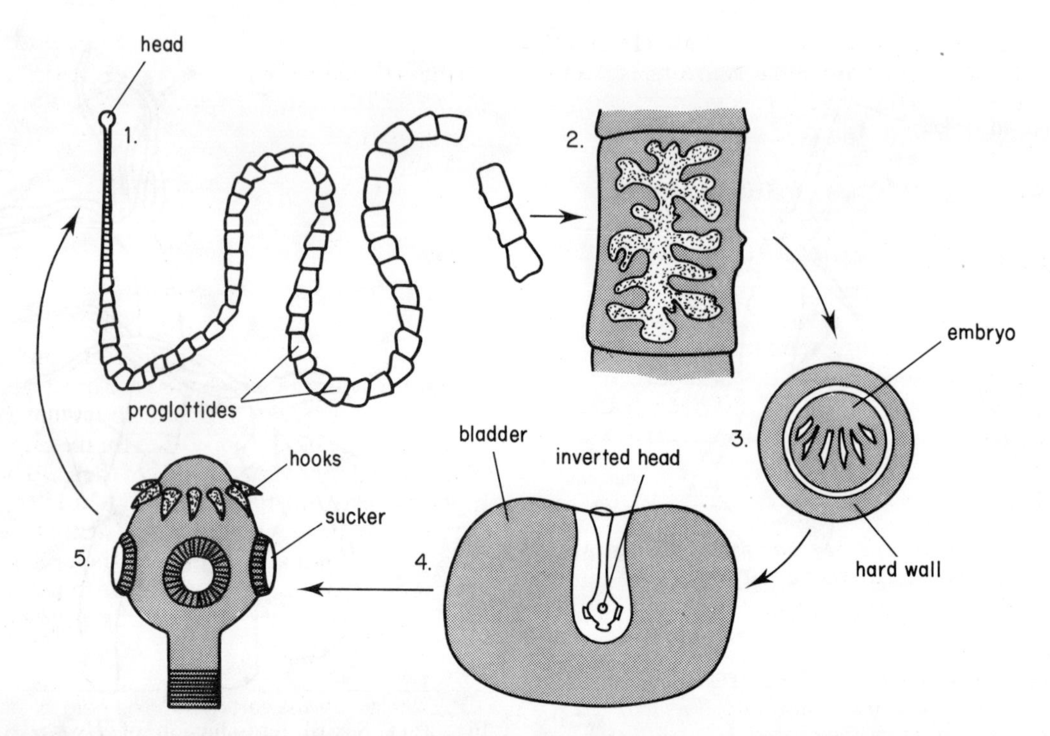

Fig. 4:13 The life history of a tapeworm: 1. Adult worm in intestine of primary host. 2. Single detached proglottis full of eggs which passes out with the faeces. 3. 6-hooked embryo which may be swallowed by secondary host. 4. Bladderworm which develops in the tissues of the secondary host. 5. Head enlarged.

anus and die. New proglottides are constantly being formed just behind the head while old ones break away from the other end when mature. As each proglottis enlarges it develops a complete set of reproductive organs and by the time it drops off it is no more than a bag of fertilized eggs. These proglottides are deposited with the faeces of the host.

Life history

All tapeworms live in two hosts during the course of their life history. The adult stage occurs in the intestine of the **primary host**, and another stage, the **bladderworm**, lives in the muscles of the **secondary host**. We will take the dog tapeworm, *Taenia serrata*, as an example, as this is a common one. The adult stage occurs in the intestines of dogs and foxes, and the bladderworm stage in the muscles of rabbits. When the proglottides, full of eggs, pass out of the dog with the faeces, the proglottides soon die and disintegrate. If the eggs come in contact with vegetation and are eaten by a rabbit, their hard coats are digested by the stomach juices of the rabbit and a microscopic embryo is released from each. This has six hooks enabling it to bore through the intestine wall into a blood vessel. The embryo is carried to the muscles where it grows into a bladderworm. This is a small, glistening, hollow sphere in which the head-end of a new tapeworm is developing.

The bladderworm will not develop further unless the rabbit is eaten by a dog or fox. In this event the head becomes everted and it becomes attached to the wall of the intestine and grows into another worm.

Whatever the species of tapeworm there is always a feeding relationship between primary and secondary hosts, otherwise the bladderworm would not be able to infect the primary host. Here are some other examples:

PRIMARY HOST	SECONDARY HOST
cat	mouse
lion	buffalo
trout	minnow
rat	rat flea

In spite of this relationship, you will realise that getting from one host to another is a very chancy affair. For example, the likelihood of the dog tapeworm eggs being swallowed by a rabbit is very slight. However, as tapeworms lay enormous numbers of eggs, this makes the event more probable.

Tapeworms in humans

It is possible for humans to acquire tapeworms through eating meat which contains bladderworms. One species may occur in pork and another in beef, but the likelihood of infection is remote in countries where proper precautions are taken. If sanitation is good (so that faeces are not left exposed), the chance that eggs will pass from an infected person to the food of pigs or cattle is very small; secondly, if meat is inspected (many countries have strict laws about this), infected meat is unlikely to reach the shops; thirdly, if the meat is properly cooked the bladderworms are killed.

Earthworms

These belong to the phylum Annelida along with marine worms, such as lugworms and ragworms, and leeches.

Most soils contain earthworms although they are most numerous in those which have a high humus content derived from such things as rotting leaves.

There are many species of earthworms in Britain, the largest being the common earthworm, *Lumbricus terrestris*, and the long worm, *Allolobophora longa*. These are much alike but the latter is a duller brown at the front end and greyer further back. A red worm with yellow rings, *Eisenia foetida*, is common in compost heaps.

Counting earthworms in a field

How many earthworms do you think there would be in a grassy field such as a playing field? If the area was a large one, e.g. one hectare (10,000 m^2), it would obviously be impossible to count them, but it would be reasonable to count the number in a single metre square and multiply by 10,000. But not every square will have the same number. So to be more accurate you should take at least five squares at random and find the average before making the calculation. Proceed as follows:

Use a 2% solution of formalin to bring the earthworms to the surface for counting. As this may temporarily damage the grass do not use your best lawn for the experiment.

Cut a piece of string just over 4 metres long so that when the ends are tied together it will exactly enclose an area of 1 m^2 when pegged out with four skewers. If the soil is dry, soak it with water first. Now pour over the area 10 litres of the 2% formalin. When the worms appear, collect them with forceps and give each a quick wash in water to remove any formalin, before putting them in a beaker. It will be about 15 minutes before they all come up. Count them, take the average for the number of squares treated, and calculate how many there would be in 1 hectare. Keep one large worm for examination; the rest should be returned to the soil.

It is possible by this method to compare earthworm populations in various kinds of soil and find the conditions they like best. Would it be better to compare numbers or total weight?

Structure

You will see from Fig. 4:14 that the body of the earthworm (like all members of the phylum Annelida) is built on the plan of a tube within a tube. The outer tube is the body wall, which is mainly muscular, and the inner one is the gut. Between the two is the body cavity or **coelom** which is filled with fluid. It acts as a shock absorber and provides something firm for the muscles to act upon. A blood system is present; annelids are the simplest animals to have one.

If you examine one of the larger earthworms you will see

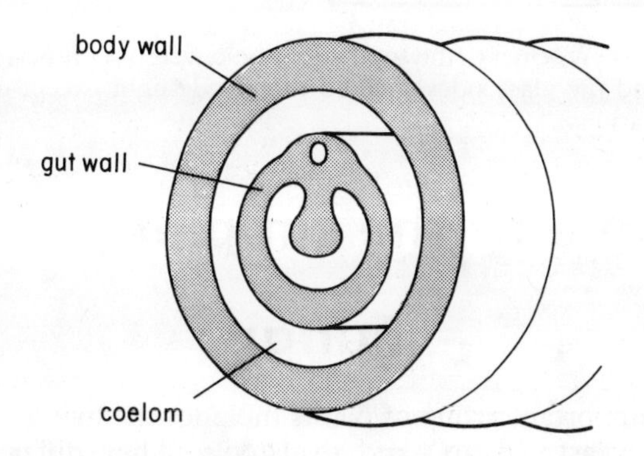

Fig. 4:14 The body of an earthworm is built on a plan of a tube within a tube.

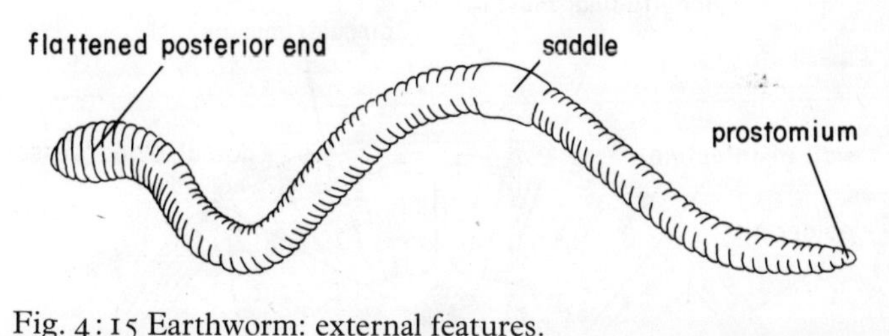

Fig. 4:15 Earthworm: external features.

that the body is divided into a great number of segments (about 150) most of which are very much alike (Fig. 4:15). If mature, it will have a **saddle** about one third of the way down from the head; this is used for making an egg cocoon. The mouth is on the underside of the head below a projecting **prostomium** (Fig. 4:16). The hind end of the body is flatter and wider, thus helping the worm to grip its burrow. The anus is right at the end. On the underside, on all but the first and last segments, there are four pairs of bristles or **chaetae**: these project backwards and help the worm to grip the soil.

1. Examine a living earthworm and note its external features (as above). Note the pink line going down the body on the upper surface; this is a blood vessel showing through the skin. Look at this vessel with the help of a lens. Can you see its walls contracting? Which way is the blood flowing?
2. Gently stroke the underside of the earthworm from the head end backwards; notice how smooth and slippery it is due to the secretion of mucilage from its skin. Now stroke it in the opposite direction: can you feel the chaetae?
3. Observe how it moves. You can see this best if you put it on a piece of coarse paper. Notice how it elongates its front end, segment by segment; this action helps it to penetrate between loose soil particles when burrowing.

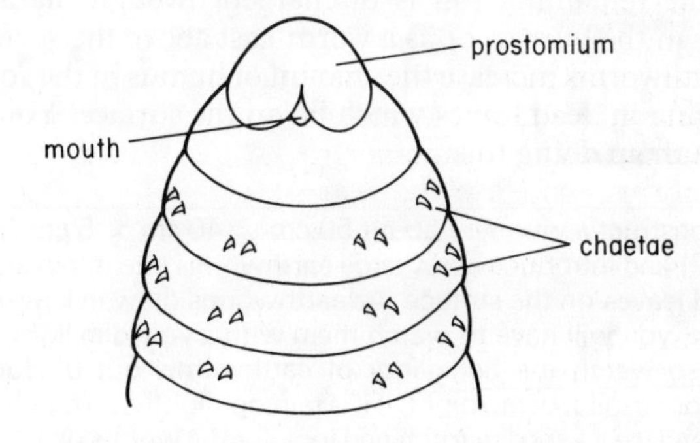

Fig. 4:16 Earthworm: ventral view of head region.

Movement

Two sets of muscles in the body wall are concerned with movement—one circular, the other longitudinal (Fig. 4:17). When the circular muscles of a segment contract and the longitudinal muscles relax it becomes longer and thinner; when the longitudinal muscles contract and the circular muscles relax it becomes shorter and fatter. Locomotion occurs when waves of muscular contraction pass down the worm. It is prevented from slipping backwards by the chaetae, which are protruded from those segments in which the longitudinal muscles are contracted.

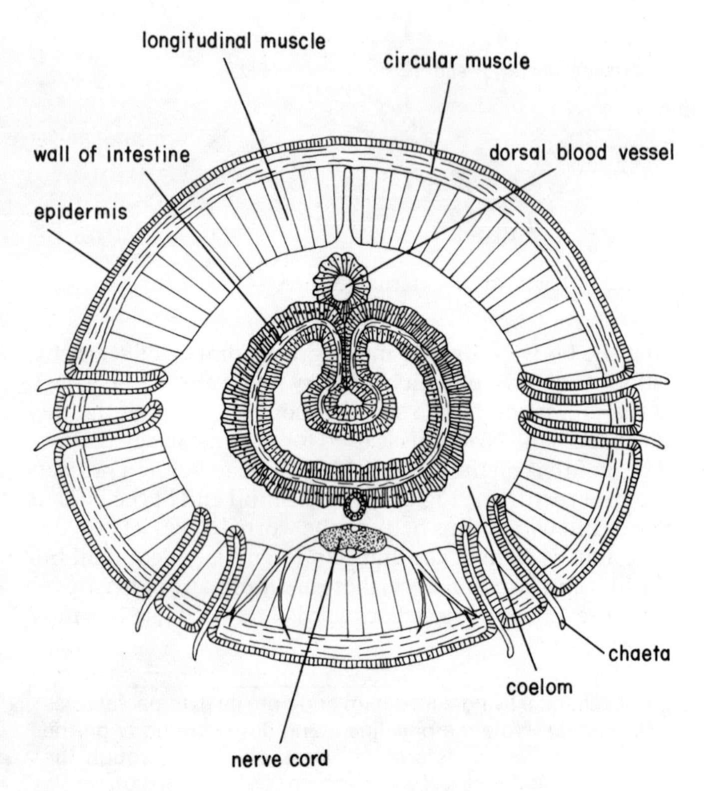

Fig. 4:17 Transverse section through the intestinal region of an earthworm.

Feeding

Earthworms feed by taking in large quantities of soil through the mouth as they burrow; it is ground up in a muscular part of the gut, the **gizzard**, and digestive juices are poured on it which dissolve the particles of humus. The soluble products of digestion are absorbed into the blood and the remaining soil is discharged through the anus, either in the burrow or as a worm-cast above the surface.

Earthworms increase the amount of humus in the soil by dragging in dead leaves which lie on the surface. You can watch them doing this.

> Construct a wormery about 50 cm × 40 cm × 5 cm (Fig. 4:18) and introduce a few large earthworms into it; put some dead leaves on the surface. As earthworms draw in leaves at night, you will have to watch them with a very dim light.
>
> Also watch the behaviour of earthworms out of doors. Choose a mild, damp night; if it is raining slightly, so much the better. Use a *very* dim torch and look for the worms on a lawn or similar place with short grass. You will see them lying on the surface of the ground with their hind ends in the burrows. Tread very lightly or you will disturb them. Observe how they search for leaves or rotting materials. Notice the effect of bringing the torch nearer to them. What happens if you tread more heavily? Touch one when it is fully stretched out and see how quickly it moves. This is its reaction if it is attacked by a badger or a little owl. Which muscles is it using when it reacts in this way?

Habits

In the daytime earthworms usually remain below ground in their burrows. The latter are often plugged with leaves which the worms have brought back during the night (on gravel paths they use piles of small stones instead). They grip these with their mouths by suction.

There are interesting investigations you could carry out on earthworms. You could test their reactions to different foods, such as onion and carrot, by scattering similar sized pieces in equal numbers on the surface of a wormery and noting if one food disappears more quickly. You could also test in the same manner whether they select leaves of different species and shapes and drag them into their burrows.

The importance of earthworms

Earthworms are of great importance in keeping soil in good condition. By their burrowing activities they make the soil looser and more porous; this improves the drainage and aeration, bringing more oxygen to the roots of plants which need it for respiration. By their feeding activities the soil is broken up into fine particles; it also becomes thoroughly mixed as they take it into their bodies at one level and deposit it on the surface in the form of worm casts, or in spaces within the soil itself. In addition, by drawing leaves into their burrows, organic matter becomes incorporated into the soil and, when this rots, it provides more nutrients for plants.

Earthworms are also a most important source of food for other animals. Moles are dependent upon them, shrews and hedgehogs take them regularly, badgers eat them in vast numbers and many birds such as thrushes and blackbirds rely on them.

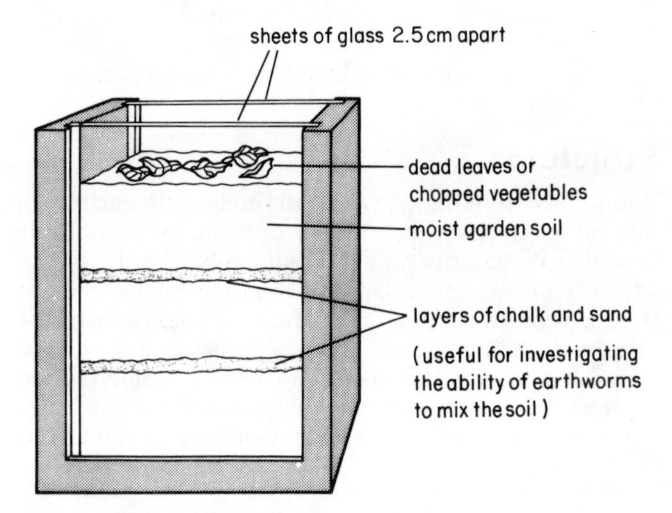

Fig. 4:18 Wormery (the top should be covered with perforated zinc and the glass sides covered to exclude light except during observations).

Some of the simpler plants

Fungi

This important group of plants includes the moulds, mildews, yeasts (p. 91) and toadstools. They differ from typical green plants in two important respects:

1. They have no chlorophyll, and so they are unable to photosynthesise. Instead, they have to feed on complex

organic substances (carbon compounds) obtained from dead or living material. If they feed on dead organic matter they are called **saprophytes** and the result of this process is the decay of the material. If they feed on living matter they are called **parasites**.

2. Their bodies are built up from long threads called **hyphae**. Sometimes these form a tangled mass rather like cotton wool called a **mycelium**, which creeps all over and penetrates into the organic matter on which the fungus is feeding. In the larger fungi such as mushrooms, the hyphae become bound together to form the solid reproductive bodies which we use as food. In many of the simpler fungi the hyphae are hollow branching tubes and many nuclei occur in the cytoplasm which lines the wall. In others the hyphae have cross walls and are thus divided into true cells. The walls of the hyphae are not made of the cellulose typical of green plants, but of a nitrogenous compound.

Fungi reproduce by means of spores which are so small and light that they may be carried almost everywhere by the slightest wind current. A mushroom is said to release half a million spores every minute for several days!

> Observe these spores by cutting off the stalk of a mushroom and inverting the cap over a piece of paper: put a cover on top to eliminate draughts and examine the paper next day. You will see a pattern corresponding to the underside of the cap composed of millions of spores (Fig. 4:19). If you try this with various toadstools you will find that the spores vary in colour according to species.

Fig. 4:19 Photograph of the inverted cap of a mushroom and the spore pattern it has produced.

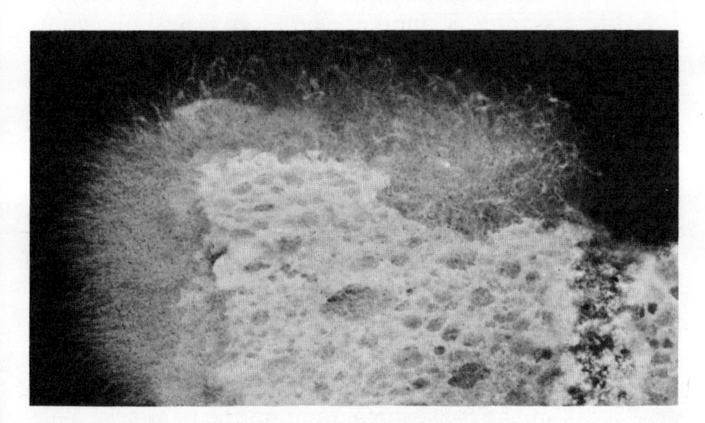

Fig. 4:20 Moulds growing on damp bread.

Saprophytic fungi

Because fungus spores are so common in the air, organic matter which is damp and exposed to the air soon goes mouldy as the fungus spores settle, grow and feed upon it. The destruction of timber in houses due to wet or dry rot is also due to saprophytic fungi. Thus fungi cause decay by feeding on the dead organic matter by a process called **external digestion** (for details see p. 101). Decay is a most important process as it causes nutrients to be returned to the soil to be used once more. Imagine what it would be like if the dead bodies of plants and animals never decayed!

> You can grow many kinds of saprophytic fungus for yourself. Take a slice of bread, moisten it with water and then sprinkle some dust on it (not chalk dust). The dust is sure to contain fungus spores. Place it under a glass bell-jar or plastic cover, put it in a warm, dark place, and examine it at intervals over the next few weeks. Note how the mycelia of the different fungi vary in colour and in the way they grow. Look out specially for a species which puts up vertical hyphae which end in black rounded heads. This will probably be a species either of *Mucor* or *Rhizopus*: either will do for examination as their structure is very similar. Examine a portion of the mycelium in a drop of water under the microscope.

Mucor

Structure and asexual reproduction

The mycelium consists of a tangled mass of hollow, branching hyphae. The cytoplasm lines the walls leaving a continuous vacuole in the centre containing cell sap which acts as a transport system for food substances in solution. Some hyphae penetrate the organic matter on which the fungus is living and feed the whole mycelium by digesting the food externally and absorbing the products. Other hyphae

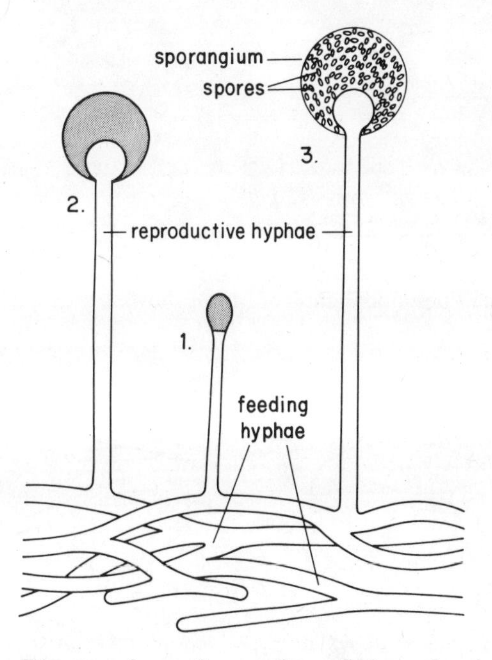

Fig. 4:21 Diagram of part of a mycelium of *Mucor* showing three stages in sporangium formation.

grow vertically and form rounded heads called **sporangia** (Fig. 4:21). Here the cytoplasm collects and breaks up into hundreds of spores. When ripe, the sporangium wall splits and the spores are released; they may be carried long distances by wind currents (in some species the spores are sticky and are carried by insects). Each spore is capable of growing into a new mycelium if it settles on suitable organic matter. This method of reproduction is asexual as no fusion of cells is involved.

Sexual reproduction

Mucor can also reproduce by a sexual method called **conjugation** which results in the formation of **zygospores**: these can withstand longer periods of drought than the asexually produced spores. Zygospores form as a result of contact between two hyphae (Fig. 4:23). The tips swell up, the wall between them dissolves, and the nuclei fuse in pairs, one from each hypha. The resulting zygospore forms a hard, rough wall round itself which is resistant to water loss and may remain dormant for months before germinating. When this happens, a single vertical hypha grows out from the zygospore and produces a sporangium at its tip. On bursting, many spores are released, each capable of producing a new mycelium.

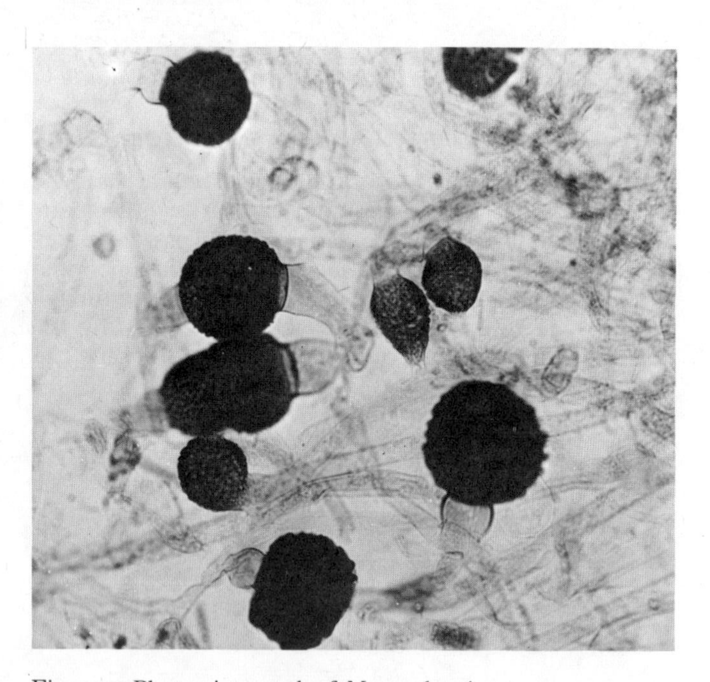

Fig. 4:22 Photomicrograph of *Mucor* showing zygospores.

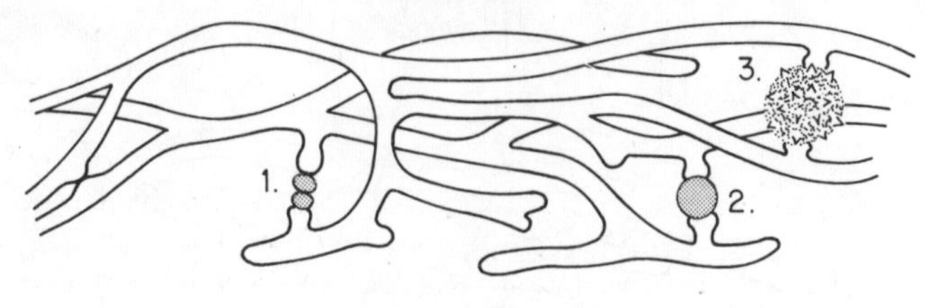

Fig. 4:23 Stages in sexual reproduction of *Mucor*. 1. Hyphal tips meet and swell. 2. Dividing wall dissolves and contents fuse. 3. Zygospore swells and forms hard wall.

Parasitic fungi

In contrast to the saprophytic fungi which feed on dead organic matter and cause decay, there are may parasitic species which live on living plants and animals and cause disease. Most plants have their own specific fungal parasites so it is necessary for the fungal spores by some means to get to new plants of the same host species. The spores are carried by wind, water, soil or insects and even on the seed coats of the infected plants. By producing vast numbers of spores, the chance of at least some reaching another plant becomes greater. Man, by growing crops in fields, makes it much easier for the fungus to spread as the plants are nearer together than in the wild.

Many fungi are of great economic importance. Some such as **rusts** and **smuts** attack cereals, e.g. maize, rice and wheat, greatly reducing their yields and sometimes destroying the whole crop.

Rusts often form patches and streaks of brown, orange or red on the leaves of the plants while smuts attack the inflorescences (flower heads), forming black or brown sooty masses which destroy the developing grain.

Another group of fungi known as **mildews** also causes widespread damage to crops. These form downy masses of hyphae and spores on the surface of leaves and fruits. One species, *Phytophthora infestans*, which causes potato and tomato **blight**, played an important part in history as it caused the complete failure of the potato crop in Ireland in the middle of the last century. This resulted in widespread starvation and caused a great number of families to emigrate, many to the United States.

The outbreak of **Dutch elm disease**, which had killed 16 million elm trees in Britain by 1980, was caused by another parasitic fungus spread by small beetles. The beetles live most of their lives in the bark only leaving as adults to mate and lay eggs on another tree. The fungus carried by the beetles produces a toxic substance which causes the transport system of the tree to become clogged up, leading eventually to its death.

Another fungus which attacks trees is the honey fungus, *Armillaria mellea*, a toadstool which spreads underground by means of long black cords made up of hyphae; these enter the roots of trees and eventually kill them.

Some fungi cause diseases in man. **Ringworm**, in spite of the name, is a disease caused by a fungus which forms a mycelium in the skin causing the hair to fall out, usually in a circular patch (Fig. 4:24).

Athlete's foot is another common fungal infection which usually attacks the skin between the toes where it finds warmth and moisture. To prevent this, it is necessary that feet should always be kept clean and the region between the toes dried after bathing. It spreads very quickly by spores, often on towels; this is a good reason for using only your own towel.

Fungal control

Many methods are used to combat fungus diseases in plants. This may be done by:

1. Carrying out breeding experiments in an attempt to produce crop varieties which are immune to attack. This has been successful for some cereals.

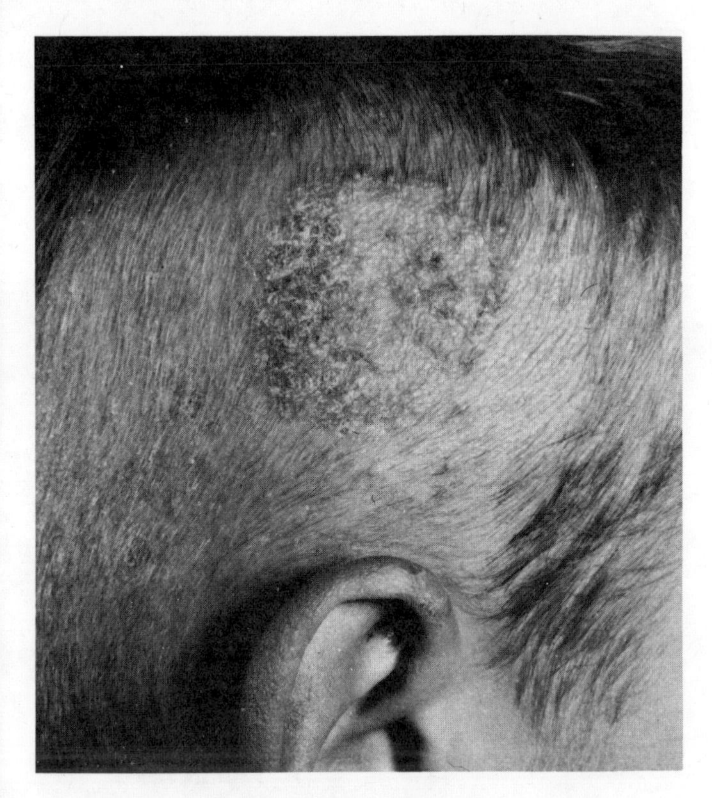

Fig. 4:24 Ringworm: a fungus which attacks human skin.

2. Spraying with chemicals called **fungicides**. These are mainly copper compounds and can be very helpful if the crop is sprayed before the disease becomes conspicuous, e.g. for potato blight.
3. Good farming techniques: a) weeding, which reduces the humidity around the crop by improving air circulation, b) by using a crop rotation (p. 110) which reduces the likelihood of soil-borne infection, c) by good manuring which makes plants stronger and more resistant.
4. Sterilising the soil by heating it strongly. This is done when raising seedlings commercially under glass to prevent infection from soil-borne fungi.

Mosses

Mosses belong to the phylum Bryophyta (p. 17). Their structure is intermediate between the algae (which have no leaves and stems) and the ferns (which have leaves, stems and roots). They have hair-like structures called **rhizoids**, which penetrate the soil, absorb water and help to anchor the plant, but these differ from true roots in having no woody conducting cells. Moss plants do not grow singly, but in clumps or mats.

Examine their structure by separating out individual plants from a clump, washing away any soil that clings to them. If possible choose one which has a capsule attached (Fig. 4:26). Note the presence of simple leaves clustered round the stem and the absence of true roots.

The capsule which projects from the tip of the moss plant produces the spores. Examine it with a lens. When young it is covered by a cap, but when mature the cap falls off and the spores are dispersed by the wind as they are very light. This only happens in dry weather because at the entrance to the capsule there is a ring of stiff teeth which close the entrance when wet, but which bend outwards when dry.

The life history of a moss (Fig. 4:27) involves two distinct generations which alternate with one another. The first is the typical moss plant with stem, leaf and rhizoids. It is called the **gametophyte** because it forms at its tip the sex organs which produce either eggs or sperms, or in some species both. Fertilization takes place in wet weather, and the fertilized egg (zygote) grows into a capsule on the end of a long stalk. This is the second generation known as a

Fig. 4:25 A common moss, *Polytrichum*.

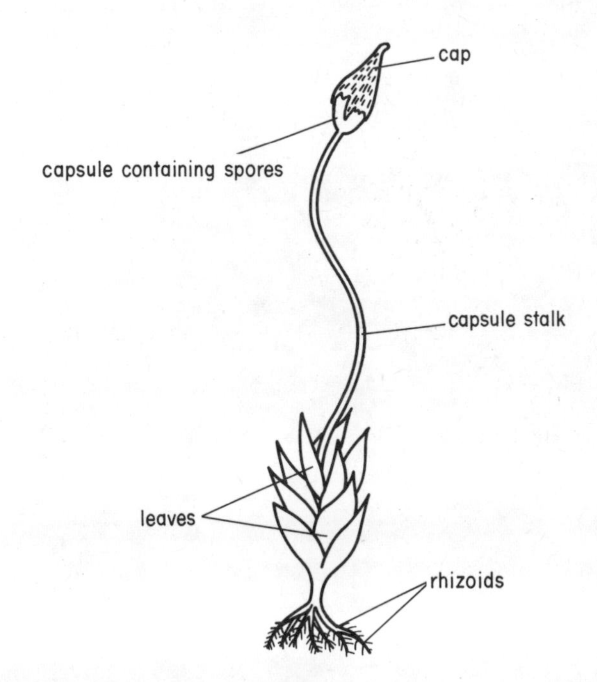

Fig. 4:26 An isolated moss plant.

sporophyte because it produces spores asexually. If the spores fall on damp soil when they are liberated, each will form a green thread-like structure from which a new moss plant will develop. This **alternation of generations** between a gametophyte and a sporophyte is characteristic of all mosses and liverworts.

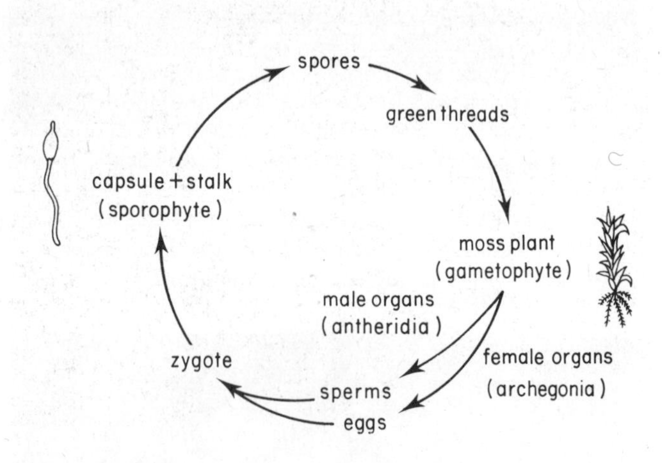

Fig. 4:27 Life cycle of a moss.

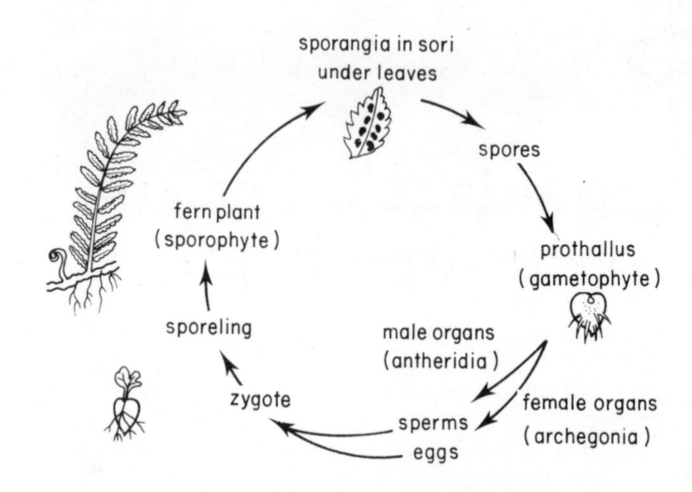

Fig. 4:28 Life cycle of a fern.

Ferns

Ferns have a rather similar life history, but in these the sporophyte is the large plant we recognise as a fern, and the gametophyte, called a **prothallus**, is a flat green structure less than 0·5 cm long which lives on the surface of the soil (Fig. 4:28). In the species illustrated the spores are formed in sporangia which are grouped together in sori on the underside of the leaflets (Fig. 4:30) but when sectioned and examined under a microscope the sporangia containing the spores may be seen (Fig. 4:31). The sori appear as brown spots to the naked eye.

Fig. 4:29 Fern plant (sporophyte).

Fig. 4:30 Leaflets of a fern frond showing sori.

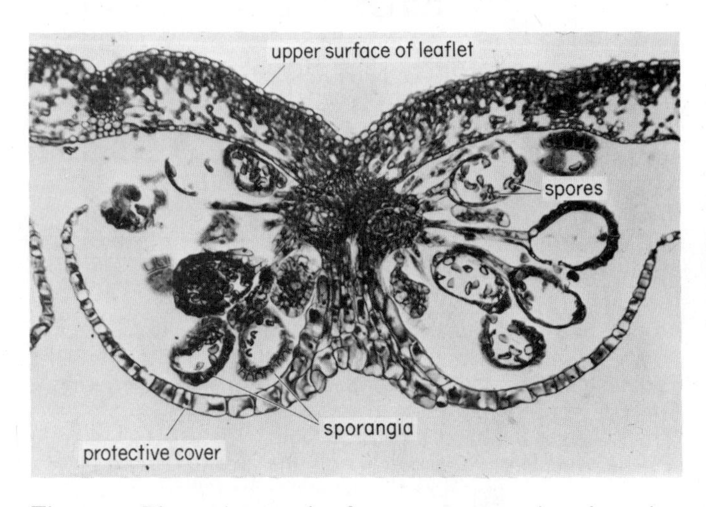

Fig. 4:31 Photomicrograph of a transverse section through a sorus showing the sporangia.

5

Organisation of the flowering plant

Plants, like animals, have their cells arranged in tissues, organs and organ systems. You can discover the general plan of structure by looking at whole plants of such species as wallflower, shepherd's purse, groundsel or buttercup. They differ in detail, but all have the same basic parts (Fig. 5:1).

We can now look more closely at the structure and functions of the different parts.

The root system

Examine some germinating mustard seeds which have been grown on damp filter paper in a covered Petri dish. They show very simple root systems consisting of a main

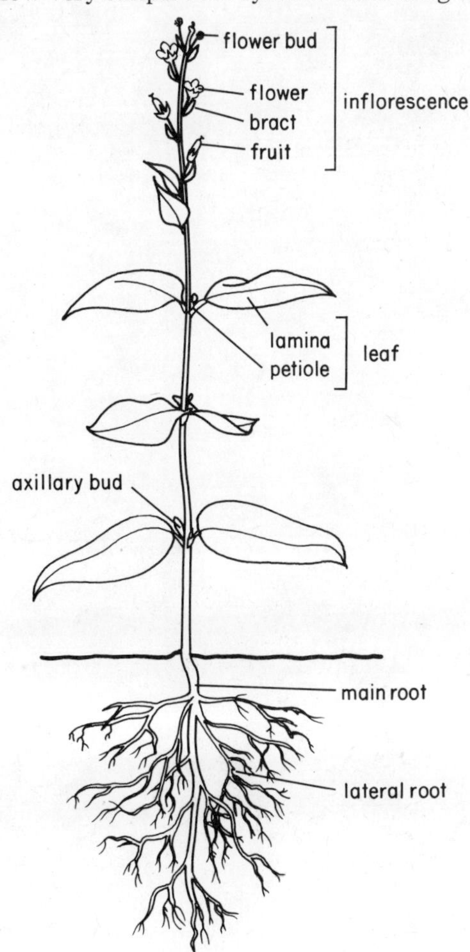

Fig. 5:1 Diagram of a whole plant to show the main structures.

Fig. 5:2 Roots of cress seedlings showing root hairs.

root with vast numbers of **root hairs** looking like cotton wool protruding from the surface (Fig. 5:2). Most root systems start in this way. Later, lateral branches arise with more root hairs near the tip of each, and further branching may occur more haphazardly to produce a very complex system. If the main root remains bigger than the rest it is called a **tap root system**, but when most of the roots appear alike and arise as a bunch from the base of the stem it is called a **fibrous root system**. The latter is characteristic of grasses (Fig. 5:3). Roots can also arise from stems and even leaves; they are then described as **adventitious**. If you pull off a piece of ivy from a tree trunk, you will find it was attached by roots growing from the stem. When gardeners take cuttings, they remove a shoot and roots grow out from the base when it is planted. Similarly, when a leaf of the African violet is planted, roots will grow from the leaf stalk.

The functions of roots

1. They anchor the plant firmly in the ground, preventing

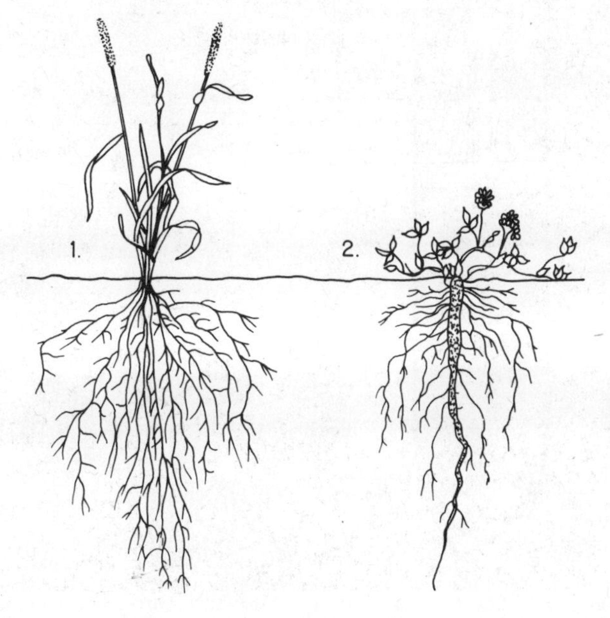

Fig. 5:3 1. A fibrous root system. 2. A tap root system.

it from being washed away by rain or blown over by the wind.

2. They absorb water and mineral salts from the soil and conduct them to the shoot system.

3. They often store food, sometimes getting very large in the process, as in carrots and parsnips.

Roots are admirably adapted for anchorage and water absorption by their pattern of branching. Both fibrous and tap root systems cover a relatively wide area around the plant, making use of the intricate air spaces between the soil particles to grow in all directions. Grasses, because of their fibrous root systems, help to keep the soil particles together, and thus prevent erosion. For this reason they are grown in such places as motorway embankments. A species called marram grass is also planted on sand dunes to help consolidate the sand. It is effective because of its network of underground stems and the adventitious roots which arise from them.

Different species tap different depths of soil, some being shallow rooting, others penetrating deeper. In this way they can live together in the same place without competing too much for the available water and salts.

The root pattern depends to some extent on the depth of the soil. When wheat grows in the very deep soil of the Ukraine its fibrous system may go down as much as 2·7 m, but in Britain this depth is impossible because of the underlying rock. Some trees growing in shallow soil send their roots parallel to the surface in all directions for 30 m or more. In deserts, they may penetrate very deeply to reach the water.

Structure of the root

We can study this better by examining thin sections under the microscope. If you examine a longitudinal section of a root tip (Fig. 5:4), you will see the **root cap**, a protective layer which prevents the delicate cells of the root from becoming damaged as the root grows between the soil particles. Behind it is a region of small cells which look alike and are capable of active division. This is the **region of cell division**. Compare these cells with the ones further back which become progressively more elongated and have large vacuoles. This is the **region of vacuolation**. Still further from the tip you will see that the centre of the root begins to look different from the rest. Here the cells have developed into the main conducting region of the root: this is known as the **stele** or **vascular system**. If you look at a transverse section through a young root, this will give you a better idea of the structure of the central stele (Figs. 5:6, 7 & 8). You should see that it is composed of different kinds of cells called **xylem** and **phloem**. The xylem is the woody portion of the root. It largely consists of elongated vessels, which conduct water and salts and are continuous with similar vessels in the stems and leaves. These vessels arise from columns of living cells placed end to end; they become long pipes when the end walls of these cells break down. The side walls are then strengthened with **lignin** (wood) and the cytoplasm eventually dies; hence they can no longer be called true cells. The phloem, by contrast, consists of living cells. Some of these, the sieve tubes, are also arranged end to end, their end-walls being perforated. This facilitates the conduction of food, made in the shoot, down to the tissues of the root.

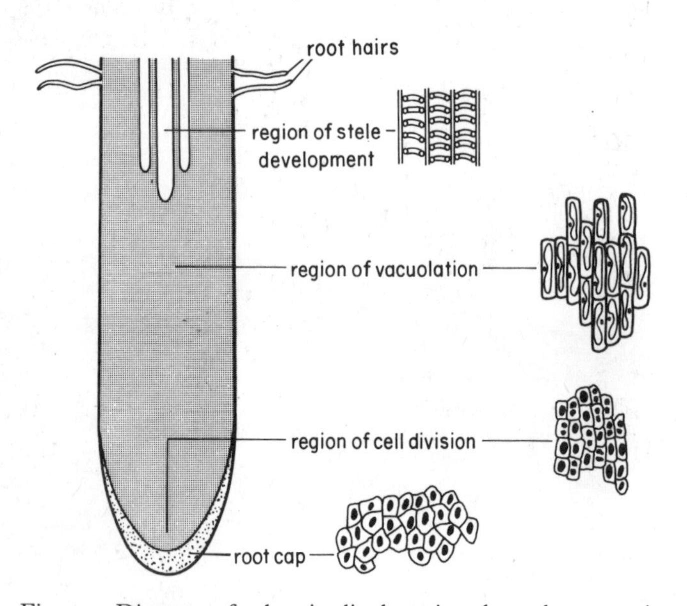

Fig. 5:4 Diagram of a longitudinal section through a root tip showing regions.

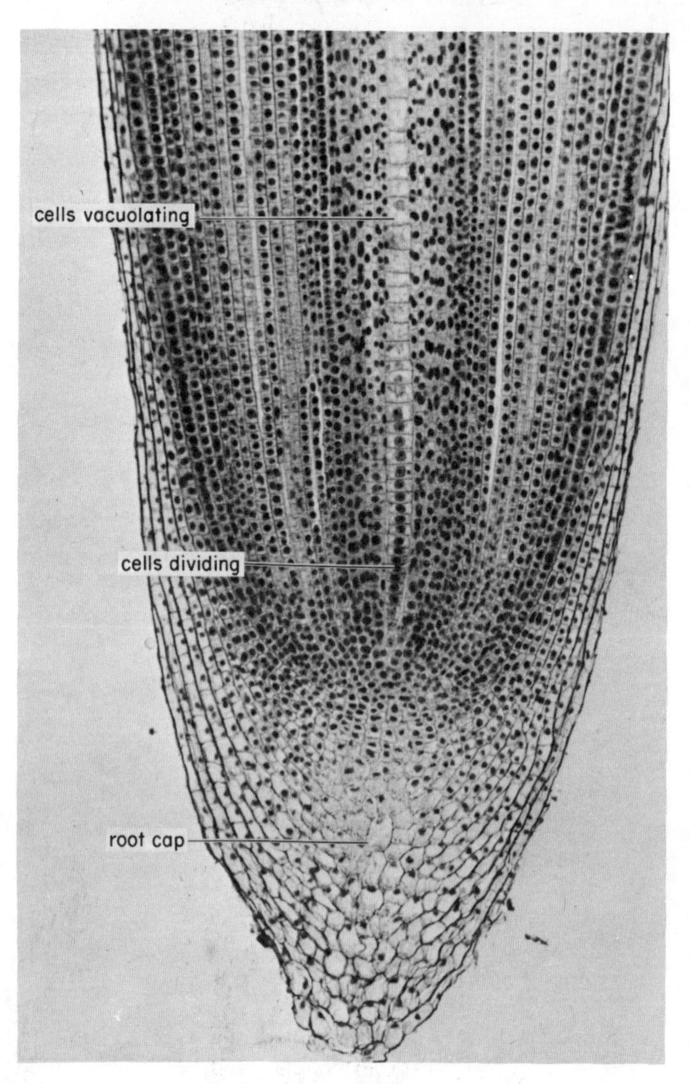

Fig. 5:5 Photomicrograph of a longitudinal section through the root tip of onion. × 60

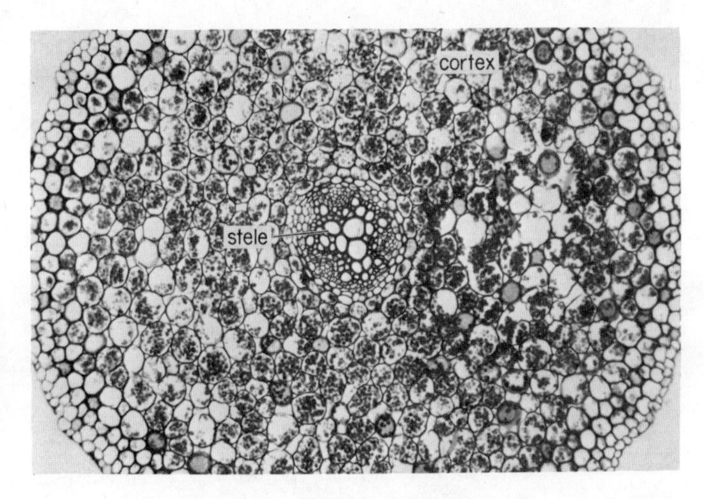

Fig. 5:6 Low power photomicrograph of a transverse section through a young buttercup root.

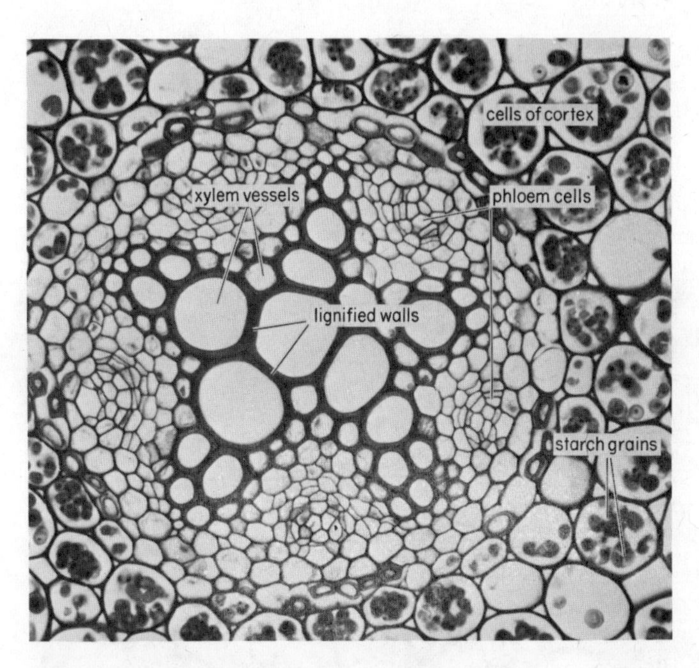

Fig. 5:8 Photomicrograph of the stele region of a buttercup root much enlarged.

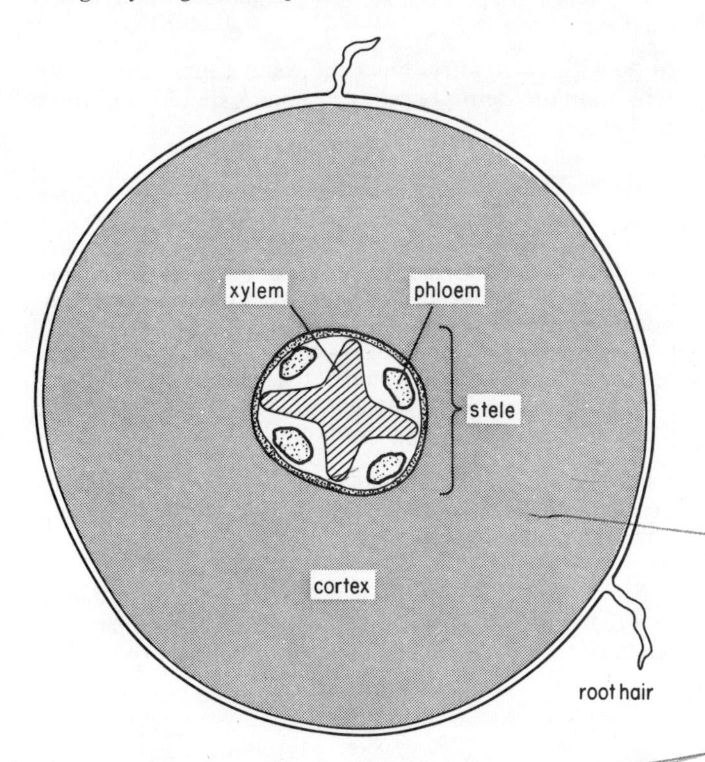

Fig. 5:7 Diagram of a transverse section through a young root.

The shoot system

The stem

A stem differs from a root in bearing buds and leaves although they are not always easy to see. It ends in a terminal bud which is the growing point of the stem. Other buds arise in the axils of the leaves. When a gardener prunes roses or fruit trees he makes a cut slightly above a bud. This stimulates the bud to grow into a new shoot in the spring.

The majority of stems are rigid and upright, but some such as honeysuckle or runner beans obtain support by twisting up other plants. Others are modified for vegetative reproduction, food storage or both. These include **runners** which creep along the ground and root at intervals to form new plants, e.g. strawberry; **rhizomes** which remain underground, e.g. iris; and **tubers** which are the swollen ends of rhizomes, e.g. potato (see p. 175). Rhizomes and tubers do not appear to be stems at first sight as their leaves are reduced to scales which eventually fall off leaving only a scar, also the buds present are often small.

The stems of herbaceous plants are green, but in shrubs and trees they become brown due to the formation of a corky bark on the outside.

The functions of stems

The main functions of the stem are support and conduction. The leaves have to be supported in such a way that they receive enough light for making food; this is why the *arrangement* of leaves on a stem is important. The flowers also need support because they need to be in the best position to be detected by insects or, if wind pollinated, to be blown by the wind. The stem also acts as the conducting system between leaves and roots, so that water and salts pass to the leaves and the food the leaves manufacture passes from them to other parts.

Structure of the stem

By examining a transverse section through the stem of a sunflower (Fig. 5:10), you will see that essentially it consists of a tough sheath of cells, the epidermis, a mass of thin-walled cells inside comprising the cortex and the pith, and a ring of **vascular bundles** (Fig. 5:11).

Each bundle consists of xylem, phloem and **cambium**. The xylem vessels, for water and salt conduction, are continuous with those in the root and leaves; they have smaller, lignified fibres around them to give added support. The phloem, as in the root, consists of sieve tubes for food conduction, and smaller, surrounding cells. Phloem cells have thin cellulose walls which might easily be crushed if it were not for the protection afforded by the xylem on the inside and a tough bunch of woody fibres on the outside.

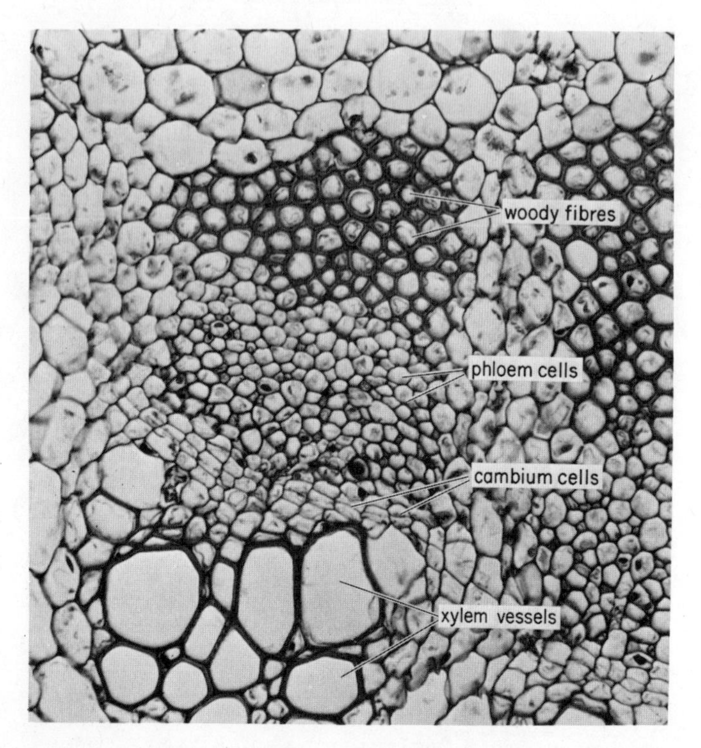

Fig. 5:9 High power photomicrograph of a transverse section through the vascular bundle of a sunflower stem.

The cambium, which lies between the xylem and phloem, consists of a thin layer of cells which are still capable of cell division (all the other cells of the stem such as xylem vessels or pith cells are permanent structures). It is by the division of these cells that a stem grows in thickness. Keeping in mind the main functions of the stem which are support and conduction, let us find out how it is structurally adapted to carry them out.

How does a stem keep rigid?

You will know that if plants are deprived of water they will droop. We can deduce from this that water has something

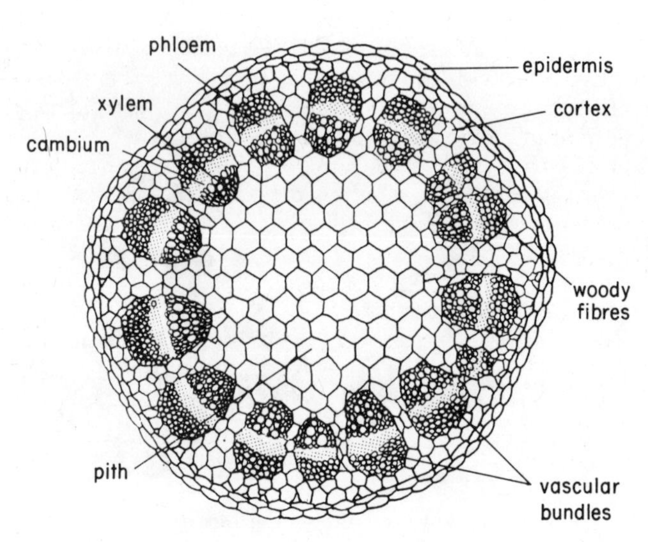

Fig. 5:10 Drawing of a transverse section of a young stem of sunflower.

to do with its rigidity. Does the water cause some sort of tension in the stem? These experiments should help you to find out:

Take a young green stem; a dandelion stalk will do. Cut it lengthwise into two and cut one strip into two again by a similar cut. Note how each strip curls and see whether the epidermis is on the inside or outside of the bend. What does this tell you about differences in tension between the epidermis and the inner cells?

Now take a piece of rhubarb stalk (this is really a large leaf stalk), and carefully cut a strip of the epidermis, leaving one end still attached. What happens when you try to replace the strip in its original position. Does it fit?

Now take a small cork borer and force it into the cut end of the rhubarb stalk to a depth of about 5 cm. Withdraw it carefully. What do you notice about the cylinder of cells you have isolated in the centre?

From these experiments you should be able to deduce that there are tensions inside these plant organs. You will see later (p. 85) that when cells are given plenty of water

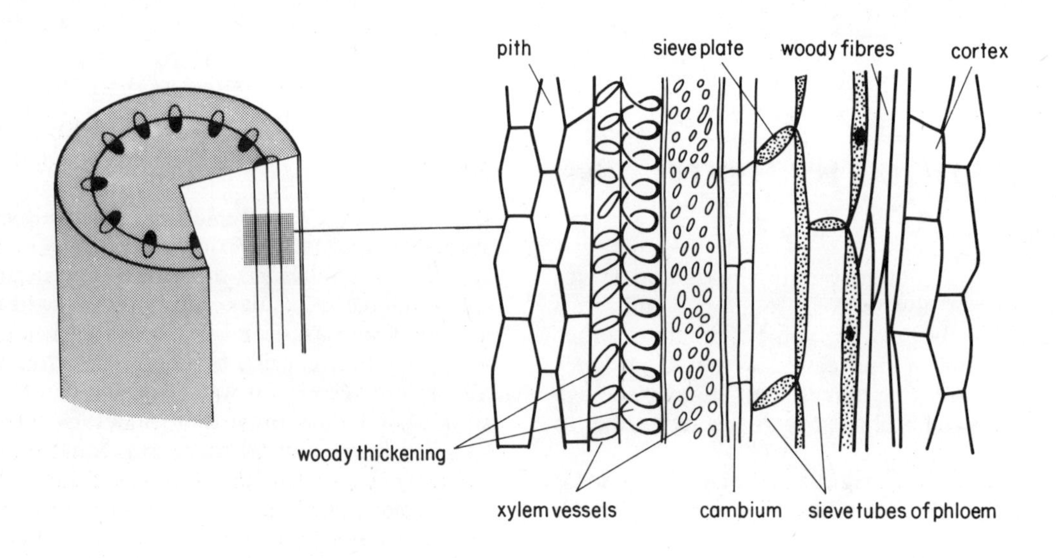

Fig. 5:11 Diagrammatic longitudinal section through a vascular bundle of a sunflower stem.

they absorb it by osmosis and swell up. This is what the pith cells do, with the result that they press against each other and against the epidermis. Because the epidermal cells are more rigid, they can only stretch a little. Thus the intake of water by the pith, and the pushing outwards of the expanding cells against the epidermis, produce a rigidity rather similar to the inner tube of a bicycle tyre when it is filled with air: it presses against the outer tube which will not stretch much, so the more air that is pumped in the more rigid the tyre becomes.

Shoots are subjected to high wind pressure which could easily bend and damage them. To withstand this bending strain, they have certain structural adaptations. Look at Fig. 5:10 once more. As wood provides greater support than cellulose, the position of the wood is significant. Can you see that the wood, which occurs in the vascular bundles, forms a ring of girders near the outside? This arrangement is excellent because it prevents too much bending; it is used for the same purpose by civil engineers in constructional work, e.g. the steel used for scaffolding is tubular. Compare this with the root structure where the lignified material is in the centre (as in a cable), an arrangement best suited to withstand a pulling strain.

The vascular bundles not only aid rigidity but contain the conducting elements for water, salts and soluble foods.

The leaf

Leaves all have a flat green **lamina** or blade, and most of them, a **petiole** or stalk. The petiole is attached to the stem and enters the leaf to form the mid-rib which gives rise to a branching system of veins. However, leaves differ from each other in shape and in many other ways (Fig. 5:12).

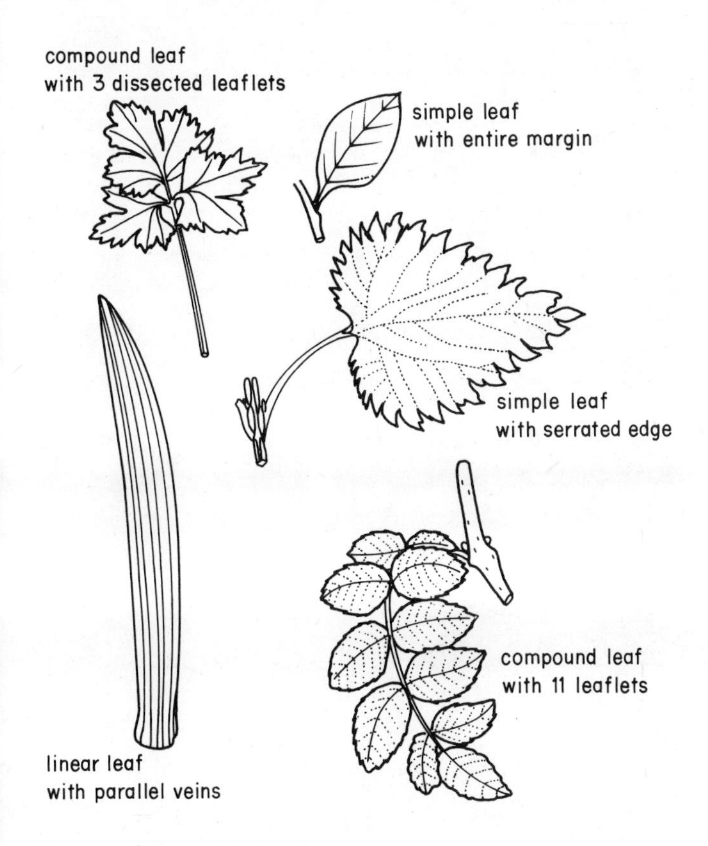

Fig. 5:12 Various kinds of leaves.

Some are hairy, others smooth. Some are in one piece (simple), others are divided into complete leaflets (compound). Simple leaves may be partially divided into segments (dissected). The edge of the leaf may be smooth (entire) or toothed (serrated). There may be a petiole present (petiolate) or it may be absent (sessile). Two green projections may occur at the base of the petiole, the stipules (stipulate), or they may be absent (non-stipulate). The veins may arise as branches from the mid-rib forming a minute network which can be seen when you hold it up to the light (reticulate veins), or the main veins may run more or less parallel to each other as in grasses (parallel veins).

Leaves also have characteristic smells and tastes. We flavour food with herbs such as mint or parsley, but some leaves are bitter or poisonous and we leave them alone. Could it be that a bad taste or smell would help to protect a plant from being eaten by animals? Can you think of any evidence to support this hypothesis?

Collect leaves from a number of different plants, selecting those which vary widely in their characteristics. Try and describe each as accurately as you can using the descriptions given above.

Functions of the leaf

There are several important processes which occur in leaves. First and foremost there is **photosynthesis**, the manufacture of carbohydrates such as sugar and starch. For this process light is needed, as well as carbon dioxide from the air, and water which comes up from the soil through the xylem vessels of root and stem.

Secondly, **respiration** occurs in all the living cells of the leaf, the oxygen needed for the purpose diffusing in from the atmosphere. Thirdly, **transpiration** takes place. This is the evaporation of water from the leaf, the water vapour passing out into the atmosphere.

In all three processes gases pass in or out of the leaf. Hence the greater the surface area the more efficient the exchange of gases will be. Thus the flat shape of leaves is an adaptation which allows quick gaseous exchange and at the same time allows a maximum amount of light to fall on them for photosynthesis.

Internal structure of a leaf

If you examine a transverse section of a leaf under a microscope (Fig. 5:14), you will see that the epidermis is a layer of colourless cells whose outer walls are thicker than the inner ones. This is due to the presence of a **cuticle** on the outside which helps to reduce evaporation of water and is a protection against mechanical injury, insect attack, or the entry of fungi and bacteria. It also helps to maintain the shape and rigidity of the leaf. In many leaves the cuticle is thicker on the upper surface than on the lower. Can you think of a reason for this?

The lower epidermis is perforated by a vast number of microscopic pores, the **stomata** (sing. stoma); in some species they occur on both surfaces. Stomata allow gases to pass in and out of the leaf and are capable of opening and closing, thus regulating this gaseous exchange (p. 99).

Between the two epidermal layers are numerous mesophyll cells which contain chloroplasts. The latter contain

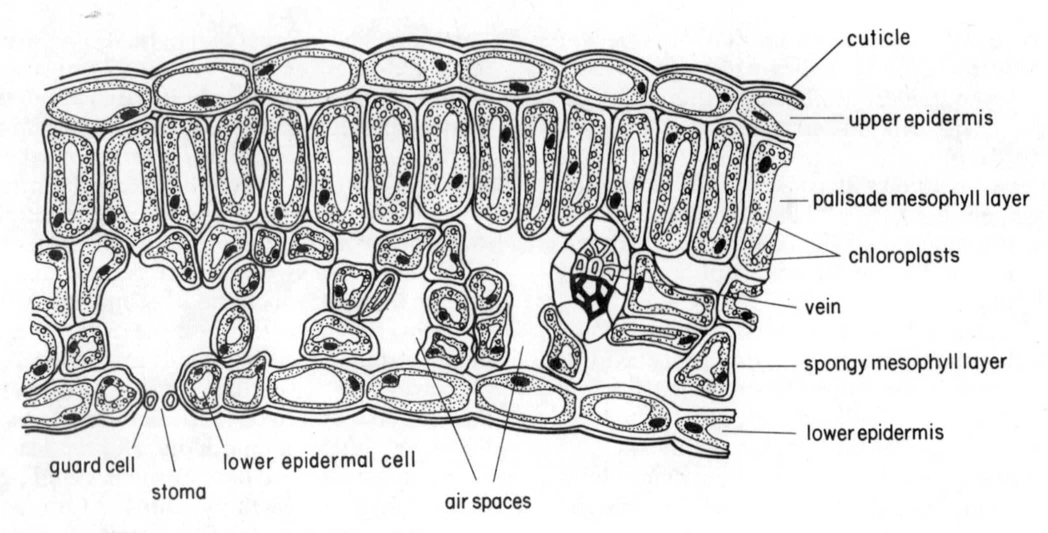

Fig. 5:13 Simplified diagram of a transverse section of a leaf.

the green substance chlorophyll, which is necessary for photosynthesis. The mesophyll cells vary considerably, those near the upper surface are elongated, crowded together and contain many chloroplasts, while those below are more irregular, have larger spaces between them and contain fewer chloroplasts. Why do you think most of the chloroplasts are near the upper surface?

The mid-rib and side veins which can be seen in cross section contain the xylem and phloem elements which are continuous with those of the stem and leaf stalk.

Thus the structure of a leaf is well adapted for its primary function of photosynthesis by having a large leaf surface to receive the light, many chloroplasts in positions where the light will reach them, a system of air spaces for conducting the necessary gases to and from the cells, and veins to bring water to the cells and conduct the sugars made in them to other parts of the plant.

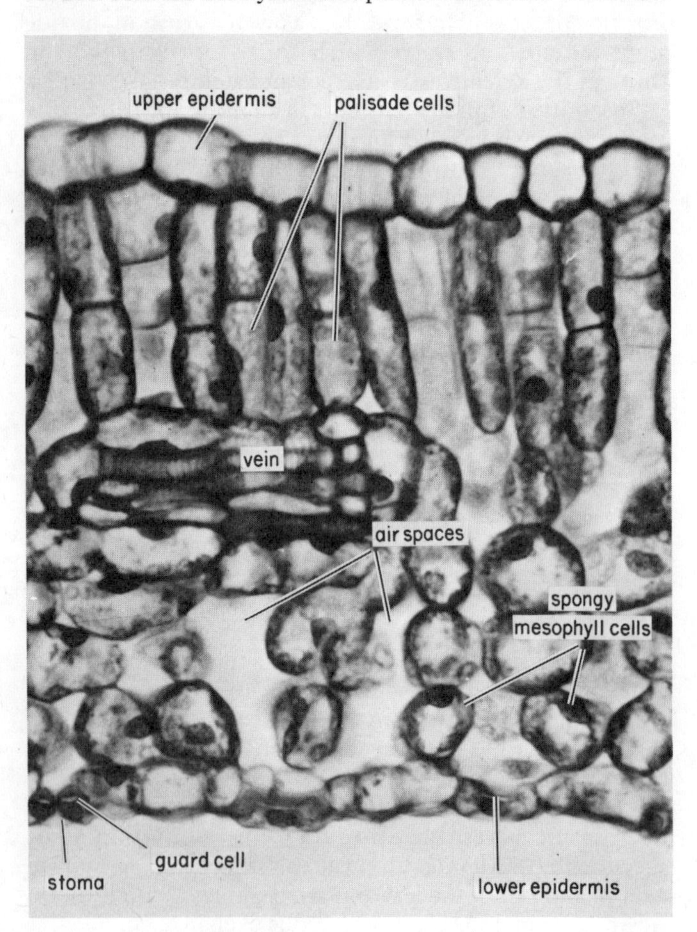

Fig. 5:14 Photomicrograph of a transverse section through the lamina of a privet leaf.

6

Insects: 1

The most successful invertebrates

Biologically speaking, insects are an extremely successful class of animals both in terms of numbers and in the distribution of their species. Nearly a million insects have already been named, i.e. more than 80% of all animal species, and many of these species contain astronomical numbers of individuals. In terms of distribution the picture is equally impressive as they are to be found in almost every kind of habitat except the sea, where only a few highly specialised insects occur.

The insects we shall study have been selected to illustrate aspects of insect life which are particularly important and interesting. Locusts and butterflies have been chosen to illustrate the basic structure of insects and their contrasting life histories. Mosquitoes and flies will be studied largely because of their important role as carriers of disease. Finally, the honey bee has been selected as an example of a species which has evolved a most complex society and which has even developed a 'language' of its own.

Insects belong to the phylum Arthropoda along with crustaceans, arachnids and myriapods. They share with them the important characteristics of having:

1. A skeleton on the outside, i.e. an **exoskeleton**.
2. A segmented body.
3. Various paired **appendages** which are jointed, e.g. legs and feelers.

The importance of an exoskeleton

The exoskeleton of an insect is called a **cuticle**. It is an excellent protection against mechanical injury. When a blue-bottle fly goes head-on into the glass of a window it is unhurt thanks to the cuticle's toughness. Yet the cuticle is very light and in many ways resembles a very strong plastic. The cuticle also keeps out fungal and bacterial spores which might otherwise damage the living tissues within. However, the chief function of the cuticle is to prevent too much water evaporating from the body, thus allowing insects to live in relatively dry places. The cuticle also aids movement as it serves as a rigid base for the attachment of muscles.

An exoskeleton also brings certain problems; these are rather similar to those which faced mediaeval knights when they wore armour. One is that any continuous hard covering impedes movement. To overcome this difficulty it was necessary for both knights and insects to have joints in parts of their armour. Nevertheless, the joints are the weak spots and in insects may serve as the point of entry for such attacking weapons as stings and poison jaws.

Growth is another problem if the armour does not stretch. The knight had no option but to acquire some new armour; the insect solves the problem by a process called **ecdysis** or moulting. So, when the pressure of growth becomes too great, all insects periodically split their cuticle, having first formed a new soft one underneath to take its place. This new cuticle is able to stretch considerably before it hardens up, consequently, after a moult, an insect looks distinctly bigger than it did a few minutes before. But ecdysis also has its hazards as the animal is very vulnerable before the new cuticle hardens, so many insects hide away during the moulting period.

Locusts

Locusts are excellent to study as they are large enough for you to see their structure easily (Fig. 6:1), you can breed them without difficulty in the laboratory and you can watch their behaviour at all stages of their life history. In addition, they are of great economic importance.

Locusts, grasshoppers and crickets belong to the order Orthoptera (straight and narrow wings). The African mi-

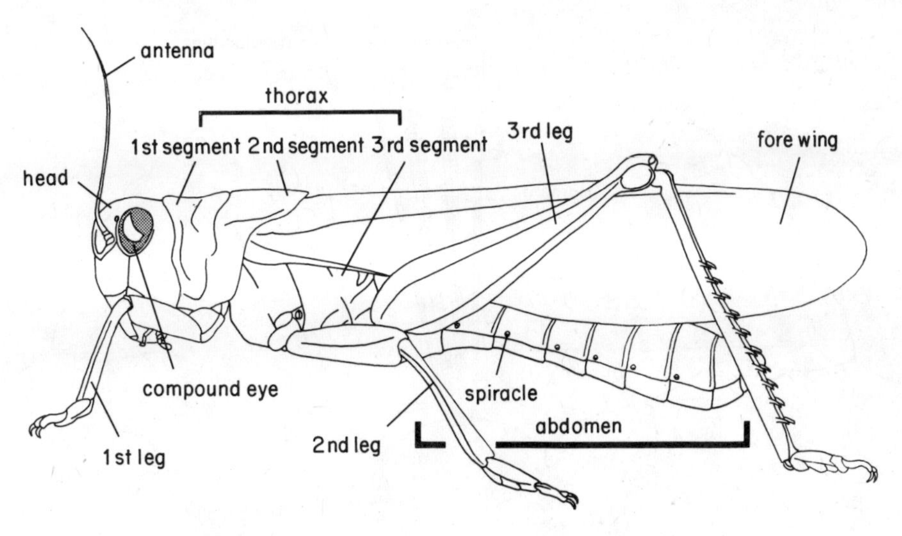

Fig. 6:1 Adult African migratory locust, side view.

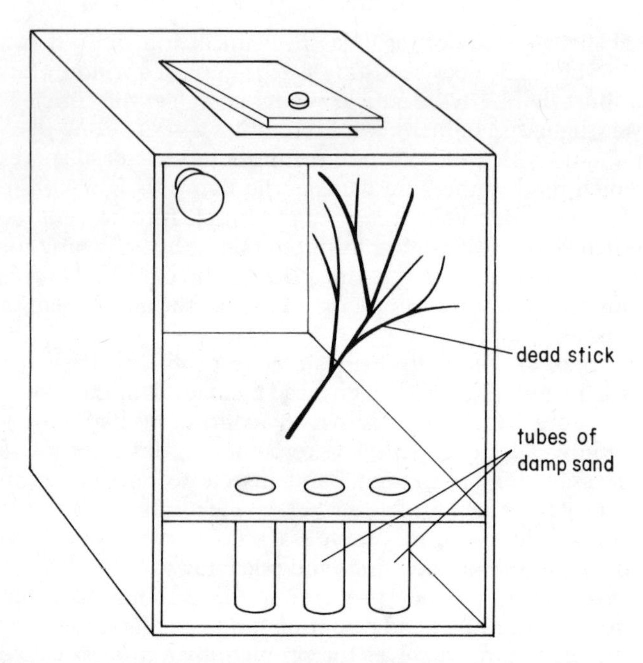

Fig. 6:2 Locust cage approximately 50 cm × 40 cm × 40 cm.

gratory locust, *Locusta migratoria*, lives in hot countries, hence it is best to keep them in heated cages.

> Set up a cage as in Fig. 6:2. You should aim at a temperature of about 34°C by day and about 28°C at night. This can usually be done by using two electric bulbs by day and one by night. Put in sticks for the locusts to rest on, and for their main food provide fresh grass each day, with wheat bran to supplement it. Clean the cage daily. When breeding, the adults will need sand in the specimen tubes for egg-laying which should be kept moist by adding a little water each day.

As locusts are very active creatures you will be able to see their structure best by examining a freshly-killed one.

External features

Locusts have a tough, but flexible exoskeleton. It not only covers the body, but also forms the main substance of the wings. The body is divided into three main parts—**head**, **thorax** and **abdomen**.

1. The head. This contains the brain. During locomotion the head is the part that goes first and so it must be well provided with sense organs to give information to the brain. For this purpose, it has a pair of jointed feelers, **antennae**, which are concerned with touch and smell, and a pair of **compound eyes** for sight. The eyes are called compound because under a lens you can see they are composed of many units. Fig. 6:3 is a section through a compound eye. Each unit points in a slightly different direction so the whole eye covers a wide field. The complete image produced is something like a mosaic of tiny dots of varying intensity, each dot resulting from a single unit.

The head is also the first part to reach food, so it contains the mouth and appendages called **mouthparts** which help to select, hold and prepare the food for entry into the mouth.

All insects have complex mouthparts, but they follow the general plan of having: a) an upper lip or **labrum**; b) a lower lip or **labium**; c) a pair of jaws, the **mandibles**; d) a pair of accessory jaws, the **maxillae**.

If you look at a locust head-on and slightly from below (Fig. 6:4) you will see the labrum which largely hides the other mouthparts. If the labrum is folded back the mandibles become fully exposed. The mandibles are extremely hard and are moved from side to side by strong muscles so that their toothed edges come together and help grind up the food. The maxillae are attached to the side of the head behind the mandibles; they are more flexible than the mandibles which they assist in breaking up the food with their toothed edges. Part of each maxilla is elongated to from a movable **palp** which is sensitive to touch and chemical stimuli. The labium acts as a lower lip and is really the fusion of two structures similar to maxillae; it has a smaller palp on either side.

2. The thorax. In all insects this is composed of three segments which are best seen from the side as a large plate, the **pronotum**, covers them dorsally. The wings are attached to the second and third segments; the forewings are tougher than the hind ones which lie below them when at rest. One pair of legs is attached to each thoracic segment and although they differ in size, each is built on the same plan of **femur** (pl. femora), **tibia** and **tarsus**, with two small additional joints where the leg is attached to the body

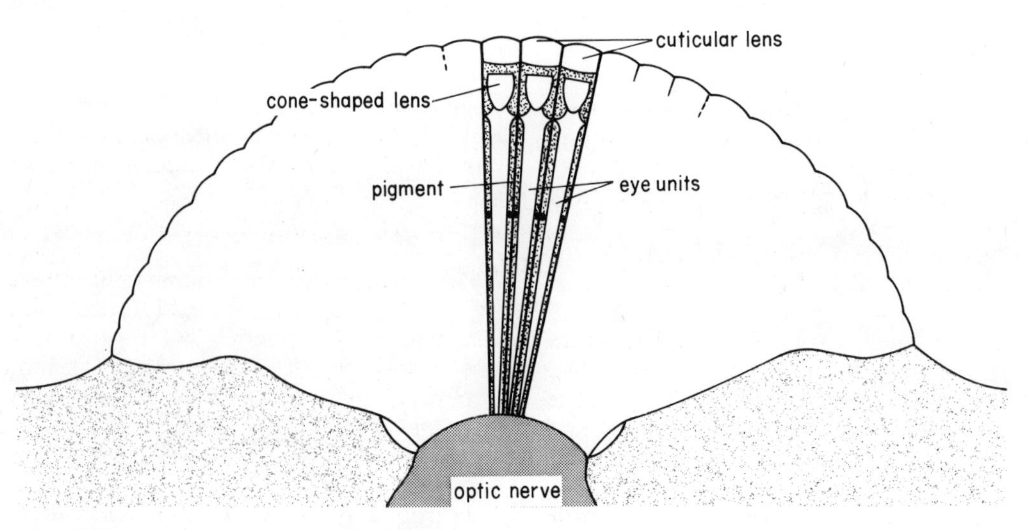

Fig. 6:3 Section through a compound eye (much simplified).

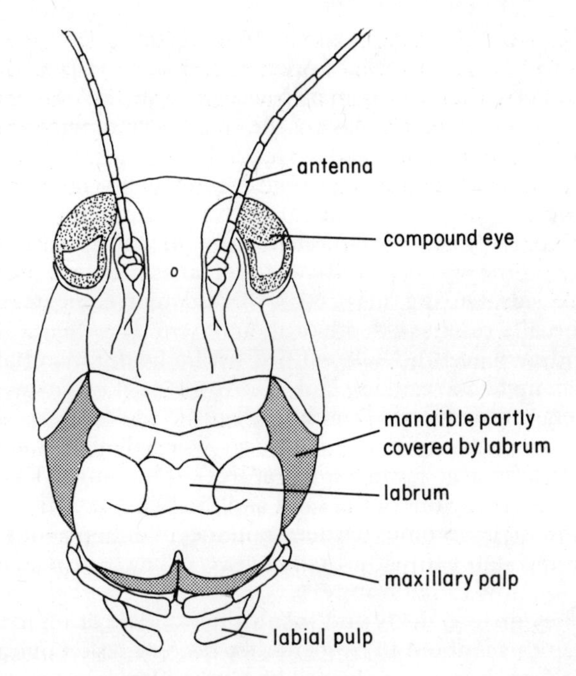

Fig. 6:4 Head of locust from the front.

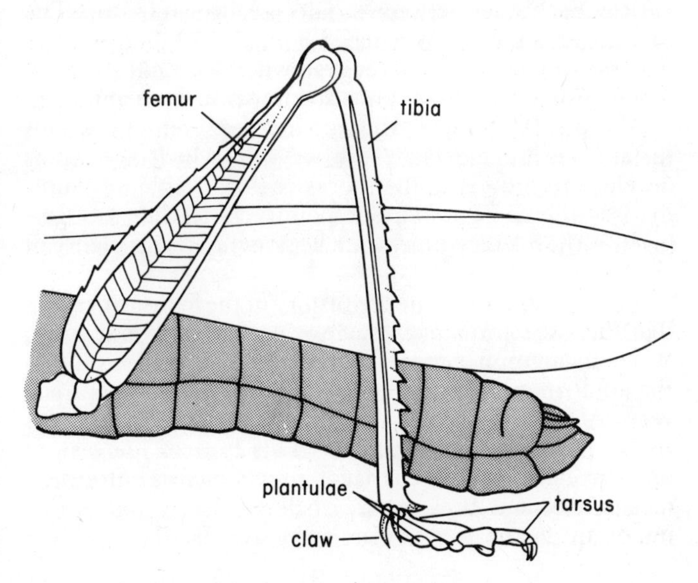

Fig. 6:5 Third leg of a locust.

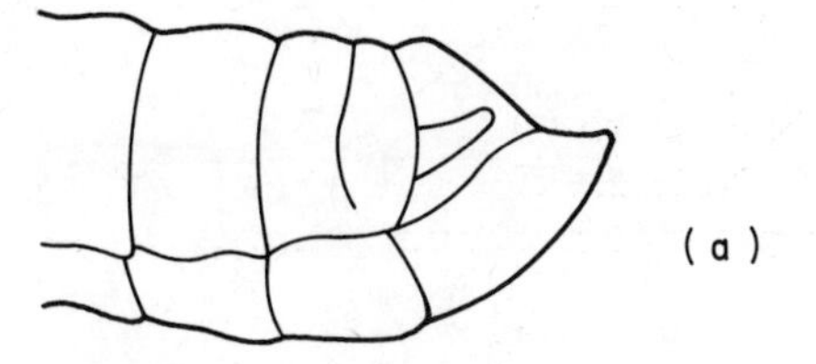

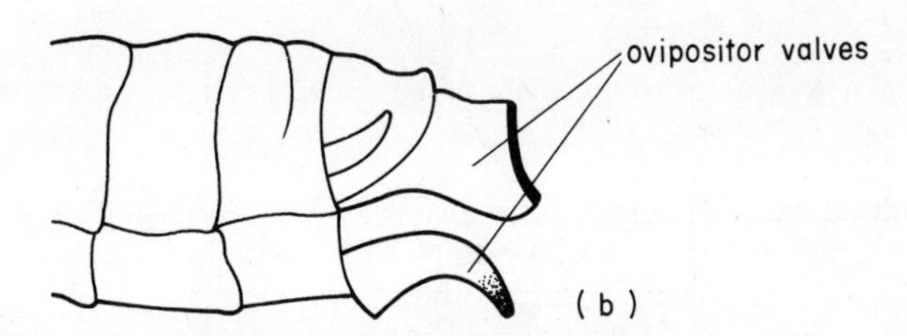

Fig. 6:6 Abdomen of: a) male b) female locust.

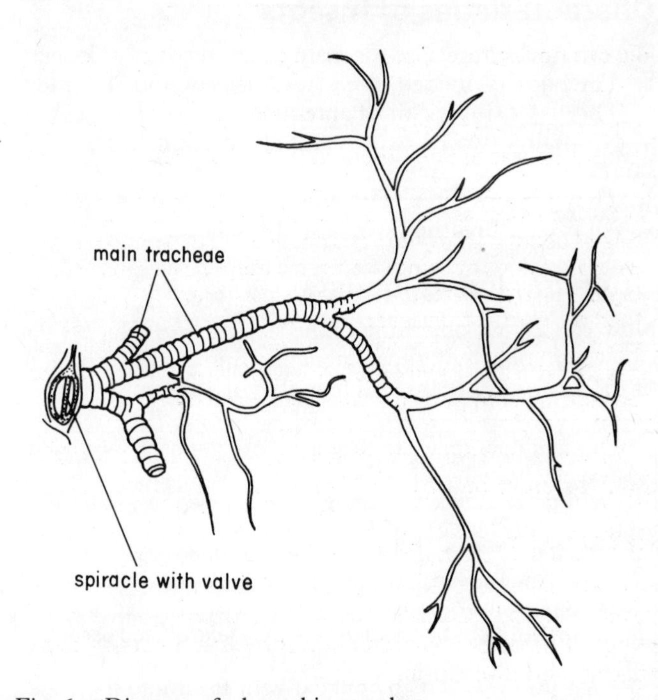

Fig. 6:7 Diagram of a branching trachea.

(Fig. 6:5). Under a lens you can see that the tarsus has five segments and ends in a pair of claws. Below them are rounded pads, the **plantulae**, which act as suckers. If a leg is moved you will find that the cuticle is softer and more flexible at the joints.

3. The abdomen. This has seven visible segments. On the first of these there is a pair of rounded membranes, one on each side. These are hearing organs. They act like our eardrums by vibrating when sound waves fall on them from the air. Locusts make sounds by a process called **stridulation**. They rub the inner side of a back leg (notice the fine spines on it) against the edge of the forewing and the sound which results is amplified because the wing acts as a resonator.

The structures at the end of the abdomen differ in the two sexes (Fig. 6:6); they are concerned with pairing and egg laying. Along the sides of the body you will see a series of pores called **spiracles**. These are the openings of the **tracheal system** which is the breathing system of insects.

The tracheal system

This is a most efficient system of fine tubes called **tracheae** (sing. trachea) (Fig. 6:7). All but the finest tracheae are supported by strengthening rings to prevent them from collapsing; they branch into finer and finer tubes which supply all the organs of the body directly with oxygen from the air. In insects the blood, which is colourless, plays no part in carrying oxygen. The general arrangement of the tracheal system is shown in Fig. 6:8. Each spiracle is controlled by a valve which is normally kept shut, but is opened periodically to let air in. The movement of air is

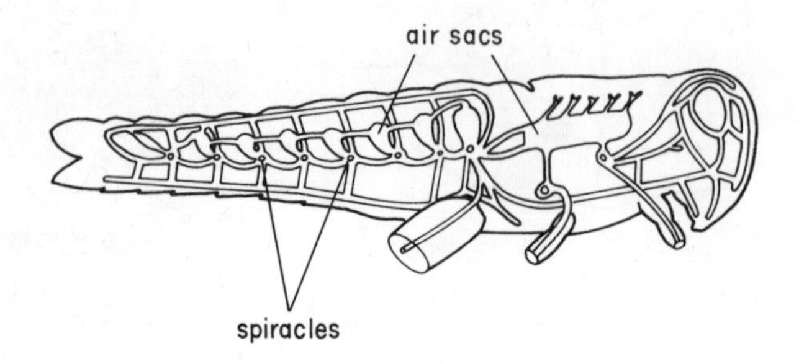

Fig. 6:8 General arrangement of the tracheal system of a locust.

helped by the alternate expansion and compression of the abdomen which acts as a pump. In locusts this is a dorso-ventral movement, but in some insects, e.g. wasps, it is a lengthening and telescoping action. You can see this well when a wasp settles on some jam.

Characteristics of insects

We can now summarise the main characteristics of insects:
1. The body is divided into a head, thorax and abdomen.
2. There is a single pair of antennae.
3. Normally, two pairs of wings are present in the adult stage.
4. They possess three pairs of legs in the adult stage.
5. All insects breathe by means of tracheae.

Studying living locusts

Now you know something of the structure of a locust you can watch some living specimens and find out the functions of their various organs and how they use them.

1 Observe how they use their legs. Are the back legs used differently from the others?
2. Why do you think the femora of the back legs are so fat and long?
3. When they climb up the side, how do they grip?
4. When they are on a twig, what part is played by the tarsus with its plantulae?
5. If they are disturbed, watch how they take off and use their wings.
6. Test their reactions by putting your hand into the cage from the top a) very slowly indeed, b) more rapidly. Remember, in the wild they are attacked by animals which move rapidly.
7. Watch how they feed. Look for the movement of the mouthparts, especially the labrum and jaws. What are the palps doing?
8. Note how they test food with their palps before feeding. Compare their reactions when: (a) a piece of filter paper soaked in dilute sucrose solution, (b) a blade of grass are held to both labial and maxillary palps.

Life history

Apart from differences in the structures at the end of the abdomen, adult males are smaller than the females and are more yellow when mature. Mating takes place when the male climbs on to the back of the female and curves its abdomen round to make contact with the end of the abdomen of the female. They remain together for some time while sperms are being passed from the body of the male into that of the female, a process known as **copulation**. The sperms are stored in a special sac near the end of the egg tube. During egg-laying the eggs move down the egg tube and are fertilized as they pass the opening of this sac.

You may be fortunate enough to see a female laying eggs in one of the specimen tubes containing damp sand. First the locust buries its abdomen into the sand, using the valves at the end to push away the sand and make penetration easier. During this process the abdomen elongates and eventually reaches a depth of about 10 cm. It is then slowly withdrawn and the hole is filled with a frothy material in which up to 200 eggs are laid. The froth quickly hardens to form a protective case round the eggs.

The eggs hatch in about 11 days when they are kept at 34°C (longer at lower temperatures). The young locusts push their way out of the sand and their colour darkens as their cuticle becomes harder; soon they will hop about the cage and start eating the bran or grass. They have no wings and are now called **hoppers**.

They grow quickly and moult their cuticle at intervals. When this is about to happen, they become less active and cling firmly to some object like a twig. The cuticle splits down the back and the locust pulls itself out of the old cuticle. Each stage between moults is called an **instar**. The first instar is the newly hatched hopper, and there are five instars in all before the last ecdysis when the adult emerges. These immature locusts are also known as **nymphs**.

Watch for differences in structures between the various instars, noting especially the ways in which the wings develop from buds in the thorax region. Table 6:1 summarises the more important points. Note that the time given is the average time when kept at 34°C with plenty of food.

We can summarise the life history of the locust by saying that there are three main stages—egg, nymph and adult. When the nymph hatches from the egg it already resembles the adult in many ways and its further development is a very gradual one. Therefore, the locust is said to exhibit **incomplete metamorphosis**. This type of life history differs considerably from that of insects such as butterflies, mosquitoes and bees, as we shall see later, and is one important factor used in classifying insects.

The locust as a major pest in the tropics

Locust swarms are a great scourge over large areas of Africa, the Middle East, India and South America. It is difficult to imagine how large these swarms can be. They have been known to cover an area of up to 1,000 square kilometres, and to contain perhaps 40,000 million locusts! Even an average swarm may cover 128–256 square kilometres and contain over 5,000 million locusts. When they fly over, the whole sky is darkened. When they settle they utterly destroy the green vegetation, crops are ruined and famine often follows in their wake. You can get some idea of the destruction they cause when you realise that an average adult of 2–3 g eats its own weight of green food every day. Multiply that weight by 5,000 million!

The African locust, *Locusta migratoria*, occurs in two distinct forms which differ both in appearance and behaviour. The first is the light coloured **solitary form** which behaves like a large grasshopper and does relatively

TABLE 6:1. STAGES IN THE LIFE CYCLE OF THE AFRICAN MIGRATORY LOCUST

Stage	*Length*	*Duration of stage*	*Main characteristics*
Egg	0·5 cm	11 days	Position of eyes visible
1st instar	0·9 cm	5	No wings. Black
2nd instar	1·2 cm	4	Wing buds just visible
3rd instar	1·9 cm	4	Wing buds point down
4th instar	2·3 cm	5	Wing buds point up
5th instar	3·2 cm	8	Wings half the length of body
Adult	5·5 cm	Several weeks	Wings longer than body

little damage, and the second is the darker swarming or **gregarious form**.

Up to 1921 these two forms were considered to be different species, but it was then discovered that if locusts of the solitary form were bred together under crowded conditions their progeny developed into the gregarious form.

In the wild the solitary form is the usual one, but swarms may arise if a season when conditions are good for the locusts is followed by a bad one. During the good season the solitary locusts breed rapidly, but if the country then becomes scorched up, the food dwindles and the locusts congregate in the few areas where there is enough water to keep the vegetation green and where reproduction can take place. Thus they become crowded together, breed rapidly and produce a vast swarm. With further reproduction, astronomical numbers may result.

You may wonder how the solitary form gives rise to the gregarious when bred under crowded conditions. The answer lies in a **hormone**, a chemical substance which is produced by a gland in the head of the locust. Under crowded conditions this hormone is secreted in larger quantities and this has the effect of speeding up reproduction and bringing about the changes in the progeny which are characteristic of the gregarious form.

For centuries man has attempted to control locust swarms. Success today lies in spotting the swarms when they are in the hopper stage and unable to fly; they can then be sprayed from the air with insecticide. Hence for control to be effective early information about the location of developing swarms is essential and then quick action must be taken to deal with them. To cope with the problem, several international control organizations have been set up. They receive the information, decide on the strategy and co-ordinate the action. In this way considerable success has been achieved, but locust swarms are no respecters of national boundaries and international co-operation is essential. In some areas this has yet to be achieved.

Fig. 6:9 Part of a locust swarm, Ethiopia.

Fig. 6:10 Maize damaged by locusts.

Butterflies

These insects belong to the order Lepidoptera (scaly wings), because their four wings are covered with tiny scales of various colours which give them their pattern.

Their life history differs from that of the locust by having a **complete metamorphosis**. This means that there is a rather sudden change between the larval and adult stages. So there are four stages in the life history: **egg** (or ovum), **caterpillar** (or larva), **chrysalis** (or pupa), and **adult** (or imago).

The term **larva** is used to describe a stage in the life history of *any* animal which lives an independent existence from the adult and differs markedly from it in appearance.

Fig. 6:11 Three closely related species of the genus *Pieris*.

Thus the caterpillar does not look like the butterfly and feeds quite differently.

The term **pupa** is given to a resting stage between a larva and an adult. Although normally inactive, great changes take place internally as larval features are lost and adult characteristics develop. These changes are not very evident externally until the pupal skin splits and the adult, in this case, the butterfly, emerges.

The species we shall study to illustrate this life history is the large white, *Pieris brassicae*. There are two other species which resemble it rather closely (Fig. 6:11): the small white, *Pieris rapae*, and the green-veined white, *Pieris napae*. You will notice that they have all been classified in the same genus because of their similarities. These three species differ mainly in size, the colour of the wing veins underneath, and the appearance and food preferences of the caterpillars. The large and small white both feed on cabbages and cause quite a lot of damage, but the green-veined white feeds only on wild plants of the cabbage family and does no damage to crops.

Large white butterflies are best looked for in gardens, allotments or fields, especially where cabbages and allied species are grown, but they may visit flowers in any situation. This species has two broods in the year—May and June, and August and September. In some years large numbers migrate from the continent to Britain and swell the ranks of the resident population.

Life history of the large white

Egg-laying

It is possible to make the females lay eggs in captivity. They should first be fed on sugar solution (p. 47), and then placed in a large muslin cage in which they can fly freely. They should be supplied with fresh cabbage leaves and the cage should be placed out of doors in the sunshine.

If you watch butterflies in a vegetable garden you will see how they choose members of the cabbage family on which to lay their eggs in preference to other plants. They recognise the plants largely by scent. Sometimes they will lay on nasturtium leaves instead of cabbages and these have been found to contain a chemical substance similar to that found in cabbages. A female usually chooses the underside of one of the younger leaves, laying the eggs, one at a time, to form a compact mass. The number depends on whether or not it is disturbed while egg-laying, but anything up to 100 can be laid.

The egg is elliptical, but drawn out at its apex, and is 1 mm high. Its flat end is glued to the leaf by a sticky secretion. When first laid it is pale yellow in colour, but soon

Fig. 6:12 Eggs of the large white butterfly.

changes to bright yellow and then to orange (Fig. 6:12). Each egg contains a supply of food, the **yolk**, on which the developing caterpillar feeds.

> Examine some eggs under the microscope and notice the pattern on their shells. Perhaps you will see the caterpillars hatch out. They do this when 6–10 days old according to the temperature. You can tell when they are about to hatch as the egg turns a dull colour with a dark region at the top; this is the head of the young caterpillar showing through. Notice how they use their jaws to enlarge the hole at the top of the egg before crawling out and then eat the remainder of the shell.

Fig. 6:13 Caterpillars of the large white on a cabbage.

The caterpillar

On hatching, the tiny caterpillars form threads of silk over the surface of the slippery leaf; this gives them something to grip. The silk is formed from a tubular **spinneret** behind the jaws; this secretes a fluid which hardens into a thread of silk when in contact with air. The caterpillars keep close together at first. They feed by scraping off the epidermis of the leaf with their jaws. Later, as they get bigger, they make large holes in the leaf, working methodi-

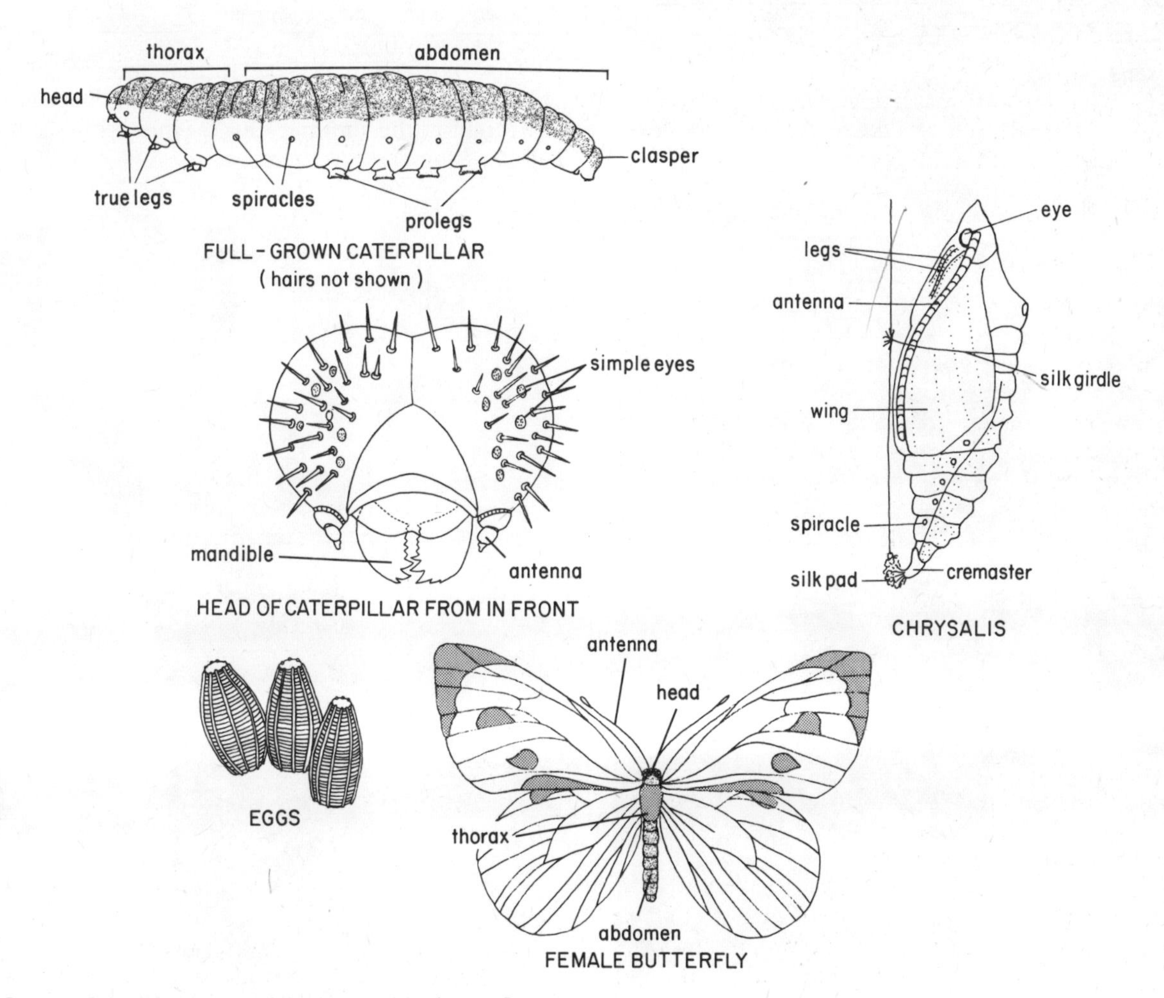

Fig. 6:14 Stages of the life history of the large white butterfly.

cally along a cut edge. The caterpillar moults four times during its growth over a period of about a month. Its colour remains the same throughout development. It has three yellow longitudinal stripes, one dorsal and one each side on a greyish green background. It is speckled with black.

1. Examine a preserved specimen of a full-grown caterpillar. Note the shiny black head, the three segments behind it which form the thorax and the nine abdominal segments behind these; also the warty skin and the sparse covering of hairs. Look at the legs and prolegs using a hand lens. The true legs are attached to the thorax, are jointed and end in a claw, while the prolegs and claspers are fleshy protuberances from the abdomen which end in pads bordered by a half-ring of minute hooks. Look also for the spiracles through which it breathes. There is one pair on most segments. Which segments have no spiracles? Look at the head from the front. Note the powerful jaws and the pair of very small antennae at the side. There are no compound eyes, but there are six simple eyes which consist of a single lens with a few light-sensitive cells. They are probably just sensitive enough to distinguish light from darkness.
2. Examine a living caterpillar to see how it moves and feeds. Note how it uses its true legs both for walking and holding the leaf while feeding, while the prolegs and claspers are used for gripping tightly. Can you see the segments extending and contracting as it wriggles along? This is rather similar to the method used by earthworms (p. 27), the same kinds of muscles being used. Note also how it uses its mandibles; how does the method compare with that of the locust?

The chrysalis
The full-grown caterpillar may pupate on the cabbages in the summer, but usually it leaves the food plant and moves rapidly over the ground until it comes to something like a fence, tree or wall up which it can climb. It then selects a place where it will be protected from extreme conditions, and spins a mat of silk into which it digs its claspers. Finally, it forms a girdle of silk round its body which helps to hold it in a fairly vertical position, head upwards. Making the girdle is quite an acrobatic feat, as the silk comes from the spinneret in the head and has to be fixed to the support on either side of the body.

The caterpillar remains quiescent for a day or so, becoming shorter and fatter. Then it starts to wriggle violently and the cuticle splits in the thorax region. Next it frees its head from the old skin and pushes the latter through the girdle by rhythmic movements of the body until the old skin drops off the end. The new chrysalis now firmly attaches itself to the silken pad with the help of its hooks and is held in its upright position by the silk girdle.

The cuticle of the chrysalis is at first soft and pale in colour, but it soon hardens and the colour changes to become more like that of its background. The mechanism for this colour adaptation is not understood, but it is known that the simple eyes of the caterpillar can detect colours, so when it settles to pupate the background colour may be detected and an appropriate chemical reaction triggered.

Examine a chrysalis with a hand lens. Note the butterfly features that are visible (Fig. 6:14). You should be able to make out the wings, legs, antennae and proboscis (feeding tube) below the skin. These structures start to form towards the end of the caterpillar phase.

During the pupal stage further internal changes take place before the butterfly emerges. This stage lasts about a fortnight in the summer, but for the second brood, it lasts all the winter.

Fig. 6:15 Large white: (left) caterpillar about to pupate (right) chrysalis.

Emergence of the butterfly
The emergence of the butterfly from the chrysalis is another instance of moulting. The cuticle splits in the thoracic region and the butterfly levers itself out with its legs, and if possible, hangs upside down. The wings are at first small and crumpled, but they are gradually 'pumped up' by blood which is forced into their veins. Once the wings have reached their full size, the blood hardens in the veins and thus gives the wings support. Butterflies bred in captivity sometimes have deformed wings because they have not been given supports from which they can hang when the wings are expanding and hardening.

Mating
The female mates soon after emergence, if the weather is sunny. The male is attracted to the female in the first place by sight. After a brief courtship flight they join their abdomens together and sperms are passed into the female's body, as in locusts.

Structure of the butterfly

The wing patterns differ according to sex, the male lacking the three black markings on the forewings (Fig. 6:11). If you examine a wing under the microscope you will see how the overlapping scales make the pattern.

The head bears a pair of knobbed antennae which are organs of touch and smell; also a pair of large compound eyes. The eyes form fairly clear images of near objects and are quick to detect any movements near them. This is because the image of a moving object is picked up by many eye units in quick succession; such a stimulus usually causes the butterfly to fly off.

How butterflies feed

As butterflies feed only on liquid food, the mouthparts are very different from those of locusts although formed from the same basic parts. The main structure is the **proboscis**

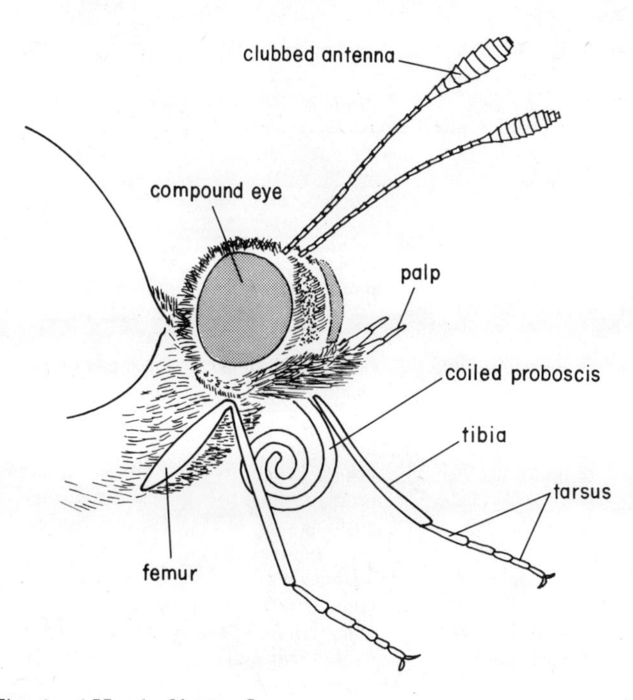

Fig. 6:16 Head of butterfly.

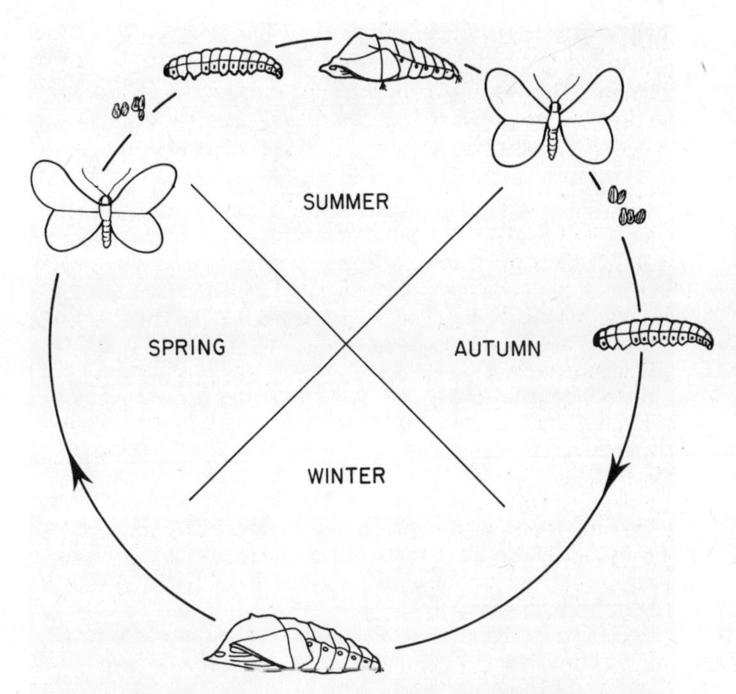

Fig. 6:17 Life cycle of the large white butterfly. This species has two broods during the year.

which is curled up like a spring under the head when not in use (Fig. 6:16). It is formed from the two maxillae which are grooved along their inner surfaces and fused together to form a tube through which the fluid is sucked up. Nectar, a sugar solution secreted by flowers, is the main food of all butterflies although some will also take the juices from rotting material such as fruit. You can watch how a butterfly uses its proboscis when it settles on a flower. The process can also be demonstrated in the laboratory in the following way:

> First place a drop of sugar solution on a piece of rough card. Wait until the butterfly folds its wings together above its back. Now hold it carefully by the top of its wings between your finger and thumb. Place the butterfly on the card with its head near the sugar solution, but do not let go yet. It may put out its proboscis immediately, but if not, put a pin into the coil of the proboscis (you will not hurt it), and draw it out until the end reaches the sugar. The butterfly will at once start sucking and you can then release your grip. Gradually the drop will get smaller as it continues to feed.

You can now refer to Fig. 6:17 which summarises the life history.

Controlling numbers

Artificial methods
The large white and, to a lesser extent, the small white are the only species of butterfly in Britain which are pests. In large numbers their caterpillars can reduce a patch of cabbages to skeletons, leaving only the stalks and leaf veins. When cabbages are grown commercially the damage is prevented by spraying them with insecticides, but it is important that a non-persistent type is used. In a garden this is unnecessary if the plants are examined *each week* during the egg-laying season for patches of eggs, which can then be crushed.

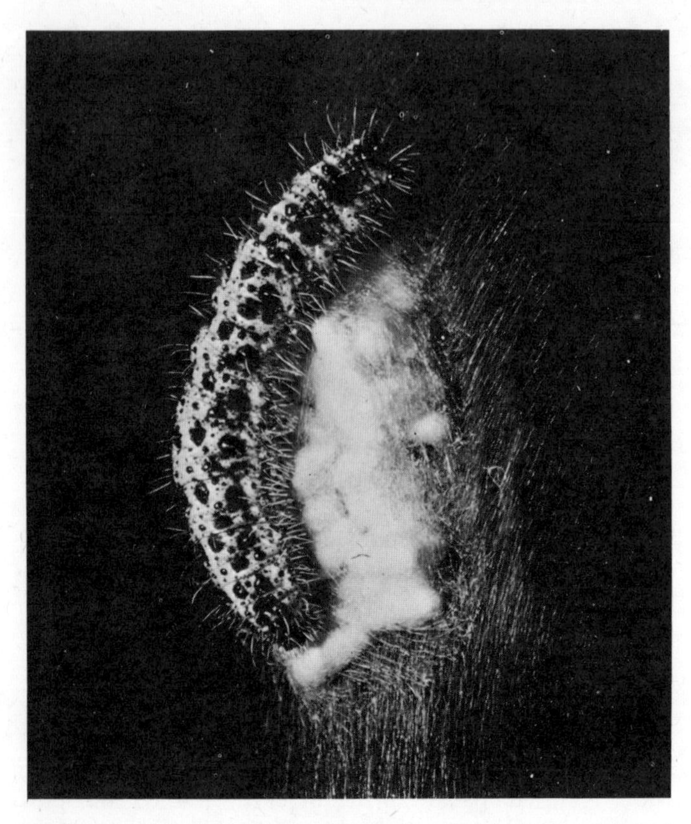

Fig. 6:18 Clusters of *Apanteles* cocoons near the body of a dead caterpillar.

Natural methods
Birds keep down the numbers of caterpillars by eating them. A species of ichneumon fly, *Apanteles glomeratus*, is also very effective. The female ichneumon has a sting-like, egg-laying tube, the **ovipositor**. Having found a caterpillar, it uses this ovipositor to pierce the skin, and then passes eggs through the ovipositor into the body of the victim. The eggs hatch into larvae which feed first on the less vital tissues of the caterpillar, which surprisingly, continues to feed and grow. When the caterpillar crawls away to its pupation site most of its organs have been consumed and it soon dies. The ichneumon larvae then eat their way through the skin and form a cluster of sulphur yellow cocoons in which they pupate (Fig. 6:18). Later, they hatch into adult ichneumons ready to attack the next generation of caterpillars.

Another insect called a chalcid wasp, *Pteranotus puparum*, lays its eggs inside the chrysalis, with a similar result. The caterpillars are also subject to diseases. So with all these hazards the likelihood of an egg eventually turning into a butterfly is not very great. It has been calculated that for every 1,000 young caterpillars only about 3 become butterflies. This is an example of how the numbers of animals are regulated by natural means.

Butterflies in Britain

There are 64 species to be seen in Britain, although some are extremely rare; others used to occur, but are now extinct. Many species are far less common than they were, and this is probably due to the increasing use of insecticides, the great reduction in marginal land, and the destruction of the specialised habitats in which they live. Some species vary greatly in number from year to year because they migrate from other countries. Some years they fly to Britain in large numbers. The painted lady comes from as far as the Sudan and the clouded yellow from the Mediterranean region. These two species cannot survive a British winter, but other migrants such as small tortoiseshells and large whites merely swell the numbers of those which are native. The numbers of any one species also fluctuate widely because of seasonal conditions; a wet season, for instance, encourages disease and so fewer survive.

TABLE 6:2. SUMMARY OF THE MAIN CHARACTERISTICS ON WHICH DIFFERENT ORDERS OF INSECTS ARE CLASSIFIED

Order	*Orthoptera*	*Odonata*	*Lepidoptera*	*Coleoptera*	*Diptera*	*Hymenoptera*
Examples	locusts grasshoppers	dragonflies	butterflies moths	beetles	flies mosquitoes	ants ichneumons bees hornets wasps
Wings	4 thin and straight	4 transparent and similar	4 covered in scales	4 front pair horny, hind pair membranous	2 hind pair modified as balancing organs	4 front pair hooked to hind pair
Mouthparts	strong jaws for biting	strong jaws for biting	proboscis for sucking	strong jaws for biting	proboscis for licking (flies), or for piercing and sucking (mosquitoes)	jaws for biting (ants and wasps), or proboscis for sucking (bees)
Life history	3 stages: egg–nymph–imago. Incomplete metamorphosis	3 stages: egg–nymph–imago. Incomplete metamorphosis	4 stages: egg–larva–pupa–imago. Complete metamorphosis	4 stages: egg–larva–pupa–imago. Complete metamorphosis	4 stages: egg–larva–pupa–imago. Complete metamorphosis	4 stages: egg–larva–pupa–imago. Complete metamorphosis

Butterflies should not be collected just for the sake of it, but breeding them is very interesting and worthwhile. This enables the caterpillars to be guarded from their natural enemies and larger numbers of the butterflies can be released. In this way you can help conserve these beautiful insects.

Fig. 6:19 The rare swallowtail butterfly is now confined mainly to the Norfolk Broads.

All butterflies have the same stages in their life history, but the details are different. Some have one brood during the year, others up to three. Some pass the winter as eggs, others as larvae or pupae, and a few hibernate in the butterfly stage. Small tortoiseshells, peacocks, commas and brimstones are the more common hibernators; they are usually the first ones to be seen in spring.

Moths

Moths are very similar to butterflies, and in practice there is no perfect way of distinguishing them. However, it is true to say for British insects that butterflies have knobbed antennae while moths have antennae of many other shapes, but they are never knobbed.

Classification of insects

From the two species studied, the locust and the butterfly, you will have noticed that although they have many characteristics in common, they differ markedly in their life history, their type of mouthparts and their wings. It is differences such as these that are used to classify insects into different orders. Table 6:2 gives details of six orders, selected because they include some of the commonest insects.

7

Insects: 2

We have already seen with the locust and the large white butterfly that insects can be considerable economic pests. Other pest species which are of great importance include the weevil beetles which get into grain stores, the beetles which attack furniture and the wooden beams of buildings, the termites of the tropics which destroy anything made of wood and even cause houses to collapse, and the clothes moths which make holes in any woollen material.

Insects as carriers of disease

Some insects are important to man because they spread diseases. Diseases are caused by a variety of organisms such as viruses, bacteria, protozoa and worms. They enter the bodies of people, animals and plants and cause various kinds of damage by releasing poisons (toxins) or destroying tissues; in this way they act as parasites. But every parasite must have a means of transference from one organism to another, otherwise the species would die out; insects often act as vehicles for carrying them, and in doing so help to spread the disease. Man has to wage war constantly on some species of insects because many of the diseases they carry are dangerous, e.g. malaria, yellow fever, cholera and plague. Insects transfer these disease-producing organisms in two main ways: by injecting them into the blood stream and by contaminating food.

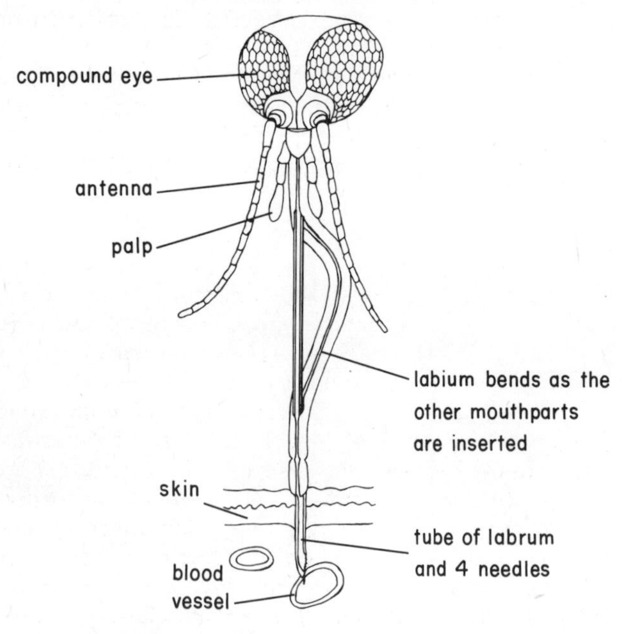

Fig. 7:1 Mouthparts of female mosquito in action.

Transference by injection

Mosquitoes

We have all been bitten by mosquitoes; let us see what happens when this takes place. The mouthparts are highly efficient tools for the task and include two pairs of needles for piercing the skin (mandibles and maxillae) and a tube for sucking up the blood (labrum). All these lie in the groove of the labium (Fig. 7:1). The latter does not enter the skin itself, but appears to guide the other parts as fingers guide a billiard cue. The tube formed by the labrum is so narrow that it would quickly become blocked up if the blood coagulated. This is prevented, however, because saliva containing an anti-coagulant is first poured down the tube; the blood can then be sucked up safely. It is the saliva which probably causes the irritation and swelling that follows the bite.

Life cycle of a mosquito

A mosquito which is easily obtainable is the common gnat, *Culex pipiens* (Fig. 7:2). Fortunately this species does not carry malaria, but its life history is similar to those that do. Gnats are frequently seen during the warmer months near water. You can find the larval and pupal stages in small patches of standing water, e.g. in water troughs, artificial ponds and in marshes or ditches.

Living mosquitoes are best examined in a specimen tube. They have only one pair of wings: for this reason they are classified in the order Diptera. In the place of a second pair of wings are two small projections called **halteres**. These act as gyroscopic organs, helping the mosquito to keep its balance while flying. You can tell the sex by the antennae. They are bushy in males and much thinner in females. Both sexes have a long straight proboscis, but only the female can suck blood; the males feed on nectar from plants.

To study the whole life history, collect a number of females and enclose them in a muslin net over a bowl of water containing some decaying leaves. The females will then lay their rafts of eggs (Fig. 7:3a) on the surface of the water. A raft contains 100–200 cigar-shaped eggs stuck together and buoyed up by air bubbles trapped between them. The eggs hatch in a few days, the larvae forcing their way through the bottom of the eggs into the water.

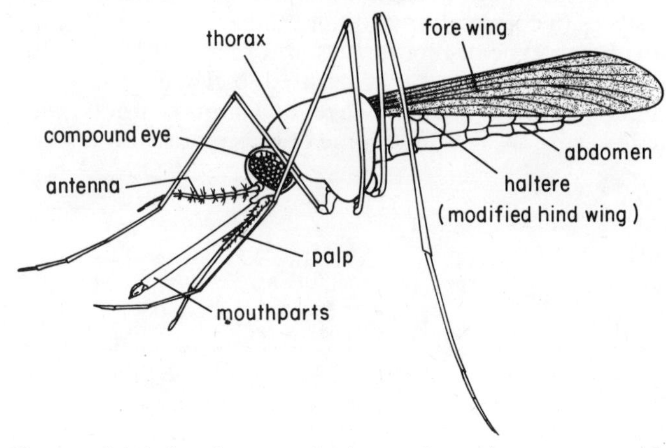

Fig. 7:2 Adult female mosquito (appendages shown on one side only).

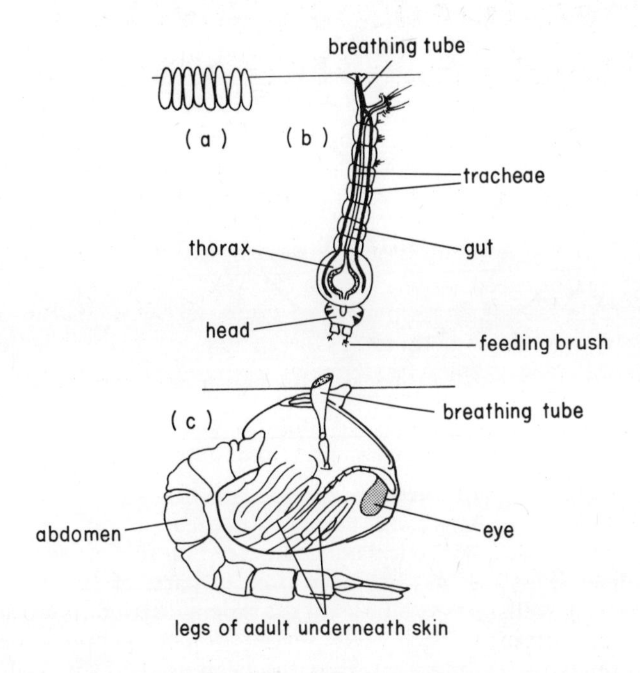

Fig. 7:3 Stages in the life history of the mosquito (*Culex*): a) egg raft b) larva c) pupa.

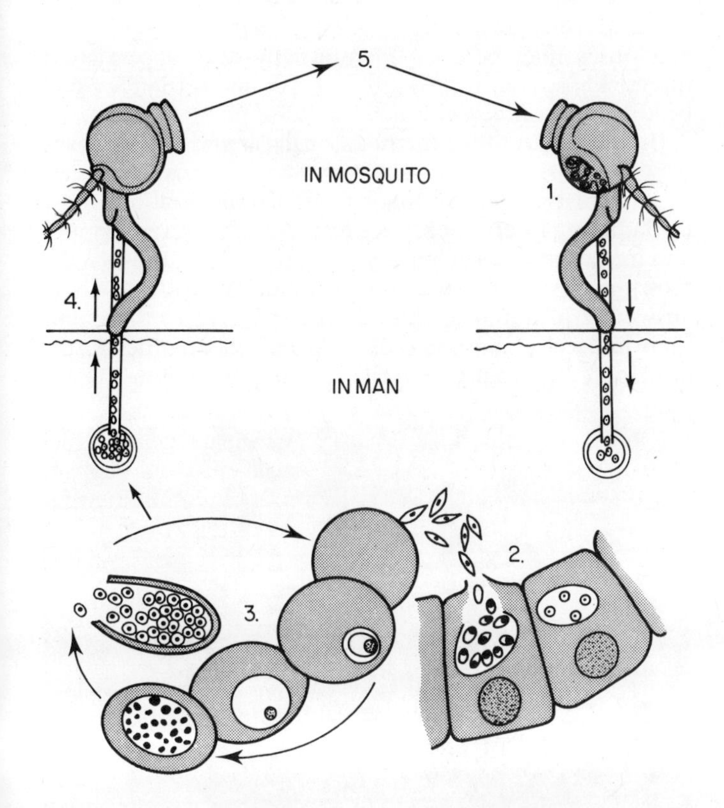

Fig. 7:4 Life cycle of the malarial parasite: 1. Parasites injected into a person's blood system. 2. Multiplication within the liver cells. 3. Red blood cells invaded and multiplication within them (process repeated many times). 4. Parasites absorbed into the body of a biting mosquito. 5. Multiplication within the mosquito's body and passage to its salivary glands.

1 Examine a larva (Fig. 7:3b) under a microscope in a drop of water on a cavity slide. Note the feeding brush on its head used for wafting the micro-organisms on which it feeds towards the mouth.

Notice the breathing tube which projects from the end of the abdomen. Inside it you will see two tubes which continue down the body on either side of the gut. These are large tracheal tubes which supply all parts of the body with oxygen. Observe how the gut shows squeezing movements which help to move the food along inside.

2. Watch some larvae in a beaker of water. Being heavier than water they have to flick their bodies from side to side to reach the surface to breathe. A larva is able to hang motionless in the surface tension film and take in air because on reaching the surface it opens the breathing tube by spreading out five small flaps. These flaps prevent water from entering the tube when submerged.

When the larvae are at the surface, move your hand above them without touching the beaker. What happens?

Again, when they are at the surface, tap the bench from below (why below?). What happens?

From your observations you should be able to deduce the stimuli to which they are sensitive and guess the possible usefulness of their reactions in avoiding being eaten by predators, e.g. wading birds.

The larval stage lasts about three weeks, during which there are three ecdyses. The fourth moult reveals the pupa underneath. Unlike the pupae of butterflies they are capable of rapid movement but, like them, they do not feed. They may be distinguished from the larvae by their top-heavy appearance (Fig. 7:3c).

Look at a pupa under the microscope. What adult organs can you see? Look for the two breathing tubes which project like horns from the top of the thorax, penetrating the surface tension film of the water when the pupa is resting. Watch some pupae in a beaker of water and compare their behaviour with that of the larvae. Are they heavier or lighter than the water?

The pupal stage lasts only a few days. The cuticle then splits in the head and thorax region and the mosquito drags itself out, using the old cuticle as a raft while it enlarges and dries its wings before flying off.

It can be seen, therefore, that the life history of the mosquito, like the butterfly, exhibits a complete metamorphosis, but it differs from the butterfly in having aquatic larvae and pupae, and in having pupae which are mobile.

Malaria

It has been estimated that 200 million cases of malaria occur each year and more than 2 million people die of it. The organism which causes it is a protozoan parasite called *Plasmodium*, of which there are several species producing various forms of the disease. Certain species of mosquito belonging to the genus *Anopheles* are entirely responsible for the transmission of the disease. The organisms are injected into the blood stream via the mosquito's saliva and, after a period in the liver cells, they attack the red corpuscles and burst out again to infect more and more cells. Toxins liberated into the blood when the corpuscles burst produce the characteristic fever of malaria. If a mosquito feeds on the blood of a person suffering from malaria

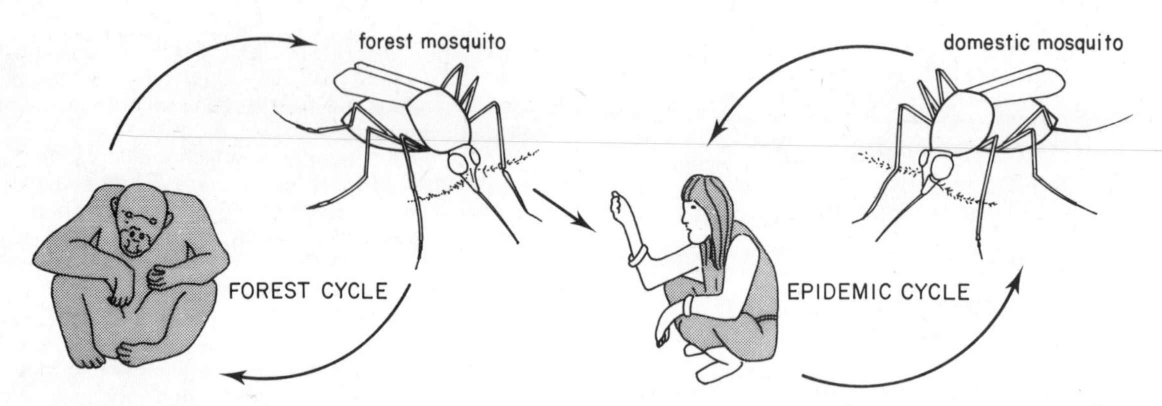

Fig. 7:5 Diagram to show how yellow fever which is carried by a species of forest mosquito from monkey to monkey may affect man and be passed on to others by a species of domestic mosquito, causing a possible epidemic.

these microscopic parasites are sucked up and undergo a complicated cycle inside the mosquito. As a result, spores are produced which end up in the salivary glands from which they may be injected into another person who may then contract the disease (Fig. 7:4).

Yellow fever

Certain species of mosquito transmit other tropical fevers. One of the most important is yellow fever, a disease caused by a virus (p. 143). This disease may be present in colonies of monkeys in both Africa and South America. It is spread from monkey to monkey by a forest species of mosquito. A person working in the forest might be unlucky enough to be bitten by an infected mosquito and contract the disease. If, however, he travelled back to a town and was bitten by a town-dwelling species of mosquito, the disease might be spread to other people and an epidemic could break out (Fig. 7:5). It is because of this danger that visitors to these areas have to be inoculated against yellow fever before they are allowed to enter or leave the country.

Methods of mosquito control

In order to eradicate the diseases which are carried by mosquitoes, these insects have to be killed if possible. Here are some of the methods of control which have been used successfully:

1. The larvae and pupae may be attacked by spraying with oil all standing water where they may be breeding. The oil spreads out to form a thin film which reduces the surface tension of the water, so that when they come to the surface to breathe, they sink and water gets into their breathing tubes. This suffocates them. By adding insecticide to the oil this treatment becomes more effective. Vast areas have been sprayed from the air in this way.

There is a particular difficulty, however, in areas where wells are used for drinking water, as the water becomes contaminated with the oil. In some parts lead-free petrol is used instead, which, after killing the larvae and pupae, quickly evaporates, leaving the water still drinkable.

2. A new method being tried is the use of a bacterium, *Bacillus thuringensis*. The bacterial culture is dried into a powder which can be mixed with water when needed and sprayed on mosquito breeding grounds. Although harmless to man, it produces a substance poisonous to mosquito larvae (and also kills blackfly which spreads a parasite causing river blindness—a disease affecting some 20 million people in Africa and parts of America).
3. Small fish may be introduced to feed on the larvae and pupae. This method has been used in parts of India for treating wells. For example, in the Salem district in South India where 11,000 wells serve 300,000 people, the minnow fish, *Gambusia affinis*, has been used. If the wells dry up, they have to be re-stocked.
4. Marshes and swamps may be drained to prevent breeding. This has been effective in many parts of the world and is particularly important in regions within mosquito-flying distance from towns and villages.
5. Houses may be sprayed internally with a persistent insecticide so that if mosquitoes enter and settle they will be killed by contact with it.

By these and other methods, malaria and yellow fever have been largely eradicated from many regions. However, in some places man has made the malarial situation worse by damming rivers to produce reservoirs for water supplies and irrigation, thus enlarging the breeding grounds for the mosquito. Also, the use of insecticides is not the whole answer as mosquitoes can become resistant to these and their indiscriminate use causes side effects on other forms of wildlife, not to mention the human population.

Fig. 7:6 Spraying houses with insecticide as part of an anti-malarial operation.

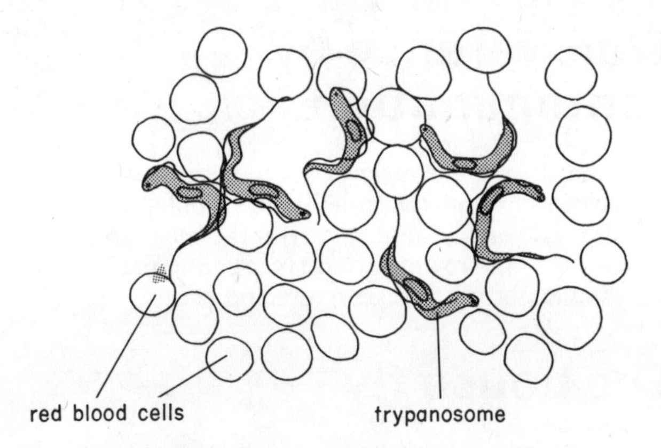

Fig. 7:7 Blood infected by trypanosomes.

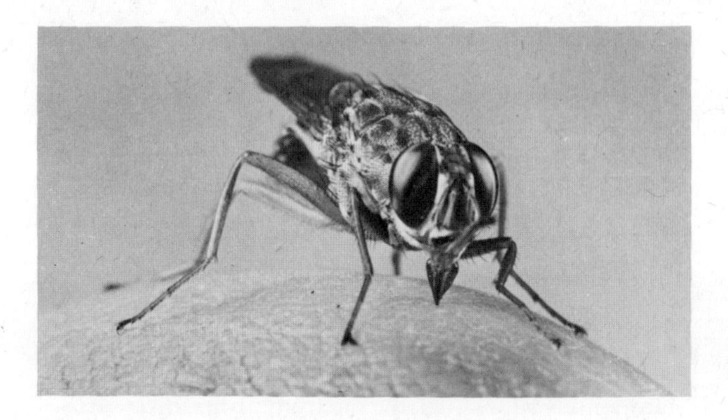

Fig. 7:8 Tsetse fly feeding on blood from a finger.

Research still in progress has shown that in certain irrigated rice fields in California the population of mosquito larvae is kept in check by flatworms which produce a slime which paralyses them on contact. So the presence of flatworms in rice fields may help to reduce malaria and other mosquito-borne diseases. In places where malaria is rife, gauze placed over doors and windows and netting over beds greatly reduces the risk of people being bitten by mosquitoes, and drugs such as paludrine and daraprim can be used both as a preventative and in larger doses as a cure. However, some strains of the malarial parasite have now become resistant to these drugs and new ones such as maloprim are recommended where this is so.

Tsetse flies

Sleeping sickness

This is a disease which is prevalent in parts of Africa. It is carried by the tsetse fly. The parasite which causes it is a protozoan called a **trypanosome** (Fig. 7:7). The trypanosomes live in the blood stream of wild antelopes and other game, and are transmitted from animal to animal through the bite of the tsetse. If people are bitten by an infected insect they may contract the disease. Some species of trypanosomes spread from the wild animals to domestic species, causing the disease called **nagana**. This makes the keeping of cattle very difficult in some parts of Africa.

Fleas

Bubonic plague or black-death

This is a bacterial disease which occurs in rodents, especially the black rat. It is carried from rat to rat by means of the rat flea. It is a great danger to humans, because when large numbers of rats die of the disease, the fleas leave the rats and may then bite people. This is what happened in the 14th century when it is estimated that 24 million people in Europe died of the disease. Today it is still present in wild populations of rodents in various parts of the world. It constitutes a potential danger if infection spreads to domestic rodents, as the fleas of these animals are more likely to come into contact with people.

Myxomatosis

This disease, which attacks rabbits, is caused by a virus which is carried from rabbit to rabbit by insects. The disease was introduced into Australia to control the rabbit population which had reached plague proportions. The introduction was made in 1950, and by the following year it had spread over an area of $2\frac{1}{2}$ million km^2, bringing about a very high proportion of deaths. Myxomatosis today is still a major factor in keeping down the rabbit population in Australia, but the rabbits now show much greater resistance to the disease than formerly.

The disease was also introduced into Europe in 1952, reaching Britain in 1953. It spread more slowly than in Australia probably because there it was carried by mos-

TABLE 7:1. SUMMARY OF SOME OF THE DISEASES TRANSMITTED BY INSECTS AND TICKS

Carrier	*Parasite causing the disease*	*Disease*
Mosquito	round worms (*Filaria*)	elephantiasis
	virus	yellow fever
	virus	dengue fever
	protozoan (*Plasmodium*)	malaria
	virus	myxomatosis (rabbits)
Fleas	bacterium (a bacillus)	bubonic plague
	virus	myxomatosis (rabbits)
Testse fly	protozoan (a trypanosome)	sleeping sickness (man)
		nagana (cattle)
Lice and ticks	bacterium (*Rickettsia*)	typhus
	bacterium (a spirochaete)	relapsing fever

quitoes and in Europe by rabbit fleas. By this means the rabbit population was greatly reduced. Outbreaks of the disease still occur sporadically, but its effect is now more limited.

Table 7:1 gives a summary of some of the diseases carried by insects and ticks (which are arachnids).

Aphids

Plant diseases

Plants are not immune from the diseases spread by insects. Many aphids which suck the sap from plants inject viruses into them at the same time and so spread certain diseases; one such virus that attacks potato leaves causes leaf-roll (see also p. 145).

Fig. 7:9 Aphids may transfer virus diseases from plant to plant when they feed.

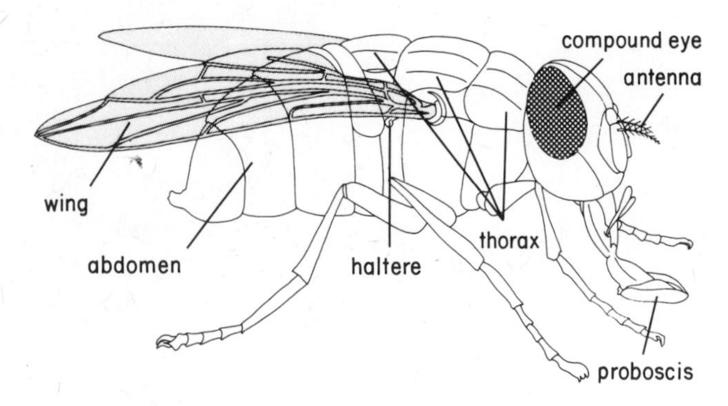

Fig. 7:10 House fly (covering of hairs not shown).

Transference by contamination

The other important means by which diseases are spread by insects is through the contamination of food. The house fly, *Musca domestica* (Fig. 7:10), is the most important of these insects, and to understand how contamination occurs you should study a fly's structure and habits.

The house fly

> Prepare two covered Petri dishes, placing a small lump of sugar in one and a drop of sugar solution in the other. Catch some house flies and after starving them for a few hours introduce one into each dish. Observe their feeding methods, using a hand lens or, preferably, a binocular microscope.

House flies can only feed on liquid foods and yet they are attracted to solids. You will see the reason for this if you look at the structure of the proboscis (Fig. 7:11). It is a

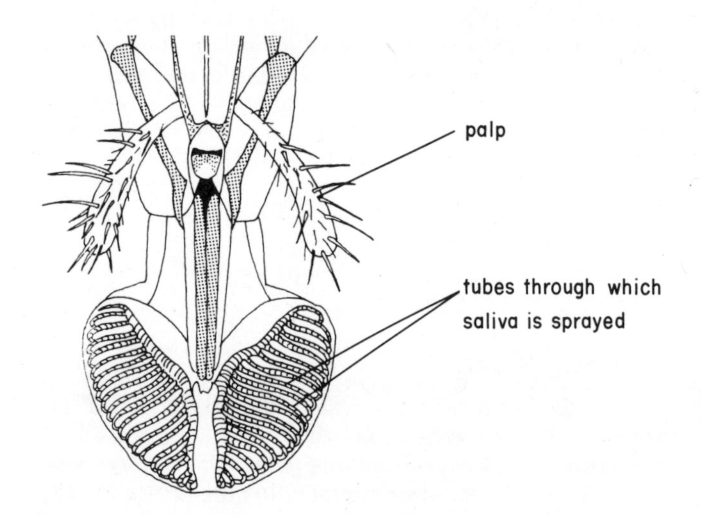

Fig. 7:11 Proboscis of house fly much enlarged.

device for spraying the food with saliva and also for sucking it up. When it dabs at some solid food it sprays saliva over it through fine tubes; the saliva, being a digestive fluid, quickly dissolves the food and the semi-digested liquid is then sucked up by the pumping action of the muscles of the proboscis. Some of this fluid may be forced out again and left behind as a vomit spot. The glass of windows and pictures may become covered with these spots if flies frequently settle on them. You may also have noticed when watching a fly feeding that, as it moves about, it touches the end of its abdomen on the food and deposits a minute drop of waste matter from its gut. Flies also have very hairy legs and bodies to which all sorts of dirt can easily cling.

Their danger to man

You can now build up a picture of how they carry disease by contaminating food. Unfortunately they visit all kinds of material, including human faeces, and if these are infected with organisms responsible for such intestinal diseases as cholera, dysentery and typhoid, they may transfer

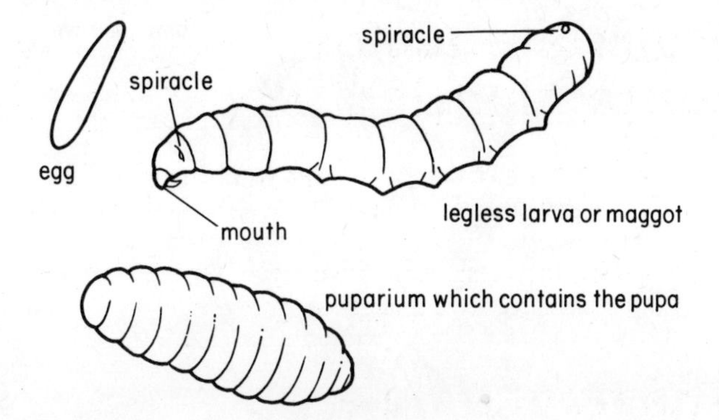

Fig. 7:12 House fly: stages in the life history.

them to food. Bacteria may be carried on the hairs of the body or legs, in the droppings, in vomit spots or via the saliva. It is obvious from this that house flies are a menace to man and need to be destroyed if possible. The best way is to prevent them from breeding. Keeping this in mind, we will now study their life history (Fig. 7:12).

Life history

Eggs are laid in large clusters in decaying material such as manure heaps; they are laid just under the surface so preventing them from drying up. They hatch very quickly (in about 8 hours at 30°C) into almost transparent, white, legless maggots. These larvae feed on the semi-liquid material and burrow into it. After several moults the maggots pupate, but unlike the butterfly and the mosquito the last larval skin is not shed, but remains as a protective coat called a **puparium**. Inside this covering each larva becomes a pupa, from which an adult fly emerges after about three days—a very quick metamorphosis.

Control

From this life history, and from your knowledge of the feeding habits of the adult flies, you will see how important it is to keep flies out of houses and prevent them from contaminating food. To this end food should be kept under covers, and house flies which get into the house should be killed with an insecticide spray. To prevent them from breeding, compost or manure heaps should be as far away as possible from houses; lids should be kept on dustbins which should be emptied frequently and town rubbish should be burnt or sprayed with insecticide. However, some insecticides can be dangerous in other ways, so it is important that the right kind is used. In some towns and cities the house fly has been virtually eliminated by using these methods, but the problem is more difficult in country districts. In some tropical countries the problem is much greater and the risk of disease is far more serious. But once more, a word of warning should be given about the indiscriminate use of insecticides; some are persistent and have dangerous side effects, and their use should be carefully controlled.

Useful insects

So far we have studied insects which are mainly harmful to man, either because they are economic pests or because they carry disease, but many insects are useful.

In the tropics, for example, termites have a bad name because of the harm they do to anything made of wood, including houses, but they are also very useful as refuse disposal agents, clearing up the forests of debris and turning the old wood into nutrients which are quickly returned to the soil.

Amongst the most beneficial insects are the many species of bees, of which the honey bee is the most important example. Honey bees are valuable to man, not merely because they produce honey, but because they bring about the pollination of a great many species of flowers without which no seeds or fruits would be formed. We will now study the honey bee, *Apis mellifera*, in more detail.

The honey bee

Honey bees are **social insects** because they live in colonies and have evolved a complex society where there is much division of labour between the individuals. A **colony** in summer contains three kinds of bee: fertile males, called **drones**, of which there are usually several hundred; a single fertile female, the **queen**, which does all the egg-laying and up to 100,000 sterile females, called **workers**, which carry out the work of the hive. In the autumn the drones are excluded from the hive and soon die, and the number of workers is considerably reduced (Fig. 7:13).

The '**nest**' is composed of a series of wax combs which hang vertically and parallel to each other leaving a small gap between each. Each comb is two cells thick; they lie back to back and nearly horizontal. These cells are made of wax which is produced by glands in the abdomen of the workers and modelled by their mandibles. The cells are of three kinds, according to the type of bee which will develop inside. Not all the cells are for breeding; those towards the outside of the nest are used for storing honey and pollen.

Life history

The queen spends her time laying eggs, often as many as 1500 a day. One egg is placed in each cell (Fig. 7:14). If it is

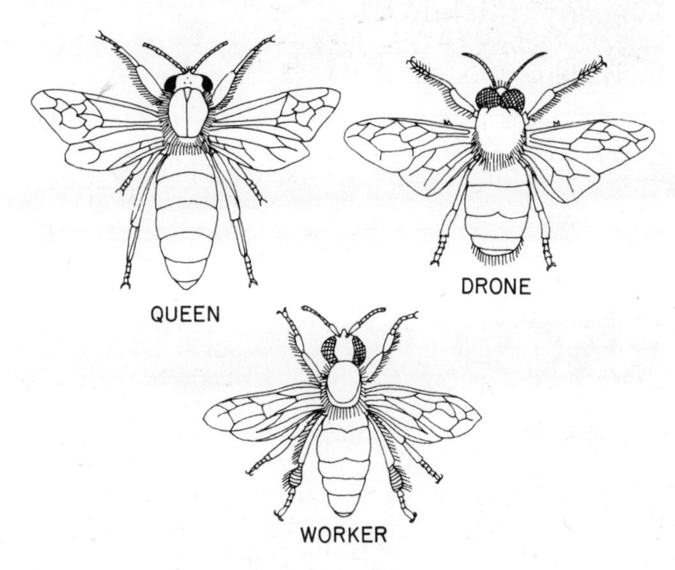

Fig. 7:13 The three kinds of individual in a honey bee colony.

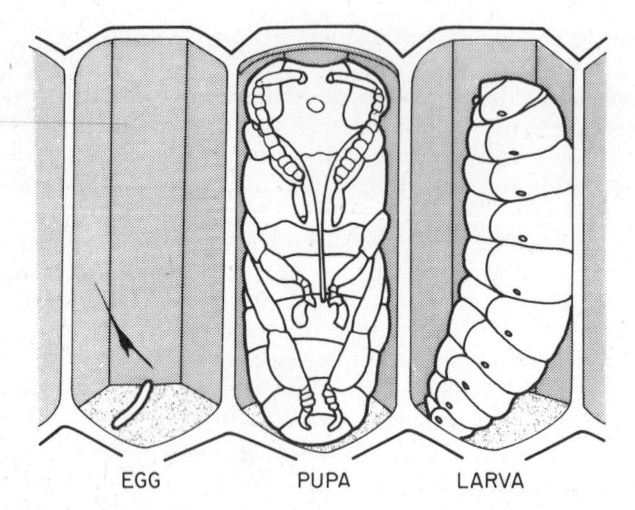

Fig. 7:14 Honey bee: section through cells showing developmental stages.

placed in a queen or worker cell it is fertilized and develops into a female (queen or worker); those laid in drone cells are not fertilized, but develop nevertheless and become drones. This unusual kind of reproduction when an unfertilized egg develops is called **parthenogenesis**.

The eggs hatch after 3 or 4 days into white legless larvae which are all fed for the next few days on royal jelly, a white milky substance regurgitated by workers from their salivary glands. This is a very rich food. Queen larvae are fed on royal jelly until they pupate, but drone and worker larvae after a few days are given a mixture of honey and pollen called 'bee bread'. It is this difference in the food supplied to queen and worker larvae that causes the queens to become mature and the workers sterile. The larvae pupate in the cells when 5 or 6 days old after the workers have sealed them in with a capping of wax. The adults emerge by biting their way through the capping after another 10 or 11 days. Thus the colony can grow very fast.

The queen may live as long as 5 years, drones about 5 weeks and workers only 4 weeks in the summer; but workers hatched in the autumn live right through the winter.

Division of labour

Workers perform many tasks which vary according to their age and the needs of the colony. At one time it was thought that they carried out a whole series of tasks in a particular sequence, but this is not so. The life of a worker may be divided into three periods. During the first three or four days the only duty seems to be the cleaning out of the cells before the queen again lays eggs in them. During the second period of about two weeks they wander about the hive a great deal, performing many tasks. These include the feeding of younger larvae on royal jelly; building more comb and capping the cells when ready; regulating the temperature of the hive by clustering over the brood if cold, or fanning with the wings to circulate cooler air if too hot; and receiving nectar from foraging bees and storing it in the cells. During the final period of life, which is usually only about a week or ten days, they forage for nectar and pollen and defend the nest from robber bees or wasps. Thus through division of labour the complex running of the hive is carried out efficiently.

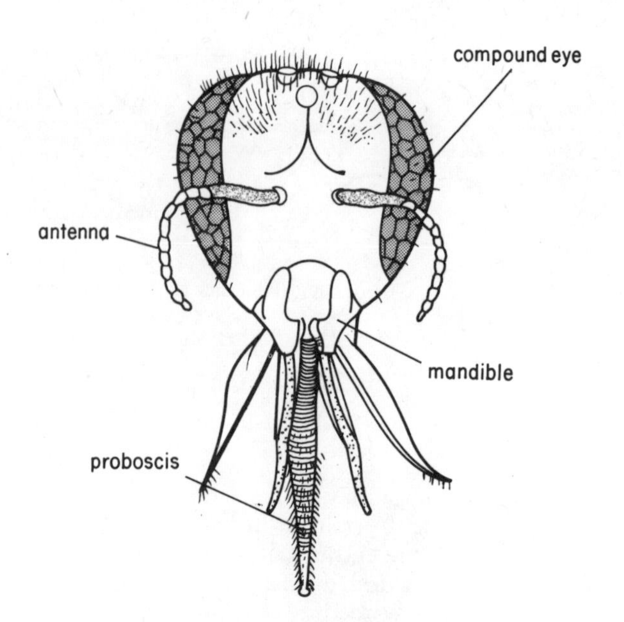

Fig. 7:15 Head of worker bee showing mouthparts.

Food and foraging

Workers collect nectar from the nectaries of flowers. Nectar is a dilute solution of sugar in water which is sucked up with the help of the proboscis (Fig. 7:15). It is carried in the crop, and when the worker returns to the hive it is regurgitated and passed on to other workers. As it passes from worker to worker it is changed by digestive juices into honey which is stored in the cells. When some of the water has evaporated from it, it is sealed off by a capping of wax for later use. Much honey is needed to keep the colony alive during the winter. Pollen is also gathered from the stamens of flowers and stored with the honey; it is rich in protein.

1. Watch bees which are visiting flowers. Notice how each probes into the nectaries with its proboscis, and the way its body becomes dusted with pollen. See if you can make out the way a bee cleans its body with its legs. If you watch by the entrance of a hive (do not stand in front!) you will see that the colour of the pollen varies; this is a clue to the flowers being visited, e.g. apple (grey-green), crocus (orange), horse chestnut (brown).
2. Catch one or two workers in a net, transfer them to a glass-topped box, and examine them under a hand lens or binocular microscope. Notice how they clean their antennae with a front leg.
3. Now remove the legs of a dead bee, examine under the microscope, and compare the structure of the fore, middle and hind legs. Notice the notch on the first leg, which is used for cleaning the antennae, the prong on the second leg for pushing the pack of pollen from the basket, and the 'brush' and 'basket' on the third leg. Can you see how the pollen is held in the 'basket'? Why does it not fall out?

Swarming

When a colony becomes large the workers construct queen cells, and when the new queens are nearly ready to emerge, swarming takes place. When this occurs, up to 20,000 workers leave the hive with the old queen. The air is thick with flying bees. The queen soon settles, often on the bough of a tree, and the workers cluster round her until

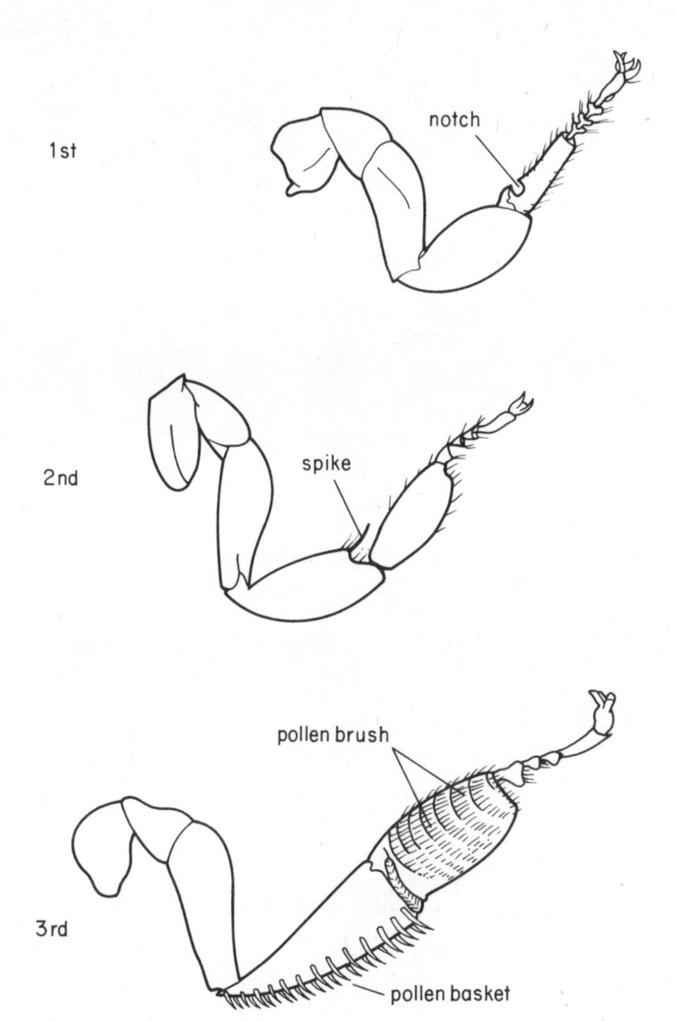

Fig. 7:16 Legs of a worker bee.

they form a solid mass of bees. They remain in this condition while scout bees look for a suitable place to build a new colony. Sometimes the scouts find an empty hive, a hollow tree or even the chimney of a house, and then the swarm flies off, takes possession of the new home, and builds a new nest. Meanwhile, in the parent colony, the first new queen to hatch usually kills off her rivals by stinging them while they are still in their cells. If more than

Fig. 7:17 Queen bee surrounded by workers.

Fig. 7:18 Swarm of bees in an apple tree.

one emerges they fight until only one survives. After a few days the successful queen, followed by the drones, leaves the hive for the mating flight. One of the drones mates with the queen while in the air. After separation the queen returns to the hive and starts to lay eggs at once.

Communication

A society works best when there are good means of communication between the individual members. Bees are able to do this largely through their senses of touch and scent. Worker bees learn the lie of the land round their hive like a map; they find the flowers both by scent and sight. What happens if a bee finds a good source of nectar—can it tell others about it? You can find out in this way.

Fig. 7:19 Bee-keeper examining a hive.

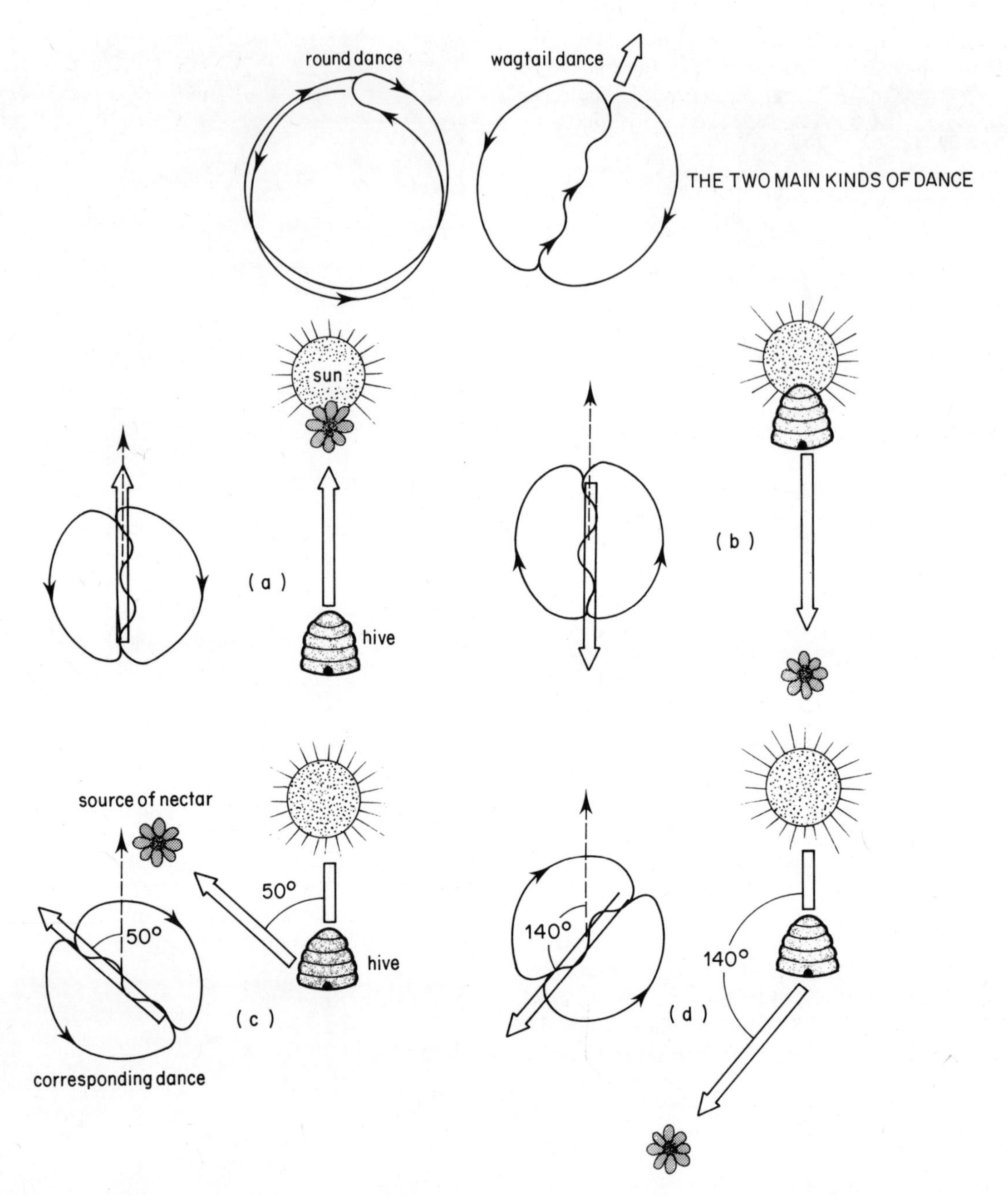

Fig. 7:20 Bee dances: a) Sun and nectar in same direction—bee runs vertically up comb. b) Sun and nectar in opposite directions—bee runs vertically down comb. c) Sun and nectar 50° from each other—bee runs at corresponding angle. d) Sun and nectar 140° from each other—bee runs at corresponding angle.

On a sunny day catch a bee in an inverted glass tumbler, preventing its escape with a piece of cardboard. While it buzzes about introduce a drop of sugar solution on to the cardboard. When the bee finds this, it will settle quietly to feed. Now remove the tumbler and place a spot of quick-drying cellulose paint on its thorax as a recognition sign. When it flies off, watch how it circles the area and observe carefully the details of its location so that it can find it again. When it has gone, add more sugar solution to the cardboard, which must be left in the same position as before. The bee should return in a few minutes. On finding there is more sugar present, will it tell others about it? Watch to see if others come to the sugar. This would be good circumstantial evidence, but it could be merely chance. To make sure, similar sources of sugar should be set up, say 10 m away. See if bees come to these too. There are many variations of this kind of experiment that can be done which you might like to follow up, e.g. what happens if you move the cardboard a few yards away while the bee has gone back to the hive. Will it find it when it returns?

The means of communication is by dancing—you can watch this if you have an observation hive, or you can see a film. On returning to the hive after finding a good source of nectar or pollen, a bee will carry out a dance on the vertical comb, the type of dance varying with the distance from the hive. The two extreme kinds are the **round dance** which indicates that the food is only a short distance from the hive, and the **wag-tail dance,** when food is more than 100 m away. In the latter the bee wags its abdomen from side to side as it makes a straight run. Follow the arrows in Fig. 7:20 to see the forms of the dances. In the wag-tail dance the *length* of the straight run varies with the distance of the food from the hive, hence the number of straight runs per minute also varies. For example, if food is 300 m away the worker makes 28 runs per minute, but when 3000 m away, only 9. The *direction* of the food is indicated by the angle of the straight run with the vertical. Running vertically upwards means that the food is in the direction of the sun; running vertically downwards means that it is in the opposite direction.

As the workers crowd round the dancer they touch it with their antennae and thus determine the direction of its run (it is dark in the hive). The dancer also regurgitates a drop of nectar which, together with the scent, indicates the kinds of flowers it has visited.

The success of insects

Why have they been so successful? Looking back on the species we have studied, certain reasons stand out; you may think of others:

1. *Size*. Their relatively small size means that their food requirements to complete their life history are also small and they can live in a great variety of places which may provide both food and shelter from enemies.
2. *Exoskeleton*. This provides them with a very efficient protection against mechanical injury, enemies, disease and loss of water.
3. *Wings*. Flight is an excellent means of escaping from enemies and it enables insects to travel long distances, spread to other habitats and find food.
4. *Excellent sense organs*. These are very varied and sensitive and provide important information about their environment; they include organs of touch, sight including colour perception, smell, taste, hearing and the detection of low frequency vibrations.
5. *Reproductive capacity*. Most insects when they reproduce have very large numbers of progeny. In this way they can multiply very quickly if conditions are suitable. Some, such as bees and aphids, can even reproduce parthenogenetically.
6. *Adaptability*. In the course of evolution they have shown great powers of adaptability, such as in the food they eat (consider their varied mouthparts) and in their many protective devices, e.g. stings and protective colouration.
7. *The ability to live together in colonies*. This only applies to a relatively small number of species, but in these it has been a very important factor in their success. The most important colonial insects are the ants, bees, wasps and termites. All these have evolved complex societies where there is much division of labour between the individuals for the benefit of the whole colony.

8

Fish

Adaptation to environment

We have already seen many examples of how the structure of animals is closely related to function. In the next four chapters we are going to study the vertebrates largely from the point of view of how their structure has become adapted to the type of environment in which they are living. The vertebrates illustrate this principle very well, because during the course of evolution there has been a gradual progression from aquatic to terrestrial types. Thus the fish are truly aquatic; the amphibians spend the early part of their lives in water, but as adults live on land; and the reptiles, birds and mammals are typically terrestrial. However, the birds have also become adapted to life in the air, and the mammals, although primarily land animals, have spread to many different habitats, some such as the whales having become adapted once more to the water, while others such as the bats have learnt to fly.

The main lines of vertebrate evolution which have taken place over a period of about 400 million years are indicated in Fig. 8:1.

Water as a medium to live in

Let us first consider the properties of water compared with air.

1. Water is a much denser medium than air. This means two things to a fish living in it: a) it offers considerable resistance to movement, b) it gives support and so reduces the weight to be carried.
2. The temperature of water varies much less than that of air. To the fish this means that it does not have to contend with such extremes of temperature.
3. Far less oxygen dissolves in water than is present in a similar volume of air. For example there are about 8 mg in a litre of water and 250 mg in a litre of air. This is an important factor in respiration.
4. Stimuli are transmitted through water rather differently than through air. For example, light soon becomes absorbed as the depth increases and sound vibrations travel faster in water.

Keep these factors in mind when studying the characteristics of fish and their adaptations to life in the water.

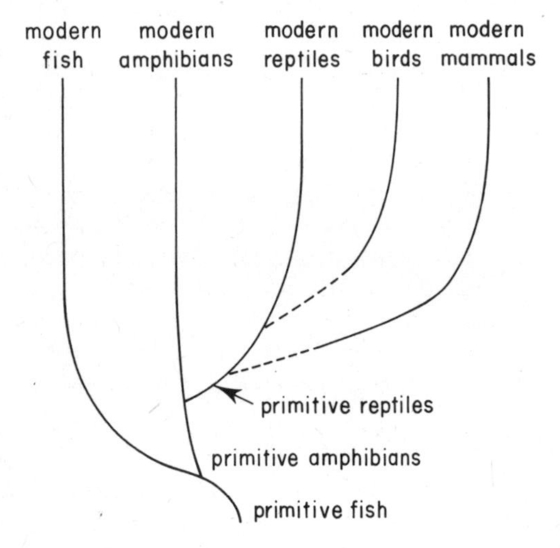

Fig. 8:1 Diagram showing the probable lines of vertebrate evolution.

External features

Examine a fish such as a herring, sprat or similar fish. Compare these features with Fig. 8:2. Note the shape, number and position of the fins. Open the mouth and see how large it is. Feel the jaws for the numerous sharp teeth used for gripping prey. Lift up the operculum and see the gills underneath. Look for the anal opening on the ventral side: the tail is made up of all the body posterior to it. See how the scales overlap and note the main sense organs, the eyes, nostrils and lateral line.

We will now consider some of these structures in more detail and show how they function and help the fish to live in water.

Shape

The shape of a typical fish is said to be **streamlined**. It is thin, pointed in front, reaches a maximum width about a third of the way back and then gradually tapers. This shape gives least resistance to the water when the fish is swimming. Ships and submarines follow the same principle.

Resistance to water is also reduced by the smoothness of the skin due to a covering of overlapping, bony scales made slippery by a covering of mucilage. The scales and mucilage are also protective, the latter being antiseptic in many species, i.e. it destroys bacteria and fungus spores which settle on it.

If some scales are examined under the microscope you will see concentric growth lines by which the age of the fish may be determined. There is a group of rings for each year of growth, the width varying with the amount of food taken during that period.

Respiration

When a fish breathes, oxygen dissolved in the water is extracted by the gills. This involves the passing of a stream of water from the mouth through the gill slits and out under the operculum (Fig. 8:3). To produce this flow of water the fish opens its mouth and lowers the floor of the mouth cavity so that water flows in. It then closes its mouth, constricts the opening to the oesophagus so that water is not swallowed and raises the floor of its mouth cavity. This forces the water over the gills and out under the operculum.

There are four gills on each side, each consisting of a curved supporting bar and vast numbers of filaments which contain blood vessels. Because the surface area of the filaments is so great the diffusion of oxygen from the water into the blood inside them is very efficient.

The bony operculum has around its unattached edge a

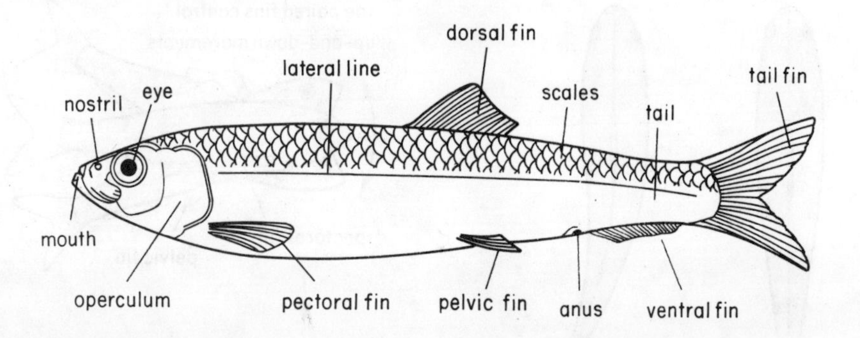

Fig. 8:2 Herring, lateral view.

flexible membrane, the opercular valve. When pressure within the mouth cavity is lowered, water is prevented from entering the gill chamber from under the operculum as the valve is pressed close to the body by the water pressure outside. This action helps to reduce damage to the gills by any solid particles in the water.

Feeding

The mouth has a wide gape which increases the chance of catching prey. In fish, e.g. herring, which feed on small organisms, the prey are sucked in with the water and prevented from being forced out through the gill slits by gill rakers on the inside of the gill bars (Fig. 8:3) which act as a sieve. So the rakers allow the water to pass but retain the food which is then swallowed.

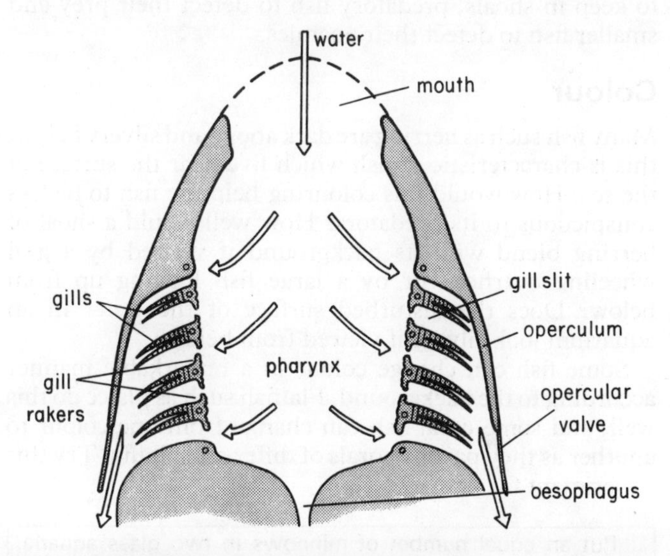

Fig. 8:3 Diagrammatic section through the head of a fish. Breathing takes place in two stages: 1. Water enters when the mouth is opened and the floor of the mouth is lowered, the opercular valve being closed. 2. The mouth is closed, the floor raised, the oesophagus constricted, the opercular valve opened and water is forced out.

Swimming

The whole body of the fish is concerned with movement, but its flexibility depends upon the species; an eel can throw its body into undulations like a snake, but a herring or stickleback cannot. However, all fish move their tails from side to side, not just their tail fins. This movement

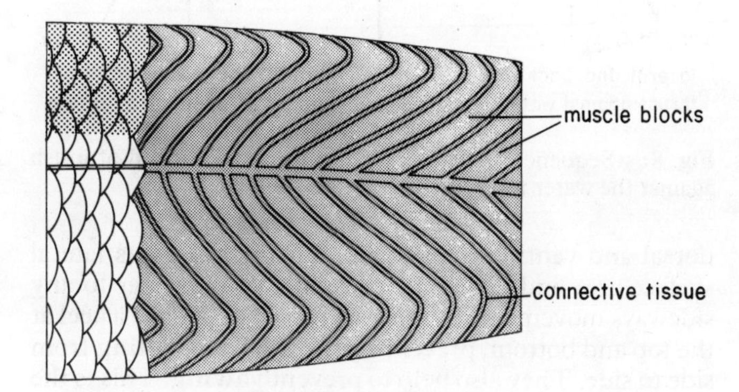

Fig. 8:4 Diagram of the tail region of a herring with the skin removed to show muscle blocks.

is brought about by the contraction of the blocks of muscle on one side of the fish in a sequence starting anteriorly, followed by a similar series of contractions on the other side.

If you remove the skin from part of the tail region of a fish you will see the underlying blocks of muscle (Fig. 8:4). When you eat fish the muscles flake off in these blocks. If you make a clean cut through the tail as in a fish cutlet you will see that the tail is composed almost entirely of solid muscle supported by the flexible vertebral column.

When the tail lashes from side to side it produces both a backwards and a lateral thrust against the water. The resistance of the water to this movement causes the fish to go in the opposite direction to these thrusts, i.e. forwards and sideways, for example to the left (Fig. 8:5). When this is repeated from the other side, the fish moves forwards and sideways to the right; the sideways effects cancel out and thus the fish moves forwards in a straight line.

The fins also help in movement, but their main use is to guide and stabilise the fish. Fins vary in number and position in different species. They are supported by fine bony rays, but are nevertheless quite flexible. The pectoral and pelvic fins (Fig. 8:6) are the only fins which are paired; they are more lateral in position, and they correspond to the fore and hind limbs of other vertebrates. The other fins are single and occur dorsally, ventrally or on the tail.

The use of the fins

The tail fin of most fishes helps the efficiency of the tail movements by increasing the surface area and therefore the thrust; the angle at which it is held also has an effect. If you watch a fish in an aquarium you will see how it keeps its

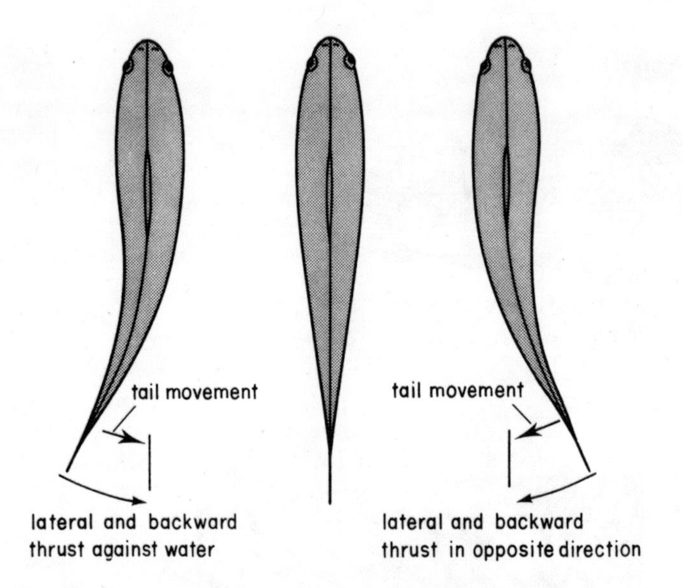

Fig. 8:5 Sequence showing how the thrust of the tail of a fish against the water results in forward movement.

dorsal and ventral fins vertical. This increases its lateral surface area and hence the resistance of the water to any sideways movement. In this way these fins act like a keel at the top and bottom, preventing the fish from **rolling** from side to side. They also help to prevent **yawing**. This is the tendency for the body to change its direction from side to side with the thrust of the tail. The paired fins have different functions. In most fish, by altering their angle they can cause the fish to swim upwards or downwards in the same way as wing flaps are used in aeroplanes. The pectoral fins are also used as brakes to adjust speed.

The swim bladder

The reason why most fish do not sink in water is that they achieve a similar specific gravity to the water with the help of a swim bladder. This is a long sac-like structure, usually silvery in colour, which lies below the vertebral column in the body cavity. Fish are able to control the amount of gas in the sac and so can adjust their specific gravity according to the depths at which they are swimming. As this adjustment takes time, what do you think would happen if a deep-sea fish was pulled very quickly to the surface on a line?

The sense organs

Most fish have large eyes. Those which live in clear water where light can penetrate easily can see well and many of them can detect colours.

The nostrils are not used for breathing as in land vertebrates, but only lead to the olfactory organs which enable the fish to 'smell' chemical substances dissolved in the water. This sense is very acute in many fish and enables species such as sharks to find their prey.

No ears are visible, but there are internal ears which can detect sound vibrations in the water; these vibrations are transmitted to the ears through the tissues of the body.

Most vibrations of low frequency are detected by the **lateral line** system (Fig. 8:7). This consists of a long canal which runs just under the skin down each side of the fish. There are pores at frequent intervals which connect the canals to the outside, and being filled with fluid, any vibrations in the water outside are transmitted to this fluid and are detected by groups of sensitive cells in the canal. Messages are then sent from these cells to the brain. Thus vibrations caused by the movements of any fish nearby can be detected by the lateral line system; this enables herring to keep in shoals, predatory fish to detect their prey and smaller fish to detect their enemies.

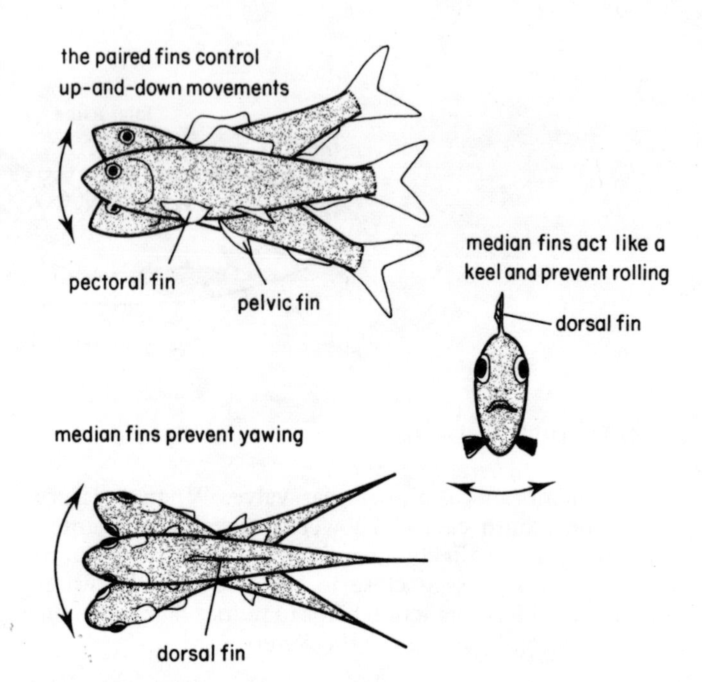

Fig. 8:6 Diagrams to show the part played by the fins in controlling a fish's movements.

Colour

Many fish such as herring are dark above and silvery below; this is characteristic of fish which live near the surface of the sea. How would this colouring help the fish to be less conspicuous to its predators? How well would a shoal of herring blend with its background if viewed by a gull wheeling overhead or by a large fish looking up from below? Does the disturbed surface of the water in an aquarium look silvery if viewed from below?

Some fish can change colour in a remarkable manner according to the background. Flatfish such as plaice do this well, and some coral fish can change from one colour to another as they pass by corals of different colours. Try this experiment for yourselves:

> Put an equal number of minnows in two glass aquaria which have no sand or gravel at the bottom. Place one on white paper and the other on black so that the paper shows through the glass. Examine the minnows after 24 hours. Have they changed colour? To make it more obvious, now mix the minnows from both aquaria together. Can you distinguish between the two sets?

Reproduction

The herring illustrates how reproduction in fish is adapted to life in water. At spawning time the herring gather in vast numbers in shallower regions of the North Sea. All the females then spawn over a period of a few hours, each

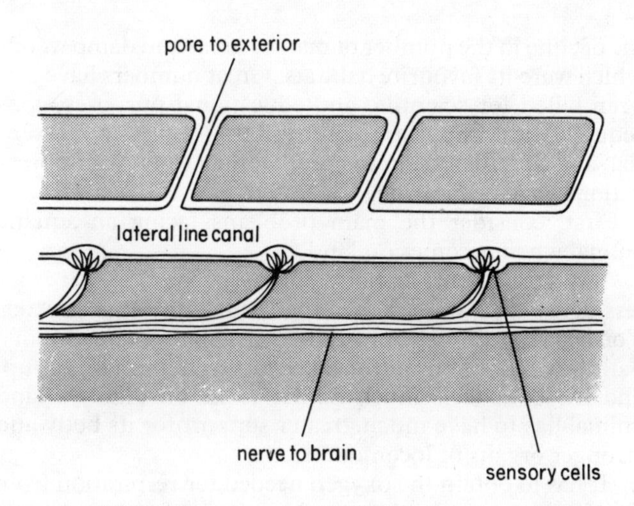

Fig. 8:7 Longitudinal section through part of the lateral line region of a fish.

laying 20–40,000 eggs which become attached to débris on the sea bottom. At the same time the males liberate the **milt**, a fluid which contains millions of sperms. Thus the water surrounding the eggs is teeming with sperms and the great majority of eggs become fertilized. This is known as **external fertilization** as it takes place outside the body of the female. The young fish soon hatch, as there is little yolk in each egg, and go about in large shoals, feeding on **plankton**, the tiny organisms which live near the surface of the sea. Other fish, such as the salmon, lay fewer and larger eggs; others such as sharks have only two or three, which are fertilized internally, develop inside the mother and hatch, before being born as relatively large fish. Refer also to stickleback reproduction, described on p. 233.

The advantage of laying great numbers of eggs is that potentially large numbers of mature fish could be produced if conditions were favourable. A disadvantage is that the young fish may be eaten in vast numbers as they are at first very small and unable to escape predation. By laying a few large eggs or producing a few young there is a better chance of their survival because they are bigger when hatched or born, but under favourable conditions the population would not expand as fast. In practice both kinds of reproduction are successful in the long term as on average enough fish in each case reach maturity to replace the parents when they die. However, commercial fishing, with increasingly efficient techniques, can upset this balance, especially when great numbers of immature fish are caught before they can reproduce. This has happened with the herring through the use of nets with too small a mesh to let the younger fish through. For these reasons fish have to be conserved by limiting the tonnage that may be taken in any one year.

Other adaptations of fish

Fish are such interesting animals that you may like to extend your knowledge about them by considering how different species are adapted to different kinds of aquatic environment. For example, would streamlining be more or less important in swift streams? You could check your conclusions by comparing the shape of fish living in fast streams, such as trout and minnow, with those in slow, such as roach, tench or bream. Can you account for any exceptions? How are fish which live on the sea bed adapted to living there? What are the characteristics of those which live at great depths?

You might like to read about the lungfish and how they are able to keep alive when the water in which they are living dries up, and more about the salmon and eel which make great journeys. Salmon spawn in upper reaches of rivers, but spend most of their lives at sea, returning to breed in the very river in which they were hatched. Conversely eels, found in ponds and rivers in Britain and Western Europe, migrate right across the Atlantic to the Sargasso Sea in order to breed. Their offspring make the three-year journey back to European rivers and even find their way to isolated ponds by travelling overland on wet nights. You will find it fascinating to read how they accomplish these feats.

9

Amphibians and reptiles

Class Amphibia

From water to land

The class Amphibia is divided into two main groups: those which have tails, the newts and salamanders, and those without, the frogs and toads. These species are present-day representatives of a stage of evolution when some of the vertebrates were beginning to leave the water and starting to colonise the land. Amphibians today can live as adults both on land and in water and show adaptations for both modes of life, but they are still dependent upon water during development.

There is a metamorphosis during their life history. This involves a change from a gill-breathing larval stage, e.g. a tadpole, into a lung-breathing adult, a frog.

It is probably because of the difficulty of being adapted as adults to both land and water that amphibians are not very numerous in the world today, although they flourished and were the dominant vertebrates 250 million years ago when they had fewer enemies. Today the majority of species occurs in the tropics where temperature and humidity are high; conditions which call for less extreme adaptations.

As an introduction to this class we will first study the common frog, *Rana temporaria* (Fig. 9:1), to see to what extent it is adapted to living on land and in water.

The common frog

This species is not nearly so common in Britain as it was, because of the draining of many ponds where it bred, and the decline in the number of marshy areas and damp woods which were its favourite habitats. Great numbers have also been killed for scientific and educational purposes; consequently it is important not to kill frogs unless it is essential, and to return them to their habitats when your observations have been made.

First, consider the main problems facing an aquatic animal when it comes on land.

1. It has to support its body during locomotion. Air is far less dense than water hence the weight carried is greater. You can see this for yourself if you let all the water out of a bath before getting out; it takes more strength to get up, and the bath feels much harder to lie on. Thus a land animal has to have much greater support for its body and stronger organs of locomotion.
2. It has to obtain the oxygen needed for respiration from the air. At first sight this should be easier because there is much more oxygen in the air than there is in the water, but there are difficulties as gills are unsuitable unless kept moist all the time.
3. It has to cope with wider fluctuations of temperature. On land the temperature ranges from well below freezing to the high temperature in the summer sun, conditions which may lead to death through freezing or through great loss of water from the body.
4. It has to have sense organs which are adapted to receive stimuli coming through the air. Light travels far better through air than through water but sound vibrations travel more slowly.

Adaptations for supporting the body and for locomotion

Unlike the fish, a frog has no tail for locomotion and instead of paired fins there are limbs.

The general arrangement of bones in the limbs is typical of all terrestrial vertebrates, including ourselves. It is said to be **pentadactyl** or five-fingered. A typical pentadactyl limb (Fig. 9:2) is arranged on a 1.2.9.5 plan. Thus there is a single bone in the upper part of the limb, two in the lower, a group of nine (some of which may be fused together) forming the wrist or ankle and five digits (fingers or toes). The

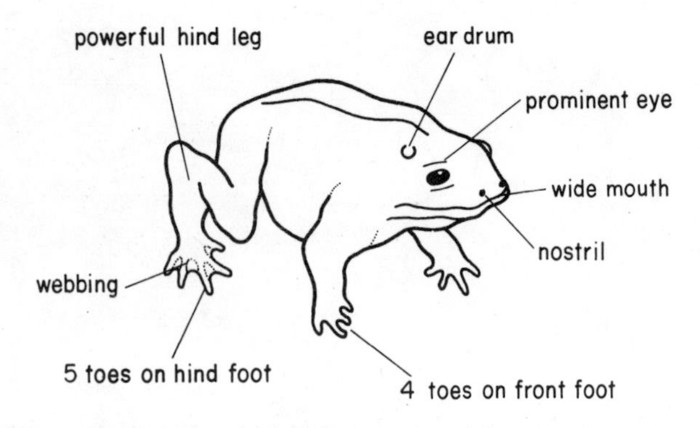

Fig. 9:1 Common frog: external features.

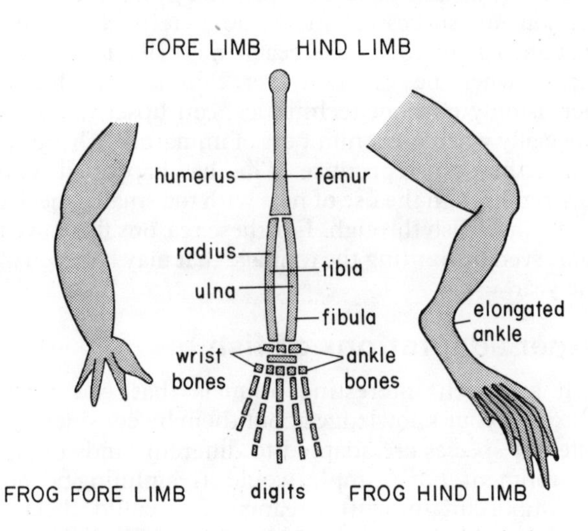

Fig. 9:2 Fore and hind limbs of a frog compared with the bones of a typical pentadactyl limb.

number of digits may have become reduced in some species, as it has been in the fore limbs of the frog, which has four.

The limbs are used both for support and locomotion. They are well adapted for these functions as the bones give rigidity and strength and the joints and powerful muscles give leverage.

The hind legs of a frog are much longer than the forelegs and have an extra portion which is an elongated ankle; this gives extra leverage when it leaps. Frogs jump considerable distances using their strong thigh muscles to straighten their hind limbs, thus pushing off hard against the ground. The fore limbs are used when walking or landing after a jump.

When swimming, the same thrusting action of the hind limbs against the water propels the frog forwards in graceful strokes, the fore limbs helping to steer it. The webbing between the toes of the hind limbs makes swimming more efficient as it increases the surface area of the thrusting feet (cf. the use of flippers by divers). So the limbs are well adapted for both life on land and in the water.

Respiration

It is possible for oxygen to pass into the blood from the air through any moist surface below which there are thin-walled blood vessels. In the frog there are three such surfaces—the skin, the inside of the mouth and the lungs.

1. *Skin respiration.* Unlike the skin of a fish, a frog's skin has no protective scales, is very thin, and is supplied with many blood capillaries. It is also kept moist when exposed to the air by the secretion of mucilage from glands in the skin, making it feel slippery. When the frog is on land, the oxygen dissolves in this mucilage and diffuses into the blood inside the capillaries while carbon dioxide passes out from the blood by the same method. When the frog is in water the same diffusion process occurs.

2. *Mouth respiration.* For this to occur, air has to be drawn into the mouth cavity. To do this, the mouth is kept shut all the time, the nostrils remain open and the frog makes rapid movements of the throat by causing the floor of the mouth to go up and down by muscular action. When the floor is lowered, pressure inside the mouth cavity is reduced and air is drawn in through the nostrils; when raised, the air is forced out again. When the air comes in contact with the moist surface of the mouth cavity the oxygen diffuses into the blood capillaries just beneath and carbon dioxide diffuses out.

3. *Lung respiration.* Periodically, while using mouth respiration the frog appears to swallow, the body swelling up at the same time. When it does this its nostrils are closed by valves so that when the floor of the mouth is raised air is prevented from going out by that route; instead it is forced into the lungs which swell up as a result. The oxygen then diffuses into the blood capillaries within the walls of the lungs.

Gaseous exchange through the skin occurs all through the frog's life, when on land, in water, and when hibernating, but this method by itself is not sufficient for an active life. On land it uses mouth and lung respiration in addition, the former all the time, the latter particularly after active movement. When the frog is floating in water you may have noticed how its nostrils are above the surface so that it can breathe as if it were on land.

> Observe a living frog in a glass container. Note its external features and how it moves its legs when jumping. Observe the throat movements associated with mouth respiration, and the closure of the nostrils and 'swallowing' action during lung respiration.
>
> Place the frog in water, note its swimming action and the position of its eyes, ears and nostrils when it is resting at the surface.

Sense organs

Chemical stimuli are detected by the olfactory organs which lie at the back of the nostrils. During respiration, chemical substances are carried to them by the air as it passes through the nostrils. These substances dissolve in the film of moisture lining the olfactory organs which become stimulated in consequence. This method contrasts with that of fish where the chemicals are dissolved in the water in which they are living and pass directly in solution to the olfactory organs.

A frog's eyes are very prominent and project above the water when it is floating. Periodically a frog will blink, but a fish cannot do this because it has no movable eyelids. Can you see why the presence of eyelids is a useful adaptation for living on land?

The eardrums, situated behind the eyes, allow the frog to receive sound vibrations both in water and in air. They are tightly-stretched pieces of skin, each being in contact with a bone which transmits the vibrations to the inner ear. The latter is very similar to that of the fish.

The greater efficiency of sight and hearing compared with fish is an important adaptation for living on land, but the position of the eyes, nostrils and eardrums on the head enables the frog to make the best use of these organs when resting almost submerged in the water. This has interesting parallels in other animals (Fig. 9:3).

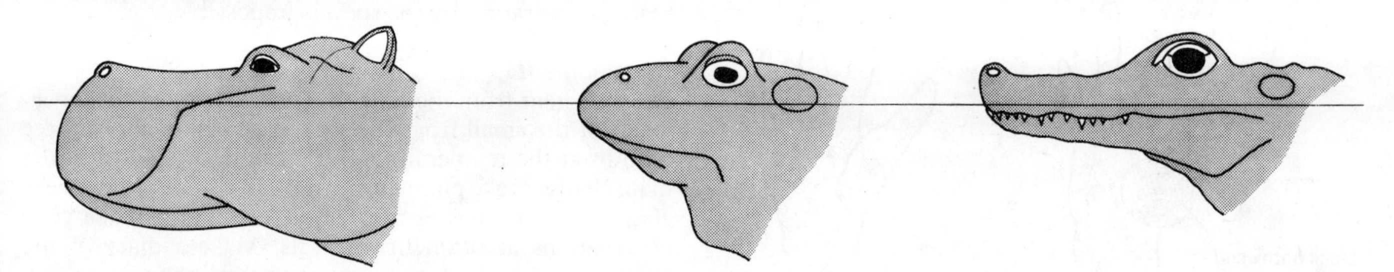

Fig. 9:3 Hippopotamus, frog and crocodile all spend much time in the water but are lung breathers. Note the similarities in the position of the eyes, ears and nostrils.

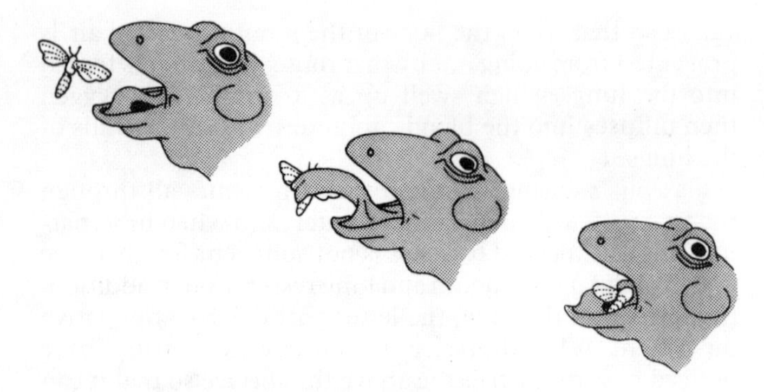

Fig. 9:4 A frog's method of catching an insect.

Feeding

Frogs feed on worms, beetles, flies and many other insects. The method of capture is adapted to land conditions, sight playing an important part in the process. It is usually when the prey moves that the frog recognises it as potential food. If the prey is slow-moving it will be picked up by the mouth with a sudden jerky movement, but flies are picked off by the tongue. The latter is bi-lobed and covered with sticky saliva and is attached near the front of the mouth. When a frog attacks a fly, the tongue is shot out, its sticky end is wrapped round the insect which in a flash is drawn into the mouth. When it swallows, the eyes are withdrawn into its head, an action which helps to push the food down the throat! (Fig. 9:4).

Life history

Just as in the course of evolution aquatic vertebrates gradually became adapted over millions of years to terrestrial conditions, so during the life history of an amphibian such as the frog we see similar changes as it develops from tadpole to adult. The main stages in the life history are as follows:

Hibernation

Amphibians, like fish and reptiles, are 'cold-blooded' or **poikilothermic**. This means they have little control over their temperature which fluctuates according to their surroundings. Thus in countries which have a cold climate they pass the winter in a torpid state known as **hibernation.**

Frogs hibernate in places where extremely low temperatures are avoided such as in the mud at the bottom of ponds or in the banks, below water-level, of streams or lakes. Occasionally a damp situation some distance from standing water is chosen. They usually go into hibernation in October and emerge in February or March according to the weather. Often, scores of frogs assemble at the same breeding place, the males making characteristic low croaking sounds.

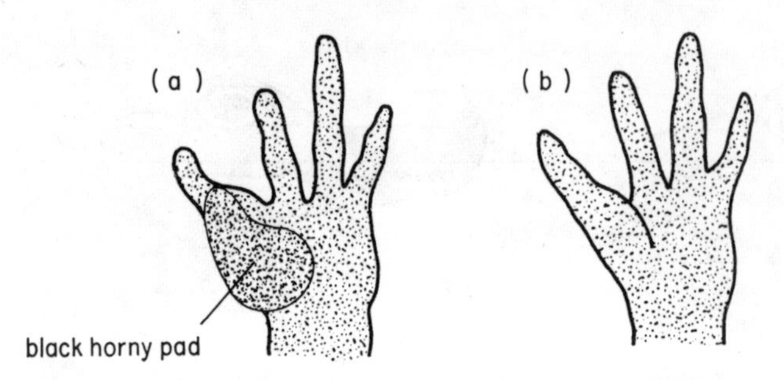

Fig. 9:5 Fore feet of frogs at the breeding season: a) male b) female.

Fig. 9:6 Common frogs pairing prior to spawning, with spawn laid by another frog.

Mating

Males are usually smaller than females and can be distinguished by their white throats; females have yellow throats. In the breeding season the males also develop a black horny pad at the base of each thumb which is used for gripping the female during pairing (Fig. 9:5). When this takes place the male climbs on to the female's back, the forelimbs being firmly clasped around the female's body, just behind the arms. They move around in this position, sometimes for several days, until egg-laying occurs. The eggs are always laid in water.

As the eggs leave the body of the female the male pours a fluid over them containing sperms; these swim to the eggs and penetrate the thin layer of jelly round each egg before the jelly swells up in the water. Fertilization takes place when the nucleus of a sperm fuses with that of an egg. Thus fertilization is external, but pairing ensures that sperms and eggs come together before the swelling of the jelly makes penetration by the sperms impossible.

Development

Development from the time the spawn is laid to the emergence of the small frogs from the water takes about three months at the temperature of the laboratory. Most of the main changes take place during the first three or four weeks after laying, so during this period, especially, make your observations at frequent intervals. Make a diary of the main events and try to relate structural changes to any changes in behaviour you notice. To rear the frogs successfully, proceed as follows:

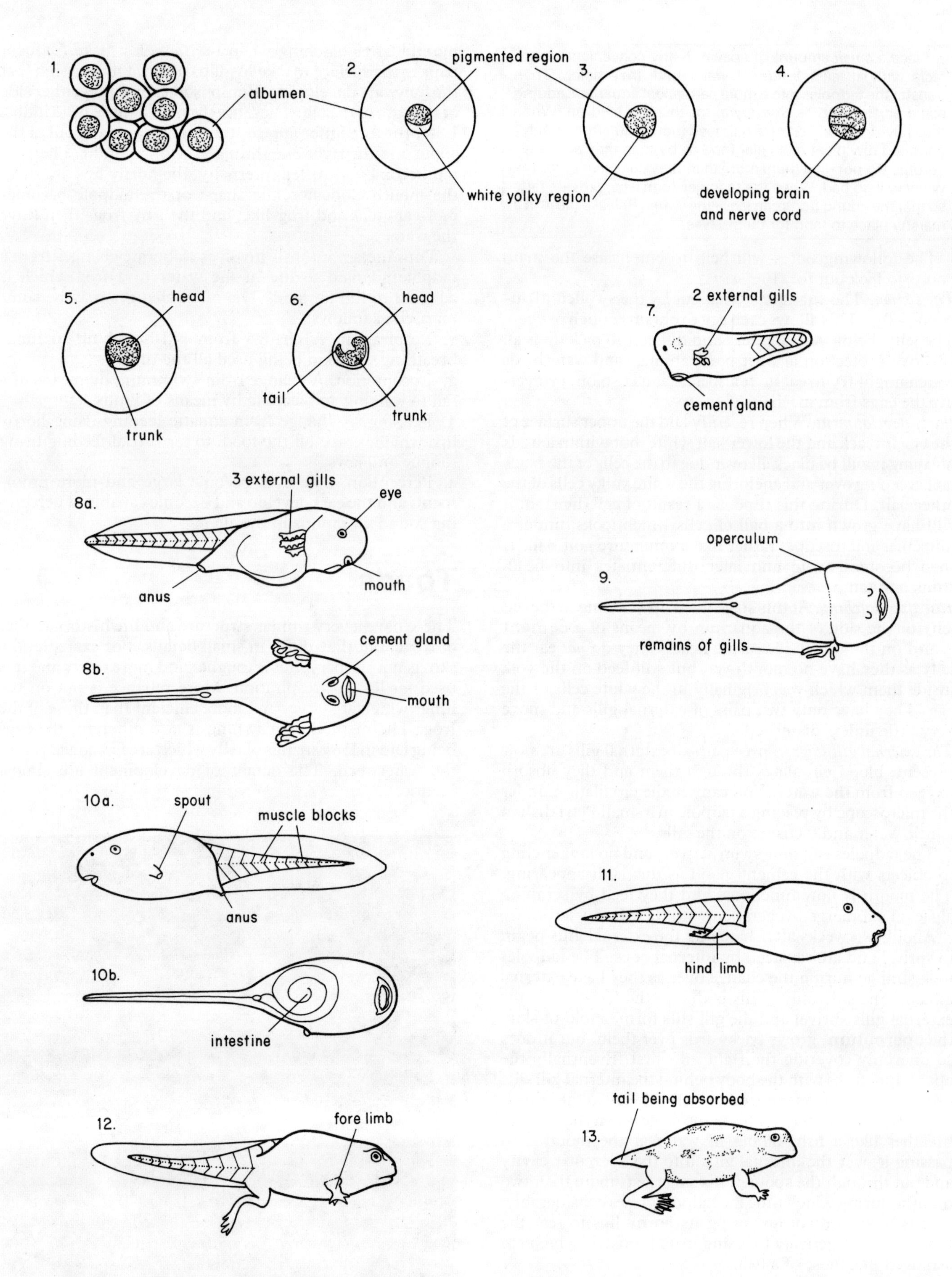

Fig. 9:7 Frog development: 1. Newly laid spawn. 2. Single egg, albumen swollen. 3.–6. Early stages in tadpole development. 7. Tadpole just after hatching. 8. External gill stage: a) lateral b) ventral. 9. Transition stage. 10. Typical larval stage with internal gills: a) lateral b) ventral. 11. Hind limbs developing. 12. Fore limbs emerging. 13. Young frog ready to leave the water.

Place a *small* amount of spawn in any convenient receptacle with at least 1 litre of water until they hatch. Then transfer the tadpoles into a more permanent aquarium, adding some algae, such as *Spirogyra*, for them to feed on. When they have lost their external gills feed them regularly on finely-chopped raw meat or similar food as by then they are carnivorous. Do not give them more than they can clear up or the water will go bad. When the tadpoles form legs, provide rafts so that the young frogs can leave the water. Release these in a marshy place to fend for themselves.

The following notes will help to emphasise the main points to look out for (Fig. 9:7).

The spawn. The eggs are spaced out by the swollen **albumen** (jelly). This allows each egg to obtain enough oxygen. The jelly, being very slippery and difficult to pick up is an effective protection against pond animals and waterbirds which might try to eat it. It also acts as a cushion, protecting the eggs from mechanical injury.

Early development. When recently laid the upper surface of the egg is black and the lower half white, but within a week of laying it will be black all over due to the cells of the black half creeping over and enclosing the white yolky cells of the other half. During this time, as a result of cell division, it will have grown into a ball of cells which looks, under a binocular microscope, rather like a miniature golf ball. It then becomes ovoid and later differentiates into head, trunk and tail.

Just after hatching. At this stage the tadpoles cling to the old jelly or the side of the aquarium by means of a **cement gland** on the underside of the head. They do *not* eat the jelly as they have no mouth yet, but still feed on the yolk inside them which was originally in the white cells of the egg. They have only two pairs of external gills and move very little unless disturbed.

The external gill stage. Three pairs of external gills are now present; blood circulates through them and they absorb oxygen from the water. You can see the circulation under the microscope by placing a tadpole in a small Petri dish in a little water and focusing on the gills.

The tadpoles can now swim actively and no longer cling to objects with the cement gland as this is disappearing. The mouth is now functional and they feed by scraping algae off submerged vegetation with their horny jaws.

About two weeks after hatching the external gills begin to shrivel and are replaced by internal ones. The tadpoles look strange during the change over as they have external gills on one side only. This is due to the fact that as the external gills shrivel and the gill slits form, a fold of skin, the **operculum**, grows backwards over them, but it does so unevenly covering the right gills first. Eventually the operculum fuses with the body behind the internal gill slits except for a small opening on the left side called the **spout.**

The typical larval stage. This is the stage when the tadpole breathes like a fish, taking in water at the mouth and passing it over the internal gills into the opercular cavity and out through the spout. This stage lasts more than two months during which time the tadpole grows considerably.

It is now carnivorous, using its horny lips to tear the food. It swims actively by using its tail muscles which are arranged like those of a fish.

Metamorphosis. This is the change from the typical larva with internal gills to the frog which breathes by means of lungs. The first sign of this change occurs about two months after hatching when the tadpoles make frequent trips to the surface to take in gulps of air. This is a sign that the lungs are developing. Bumps soon appear on either side of the base of the tail. These are the first signs of hind limbs. Later the fore limbs appear, the left one growing out of the spout and the right one through the operculum. Then the tail becomes absorbed internally, the horny lips are shed, the mouth elongates, the shape of the tadpole becomes more angular and frog-like, and the tiny frogs then leave the water.

Thus metamorphosis involves elaborate changes from a tadpole adapted to life in the water to a frog which is adapted to life on land. The main changes may be summarised as follows:

1. Respiration. A change from gill-breathing to lung-breathing, the skin being used all the time.
2. Locomotion. A change from swimming by means of a tail to walking and leaping by means of limbs.
3. Feeding. A change from aquatic feeding using horny lips which scrape off the food, to terrestrial feeding using tongue and jaws.
4. Perception. The eyes become large and more prominent, and the ears develop and become capable of perceiving sound vibrations in the air.

Toads

These have a very similar structure and life history to that of frogs, but they differ in small details. For example, the skin is not slippery, but is tougher and more warty and it is used far less for respiration. More reliance is put on the lungs which are larger and more efficient than those of the frog. The method of spawning is also different, the eggs being laid in long strings of jelly which are festooned round the waterweed. The details of development are almost identical.

Fig. 9:8 Common toad with spawn.

Newts

In many ways newts are much less specialised amphibians than frogs and their structure is more intermediate in character between fish and reptiles. For example, they retain their tail when adult and move their body and limbs in a simpler way. The frog, by contrast, has become highly specialised in structure due to its peculiar method of leaping when on land. This accounts for its short rigid body, the loss of the tail and the elongation of its hind limbs. Keeping these points in mind, it is interesting to keep newts in an aquarium where their whole life history may be observed and their behaviour compared with that of the frog.

Resting places above the surface of the water should be provided for the adults and precautions taken to prevent them climbing out. They should be fed on small live worms.

Fig. 9:9 Smooth newts in breeding condition. The male (bottom) has a prominent crest.

Adult newts may be found in ponds and canals between March and June as this is their breeding season. At other times they are more terrestrial in their habits. In all three British species, the warty, smooth and palmate newts, the males, when in breeding condition, are more brightly coloured than the females and bear a conspicuous dorsal crest (less obvious in palmate newts).

During courtship the male attracts the attention of the female by displaying its brightly coloured belly and by vibrating the tail which is curled round parallel to the body The male then lays a minute jelly-like packet of sperms, a **spermatophore**, which the female picks up by means of the **cloaca** (a ventral pouch which receives the openings of the food and reproductive canals). Inside the female's body the jelly dissolves and the sperms are freed and fertilize the eggs internally before they are laid.

When egg-laying, the female clambers over water weed and lays the eggs singly on a leaf; the leaf is then bent over so that the egg lies between the two parts of the leaf like the filling inside a sandwich.

When the eggs hatch, the larvae are tiny, transparent, fish-like creatures with feathery external gills. These gills persist until metamorphosis. The larvae are carnivorous and feed on small crustaceans such as water fleas.

Metamorphosis takes place in the late summer. Lungs develop and the larvae come to the surface to take gulps of air. Limbs slowly become visible, the front ones appearing first.

By the time the newts leave the water, their external gills have been absorbed and they become dependent upon air respiration as in frogs.

Class Reptilia

This class includes the snakes and lizards, crocodiles and alligators, turtles and tortoises. Reptiles may be considered true land animals as they can be independent of water as a medium in which to live, in spite of the fact that some, such as turtles and crocodiles, spend much of their life in water. Their main adaptations for living on land are:

1. The lungs have enlarged sufficiently to supply enough oxygen for all their needs, hence the skin is no longer needed as a respiratory surface. This allows the skin to have a hard protective covering of epidermal scales, i.e. scales formed from the outer layer. This is characteristic of all reptiles.
2. Reptiles have solved the problem of having an aquatic larval stage and being dependent upon water by laying relatively large eggs with protective leathery shells which reduce the water loss. As the embryo develops, membranes are formed which completely enclose it in a bath of fluid which acts like a minute pond of its own. Birds and mammals also use similar membranes.

Reptiles are of great interest, but it is only possible in this book to indicate briefly some of their remarkable adaptations.

Body temperature

Like amphibians they are poikilothermic, their temperature changing with that of their surroundings. When on land they are subject to wide fluctuations, and these affect the speed of their physiological processes. Thus when they become hot these processes are speeded up and they are able to be very active, when cold they become sluggish, and when very cold, quite torpid. Hence those which live in places where winters are cold are forced to hibernate.

Reptiles have many adaptations which enable them to avoid extreme temperatures. Desert reptiles, for example, avoid great heat during the day by retiring to a burrow, burying themselves in the sand or sheltering under a boulder. They do most of their feeding at night when it is cooler.

Crocodiles are able to keep a remarkably constant temperature by keeping to a regular pattern of behaviour. At night they are mainly aquatic and are warmed by the water; before dawn they haul themselves out and then bask in the morning sun. As the temperature rises they cool themselves by opening their mouths; this allows the water to evaporate from the mouth surface, taking heat from the body as a result. During the hottest part of the day they often retire to the shelter of trees, but return to bask in the sun in the late afternoon. By behaving in this way, their temperature only fluctuates a few degrees during a 24-hour period.

Feeding

Reptiles have evolved some very unusual feeding adaptations. Chameleons which are related to lizards and are adapted for life in trees and shrubs have developed a highly specialised tongue with a sticky end which can be shot out like a released spring by muscular action. This enables them to pick off an insect as far away as 10 cm, a distance not much less than the length of the body!

Snakes have specially adapted jaws which allow them to swallow eggs or prey much larger in diameter than themselves. To do this the two bones comprising the lower jaw are freely-movable units attached together in front only by an extensible ligament, and at the back they are not rigidly jointed to the skull but are connected to it by another bone which acts like an extensible hinge. This device allows large objects to be swallowed. The rows of backwardly-projecting teeth also help the swallowing process.

Some species of snakes such as boas and pythons twist their bodies round their prey and suffocate them by the pressure of their coils before swallowing them. Others, such as adders and cobras, immobilise their victims with poison using certain teeth which have become modified as poison fangs. In the adder the two fangs are hollow like hypodermic needles, and poison from glands in the head is squirted through them into the wound made when it strikes the prey. One advantage of feeding on relatively large prey is that the snake does not need to feed as frequently as other reptiles.

Locomotion

Methods of locomotion vary greatly. The most basic method is that used by lizards. This is very similar to the walking gait of the newt, but is quicker, and the body is lifted higher off the ground.

Snakes and certain lizards such as slow-worms have lost their legs during the course of evolution, but by the undulations of the body and the gripping power of the overlapping scales on the ventral side, they are able to move fast. This legless condition also enables snakes to burrow into the ground and thus find shelter. The flexibility of the snake's body also allows many species to climb very effectively; in this way they can reach the nests of birds and feed on their eggs.

Turtles have limbs modified as flippers for swimming and they are so well adapted to an aquatic life that some species only come on land in order to breed.

Reproduction

Reptiles all have internal fertilization. Usually they lay eggs. The eggs are relatively large because they contain much yolk; they also have tough, leathery shells. When the young hatch they are large enough to move actively and fend for themselves.

Successful incubation of the eggs is dependent upon their receiving both warmth and moisture. Various devices are used to keep the conditions suitable. Grass snakes often lay their eggs in heaps of rotting vegetation where extra heat is provided by bacterial action during decay. Turtles come ashore and lay their eggs in the sand at a depth where the eggs receive enough heat, but not too much, from the sun. Crocodiles select sandy situations near rivers and time their egg-laying to coincide with the onset of the dry season, thus avoiding the risk of flooding.

Crocodiles also show parental care. For three months the female zealously guards the nest against such marauders as monitor lizards and marabou storks. When ready to hatch, the young make loud croaking noises which act as a signal for the mother to scoop away some of the sand so that the young can emerge successfully. The young may then be carried to a nursery area where the water is shallow and small prey are abundant. They are gently carried by both parents, a few at a time, in a pouch inside the mouth. The mother continues to protect them for some weeks.

Some reptiles (e.g. adders and some species of lizards) do not lay eggs, but retain them in their bodies until they hatch. It is usual for them to bask in the sun, the extra heat helping to incubate the eggs inside them. The young are born as active little creatures quite capable of fending for themselves.

10

Birds

Life in the air

Birds are terrestrial animals which have taken to a largely aerial existence. Hence they are basically adapted for life on land, but many of their characteristic features are adaptations for flight. Let us consider the main characteristics of birds and see to what extent they are related to life in the air.

Fig. 10:1 Whinchat alighting.

Homoiothermy or warm-bloodedness is a characteristic they share with mammals. It is the ability to keep the temperature of the body fairly constant in spite of fluctuations in the environment. Homoiothermy is useful both for a terrestrial and an aerial existence and it enables birds and mammals to live in both extremely cold and hot places.

The structural characteristics of birds are best examined on a dead bird such as a pigeon and by looking at its skeleton (see p. 73), but it is also essential to watch living birds to see how these structures are used. It is also interesting to compare the design of birds and aeroplanes as you study their structure.

Shape

The body is streamlined and the feathers overlap in a particular direction to keep the surface smooth. Only the legs disturb the streamlined effect and in flight these are tucked up close to the body like the undercarriage of an aeroplane.

Legs

These differ somewhat from the typical pentadactyl limb (p. 64) by having an extra long bone and joint. This is caused by the fusion and elongation of certain bones giving greater length to the leg. Long legs enable a bird to jump

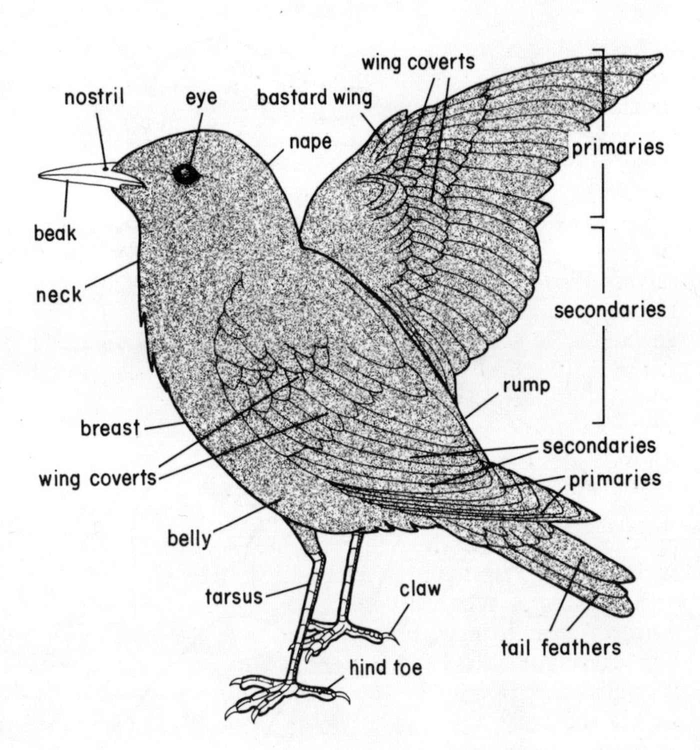

Fig. 10:2 External features of a bird.

high enough in the air at take-off to prevent it hitting its wings on the ground with the first downstroke. Wading birds have particularly long legs which help them when feeding in shallow water.

The lower part of the leg and the toes are covered with scales which are very similar to those of reptiles from which birds originally evolved.

Most species have four toes on each foot, but they differ considerably in position and structure according to the birds' habits. Pigeons are typical of perching birds in having three toes in front and one behind; this arrangement enables them to grip a branch very tightly. If you bend the leg of a fresh chicken, for example, as if it were squatting, you will see how the toes bend and grip. This is the reason why a bird does not fall off its perch when it goes to sleep. Look at Fig. 10:3 for other adaptations of feet.

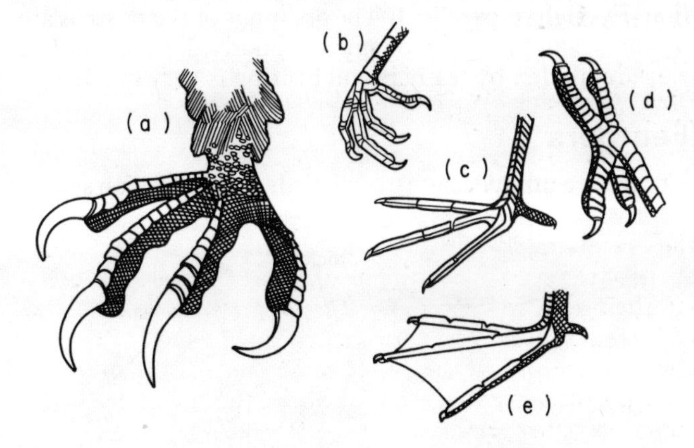

Fig. 10:3 Adaptations of birds' feet. a) Grasping and killing prey e.g. eagle. b) Perching e.g. starling. c) Walking on mud e.g. heron. d) Gripping on to bark e.g. woodpecker. e) Swimming e.g. duck.

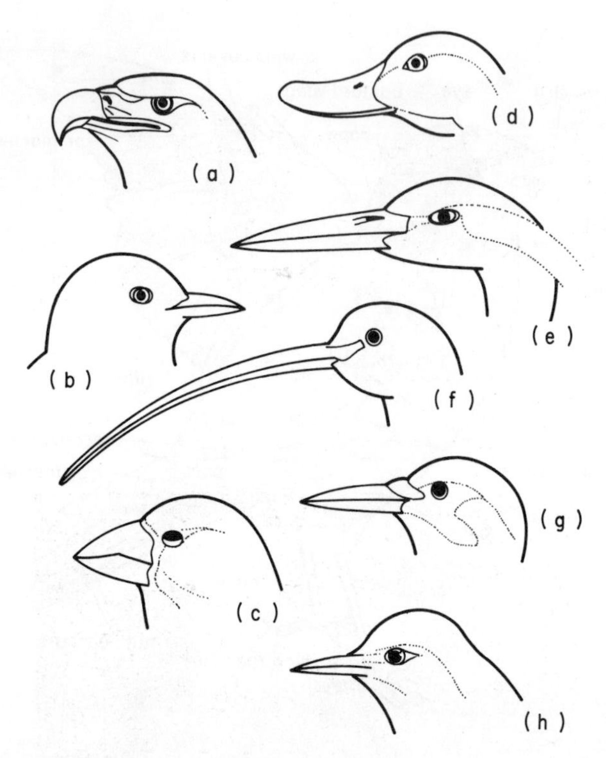

Fig. 10:4 Adaptations of birds' beaks for feeding. a) Flesh e.g. eagle. b) Varied diet e.g. blackbird. c) Seeds e.g. finch. d) Water plants and animals e.g. duck. e) Fish e.g. heron. f) Animals in mud e.g. curlew. g) Insects in wood e.g. woodpecker. h) Insects e.g. warbler.

Beak

This is a projection of the bones of the skull and consists of two mandibles covered with a horny material. The shape varies considerably according to the typical food of the species. You will see other examples in Fig. 10:4. By looking at the beak of any species you come across you should be able to guess the type of food it eats. Try this and check from a bird book whether you are correct.

Sense organs

The eyes are large relative to the size of the head. In addition to the two eyelids normally found, there is a third, the **nictitating membrane**, which sweeps sideways across the eye between the other two giving added protection. Eyesight is excellent. The openings of the nostrils are on the beak. The sense of smell is poorly developed. The ears are hidden by feathers and hearing is very good.

Feathers

These are a unique feature of all birds. Like reptilian scales, they are derived from the epidermis of the skin and are made of **keratin**, the substance which also forms the claws. The largest feathers of wing and tail are called **quills**, the smaller ones which cover the body and fill the gaps between the quills are the **coverts**, and underneath there are the **down** feathers. The air trapped next to the body by the feathers acts as an insulator, preventing loss of heat from the body.

A quill (Fig. 10:5) has many barbs which make up the vane on each side of the mid-rib or **rachis**. The barbs cling together because each barb has barbules coming from it on both sides: on one side the barbules bear hooks and on the other there is some device such as a ridge which causes the hooks to catch as they slide across each other (Fig. 10:6). This gives the feather a wind-resisting surface.

Coverts are like miniature quills in structure, but down feathers are fluffy because their barbules have no hooks to keep them together.

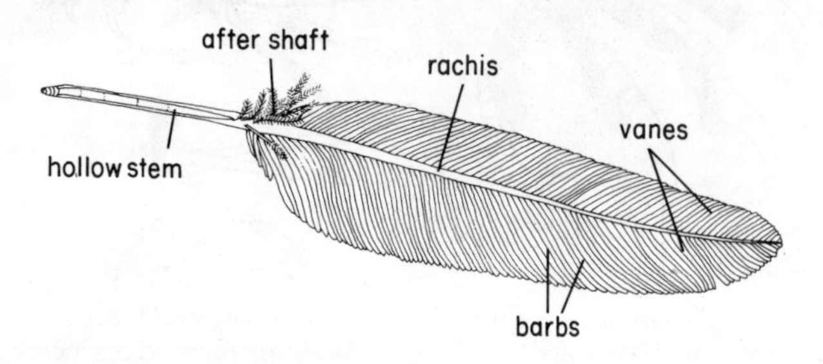

Fig. 10:5 Flight feather (quill).

1. Examine a quill, a covert and a down feather and note their main features. Notice how light a quill is and how much of a pull is necessary to cause the barbs to separate. Now see if you can repair the break by stroking the barbs between finger and thumb from the rachis end outwards. Dip the quill momentarily in water and notice if it is water-resistant.
2. Cut out a small portion of the vane, separate a few barbs by pulling gently, and examine under the microscope to see the barbules with their hooks and ridges.

A bird maintains its feathers in good condition by **preening**. First it twists its neck round and probes near the base of the tail where the preen gland is situated to obtain a little oil on the bill; next it spreads the oil on the feathers; then with its beak it smooths the barbs until they are all in place.

Wings

These are modified fore limbs which have become adapted to flight. If you examine one you will see that the undersurface is slightly concave: this makes the downstroke more effective. The leading edge is thicker than the trailing edge, just as it is in the wing of an aeroplane.

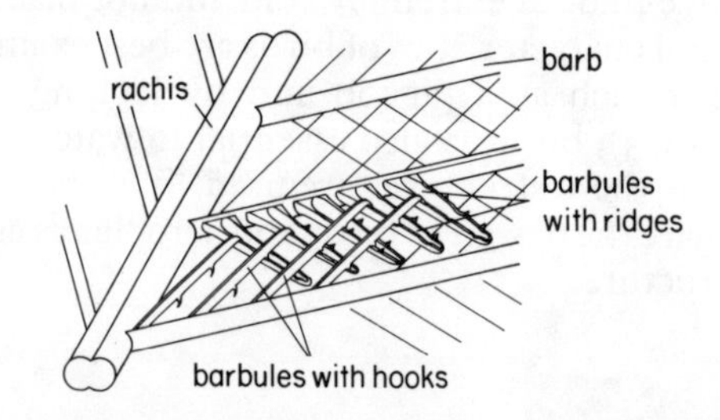

Fig. 10:6 Diagram of part of a flight feather much enlarged.

The wings of different species vary greatly in shape. Is the shape in any way connected with speed of flight or rapidity of wing beat? You could come to some conclusions about this by comparing the silhouettes of fast fliers such as swifts and hawks, heavy birds such as herons and swans and those which glide a lot such as gulls and eagles. Are there any similarities with the shapes of aircraft wings? Consider especially those aircraft which are built primarily for speed, for carrying heavy loads or as gliders.

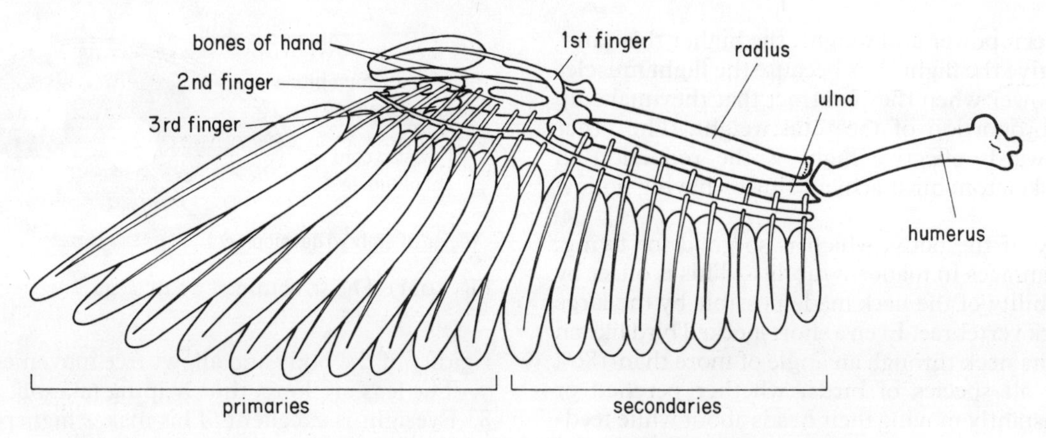

Fig. 10:7 Skeleton of a bird's wing showing the arrangement of the flight feathers (coverts removed).

The large surface area of the wings is partly due to the feathers and partly to the elongation of the limb bones. In some birds the wing span is enormous, e.g. up to 4 m in the wandering albatross. A fossil bird, recently discovered, had a wing span of 8 m!

Skeleton and muscles

Examine the skeleton of a wing (Fig. 10:7). The arrangement of the bones is not in a perfect pentadactyl plan although it has been adapted from it. Not all five fingers are present and two of the hand bones have fused with those of the wrist. The largest flight feathers, the **primaries**, are attached to the wrist and hand region, while the **secondaries** are attached to the ulna.

Look at the general form of a bird's skeleton (Fig. 10:8) and see how it is related to flight. Compared with a mammal skeleton, such as our own, the body is very rigid due to the fusion of some of the parts. This is because the great flight muscles have to be attached to something very rigid to bring about effective movement of the wings. These flight muscles are attached to the sternum which has a deep ridge down it like a keel. The muscles fit snugly on either side of it. There are two large muscles on each side: the **major pectoral** is the larger and the **minor pectoral** lies beneath it. They are both firmly attached, with one end fused to the sternum and the other to the humerus.

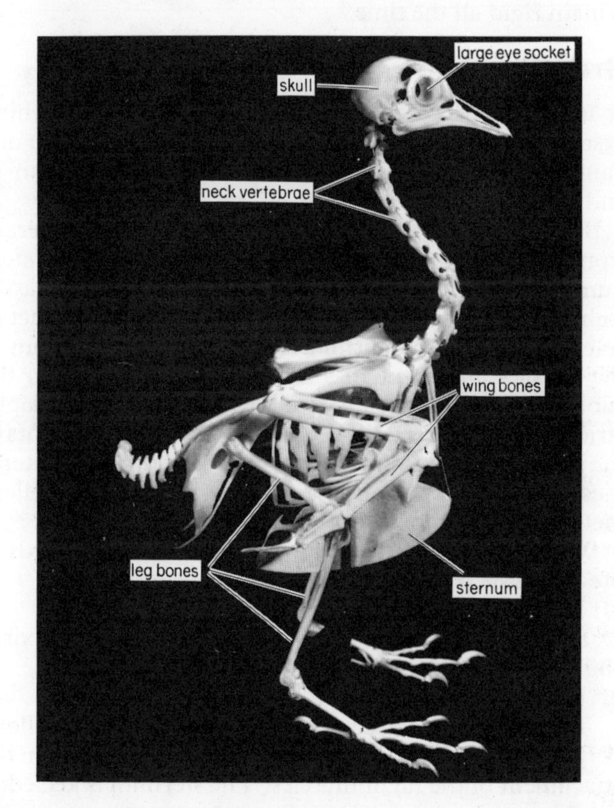

Fig. 10:8 Skeleton of a pigeon.

Flight

When the major pectoral muscle contracts it forces the wing downwards and slightly backwards, then the minor pectoral contracts and brings the wing up again. The difference in action is due to the position of insertion on the humerus of the tendons of the two muscles, the major acting from below and the minor from above. Look at Fig. 10:9 to see how this happens.

You might think that the upstroke during flight would counteract the lift given to the downstroke, but this is not so, because in the downstroke the flight feathers overlap and lie flat and no air passes between them, but in the upstroke their angle is altered so that the air does pass through. During flight the angle of the flight feathers alters during the downstroke, the primaries providing most of the backthrust and the secondaries mainly the downthrust.

In most species the flight muscles are about one-sixth of the total weight, in fast fliers, like pigeons, even more. Effective flight in both birds and aeroplanes depends upon

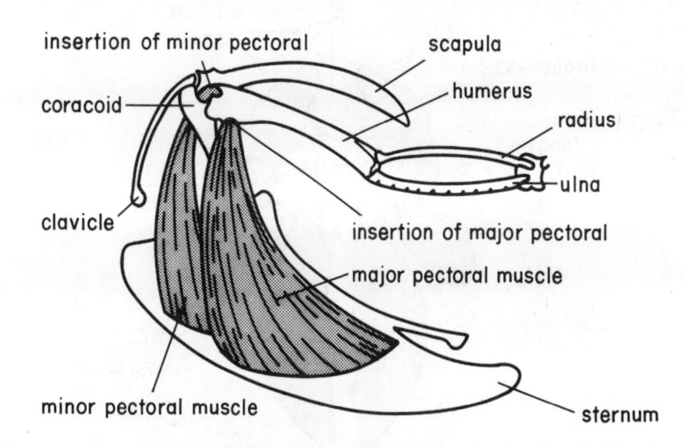

Fig. 10:9 Diagram showing how the flight muscles of a bird are attached.

the ratio between power and weight; the higher this ratio, the more effective the flight. It is because the flight muscles produce the power when they contract that they make up such a high proportion of the total weight. The other adaptation towards effective flight is the reduction in weight of the skeleton; birds achieve this by having hollow bones.

The rigidity of the body, which is so good for flying, brings disadvantages in manoeuvrability. This is offset by the great flexibility of the neck made possible by the large number of neck vertebrae. Even a short-necked bird like an owl can twist its neck through an angle of more than 180°! Observe how all species of birds, whether perched or flying, are constantly moving their heads about while feeding, preening or watching out for enemies, but their bodies remain rigid all the time.

Breathing

It is largely because of the efficiency of their breathing system that birds do not get out of breath. If you see one panting it is because it is too hot; by passing cooler air in and out rapidly, it cools itself.

Birds have lungs but the breathing method is different from that of mammals (Fig. 10:10). When the bird's sternum is lowered air is sucked down the windpipe and passes right *through* the lungs into a series of air sacs which act as reservoirs; when the sternum is raised the air passes out of the air sacs and through a series of minute tubes in the substance of the lungs. Oxygen is extracted by blood capillaries which surround the tubes. In flight the movements of the sternum correspond to the wing beats, thus the faster it flies the more oxygen it can absorb. This is an excellent method because no stale air is left in the lungs.

We can now summarise the main adaptations of birds to life in the air:

1. The fore limbs are modified as wings.
2. Feathers are present which are light, strong and flexible and which provide a large surface area.
3. The shape is streamlined.
4. The skeleton is extremely light, the bones being hollow.
5. The body is rigid, thus providing a firm base for the attachment of the flight muscles. The sternum is keeled to aid this attachment.
6. The neck is long and flexible. This compensates for the rigidity of the body and allows free movement of the head.
7. The legs are long, thus helping take-off.
8. Eyesight is excellent. This makes highspeed aerobatics and precise landings possible.
9. The efficient respiratory system and the homoiothermic condition both help to increase the efficiency of flight.

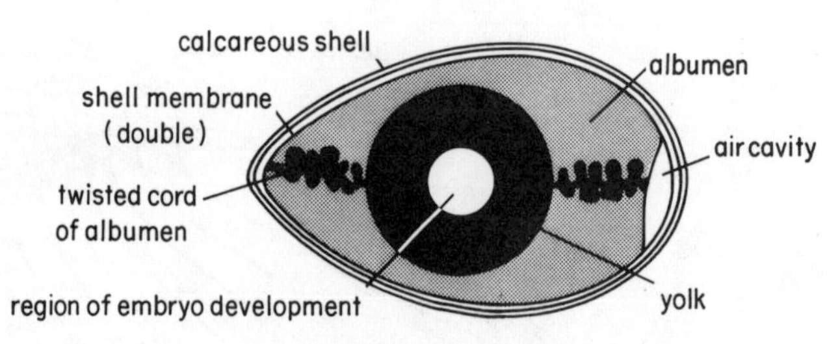

Fig. 10:11 The structure of a hen's egg.

Reproduction

All birds have internal fertilization and lay a small number of relatively large eggs (Fig. 10:11). Each has a hard protective shell which is calcareous and porous so that air can pass through. They are laid in a nest which helps to protect them, and are kept at a temperature usually higher than that of the air. This is achieved by a process of incubation when the parent keeps them close to its warm body, usually by sitting on them. Incubation also helps to prevent too much evaporation of water from the egg. In an artificial incubator for rearing chicks air has to be kept very damp for the same reason; otherwise they will die.

During development the chick embryo inside the egg becomes covered with a membrane, the **amnion**, which protects it and encloses it in a bath of water. Blood vessels spread out over the yolk surface and the blood transports the food in solution from the yolk to the embryo. As the chick grows the yolk sac becomes smaller and by the time it hatches almost all this food is used up and the chick is able to feed through its beak. When a domestic chick hatches (after 21 days) it is covered with down feathers and can move about almost at once, but in most species of birds the young are naked and helpless. However, they soon acquire feathers and after a few weeks are able to fly. During the period at the nest they are fed by the parents and grow very rapidly.

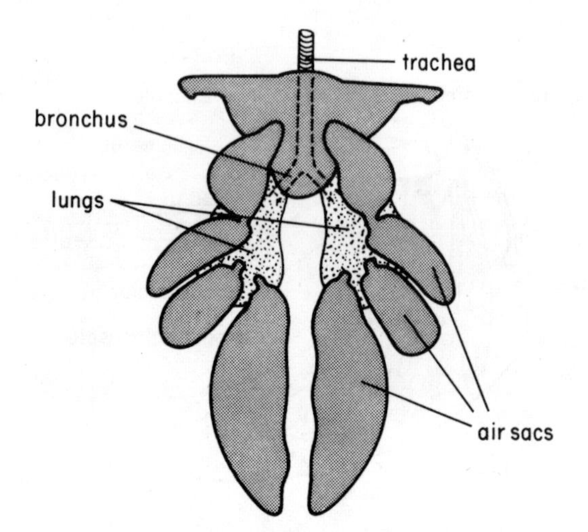

Fig. 10:10 Respiratory system of a bird.

Fig. 10:12 A kestrel feeding her young.

Looking after the young until they are independent is known as **parental care**; it is a very important factor for survival. Compare this with what happens in the majority of fish, amphibians and reptiles, where the young have to fend for themselves after hatching. Can you see any connection between parental care and the laying of only a few eggs?

> Observe the behaviour of a common species at the breeding season. By putting up nest boxes some months before nesting time, it is likely that they will be occupied and this makes observations easier. Here are some of the things to look out for:
>
> Note when nesting commences, what materials are used, when egg-laying starts, when the full clutch is laid, whether incubation is carried out by one parent or both, how often the young are fed, what food is brought, how the nest is kept clean, when the young fly and whether they keep with their parents or not. It is essential to disturb the birds as little as possible, or they may desert their nest.

Breeding adaptations

Different species show various kinds of adaptations which make reproduction more effective.

Egg shape may be significant. If eggs are round, like an owl's, they will tend to roll together in the nest if displaced, but birds which nest on cliff edges, such as guillemots, have pear-shaped eggs which do not roll off the ledge so easily. Eggs also differ greatly in colour; those of birds nesting in holes where little light penetrates are often white, and so are more easily seen by the parent when it returns to incubate them, while those of ground-nesters such as plovers and terns are mottled to resemble the surroundings.

The plumage of some species such as ducks and pheasants differ markedly between the sexes; these birds nest in the open and usually only the female incubates so her colouring shows excellent camouflage. Others which nest in holes, such as kingfishers, show no differences in plumage between the sexes.

Territorial behaviour

Most birds have **territories**. These are areas which are defended against other members of the same species. They may be defended by a male, a female, a pair or even a group of birds, for the whole year, or only during the breeding season. The main function of a territorial system is to ensure that the birds of one species are spaced out so that each can obtain enough food for itself and for bringing up its young. There is usually a correlation between the size of the territory and the availability of the right kind of food within it for that species. Thus some birds, such as robins and warblers, have territories as small as 2000 m^2, while eagles may dominate an area of more than 70 km^2.

Birds which nest colonially such as gannets and gulls have individual territories not much larger than the pecking distance between the nest owners, but these territories are not concerned with food supplies as food is obtained by ranging over considerable distances. The great advantage of nesting together is that the entire colony can act together to defend the whole breeding ground.

Territories are defended and the boundaries between adjacent territories are determined by a variety of threat postures and sounds. You can observe this for yourself very easily by watching robins. You will notice how a robin will sing from a certain tree or bush. The song serves to warn other robins that the territory is occupied and will be defended. Singing is an important aspect of aggressive behaviour; birds sing *at* each other, just as some people when they quarrel, shout at each other.

Boundaries between territories are at first vague, but they become more exact as a result of border clashes. If robin A trespasses on to B's territory, B will try to chase it off. If A holds its ground, B will fly up to it and display its red breast with a threatening posture. Usually this is enough to cause A to fly back to its territory, but if this does not happen a fight will develop. Fights to the death, however, are extremely rare. If, on the other hand, B trespasses on to A's territory, it is A which shows most aggression and hence B retreats. It appears that a bird's degree of aggression becomes less the further it strays from its own territory, thus the boundary is the region where both birds show equal levels of aggression. By first noting the places from which the neighbouring birds sing and then watching where scuffles and displays take place, it is possible to map out the territories.

Bird movements

Birds are said to be **resident** if they occur all the year in one area. In places where the seasons of the year are markedly different, the resident birds have to be very adaptable to the changing conditions. Often they have to change their feeding habits. Some species, however, make use of their power of flight to leave the region where they have bred and **migrate** to other parts where conditions are more suitable.

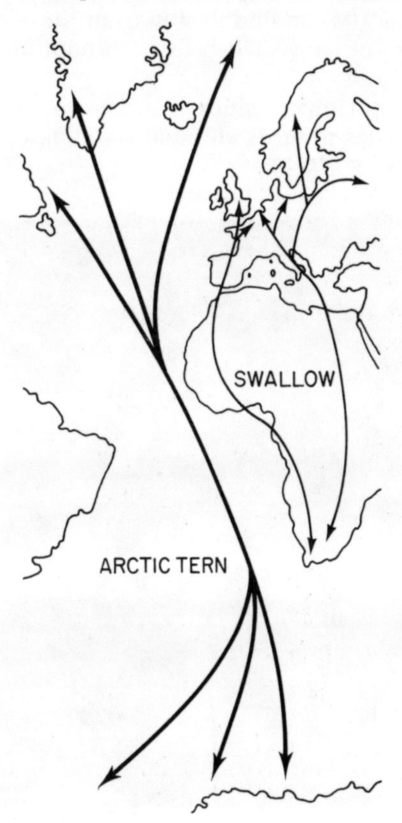

Fig. 10:13 The migration routes of the Arctic tern and swallow.

Migrants such as swallows, martins, flycatchers, nightjars and many species of warbler are insectivorous. They breed in Europe where there is plenty of food for them in the summer, but migrate to Africa in the autumn where an abundance of insects is available. Others such as redwings, fieldfares, bramblings and many species of duck and geese breed during the short summer in the northern tundra, but spend the winter in Britain and the more southern regions of Europe where the climate is mild enough for them to find sufficient food.

The most spectacular migrant is the Arctic tern which breeds within the Arctic circle and then migrates over some 16,000 km of ocean to spend the southern summer in the Antarctic.

Much of our knowledge about migration has been gained from the practice of **ringing**. Light metal rings, suitably numbered, are placed on the legs of fledglings and adults, and details of age, sex, date and locality are recorded for each ring used. The birds are then released. These data are then sent to the headquarters of the ringing scheme for reference, in the event of the ring being reported subsequently. Many thousands of birds have been ringed in different countries and the information has been used to map out the migratory routes. Additional information has been obtained by watching migration with the help of radar, but this is only successful for relatively short distances.

How do birds navigate over these vast distances?

It seems almost miraculous that a swallow can find its way back to the same barn where it nested the previous year after spending the summer in South Africa, or that a young martin or cuckoo can find its way from Europe to Africa, which it has never visited before, without any guidance from adult birds.

There is still much which is unknown, but there are important clues towards an understanding of how some species find their way.

Some birds orientate by the sun. This has been proved experimentally by keeping birds in large cages and altering the apparent position of the sun by means of mirrors. The birds adjusted their position according to the angle of movement of the mirror. It was also found that if the sky was totally obscured there was no orientation at all. What was more astonishing was that they were able to compensate for the natural 'movements' of the sun during the day, because if kept under an artificial sun they orientated at different angles to it according to the time of day; in other words they treated the artificial sun as if it were moving!

Many species migrate at night and orientate according to the pattern of stars. This has been demonstrated by putting some migrants in a planetarium. When the star pattern in the dome was rotated, the birds changed their position in a corresponding manner!

Experiments using small magnets attached to birds have shown that this can disturb their powers of orientation. It is thought that they can navigate in relation to the earth's magnetic field and minute particles of magnetite have now been discovered in the brain and muscles of birds which could act as tiny compasses.

You can now see why under heavy overcast conditions migrants may lose their way. They may also be blown hundreds of miles off course and perish in vast numbers.

Migration is a fascinating subject to read more about.

Fig. 10:14 A lesser spotted woodpecker.

Projects on birds

Here are some suggestions:

1. Try to find out which of the three main senses birds use most. For example, can they discover strong-smelling food which is covered up? Can they see colours? When a blackbird or thrush searches a lawn for worms does it see them, hear them, or smell them?
2. Erect a bird table in a quiet place where it can be viewed from a window. Find the preference foods of the species which visit it by putting out a selection of different kinds.

   Make notes on the feeding techniques of the various species. Note any aggressive behaviour between different species. Are some more timid than others? How does their behaviour affect the amount of food each species obtains? Is any particular species earliest to start feeding in the morning or last to finish in the evening?
3. Make a bird bath by digging a shallow hole and lining it with polythene sheeting. Place stones around the edge to anchor the sheeting firmly. Note how the various species that visit it use their beaks when drinking. Note also their bathing techniques. Are their bathing habits correlated with feeding times, temperature or other factors?
4. How fast do birds fly? Time different birds with a stop-watch as they pass between two fixed points such as between two trees or two hedges. Then measure the distance flown and calculate the speed. How would you allow for the effect of wind?
5. Visit different habitats such as a housing estate, wood, marsh or sea shore and identify with the help of a bird book the birds which are characteristic of each. Look out for any adaptations which help them to live successfully in these particular habitats.

11

Mammals

Fig. 11:1 The European badger.

Spreading to all habitats

Mammals represent the peak of vertebrate evolution, although the number of species living today is only about 5,000. They range in size from a species of shrew about 5 cm long and weighing 2·8 g to the mighty blue whale of up to 130 tonnes.

Like the birds, mammals are homoiothermic. This ability to keep their temperature reasonably constant irrespective of their surroundings has enabled them to spread to the coldest and hottest places of the earth. They have also shown great powers of adaptation to varying habitats and different ways of life.

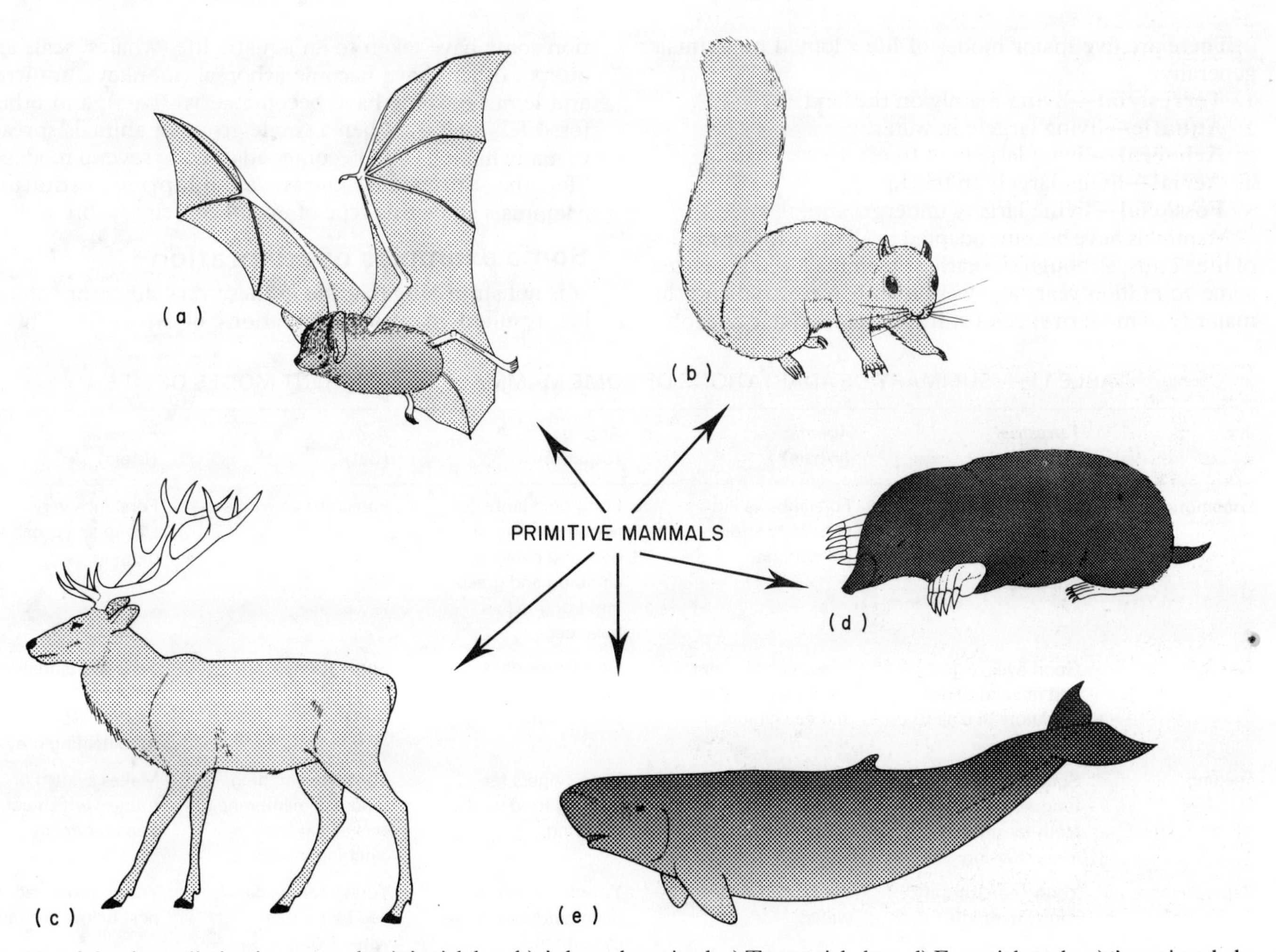

Fig. 11:2 Adaptive radiation in mammals. a) Aerial: bat. b) Arboreal: squirrel. c) Terrestrial: deer. d) Fossorial: mole. e) Aquatic: whale.

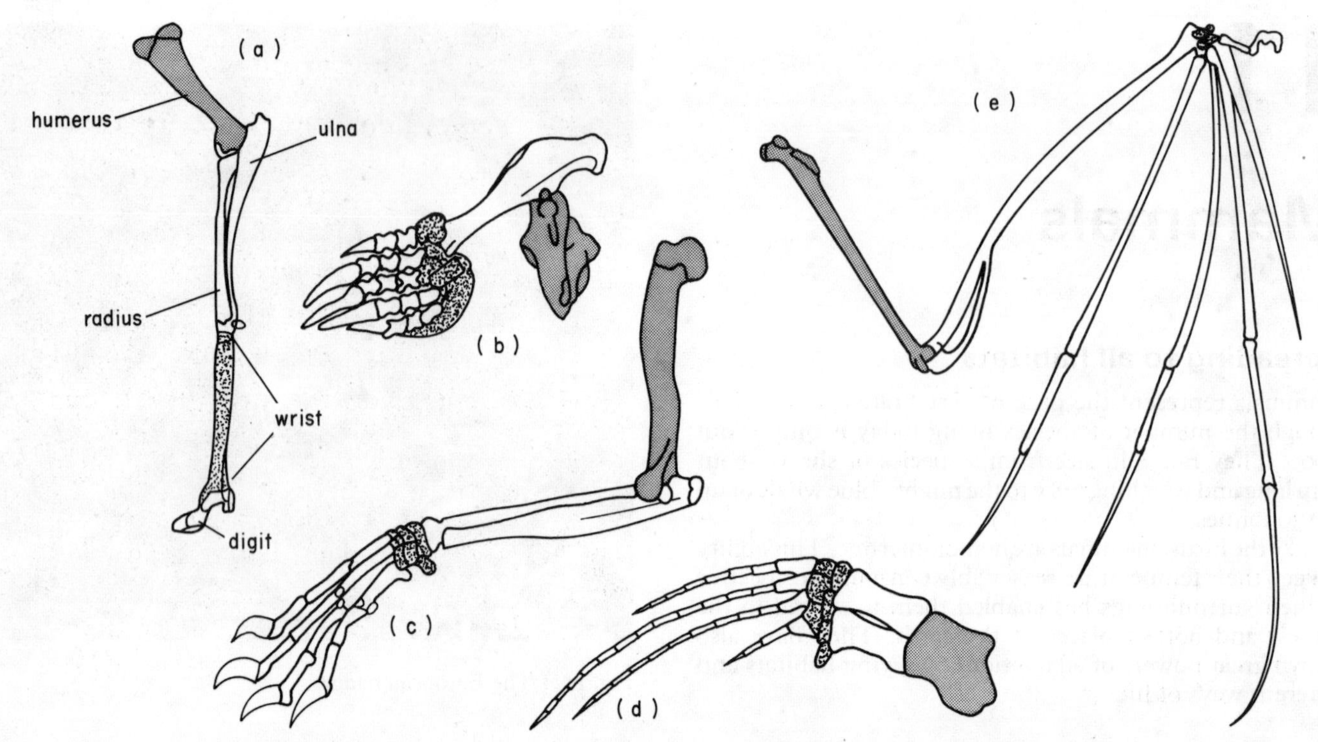

Fig. 11:3 Diagram showing how the fore limbs of certain mammals although based on the same plan have become adapted for different purposes. a) Deer: running. b) Mole: digging. c) Squirrel: jumping and climbing. d) Whale: swimming. e) Bat: flying. (Shading indicates comparable bones.)

There are five major modes of life adopted by animals generally:

1. **Terrestrial**—living mainly on the land.
2. **Aquatic**—living largely in water.
3. **Arboreal**—living largely in trees.
4. **Aerial**—living largely in the air.
5. **Fossorial**—living largely underground.

Mammals have become adapted to all these major modes of life. Thus, although the earliest mammals which existed some 70 million years ago were mainly terrestrial and the majority of modern species still are, in the course of evolution some have taken to an aquatic life (whales, seals and otters), others have become arboreal (monkeys, squirrels and lemurs), some have become aerial (bats), and others fossorial (moles). When a single group of animals spreads to many habitats and becomes adapted to several modes of life, the process is known as **adaptive radiation**. Mammals show this type of evolution remarkably well.

Some examples of adaptation

It is not surprising that life in these very different habitats has resulted in major adaptations of structure. This is

TABLE 11:1. SUMMARY OF ADAPTATIONS OF SOME MAMMALS TO DIFFERENT MODES OF LIFE

	Terrestrial (Deer)	*Aquatic* (Whale)	*Arboreal* (Squirrel)	*Aerial* (Bat)	*Fossorial* (Mole)
Locomotion	Long legs, runs on its toe-nails (hooves); good for gripping.	Forelimbs as flippers. Very strong tail with fin.	Long hind limbs for jumping. Long fingers and claws for climbing and grasping. Long tail as balancer.	Forelimbs as wings.	Forelimbs very strong and broad for digging.
Senses	Good eyes, ears and nose to detect predators in time to escape.	Uses type of sonar for communication and navigation.	Good eyesight for judging distances when leaping	Sonar for navigation in dark.	Sensitive vibrissae to detect vibrations in ground. Good nose for detecting prey.
Feeding	Special type of tongue, mouth and teeth for grazing and browsing.	Enormous mouth. Baleen for filtering plankton.	Long fingers for holding food while gnawing.	Sonar for catching food; tail membrane acts as net for catching insects.	Makes system of tunnels which act as traps for its prey.
Reproduction	Young can run very soon after birth.	Young can suckle in water.	Young can climb even before their eyes open.	Young can be carried by mother when flying.	Young protected in nest below ground.

particularly true of the limbs, as each mode of life requires a modified means of locomotion. It is remarkable that the characteristic pentadactyl limb has become adapted so that it may be used efficiently, for example, by deer for fast running on land, by whales as flippers for swimming, by squirrels for jumping and gripping boughs, by bats for flying, and by moles for shovelling earth and burrowing in the ground. By studying the skeletons of their limbs (Fig. 11:3) you will see how each has been modified from the primitive pentadactyl arrangement. The adaptations that have taken place have not only affected the limbs, but other parts too. For example, different senses become more important than others in different habitats, and feeding and reproduction provide special problems. Some of these adaptations are summarised in Table 11:1 and you could add to them considerably by reading more about the species concerned. We will study certain aspects in more detail.

The senses of mammals

In birds, sight is the most important of the senses although hearing is also very acute; the sense of smell is very poorly developed. This is largely because birds are mainly active during the day (**diurnal**). The majority of mammals, however, are most active at night (**nocturnal**), or during the transition period around dusk and dawn (**crepuscular**). Consequently, many mammals rely more on scent and hearing. The exceptions, as you would expect, are the diurnal ones. It follows from this difference between birds and mammals that when studying birds in the field you need to keep very still or watch them from a hide, but with many mammals it is often more important to be down wind so that your scent is not detected.

Sonar

Bats have a special problem due to their specialised mode of life. They are nocturnal or crepuscular, and many of them feed on insects such as moths, beetles and mosquitoes. They have to catch their prey while on the wing and at the same time avoid damaging themselves against boughs and other obstructions in their way. They do this by a remarkable echo-locating system known as **sonar**. This involves the giving out of rapid pulses of ultrasonic sounds which rebound off any object in the bat's flight path and are picked up by the extremely sensitive ears of the bat. This method of detecting objects is so effective that if bats are released in a room where extremely fine wires have been strung across no more than a wingspan apart, they will avoid striking them even in complete darkness.

In some bats the sound pulses are emitted through the mouth at a rate of about 10 per second when cruising, but if an insect such as a moth is detected the rate is increased to more than 100 per second which enables the bat to investigate it more carefully and pin-point its position. In the leaf-nosed bats the sounds are emitted through the nostrils which serve to concentrate them in a narrow beam like a bullet in a rifle barrel, thus increasing the efficiency of the technique.

The ears of bats are relatively very large and are admirably adapted for receiving the echoed sounds. In addition, they have a projecting structure within the ear, the **tragus**, which helps to focus the sounds on to the ear drum. The inner ear is also highly developed to receive sounds of

Fig. 11:4 Long-eared bat.

extremely high frequency, and by some means not yet understood, bats are able to hear the echoes of their own sounds even when subjected to much louder noises at the same time. They can also avoid hitting against each other when flying in great numbers together in the confines of a cave, although each is using its own sonar at the same time. Can they, perhaps, recognise their own voices?

The aquatic mammals, such as whales and dolphins, have similar problems of finding food and avoiding objects, but in water rather than in the air. This is particularly true in turbid water or at night. The toothed whales and dolphins use a sonar device similar in principle to that used by bats. As with bats, it has been possible to record these ultrasonic sounds by using extremely sensitive electronic equipment. When they are played back at a reduced speed, they are audible to the human ear as a series of clicks.

Whales also have the problem of communicating with each other over great distances of apparently featureless ocean. They do this by uttering a wide range of whistles, groans, grunts and trills which are conducted through the water over very considerable distances and are apparently meaningful to other whales. It is even thought that individual whales may be recognised by others by their signature tunes!

The characteristics of mammals

Let us now consider some of the features which distinguish mammals from other vertebrates, and which have been largely responsible for their success.

Hair

All mammals have some hair although some species such as whales only show a few bristles as adults. In fur-bearing types hairs are of two kinds, the long **guard hairs** and the thick felt-like mass of finer hairs forming the **underfur**. The hair traps air next to the skin, insulating the body from too much loss of heat. In otters and polar bears, for example, the underfur is so thick that when they are swimming the water cannot penetrate and the skin remains dry.

Hair has other functions beside preventing loss of heat. Some hairs are modified as special sense organs of touch, e.g. the **vibrissae**, or whiskers. In otters these vibrissae are particularly sensitive; they can detect the vibrations in the water made by a fish swimming, enabling the otter to locate the position of the fish and catch it even in dark or turbid waters.

In hedgehogs and porcupines the hairs have become modified as spines or quills and are defensive in function. If you examine a hedgehog you will see how the spines of the back and sides merge into the normal hairs of the underside. The horns of the rhinoceros are also made of hair, not bone as you would expect. They consist of a mass of hair impacted together to form a very strong and formidable weapon.

Adaptive colouration

The colour of hair in different mammals is often protective, enabling the animal to blend with its surroundings. In the majority the fur shows a gradation of tone, being darkest on the back and lightest on the belly, a condition known as **counter-shading**. As dark colours reflect less light than pale, and most light comes from above, counter-shading tends to equalise the amount reflected from all parts and so gives an appearance of flatness which makes the animal far less conspicuous. Thus a rabbit is difficult to see when it is feeding in a field, but a dead one, lying on its side, is immediately visible.

In other mammals the outline is broken up by stripes or spots as in tigers and leopards, or by having a single conspicuous dark line along the side as in many antelopes. This type of camouflage is known as **disruptive colouration**. It is very effective because it takes the eye away from the outline of the animal which gives it its familiar look.

Other mammals which live in places where there is one main prevailing colour often have fur of the same hue. Desert mammals are usually some shade of yellowish brown and those in snowy regions are white. This gives an advantage both to a predator when hunting and to the prey when being hunted. Polar bears and the young seals on which they feed are both white, and the same applies to the Arctic foxes which hunt the Arctic hares.

Some mammals change the colour of their coat according to season. This is true of the northern race of the stoat, which in winter has a white coat and is known as **ermine**, but in summer has one that is reddish-brown. Arctic foxes and Arctic hares undergo a similar change, but the new coat is acquired differently. In the fox the dark tips of the hair wear off as winter approaches leaving only the white basal parts which continue to grow; it then moults this coat in the spring, replacing it with the summer one which is blue-grey. Arctic hares, on the other hand, effect the changes by having several moults.

One of the strangest examples of hair colouring is in some of the South American sloths which live high up in the trees; they have green hairs. This is because microscopic green algae are living in the groove which runs down each hair. The general effect is to make them most inconspicuous.

Viviparity

All but the platypus and echidna (which lay eggs) bear young, and are thus said to be **viviparous**. This differs from the viviparity of other vertebrates we have mentioned such as sharks and adders, because in mammals the young are fed before birth through an organ called the **placenta** from which food is extracted from the mother's blood. When fish and reptiles bear young, the developing animals feed only from the yolk inside the egg, and the mother merely keeps the eggs inside her body until they hatch; they are then born. This method is distinguished by the term **ovoviviparity**. Marsupial mammals, e.g. the kangaroo,

Fig. 11:5 A sow suckling.

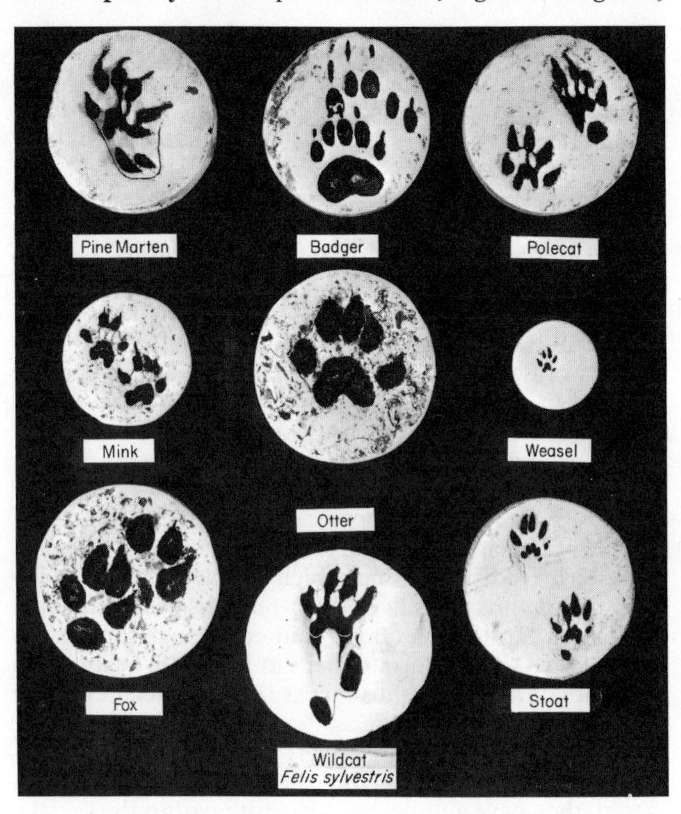

Fig. 11:6 Plaster casts of the footprints of some mammalian carnivores.

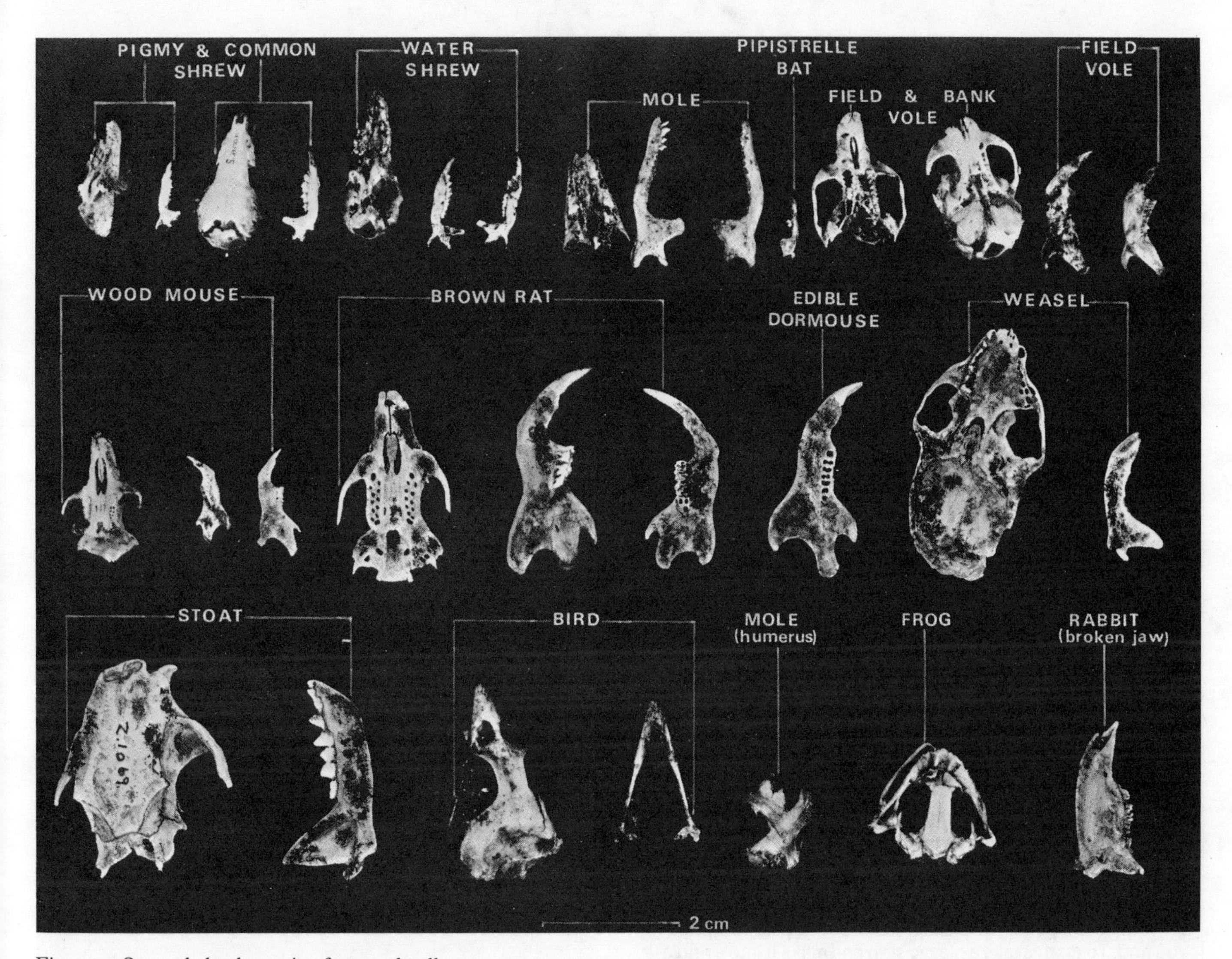

Fig. 11:7 Some skeletal remains from owl pellets.

are somewhat intermediate between these two types as the young are born in a very immature state. They are then suckled inside a pouch or **marsupium** which also protects them.

Secretion of milk

All female mammals secrete milk and suckle their young after birth. In all but the platypus and echidna the milk is secreted from **mammary glands**, also called breasts or udders in certain species. These vary both in number and position. In man and elephant there is only one pair which is thoracic, in others such as cats, cows and pigs there are two or more pairs which are abdominal. Those mammals which have large litters tend to have more. Suckling enables the mother to provide food for its young until it is **weaned**, i.e. until it can feed on solid food.

A glandular skin

Hairs, like feathers, need to be kept supple and waterproof. Mammals have minute oil glands in the skin which lubricate the base of each hair. They also possess sweat glands.

Teeth

Mammalian teeth, unlike those of the lower vertebrates, develop in sockets in the jaw. Usually they are of four kinds: incisors in front for cutting or gnawing, behind them the canines for killing the prey or tearing the food, and further back the premolars and molars for chewing or grinding. The dentition of different species varies very much according to the type of food eaten, a feature which is used in the classification of the different mammalian orders. We shall study teeth in detail in Chapter 17.

Other mammalian features

There are other characteristics which mammals share with other vertebrates, but which are developed to a greater degree. One of these is the brain which reaches its highest development in the whales and dolphins, and in apes and man. This has enabled mammals to learn more about their surroundings, develop better memories, and behave with greater intelligence than other vertebrates.

Parental care has also been developed to a far greater degree. Not only do they feed and protect the young in the uterus before birth and suckle them after birth, but they also protect them during the growing-up period. Consider the essential role which parental care plays in our own lives! How would we manage without being given shelter, food, clothes and educational facilities?

Projects you could carry out on mammals

1. Study a species of mammal for yourself, e.g. dog, cat or mouse, or wild ones such as rabbit, fox or badger. Consider first its mode of life, where it lives, what it feeds on and its general habits. Then try to discover in what ways it is adapted to live in such a habitat, feed, move and breed in the way it does. Whatever species you choose, whether it is a cat living in a city suburb, or a badger in a wood, you will find it is well adapted both in structure and behaviour to its mode of life.

2. Collect as many kinds of animal hairs as you can and try to identify their owners by the colour, length, curliness, etc. Hair often gets caught in the barbed wire round fields. Where there is a barbed wire fence bordering a wood, look especially at the lowest strand where an animal path leaves the wood. Brambles and briars also collect hairs; look carefully at those bordering animal paths.

3. Observe the behaviour of some of the small mammals such as mice and voles. Find out which species occur in your garden or in a hedgerow by attracting them with food. Just before dark put out small piles of oat grains or vegetable scraps in sheltered places where birds will not find them. Put more down each night and examine each morning. You will probably find that the food disappears quite regularly after a few nights. Discover what is eating the food by watching at night using a torch with a red filter to help you. (Many mammals are insensitive to red light.) Make detailed notes of what you see.

4. Look out for mammal tracks in snow or in muddy places. Look especially in the mud of woodland paths, by rivers, and near the edges of ponds and lakes. Identify the tracks with the help of a reference book. In dry weather you can make your own mud at strategic places and peg down some clean paper near it for the animals to tread on after going through the mud.

Take any opportunity when snow is on the ground to follow the tracks of mammals. In this way you can discover much about the habits of foxes, rabbits, badgers and many others.

5. Make a collection of plaster casts of the best footprints you find. To do this take a plastic strip and make it into a ring a little bigger than the print, fastening the ends together with a paper clip. Press the ring gently into the ground round the print and pour in a runny mixture of plaster of paris and water, made up on the spot. You can remove it carefully after about ten minutes when the plaster should have set, but it should not be cleaned up until later. To do this, place the cast under running water to remove the mud. Be sure to label it on the back with date, locality and, of course, the scientific name of the mammal.

6. Find out what small mammals exist in your neighbourhood by examining owl pellets. Owls are expert at catching them, but having digested the soft parts they eject the bones and fur in a pellet through the mouth. These may be found under places where owls roost. Soak the pellets overnight, and then pick out the skulls and bones with forceps. Identify their owners with the help of a reference book. In this way you may find the skulls of various species of shrews, mice, voles and even an occasional mole or bat. The parts of the skeletons can be cleaned in water and bleached by soaking in hydrogen peroxide solution and then mounted on cards.

7. Another method of finding out about the distribution of small mammals is to examine the contents of old bottles which have been thrown away near lay-bys and similar places by careless people. Small mammals often enter these bottles, are unable to get out and so die. Wash out the contents of the bottle into a dish and look for any skeletons. The bones can be cleaned, bleached and mounted as in 6. Put the bottles in a litter bin!

12

Diffusion and osmosis

In previous chapters we have been largely concerned with the study of whole organisms: their great diversity, their life histories, modes of life, and the ways in which they are adapted to living in various environments. We have also seen how all these organisms, whether plant or animal, large or small, carry out certain functions which are common to all living things; they all feed, respire, excrete, respond to stimuli, grow and reproduce. We shall now study more deeply these functional aspects of biology—how the body of an organism really works. For this we shall concentrate mainly on man and flowering plants: man, because he is the most important and interesting member of the animal kingdom to us; flowering plants, because we are utterly dependent upon them for life, and need to understand the contrasting ways in which they carry out their life functions.

As an introduction to this study, it is necessary first of all to understand two physical processes which are of vital importance to all living organisms. These processes, **diffusion** and **osmosis**, are concerned with the manner in which water and other substances move within the tissues of organisms.

Diffusion

Consider a solution of sugar in water. It consists of water molecules of extremely small size and sugar molecules of rather larger size. All these molecules are moving rapidly and in a random manner, hence they are evenly distributed within the liquid.

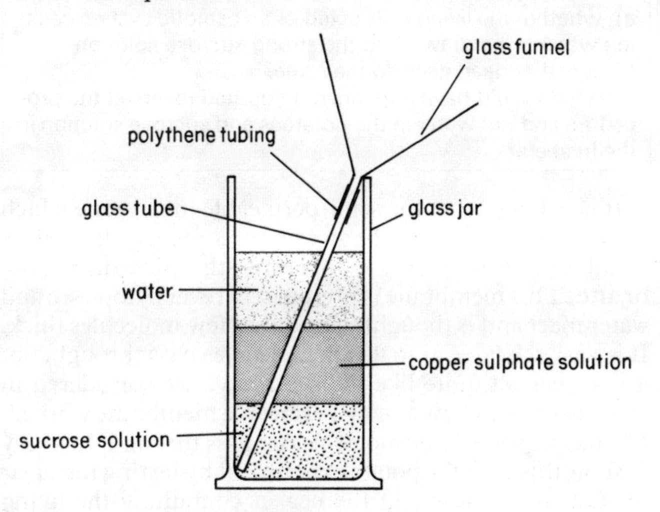

Fig. 12:1 Apparatus to demonstrate diffusion.

When there is a greater concentration of sugar in one part of the solution, as would happen when a lump of sugar slowly dissolves in it, more sugar molecules tend to move away from the region of high concentration than return to it, until eventually they become uniformly dispersed once more. This movement of molecules (or the ions into which they dissociate when in solution) from regions of high concentration to low, until uniform dispersal results, is called **diffusion**. The difference in concentration along the path between two regions is known as the **diffusion gradient**. Diffusion occurs in both gases and liquids.

You can observe diffusion taking place by using a substance such as copper sulphate which forms a coloured solution in water. The density of the colour is an indication of its concentration.

> Take a gas jar one quarter full of water. Very carefully introduce copper sulphate solution using a glass funnel and tubing as shown in Fig. 12:1. Copper sulphate solution has a higher density than water and provided that it is introduced very slowly it will form a coloured layer underneath the water. Repeat the procedure using a sucrose solution. This has a higher density than copper sulphate solution and so will form a clear layer at the bottom of the jar.
>
> Hold the glass tube in position and carefully remove the funnel. Place a finger firmly over the top of the tube and withdraw it slowly from the gas jar.
>
> You should have bands of clear liquid above and below a layer of coloured liquid. Keep the jar undisturbed for several days. Observe any diffusion of the coloured liquid into the clear layers.

Diffusion is a very important process in living organisms as substances such as oxygen, carbon dioxide, salts and sugar will tend to move from regions of high concentration to low. For example, if cells are constantly using up a gas such as oxygen, the concentration of oxygen there will be lowered, hence more oxygen will diffuse from places of higher concentration to take its place. In this way a constant supply to the cells can be kept up. We shall come upon many other important examples of diffusion in other chapters.

Osmosis

An important variation of diffusion may occur when there is some kind of barrier separating the solution of high concentration from the low. What happens depends on the kind of barrier. If it is made of something such as glass, or the cork in the bark of a woody stem, no molecules can move through it, hence it is said to be **impermeable**. If it is made of a porous material with pores sufficiently large to let the largest molecules through, diffusion will occur through it as if it did not exist; it is then said to be fully **permeable**. The cellulose walls of plant cells are like this. However, if it contains pores which are of molecular size, it could be that small molecules such as water will be able to pass through, while larger ones such as sugar will not. Such a substance is said to be **semi-permeable**. In other words, by being selective in its action, the membrane modifies the diffusion process.

Visking tubing is an artificial semi-permeable membrane which is convenient to use for studying the process.

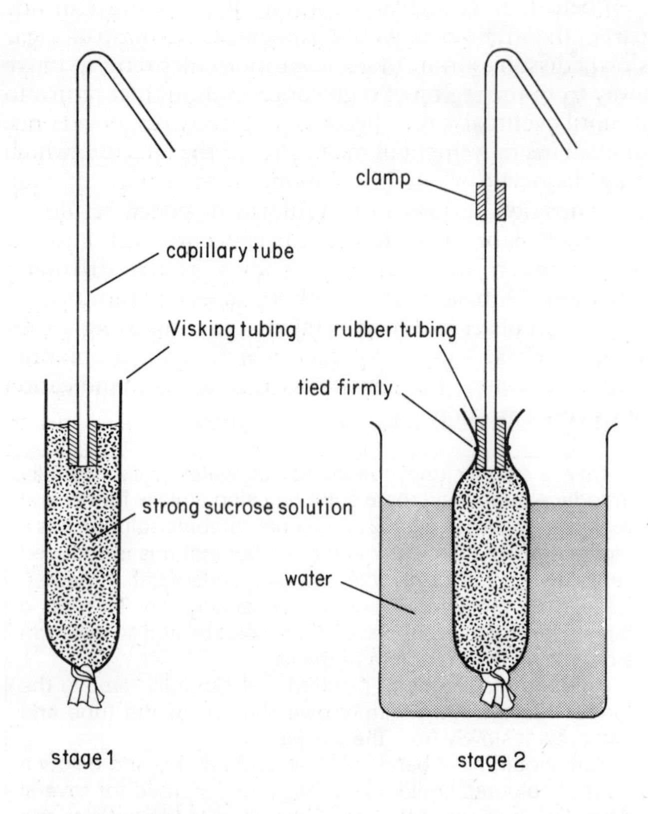

Fig. 12:2 Osmometer.

Set up the apparatus shown in Fig. 12:2 using a piece of Visking tubing about 10 cm long. First wet it and then tie one end in a knot. Half fill it with strong sucrose solution and insert the bent capillary tube. Now remove the air by squeezing the Visking tube tightly just below the level of the solution and tie firmly to make a watertight joint with the capillary tube. Wash the outside of the bag to remove any spilt sugar solution and suspend it in a beaker of water. Mark the level of the sugar solution in the capillary and observe any changes over a period of half an hour.

It is possible to explain any rise in level by saying that water has passed from the beaker through the Visking tubing into the sucrose solution thus increasing its volume. If you think of this in terms of movements of molecules (Fig. 12:3) you can say that water has passed from a region of 100% water through a semi-permeable membrane into a sugar solution where the water is less than 100%; thus the water molecules have obeyed the diffusion law by going from high concentration to low, but the sugar molecules have been prevented from doing so because they could not pass through the pores of the membrane. This process is called **osmosis** and it also takes place when a semi-permeable membrane separates a strong solution of sucrose from a weaker solution—it does not have to be pure water. More water molecules pass from the weaker solution to the stronger than in the opposite direction.

From your experiment you will have noticed that the intake of water causes a rise in the capillary tube against gravity; it causes a pressure sufficient to force it out at the top. This is called the **osmotic pressure** of the solution. Its value is equal to the pressure that would have to be exerted in the opposite direction on the column in the capillary tube to prevent it from rising. The stronger the solution, the greater is the osmotic pressure.

Osmosis in living cells

We can now find out whether osmosis occurs in living material and if so whether the semi-permeable membrane is living or non-living. For this we will compare the action of living and dead potato tissue.

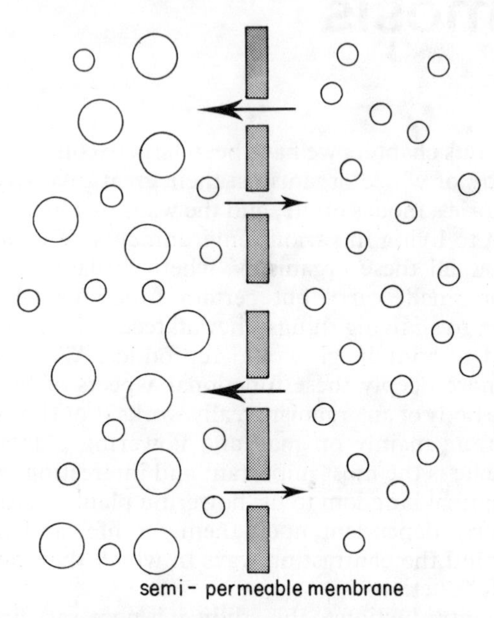

Fig. 12:3 Diagram to illustrate the action of a semi-permeable membrane. (Small circles = water molecules, large circles = sugar molecules.) Note that more water molecules pass from **right** to **left** than in the reverse direction.

1. Peel a large potato, cut it into two halves and remove the central tissue, without piercing it, to form two cups (Fig. 12:4).
2. Boil one half in water for about 2 minutes to kill the cells and then suspend each half in a beaker.
3. Add enough strong sucrose solution to each to half-fill the cups and put enough water in both beakers to come up to the same level. After 24 hours note the levels in both potatoes compared with that in the beakers.

From your observations you should be able to conclude:
a) whether the living cells acted as an osmotic system causing water to be drawn into the strong sucrose solution,
b) whether dead cells do the same,
c) what would have happened if you had reversed the procedure and put water in the potatoes and sucrose solution in the beakers?

It is believed that the semi-permeable membrane which causes living material to be an osmotic system is the surface membrane of the living cytoplasm—the **plasma membrane**. This membrane is present wherever cytoplasm and water meet and is thought to be only a few molecules thick. It is very selective as to which molecules pass through, but it does *not* act quite like the rigid sieve we considered in order to explain how a semi-permeable membrane worked. In practice some large molecules do pass through. It is easy to show this with the potato experiment by testing for sugar (p. 112) in the water in the beaker containing the living material after a few days. What probably happens is that

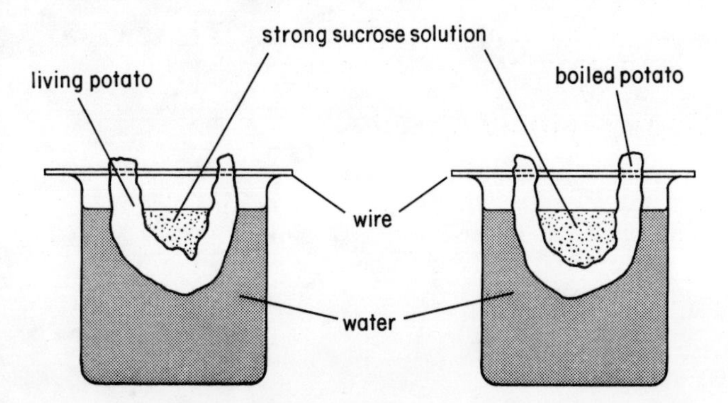

Fig. 12:4 Experiment, when first set up, to find out whether living material acts as an osmotic system.

the larger molecules pass through much more slowly than the water molecules, so the osmotic effect is the same. Other theories to explain osmosis have been put forward, but these will not be considered here.

If you think back to your observations on living plant cells (p. 4), you will see that each cell is an osmotic system (Fig. 12:5). The cytoplasm surrounds a vacuole containing water in which substances such as salts and sugars are dissolved, and both internal and external surfaces of the cytoplasm will act as semi-permeable membranes. What would happen if cells such as these were surrounded by solutions of various strengths? It is difficult to observe the effect in some cells as the cytoplasm and cell sap are colourless, but you could use cells where the sap is coloured, as it is in the epidermal cells of plant organs which appear red.

1. Strip a small piece of epidermis from the reddish area of a rhubarb petiole and mount it on a slide in a drop of water with the outside uppermost. Examine under the microscope and note the red content of some of the cells.
2. Remove the water with blotting paper and add a few drops of strong salt or sugar solution and note what happens to the red content of the cells.
3. If you observe a change, remove the strong solution once more and replace with water.

Is there any further change? Do your observations confirm that living cells act as osmotic systems?

The changes you should have seen may be explained more fully by saying that when living cells are placed in a **hypertonic** solution (one stronger than that of the cell sap), water will be drawn out from their vacuoles and the cytoplasm, which is somewhat elastic, will come away from the cell wall. At the same time the bathing solution will pass through the cell wall, because it is permeable, and fill the gaps. In this condition the cell is said to be **plasmolysed**. When the cell is then surrounded by water or a **hypotonic** solution (one weaker than the cell sap), water will be drawn into the vacuole which will expand until the cytoplasm is forced back against the cell wall. But the cell sap will still be stronger than the fluid outside, so more water will be taken in and a pressure will be exerted outwards against the cell wall causing it to stretch. The cell is now said to be **turgid**. This will continue until the increasing resistance of the cell wall to this stretching equals the osmotic effect of the water being drawn in.

Bearing these principles in mind, carry out this experiment:

1. Cut from the rhubarb petiole used in the last experiment two strips of tissue (with any epidermis removed) about 8 cm long and roughly 0·5 cm square in section. Use a ruler to guide the scalpel when making the cuts. Trim the ends so that they are of equal length and note the exact measurement.
2. Place one strip in strong sugar (or salt) solution and the other in water and leave for half an hour.
3. Measure each strip again carefully. Is there any difference in the lengths? Feel the two strips and compare their rigidity.

Is there any difference between them? Can you explain any difference in terms of osmotic action? (This experiment can be done equally well using cylinders cut from a large potato by means of a cork borer.)

You will now realise from your experiments that it is osmosis which causes water to be drawn into plant cells causing them to swell and press against each other so bringing about the rigidity of stems and leaves (p. 36). It is also by osmosis that water is drawn into the root hairs of plants from the soil.

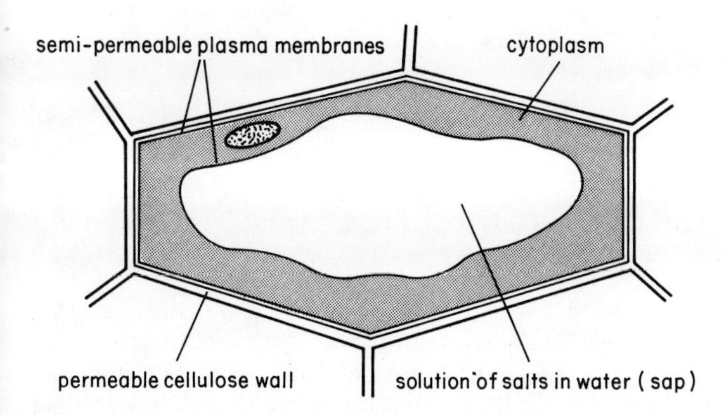

Fig. 12:5 A living cell—an osmotic system (membranes shown diagrammatically).

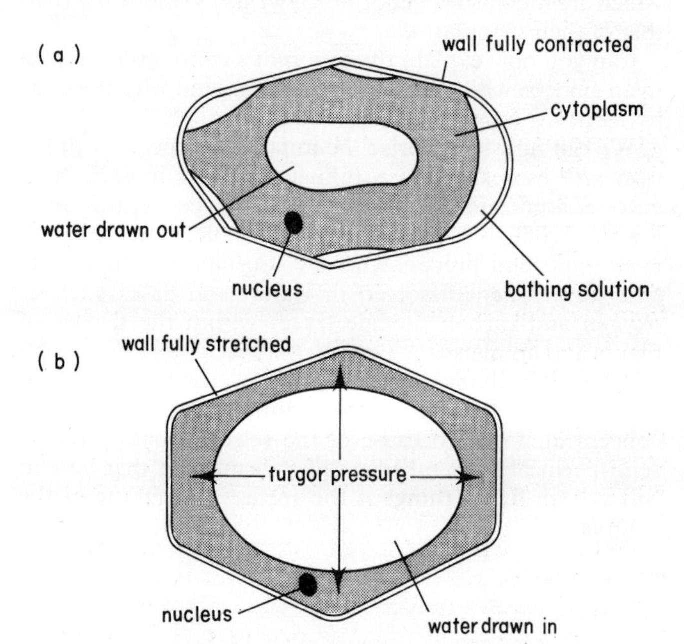

Fig. 12:6 a) Cell in hypertonic solution—plasmolysed. b) Cell in hypotonic solution—fully turgid.

What happens to animal cells under similar circumstances, as they have no tough cell walls to prevent them from expanding? You could find out by subjecting cells such as red blood corpuscles to solutions of different strengths. It is easy to obtain the material as each drop of blood contains about 5 million red corpuscles. These corpuscles normally float in a fluid plasma which contains various salts and sugars in a concentration which gives it the same osmotic strength as would be obtained by dissolving 0·85 g of salt in 100 cm^3 of water (0·85% solution). Solutions of equal osmotic strength are said to be **isotonic**. So for this experiment we shall need:

1. A 0·85% solution of sodium chloride in water which is isotonic.
2. A 2% salt solution which is therefore hypertonic.
3. Distilled water which is hypotonic.

1. Place 1 cm^3 of each of the above solutions into 3 small specimen tubes, appropriately labelled.
2. To obtain the blood sterilise the top of a finger with surgical spirit. Draw more blood into your finger by swinging your arm round, holding it downwards and wrapping a handkerchief tightly round the base of the finger. Now make a quick jab with a sterilised lancet and draw up a few drops of blood into a fine pipette which has also been sterilised.
3. Place one drop into each of the specimen tubes and shake gently.
4. After a few minutes remove a drop from the isotonic solution with a clean glass rod and examine under the high power of the microscope. Note the shape of the corpuscles and remember this is your control. Now do the same for the hypertonic and hypotonic solutions. Observe what happens to the corpuscles in each case. If the cells act as osmotic systems, in the hypertonic solution water will be drawn out and the cells will shrink. In distilled water the cells will swell and eventually burst releasing the red pigment. In the isotonic solution there will be no change.

From this experiment you should appreciate the importance of not allowing animal cells to be bathed in fluids which are much stronger or much weaker osmotically than that of their cytoplasm.

Can you now explain the action of a contractile vacuole in an amoeba which lives in pond water, and why amoebae living in sea water do not have one?

We can now summarise the important aspects of diffusion and osmosis. When diffusion occurs molecules in gases or liquids pass from regions of high concentration to low until there is an even dispersal of molecules. Diffusion is an important process whereby substances such as salts and sugar when dissolved in water, and gases such as oxygen and carbon dioxide travel within the tissues of plants and animals.

Osmosis in living systems only concerns the passage of water. The water passes from dilute solutions to more concentrated ones because of the selective nature of the semi-permeable membrane. The membrane that acts in this way in living things is the surface membrane of the cytoplasm.

13

Obtaining energy

All animals and plants need energy to carry out the vital processes going on in their bodies. We need energy to move, to talk, to think, to grow; we need it when asleep or awake, from conception to death. Every living cell in our bodies needs energy. How do organisms obtain this energy? The short answer is through **respiration**. This essential life process may be defined as the liberation of energy from food. To find out how this happens let us first make a comparison between respiration and a process we all know something about—**combustion** (burning). When a fuel such as coal, wood or gas burns, energy in the form of heat and light is liberated. This process only takes place when oxygen is present, and as a result the fuel is used up and two of the products which pass into the air are carbon dioxide and water vapour. The equation for combustion can be written:

$$\text{fuel} + \underset{\text{oxygen}}{O_2} \rightarrow \underset{\text{carbon dioxide}}{CO_2} + \underset{\text{water}}{H_2O} + \text{energy.}$$

You will know that certain foods we eat are called energy foods; sugar is a good example. If sugar is burnt in a crucible in the presence of a lot of oxygen it behaves like a fuel: it bursts into a bright flame, gives out a lot of heat, and carbon dioxide and water are formed. Is respiration something like this?

To find out we shall have to consider five experiments on living material, each concerned with one aspect of the above equation. We must show that:

1. Energy is released.
2. Carbon dioxide is given out.
3. Water is formed.
4. Food is used up in the process.
5. Oxygen is used up.

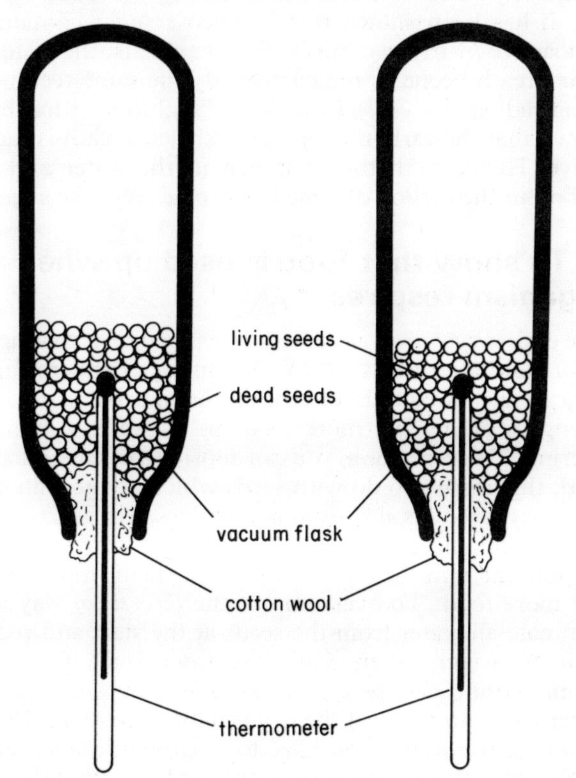

Fig. 13:1 Apparatus to determine whether heat is released from living material.

A controlled experiment

In biological experiments it is usually essential to have a **control** to ensure that we interpret the results correctly. This involves setting up two almost identical experiments, kept under similar conditions except for the one factor to be investigated. By comparing the results in the experiment and the control we then know that any difference is due to that one factor. The following experiments will illustrate this principle.

1. When organisms respire, is energy released in a form we can detect?

Set up the apparatus as in Fig. 13:1. Both flasks are alike except that one vacuum flask contains living organisms and the other, the control, dead ones. Prepare the living seeds by soaking a handful of oats in water for 24 hours. Kill a similar sample by soaking in 10% formalin. Rinse the living seeds in a dilute solution of disinfectant such as TCP which will kill off any bacteria on the surface of the germinating seeds, but leave the seeds unharmed. This is to eliminate any bacterial effect on the result.

Compare the temperatures each day for several days. What do you conclude from your results?

Think out why it is better to have a *lot* of oats in each flask; why cotton wool is used instead of a cork; why vacuum flasks are used instead of glass flasks.

If you put your hand into a heap of lawn mowings a few hours after they have been cut, the heap feels warm. This is another example of heat being given out by living organisms, in this instance by bacteria causing decay.

2. Is carbon dioxide given out when an organism respires?

There are two good tests for carbon dioxide. One is to pass it into limewater which then turns milky, the other is to pass it through bicarbonate indicator which gradually changes colour as the concentration of carbon dioxide increases. (This is because the solution becomes more acidic.) The colour change is from orange-red to yellow. First, use one of these tests to discover if you give off carbon dioxide yourself when you breathe out.

Set up the apparatus shown in Fig. 13:2 and breathe *slowly* in and out. As you breathe in, the air passes through the left-hand tube and when you breathe out it goes through the right. (Notice the position of the glass tubing which controls the direction of airflow.) Why is there an equal

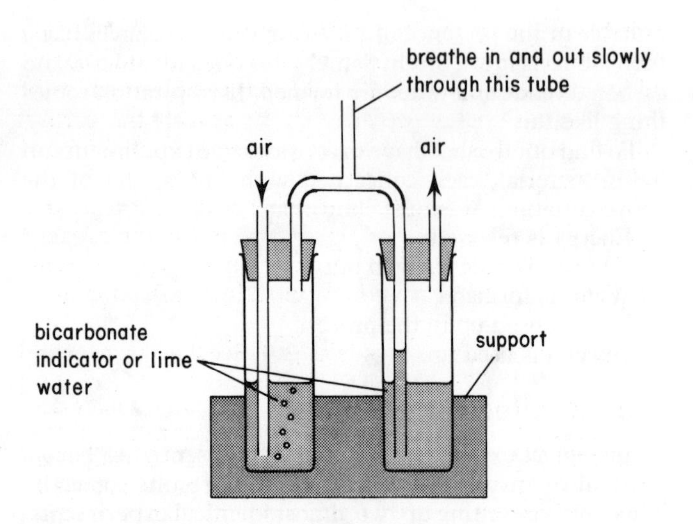

Fig. 13:2 Method of testing whether we give out carbon dioxide when we breathe.

volume of indicator in each tube?

Note any change of colour and conclude whether or not the exhaled air contains more carbon dioxide than the inhaled air.

Do other organisms breathe out carbon dioxide? To find out, use the apparatus shown in Fig. 13:3. Various members of the class could use different material such as woodlice, beetles, locust nymphs, germinating peas, chopped-up potato, etc.

Rinse out two boiling tubes with distilled water followed by a little bicarbonate indicator. Label them A and B.

Add 2 cm^3 of indicator to A and cork it at once (avoid breathing into it). This acts as a control.

Now add 2 cm^3 of indicator to B and put in the cork with the rod and perforated zinc disc attached; introduce the living material before fixing the cork tightly.

Shake the solution at intervals, but do it gently so as not to splash the living material. Note if the colour changes compared with A.

Do all the specimens give out carbon dioxide?

This apparatus can also be used to compare the *rate* at which carbon dioxide is given out when the temperature is altered. You could work out the details of how this could be done remembering that when you alter the temperature everything else must be kept exactly the same if the comparison is to be accurate. You will need a third tube C containing bicarbonate indicator solution which you have already blown through to produce a standard yellow colour to acts as an end-point.

3. Is water given out when an organism respires?

It is easy enough to test your own breath for water vapour with thoroughly dried cobalt chloride paper (which turns from blue to pink), and you know how this water vapour condenses in cold weather so that your breath becomes visible, but it is difficult to prove that this water actually results from the breakdown of the food, as all living things contain water anyway. However, scientists have proved this by using radioactive isotopes. These are atoms which

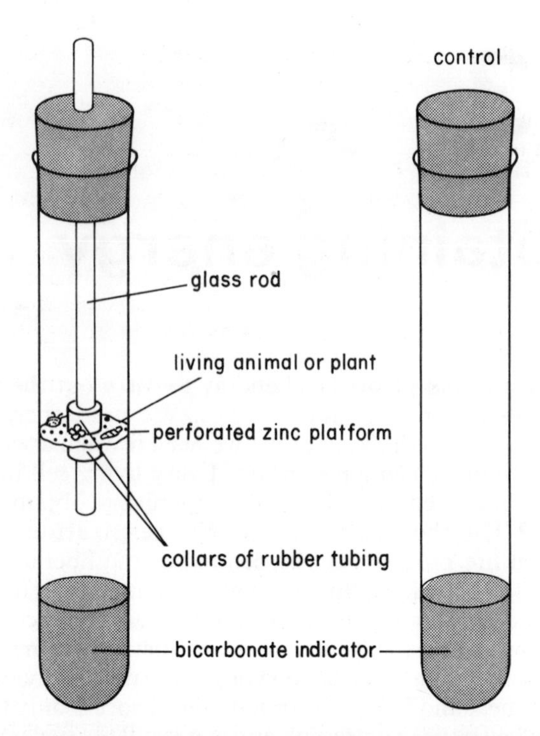

Fig. 13:3 Apparatus to test whether living things give out carbon dioxide.

can be incorporated into the molecule of a compound such as sugar which is then said to be 'labelled'. The atom can now be detected wherever it goes, because although it behaves like a normal atom, it also gives off bursts of radiation. These can be detected by electronic Geiger-Muller tubes which are connected to a machine which counts any bursts of radiation detected by the tubes. In this way it has been shown that if glucose sugar containing radioactive *hydrogen* atoms is fed to animals, the water in their breath becomes radioactive. By the same technique using radioactive *carbon* atoms in the glucose it has been shown that the carbon dioxide in exhaled breath is radioactive. Hence both the hydrogen in the water and the carbon in the carbon dioxide have come from the sugar.

4. To show that food is used up when an organism respires

The easiest way to show something is used up is to weigh it before and afterwards. But with a living organism this is difficult as its weight is altering all the time because a) it is taking in or making more food, b) its water content is altering. To overcome a) you can deprive it of any source of food; this is easy to do with seeds which are germinating because the food is already stored in the seeds, and to make them germinate you need only give them water. Also, by keeping them in the dark you prevent them from making any more food. To overcome b) the *theoretical* way is to eliminate all water from the seeds at the start and record their dry weight; then give them water to allow them to germinate and after several weeks of growth remove all the water again and record the dry weight once more. But to eliminate the water, you have to put them in an oven at 100°C until the weight is constant, and if you do this you kill the seeds! That is the problem! You can only overcome this by taking two lots of seeds which are equal in number and weight. You assume that the water content of both lots is the same and reduce one lot to dry weight. Proceed as follows:

1. Take 10 broad beans, put 5 on one pan of a balance and 5 on the other and change them around until their weights are equal.
2. Put one set in an oven and reduce to dry weight (you assume this equals the dry weight of the other). To obtain an accurate dry weight reading, cool the seeds in a desiccator before weighing them otherwise they may absorb water from the air. Return the seeds to the oven and repeat the procedure until you obtain a constant reading. Make a note of this weight.
3. Plant the second set in a bowl of wet sawdust and leave in the dark for at least two weeks so that they will grow and respire.
4. You now have to dig up the plants without breaking the roots, and remove all the sawdust which adheres to them. Do this by first soaking the bowl in water and then washing each plant gently under the tap.
5. Put the plants in an oven and reduce to dry weight.

Compare this weight with the dry weight of the first set of seeds. If it is less, what do you conclude?

The experiment is based on the assumption that the two lots of seeds were alike. Would this assumption have been more justified if you had taken many more seeds? (Compare your results with those of the whole class.)

5. To show that oxygen is used up when an organism respires

The experiment (Fig. 13:4) is based on the principle that if oxygen is used up there should be a reduction in the volume of air in the boiling tube. But we have already shown that living organisms give off carbon dioxide, so this additional quantity of gas could mask the reduction of oxygen. It is therefore necessary to absorb the carbon dioxide as soon as it is formed by putting in a small tube of soda lime.

The second boiling tube containing dead seeds is needed to act as a control, because respiration is a living process. By having tubes of the same capacity any change in volume due to temperature variation would affect each equally.

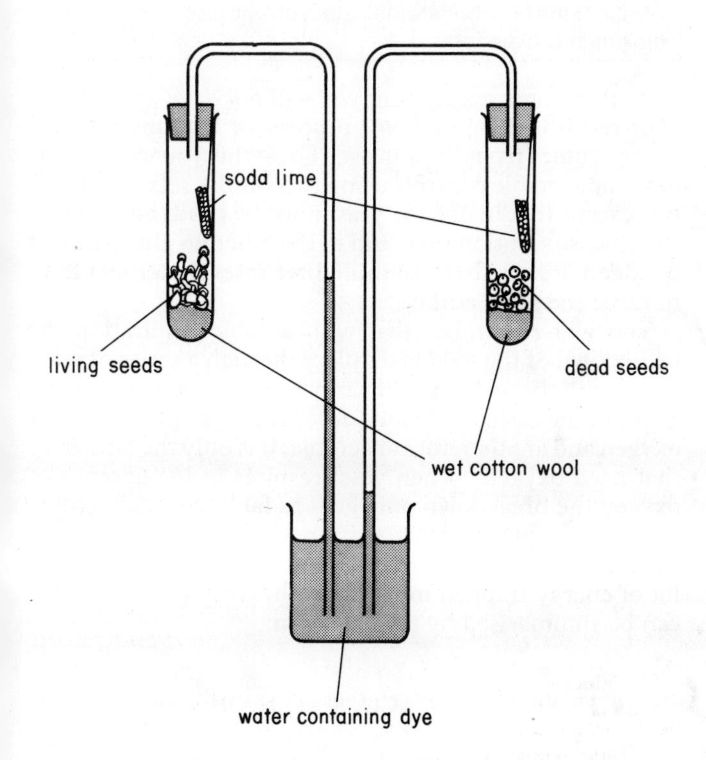

Fig. 13:4 Experiment to find out whether oxygen is absorbed when an organism respires.

Set up the apparatus as in Fig. 13:4. See that the levels in the two glass tubes are the same at the start. Leave for 3 or 4 days and note any change in level in the two tubes. If there is a change you still have to prove that the gas absorbed is in fact oxygen. To do this remove the cork from the first boiling tube and slowly lower a lighted taper into it. It will go out at once if little or no oxygen is present. Test the control in the same way.

From these experiments we can now conclude that when an organism respires the process appears to be comparable with that of burning sugar in oxygen.

$$\text{sugar} + \text{oxygen} \rightarrow \begin{matrix}\text{carbon}\\ \text{dioxide}\end{matrix} + \text{water} + \text{energy.}$$

As organisms release energy mainly from sugars and fats, and glucose sugar is one of the most important, we can write the respiratory equation more fully:

$$\underset{\substack{\text{glucose}\\ \text{(containing}\\ \text{trapped energy)}}}{C_6H_{12}O_6} + \underset{\text{oxygen}}{6O_2} \rightarrow \underset{\substack{\text{carbon}\\ \text{dioxide}}}{6CO_2} + \underset{\text{water}}{6H_2O} + \underset{\text{(released).}}{\text{energy}}$$

But there must be some difference between combustion of sugar by burning and respiration in living cells, because:
1. In the laboratory, sugar must be brought to a high temperature before it burns and liberates its energy. If this happened in our cells we would become charred!
2. Once the sugar starts burning we cannot stop it easily, but living cells seem to be able to control the process.
3. Although cells contain a lot of water, respiration goes on within them; outside the body water normally stops combustion from taking place.

Therefore we must conclude: a) that respiration is a very different process from combustion, although the same substances are used and the end products are similar; b) that a living cell is somehow able to control the release of energy; c) that the reaction takes place at normal temperatures and in a watery medium.

The full explanation of how this happens is very complex, but some of the important points are these:
1. The sugar is broken down into carbon dioxide and water by a complex series of reactions involving a number of intermediate compounds.
2. Some of these reactions release energy, but always in small quantities.
3. The reactions occur at normal temperatures because each is controlled by an **enzyme**.
4. The energy is liberated within the cell when it is needed for some vital process.

Enzymes

These are complex chemicals which are made by the cytoplasm of the living cells; they control the chemical reactions. Enzymes act as **catalysts**, i.e. they speed up a reaction, which otherwise would be extremely slow at normal temperatures, and at the end of the reaction they remain unchanged, so they can be used again and again. Each reaction is catalysed by a different enzyme.

The way an enzyme works is rather like the way a key opens a lock. The key has to be of the right pattern to perform the action, but once it has acted it can be taken out and used on any other similar lock. An enzyme, in the same way, has a particular molecular shape which only reacts with a substance whose molecule it fits. Thus the series of reactions bringing about respiration are controlled by a series of different enzymes.

Forms of energy

Some energy is released in the form of **heat** as we have already seen; occasionally it is in the form of **light**. Female glow-worm beetles, for example, give out light which enables the males to find the females prior to mating. Many deep-sea fish have luminous organs, and certain bacteria and fungi may be seen in the dark because of the light energy they emit. But the energy required by living cells, for such processes as the contraction of muscle, needs to be in a different form. It must be readily released when required. How does this happen?

When energy is released from the food during respiration, some of it is transferred to a special substance found in all cells called **ATP** (adenosine *tri*phosphate). One of three phosphate groups which form part of the ATP molecule is easily removed to form **ADP** (adenosine *di*phosphate), thus releasing energy from the molecule. Energy from the breakdown of food is used to recombine the ADP and phosphate to form ATP once more.

1. ATP ⟶ ADP + phosphate + energy.
2. Energy + phosphate + ADP → ATP.
   (from food
   breakdown)

By using a specially prepared ATP solution it is possible to demonstrate this action by putting a drop on a length of fresh muscle fibre from lean meat—it will immediately contract. Thus the **chemical** energy stored in the ATP has been turned into **mechanical** energy, i.e. for doing work. When ATP releases its energy not all of it is used for contracting the muscle. Some is lost in the form of heat; that is why, when you use your muscles a lot, you become hot.

To summarise, we can now say that in the living cells of both plants and animals energy is released from the food with the help of oxygen, and carbon dioxide and water are formed as waste products. This method of energy release is known as **aerobic respiration**.

Anaerobic respiration

The next question is whether energy can be released from food without oxygen, i.e. **anaerobically**. You can find this out by using dried or baker's yeast. Looked at under the microscope it is seen to consist of a vast number of single

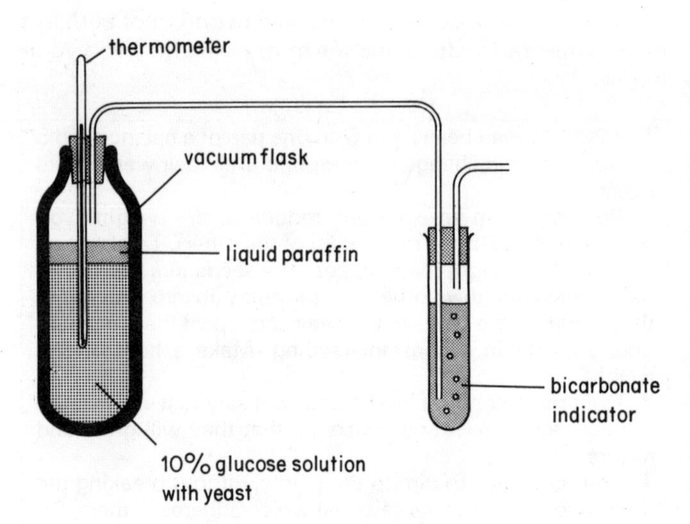

Fig. 13:5 Apparatus to test whether yeast liberates energy and carbon dioxide in the absence of oxygen.

cells. It is in fact a living fungus.

The principle of this experiment is to supply the yeast with glucose to act as food, but to deprive it of oxygen and see if energy and carbon dioxide are released.

> Boil some 10% glucose solution to remove any oxygen dissolved in it, *cool* and pour into a small vacuum flask until about three-quarters full. Add a few grammes of yeast and pour a layer of liquid paraffin on top to prevent the mixture from coming in contact with atmospheric oxygen. Set up the rest of the apparatus as in Fig. 13:5. By taking the temperature at intervals over the next few days you can find out if energy has been released, and by noting any change in the colour of the bicarbonate indicator you will discover if carbon dioxide has been formed.

By this single experiment you will not have proved that your result is due to a *living* process, or that any released energy comes from the glucose. To do this you would have to set up suitable controls using two similar sets of apparatus, except that in one the yeast must be dead (boil it up in the glucose solution first) and in the other no glucose must be added. Remember that in all three cases the temperature must be the same at the start.

You will remember that with aerobic respiration the breakdown of the food takes place through a series of reactions, with the formation of various intermediate products. Some of the earlier reactions do, in fact, take place without oxygen and are therefore anaerobic; it is only the later ones that need oxygen. When yeast respires in the absence of oxygen the breakdown only goes as far as these anaerobic reactions, so much less energy is released and the glucose is turned mainly into ethanol—a substance which has still a lot of energy trapped in it. Anaerobic respiration in yeast can be summarised by this equation:

$$\underset{\text{(containing trapped energy)}}{\overset{\text{glucose}}{C_6H_{12}O_6}} \xrightarrow{\text{enzymes}} 2CO_2 + \overset{\text{ethanol}}{2C_2H_5OH} + \underset{\text{(released)}}{\text{energy}}.$$

Although the amount of energy released anaerobically is small compared with the aerobic method, it is enough to

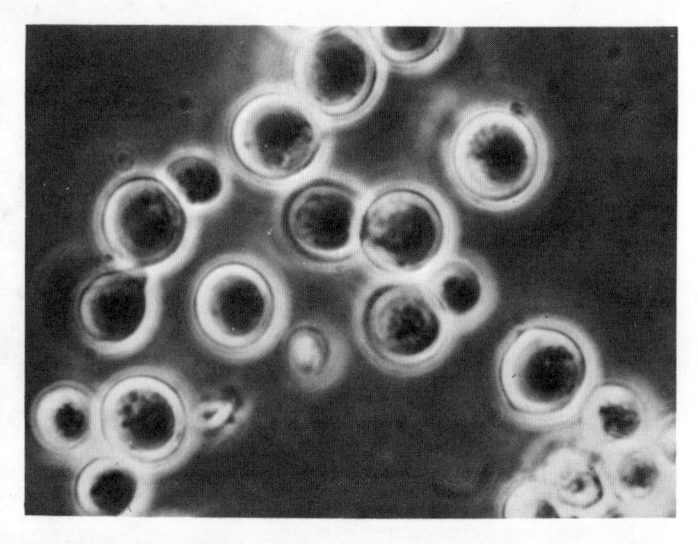

Fig. 13:6 High power photomicrograph of budding yeast cells.

keep organisms like yeast and some bacteria active and most plants can survive for short periods by this means if the oxygen is temporarily cut off.

The formation of ethanol from sugar by yeast is known as **fermentation**. It is the basis of brewing and wine making. To make beer, yeast is added to malt (crushed germinating barley in water) and the ethanol is formed from the sugar in the malt; hops are added to give it a bitter taste. A wine is formed when sweet juices, usually from fruits, are allowed to ferment. Natural yeasts occur on the skins of fruit, e.g. the bloom on grapes, so when grapes are crushed the sweet juices become mixed with the yeast and fermentation takes place.

Yeast is also used in bread-making to make the dough rise. When yeast with a little sugar solution is mixed with the dough (flour, water and a little salt), it respires anaerobically and the carbon dioxide given out, being a gas, 'blows up' the bread.

You should now be able to explain why:

1. People who make home-made wine must not cork the bottles too soon.
2. A baker must allow time for the dough to 'rise' in a warm place before he puts it in the hot oven.
3. Holes are sometimes found inside a loaf.

Muscles and energy

When we undertake strenuous physical exercise, it is not long before our muscles begin to ache and we become fatigued. One way in which physiologists have been able to study this process is by using a special kind of bicycle (Fig. 13:8). It is stationary and so enables the researcher to conduct tests on a volunteer more easily. The amount of effort needed to keep up a constant 'speed' can be increased by the use of electromagnetic brakes, so a state of near exhaustion can be reached quickly.

When blood samples are taken from a volunteer before and after a period of vigorous exercise, it has been found that a substance called lactic acid accumulates in the blood.

Fig. 13:7 Finish of the 100 m men's final, Olympic Games 1980. The oxygen debt built up during the race is repaid during the deep breathing that follows.

Fig. 13:8 Research bicycle used to study the physiology of vigorous muscular action.

But during a subsequent period of rest the amount of lactic acid is gradually reduced to its original level.

When there is a shortage of oxygen, lactic acid is an intermediate product in glucose breakdown, and some energy is released in the process. But when there is plenty of oxygen the reaction goes further; some of the lactic acid is completely oxidised to carbon dioxide and water with the liberation of more energy, and the remainder is reconverted into glucose.

During a 100 m race a well-trained athlete can hold his breath all the time—it is not until afterwards that he pants. In this case, the muscles are using the energy released during the anaerobic breakdown of glucose. It is not until afterwards that the athlete obtains the oxygen needed in order to remove the lactic acid. Therefore, when we undertake strenuous exercise we build up what is called an **oxygen debt** which has to be repaid later. In a longer race athletes have to breathe all the time, so some lactic acid is removed while they are running, and they can go on for longer before becoming exhausted. The presence of lactic acid in the blood is the main cause of muscle fatigue, but if the body is rested for long enough the tiredness goes.

14

Gaseous exchange

Respiration and the size of organisms

Every organism must have a surface through which oxygen and carbon dioxide can pass and also a means of taking these substances to or from the cells. The methods used vary according to the size of the animal and whether it lives in water or on land. Aquatic organisms utilise the oxygen which is dissolved in the water, terrestrial ones obtain it from the air.

Because microscopic, single-celled organisms such as amoeba are so small, the surface area is large compared with the volume, and oxygen can easily diffuse from the water through the surface into the cytoplasm. Similarly, carbon dioxide can diffuse out in the opposite direction. As organisms increase in bulk the ratio of surface area to volume changes.

You can see this is so by calculating the surface area to volume ratio for different cubes with sides, say of 1 cm, 2 cm and 4 cm respectively (a cube has six equal sides).

Side (cm)	Surface area (cm^2)	Volume (cm^3)	Surface area: Volume ratio
1	$1 \times 1 \times 6 = 6$	$1 \times 1 \times 1 = 1$	6:1
2	$2 \times 2 \times 6 = 24$	$2 \times 2 \times 2 = 8$	3:1
4	$4 \times 4 \times 6 = 96$	$4 \times 4 \times 4 = 64$	1·5:1

Thus increase in size brings with it two difficulties:

1. More oxygen is required, but the surface for gaseous exchange becomes relatively less and is therefore inadequate.
2. Many of the cells are some distance away from the surface and so are less likely to receive oxygen because it may be used up on the way. Thus larger animals must somehow increase their surface for gaseous exchange and have a means of taking the oxygen to the cells deep in the body (Fig. 14:1).

Simple animals like flatworms solve the problems by being flat. In this way they have a large surface/volume ratio and all cells are near the surface. Most animals, however, have a blood system which helps in two ways. First, through being in close contact with the respiratory surface, blood increases the efficiency of gaseous exchange by quickly removing oxygen which has diffused in, so increasing the rate of diffusion (p. 83). Secondly, it acts as a transport system, taking the oxygen to the cells.

Insects, as we saw in Chapter 6, have a unique method of bringing oxygen directly to the cells by means of an elaborate system of branching tubes, the tracheal system. This overcomes the necessity for having a blood system to transport the oxygen. It is a very successful method for small animals but is impracticable for large ones.

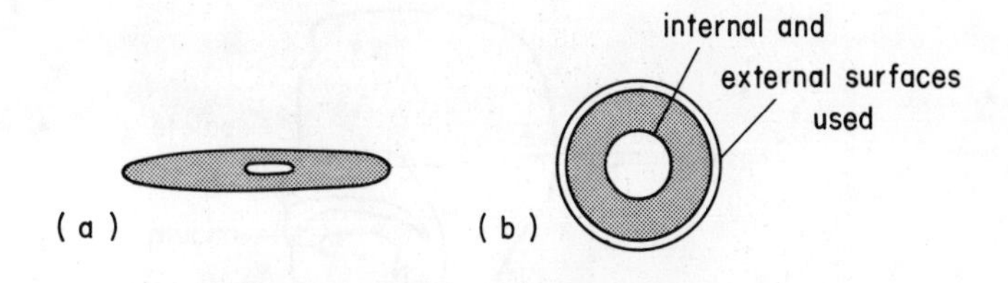

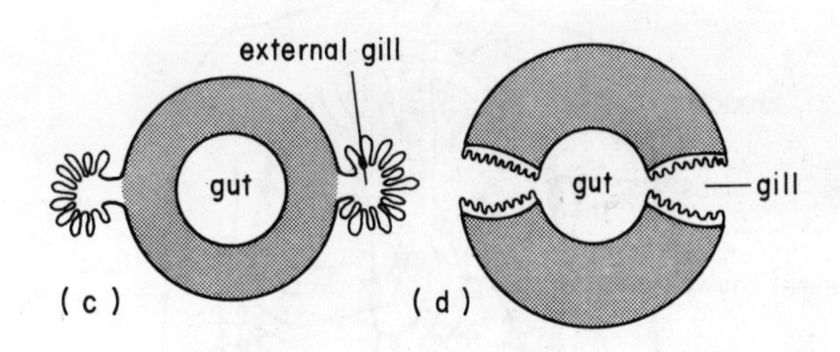

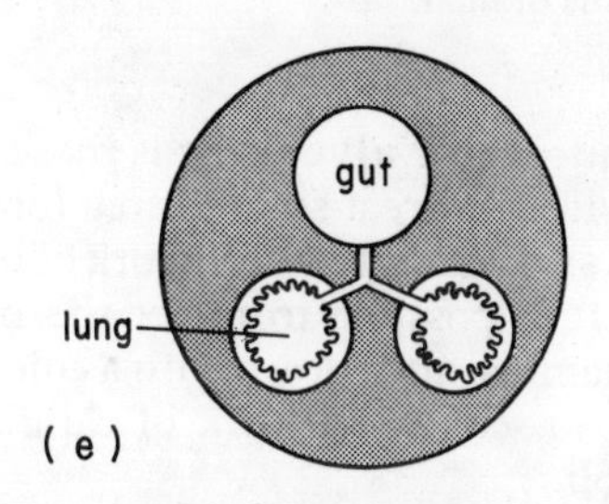

Fig. 14:1 Schematic diagrams of sections through various animals showing how the surface area has been increased for gaseous exchange. a) A flat body e.g. flatworms. b) A hollow body e.g. *Hydra*. c) Having external gills e.g. tadpole. d) Having internal gills e.g. fish. e) Having lungs e.g. mammal.

For animals like earthworms which are relatively sluggish, and so need comparatively little energy, the skin, with its special blood vessels, provides enough surface for gaseous exchange. But with larger, active animals this is not enough, so they have special respiratory organs which provide a greatly increased surface in contact with the water or air; these organs are called **gills** and **lungs** respectively.

The gill filaments of fish provide a very great surface area compared with the volume of the fish and each filament contains blood vessels very near the surface (p. 60). What is more, the water containing the oxygen is actively pumped over the filaments making the diffusion of gases even more efficient. Air-breathing vertebrates possess lungs which provide *within* the body a large surface area in close contact with blood vessels. The air enters the lungs from the outside through a system of tubes. During the course of evolution there has been a tendency for the internal surface of the lungs to increase, thus enabling the animals to become larger and more active.

Breathing organs in man

Fig. 14.2 shows the main structures concerned with respiration.

The two lungs, which are lobed, are spongy elastic organs filling up most of the chest cavity. They look solid, but in reality consist of a mass of tubes called **bronchioles** which branch more and more and become finer and finer

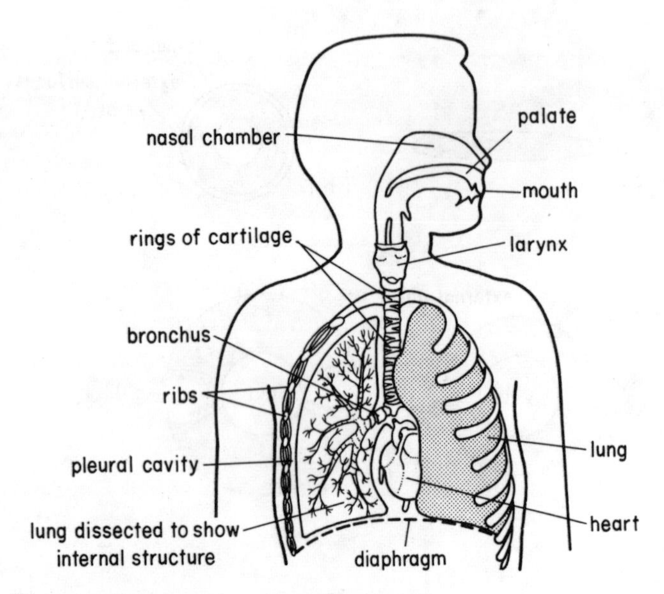

Fig. 14:2 Respiratory organs of man.

until they end up as blind-ending **alveoli**. It is these millions of alveoli that provide the great surface area for gaseous exchange, as each is surrounded by a network of blood capillaries (Fig. 14:3). It is possible to inject the blood vessels of the lung of a mammal after death with a coloured latex. This enables you to see the network of capillaries associated with the alveoli.

The lungs are in direct contact with the atmosphere through the nostrils, nasal cavity, larynx, trachea (wind pipe) and the two bronchi. Each lung is enclosed by two transparent membranes, the **pleurae**. The inner one covers the lung itself and the outer is attached to the chest wall. In between there is a little fluid which allows the two membranes to slide over one another easily when the lungs move. If the pleurae are damaged by bruising or disease

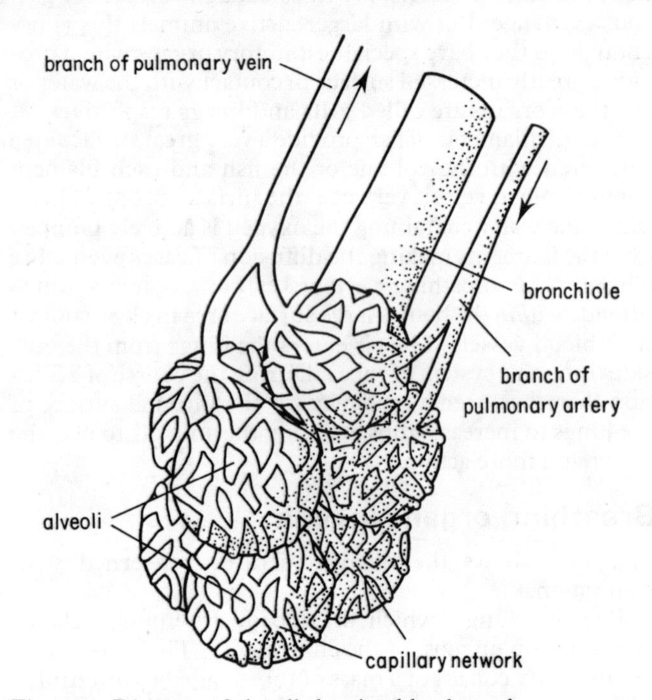

Fig. 14:3 Diagram of alveoli showing blood supply.

they secrete more fluid into the cavity between them, causing pain and difficulty in breathing. This condition is known as **pleurisy**.

The chest cavity is air-tight and separated from the abdomen by a dome-shaped structure, the **diaphragm**. The sternum, ribs and backbone (Fig. 14:4), together with the muscles of the chest wall, provide a tough box which surrounds the lungs and heart and affords them essential protection.

The breathing mechanism

The basic principle can be explained through a model (Fig. 14:5). Air enters the lungs (in this case balloons) only if the pressure inside them is momentarily reduced to below that of the atmosphere. As the lungs are thin-walled, this can be brought about by reducing the pressure in the space around them so that they expand. To reduce this pressure, the volume of the space surrounding the lungs must be

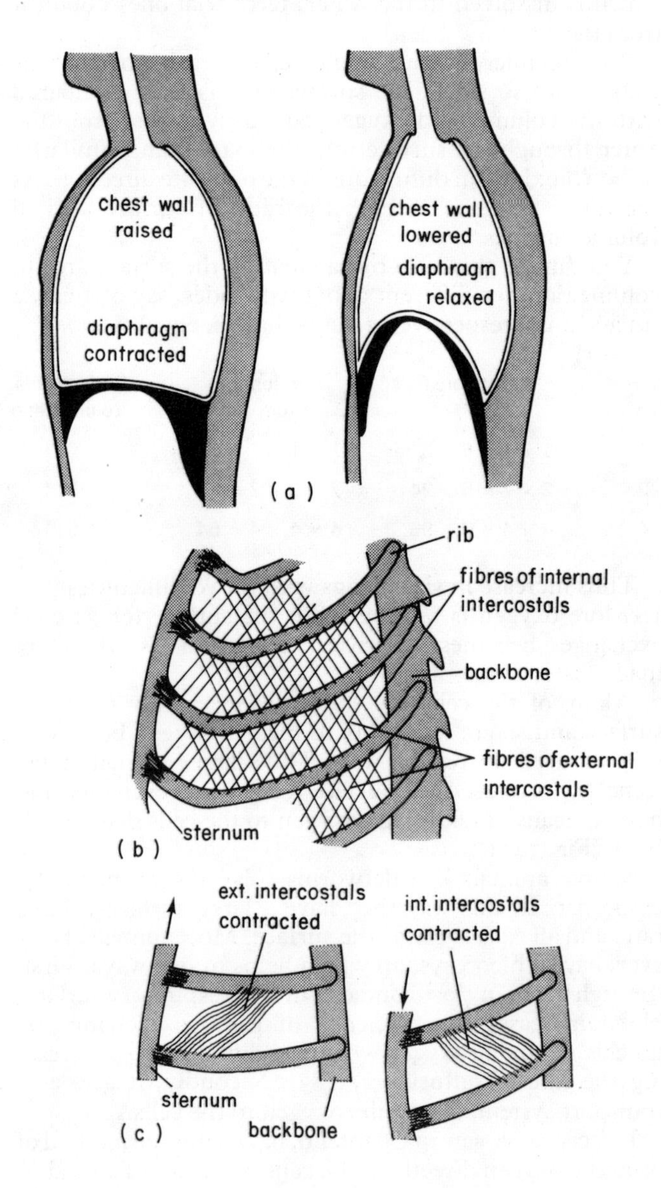

Fig. 14:4 Diagram to show the mechanism of breathing: a) Lateral view of thorax: left—breathing in, right—breathing out. b) The position of the external and internal intercostal muscles. c) The action of the intercostal muscles.

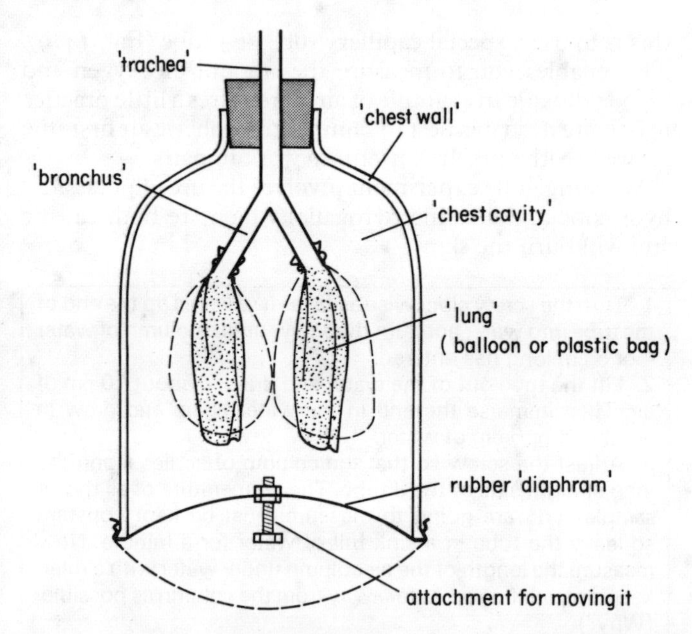

Fig. 14:5 Model thorax to demonstrate the action of the diaphragm during breathing.

increased. This can be achieved in the model by pulling down the rubber sheet. In man it happens in two ways: by lowering the diaphragm and by moving the ribs upwards and outwards. The model illustrates only the movement of the diaphragm.

The diaphragm consists of a central non-elastic membrane and an outer ring of muscle which curves downwards to join the body wall. When you take in a breath (inspiration), the radially arranged muscles of the diaphragm contract causing the membrane to be lowered. (The liver and stomach are pushed down slightly as a result.) At the same time, the rib cage is moved upwards and outwards by the contraction of the external intercostal muscles which are arranged obliquely between the ribs (Fig. 14:4). This is made possible because each rib is not fixed rigidly to the backbone and at the sternum end the bone is capable of bending.

Breathing out (expiration) is a more passive process, because it involves the relaxation of the diaphragm and external intercostal muscles. The lungs, being elastic, shrink when the pressure around them increases. In addition, there are internal intercostal muscles running at right angles to the external intercostals which contract and help to lower the rib cage. During expiration, abdominal muscles are also used to push the liver and stomach up against the diaphragm, thus forcing it further into the chest cavity.

The 'iron lung'

In cases of diseases which affect the functioning of the respiratory muscles (as sometimes happens with poliomyelitis) the patient's breathing movements have to be brought about artificially. This is done through an 'iron lung'. The patient is completely enclosed in a large metal box, except for his head, and the air pressure in the box is lowered and raised rhythmically. This has the effect of increasing and decreasing the volume of the whole chest. Thus when the pressure is reduced around the patient, the chest and the

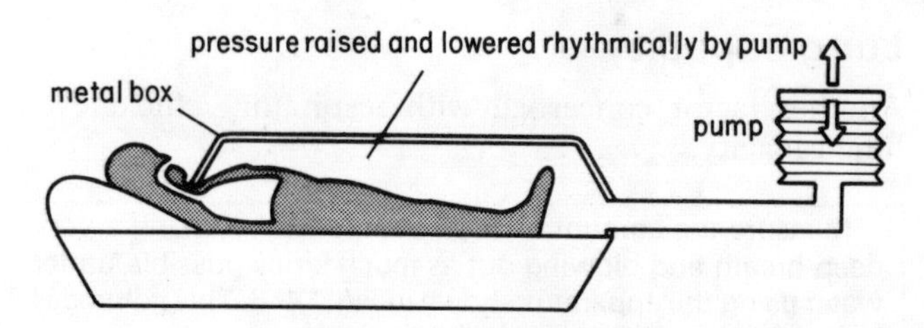

Fig. 14:6 Diagram showing the principle of the 'iron lung'.

lungs expand and so air enters the lungs. When the pressure is increased around the patient air is expelled (Fig. 14:6).

Artificial respiration

In cases where a person's breathing has stopped, for example through a drowning accident, the only hope of resuscitation is through the immediate application of artificial respiration. The patient's head is pressed well back to keep the trachea open; the mouth is opened and air is blown directly into the lungs from the rescuer's mouth. The patient's nostrils must be kept closed to prevent air from escaping (Fig. 14:7).

Control of the rate of breathing

Although our breathing movements are controlled by automatic nervous responses, the breathing rate varies according to circumstances. For example, when we take vigorous exercise, more of our food reserves are oxidised and so more carbon dioxide is produced. The latter alters the composition of the blood (it becomes more acid), and sensitive cells in the hind brain immediately react to this by increasing the rate at which nervous stimuli are sent to the intercostal and diaphragm muscles so that our breaths become quicker and deeper. The extra carbon dioxide is therefore quickly removed, the acidity of the blood decreases, and so the rate of breathing automatically slows down to its normal rhythm.

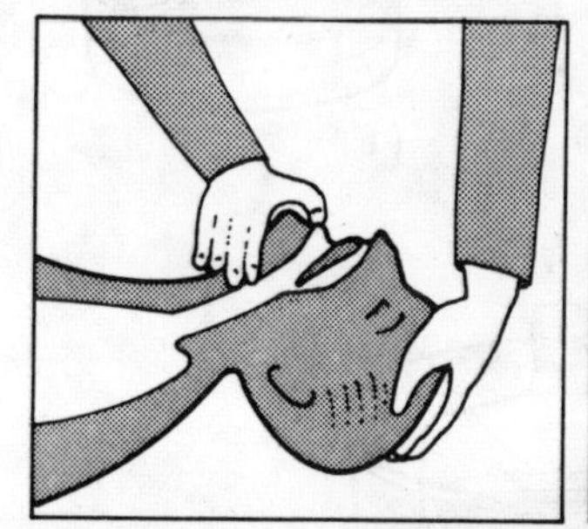

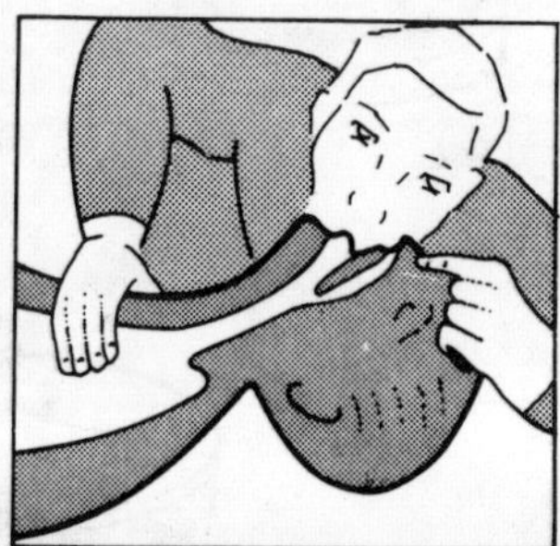

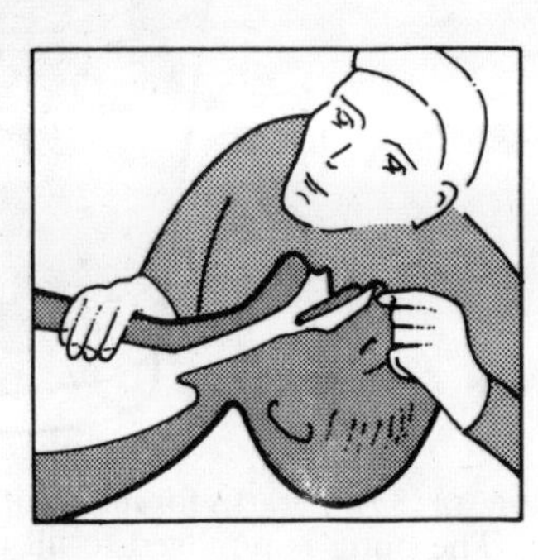

Fig. 14:7 The mouth-to-mouth method of applying artifical respiration (see text).

Lung capacity

Another factor concerned with respiratory efficiency is lung capacity.

Measure the amount of air in your lungs by taking a very deep breath and blowing out as much air as possible under water using the apparatus shown in Fig. 14:8. This volume is known as the **vital capacity** of the lungs.

When the whole class has done this draw up a table showing the vital capacity of each member in one column and the normal rate of breathing in another. Other factors can also be compared with these measurements, such as chest expansion, and even athletic ability. Have good sprinters and long-distance runners in your class rather similar characteristics in these respects?

When you have breathed out all the air you can there is still about $1\frac{1}{2}$ litres of air locked up inside the alveoli so you can now work out the **total capacity** of your lungs.

Gaseous exchange

Because oxygen is being used up by the body and carbon dioxide given out, we would expect certain differences between inspired and expired air. One way of investigating

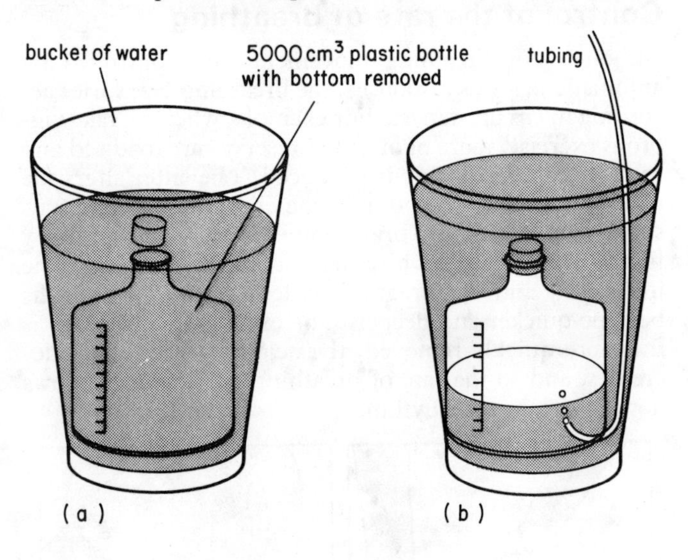

Fig. 14:8 Apparatus for measuring the vital capacity of the lungs: a) The bung is removed to fill the bottle and is then replaced. b) The bottle is held loosely to prevent tipping while air is exhaled through the tube. c) The volume of air is measured after levelling.

this is to use a special capillary tube or J tube (Fig. 14:9). This enables you to measure the amount of oxygen and carbon dioxide in a sample of air. It requires a little practice to operate it, so it is best to sample atmospheric air first (the air we breathe in). It is useful to work in pairs.

Warning! The experiment involves the use of potassium hydroxide and alkaline pyrogallol which are both caustic and will burn the skin.

1. Turn the screw clockwise as far as it will go. Dip the end of the tube into water and turn the screw until a column of water 4 or 5 cm long has entered.
2. Lift the tube out of the water and draw in about 10 cm of air. Then immerse the end in the water again and draw in another 4 or 5 cm of water.
3. Adjust the screw so that the column of air lies along the long straight side of the J tube. The temperature of all the air samples you are going to measure must be kept constant so leave the tube in a sink full of water for a minute. Now measure the length of the air column under water with a ruler, keeping your fingers as far away from the column as possible. (Why?)
4. Turn the screw until the sample is within 2–3 mm from the end of the tube (no further!) and draw in about 5 cm of concentrated potassium hydroxide solution carefully. This will dissolve any carbon dioxide with which it comes in contact. Draw the potassium hydroxide backwards and forwards several times over the first part of the tube so that a thin film of it comes in contact with the column of air. Any carbon dioxide in the bubble should now be absorbed and the bubble will be reduced in length.
5. Expel the potassium hydroxide carefully, making sure the air sample remains in the tube. Draw the air back again into the long straight part of the tube, immerse in water for a minute and then measure the length of the air column.
6. Repeat procedures 4 and 5, but this time draw in some alkaline pyrogallol which will absorb oxygen.
7. Before taking any further samples rinse the J tube thoroughly, first in dilute acid and then water. (Why?)
8. Now analyse a sample of exhaled air. It can be collected from a test tube as in Fig. 14:10. Take a fairly deep breath. Blow slowly down the rubber tube to remove the residual air inside it. While still blowing slowly place the end of the tube under the mouth of the test tube.
9. Take a sample of the exhaled air into the J tube and analyse as before.

Now calculate the percentage of carbon dioxide and oxygen as follows: Suppose, for example, the column of air was 10 cm long to begin with and the reading after treatment with potassium hydroxide was 9·6 cm and after treatment with alkaline pyrogallol 7·9 cm.

Reduction in volume with potassium hydroxide is $10{\cdot}0 - 9{\cdot}6 = 0{\cdot}4$.

Reduction in volume with alkaline pyrogallol is $9{\cdot}6 - 7{\cdot}9 = 1{\cdot}7$.

Therefore the percentage of carbon dioxide in the sample is:

$$\frac{0.4}{10.0} \times 100 = 4\%.$$

Similarly, the percentage of oxygen is:

$$\frac{1.7}{10.0} \times 100 = 17\%.$$

What conclusions can you draw from your results? Compare them with the following table which gives the average percentage composition:

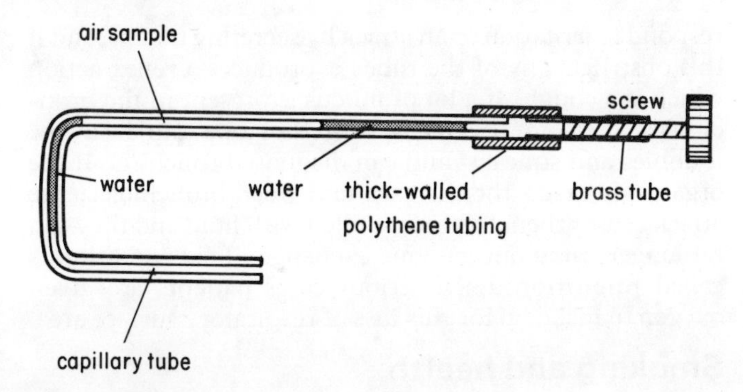

Fig. 14:9 J tube.

	Atmospheric air	*Alveolar air*	*Exhaled air*
Nitrogen	79·01	80·7	79·5
Oxygen	20·96	13·8	16·4
Carbon dioxide	0·03	5·5	4·1

With quiet breathing we only exchange about 500 cm^3 of air. Not more than about 350 cm^3 of this actually reaches the alveoli because the rest fills the trachea and bronchi. Thus it follows that the amount of fresh air coming in has comparatively little effect on the composition of gases in the alveoli. Exercise, however, makes us breathe more quickly and deeply and more of the stale air is removed.

Evidence that the blood is involved in gaseous exchange can be provided by analysing the amount of the various gases in samples of blood entering and leaving the lungs. The following table gives the volume of each gas present in 100 cm^3 of blood:

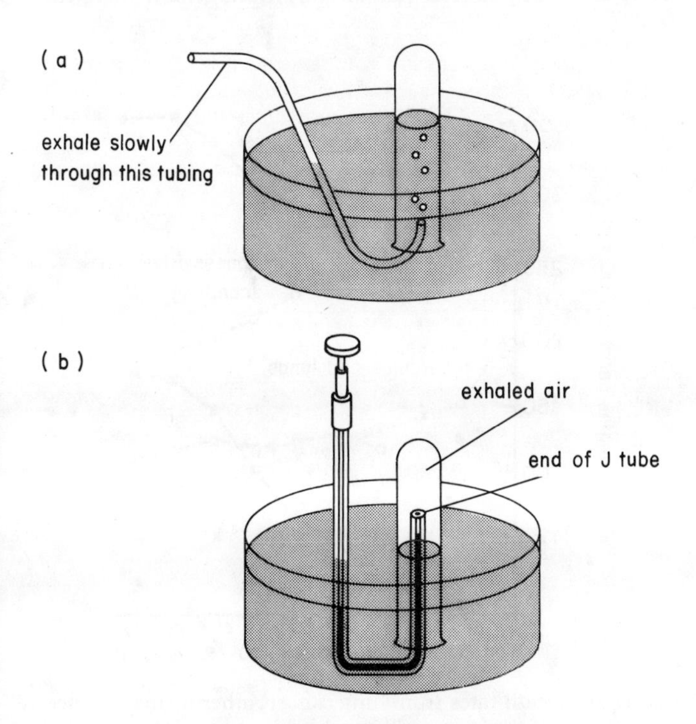

Fig. 14:10 Method of obtaining a sample of exhaled air using a J tube: a) Breathing into a test tube of water. b) Inserting the J tube.

	Blood entering the lungs	*Blood leaving the lungs*	*Change*
Nitrogen	0·9 cm^3	0·9 cm^3	—
Oxygen	10·6 cm^3	19·0 cm^3	+8·4 cm^3
Carbon dioxide	58·0 cm^3	50·0 cm^3	−8·0 cm^3

The exchange of gases at the lung surface

This takes place between the air and the blood by diffusion (Fig. 14:11). The process is very efficient because the internal surface area provided by approximately 700 million alveoli in the two lungs totals over 70 m^2 and the capillary network provides an area of over 40 m^2. Compare these figures with the total surface area of the walls, floor and ceiling of a medium-sized room. In addition, the diffusion pathway, i.e. the distance between the air in the alveoli and the blood in the capillaries, is only about 1 μm. Also, because the alveolar walls are always moist, oxygen can dissolve more easily. In the capillaries the oxygen is taken up by a red pigment, **haemoglobin**, contained within the red blood corpuscles. In this way blood can carry more oxygen than in simple solution.

The respiratory system and health

The respiratory system is open to the external environment via the nose and mouth, and is therefore a source of entry for pathogenic (disease-causing) organisms and atmospheric pollutants which might damage the delicate tissues of the lungs.

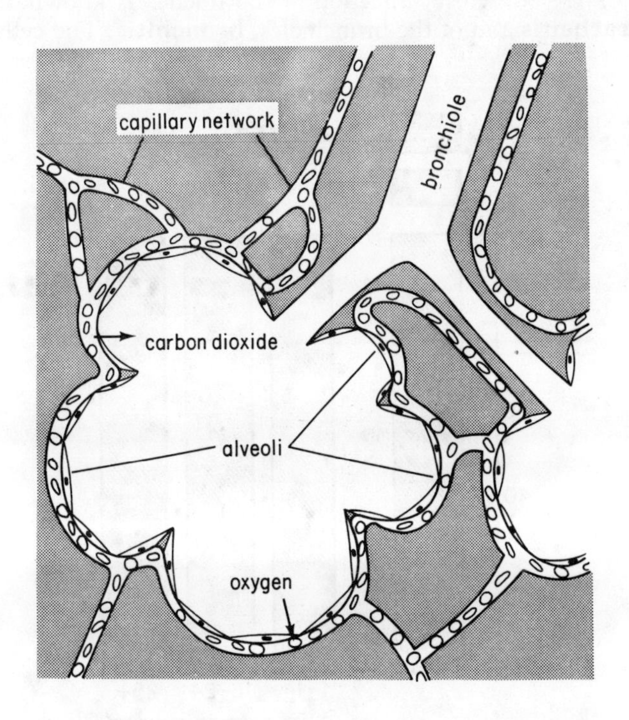

Fig. 14:11 Diagram of a section through the alveoli of the lung showing the diffusion pathway for gaseous exchange between lung and blood capillaries.

The system has its natural defences against these invaders. The nostrils and the hairs within them are coated with a slimy secretion, **mucus**. The hairs act as an efficient sieve for larger particles; after blowing your nose when you have been in a dusty or sooty atmosphere you will see plenty of evidence for this. However, if you breathe through your mouth you derive no benefit from this protection. In addition, particles may stick in the mucus secreted by cells which line the trachea and bronchioles. Most of these cells have cilia which are in constant motion, causing the mucus to pass slowly up the tubes to the back of the throat where it is swallowed unconsciously. The particles are carried with the mucus as on a conveyor belt, away from the lungs.

These defences are not always sufficient. People who are constantly inhaling gritty particles are liable to damage their lungs badly and develop various diseases such as **silicosis** and **asbestosis**. In these situations, the workers should wear masks to avoid breathing in the harmful particles.

Atmospheric pollutants also have a major effect on health. During the combustion of fuel for generating electricity vast quantities of smoke, grit, sulphur dioxide and other substances are passed into the atmosphere. In addition the exhaust fumes of cars add poisonous carbon monoxide and oxides of nitrogen and lead. All these substances may affect the respiratory system and lead to ill health.

All parts of the respiratory system may be attacked by bacteria or viruses. Usually the body defences are sufficient but, if not, the micro-organisms invade the tissues and cause inflammation and disease. If this occurs at the back of the throat (pharynx) a sore throat or **pharyngitis** develops, if in the larynx, **laryngitis**, which may cause you to lose your voice; infection of the trachea is known as **tracheitis** and of the bronchioles, **bronchitis**. The cells respond to irritation or an attack by secreting mucus, and if this obstructs any of the tubes it produces a reflex action which is a cough. If a lot of mucus is present in the bronchioles, a doctor by using a stethoscope can hear the 'bubbles and squeaks' and can diagnose bronchitis. If the organisms reach the air sacs and finer bronchioles, the attack causes them to become filled with fluid and they can no longer carry out gaseous exchange. This condition is called **pneumonia**. In serious cases patients are given oxygen to make up for this loss of respiratory surface area.

Smoking and health

Everybody should now be aware that smoking is a serious danger to life and health. A report from the Royal College of Physicians (1977) stated that on average the habitual smoker shortens his life by about $5\frac{1}{2}$ minutes for each cigarette he smokes.

Over 2,000 different substances have been found in cigarette smoke. They fall into four main groups:

1. Chemicals which are irritants to the respiratory system.
2. Tar, which contains cancer-causing substances called **carcinogens**.
3. Carbon monoxide and other poisonous gases.
4. Nicotine which affects the nervous system.

As an irritant, cigarette smoke causes extra mucus to be produced by the cells lining the bronchial tubes, while the nicotine content paralyses their ciliary action. Thus the mucus has constantly to be coughed up, resulting in 'smokers' cough' which is the first stage of **chronic bronchitis**.

Chronic bronchitis is a major health hazard and accounts for 30,000 deaths each year in Britain. Of every 8 men suffering from chronic bronchitis, 7 are smokers. It is also brought on by other air pollutants such as the smoke from factory and domestic chimneys and the exhaust fumes from cars. Legislation has already done much to reduce

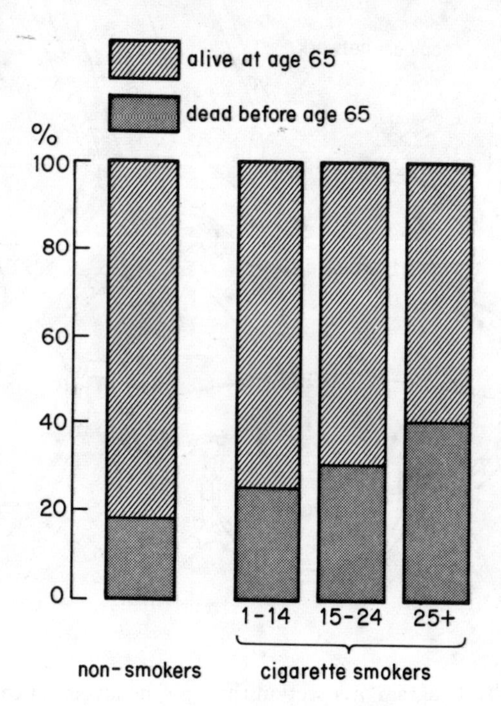

Fig. 14:12 Proportion of men aged 35 who will die before the age of 65 according to their smoking habits.

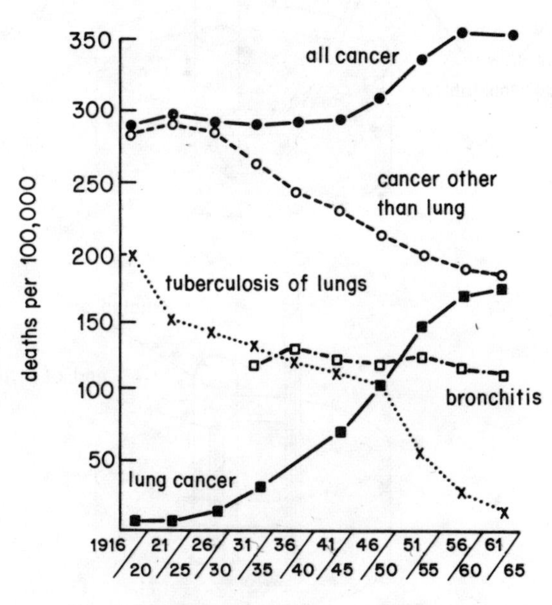

Fig. 14:13 Death rates from lung cancer, other forms of cancer, tuberculosis of the lung and bronchitis in men aged 45–64 from 1916–1965. The down curve for tuberculosis is due to improved treatment.

smog (e.g. smokeless zones in cities), but exhaust fumes still remain a major menace.

Children who smoke have been shown to damage their lung function. If the habit is continued the damage becomes permanent with serious shortness of breath in later life. Fortunately, if a child smoker gives up, the lung damage is reversible, with no long-term effects on health.

But the most dramatic consequence of smoking is **lung cancer**. It is caused by carcinogens in the tar. Unfortunately this is a disease which usually develops thirty or more years after the smoking habit has been acquired and then it is often too late to do much about it. Cases of lung cancer are increasing all the time. In the U.S.A., 2,500 people died of it in 1930, but by 1967 the number had increased to 50,000. In Britain, 30,000 men and 7,500 women died of it in 1974, about half of them before the age of 65, but these numbers are now beginning to fall as more people stop smoking. The group of people in which lung cancer deaths have gone down most in recent years has been the doctors: they know the dangers!

Another danger is that cigarette smoke contains carbon monoxide. This highly poisonous gas interferes with the capacity of blood to carry oxygen (p. 152). In heavy smokers it may lead to a loss of oxygen-carrying capacity of up to 15%. In consequence, pregnant women who smoke a lot tend to have smaller babies (on average 200 g lighter) than those who do not smoke. Also the risk of spontaneous abortion and stillbirth is increased. It is estimated that 1,500 babies in Britain were lost in 1970–71 through their mothers' smoking habits during pregnancy.

Nicotine in cigarette smoke is not only a drug which results in dependence (p. 228), but it also has important physiological effects. Along with carbon monoxide it is known to be an important contributory factor in bringing about **coronary thrombosis** (heart attack) which caused 164,000 deaths in Britain in 1974. Although many people know that cigarette smoking can kill by causing lung cancer, few realise that it contributes to $2\frac{1}{2}$ times as many deaths from diseases of the heart and blood vessels.

The seriousness of the effect of smoking can be gauged by the fact that people are being killed by cigarettes in Britain alone at the rate of one every 15 minutes. This is four times the rate of deaths through road accidents. Smoking 20 cigarettes a day increases by 20 times the chances of contracting lung cancer, compared with the non-smoker. One in 8 people who smoke 40 cigarettes a day contract the disease (Fig. 14:13).

With these facts available you may well wonder why people smoke at all. The reason is basically because nicotine is a drug of dependence. Nearly all smokers will tell you that they started because they wanted to appear tough or grown up, to be sociable, because their friends smoked, or more often just to see what it was like. They became dependent long before they realised and then found it very difficult, if not impossible, to drop the habit.

Gaseous exchange in plants

When we investigated aerobic respiration in the last chapter we used both animal and plant material according to convenience. This was done because plant and animal respiration are basically the same. It is only in the means by which oxygen reaches the cytoplasm that differences occur.

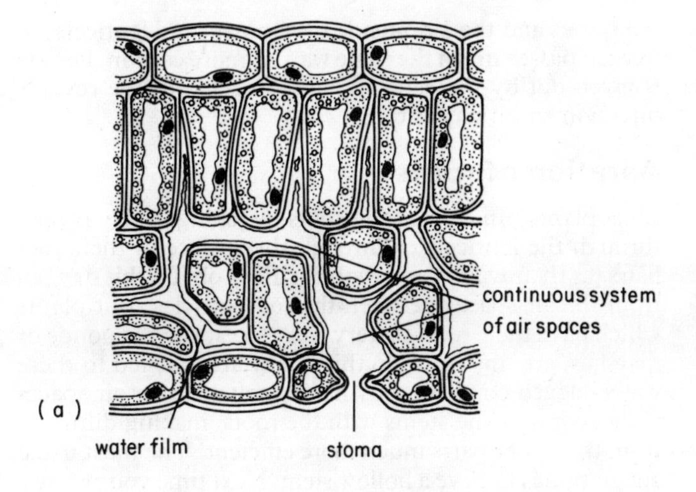

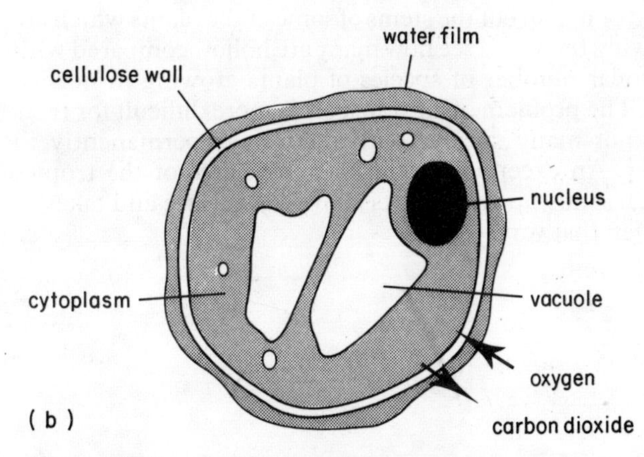

Fig. 14:14 The leaf as a respiratory organ: a) A transverse section showing system of air spaces. b) A single leaf cell showing diffusion pathways.

Respiration takes place in every living cell, so there must be a means of oxygen entry into the plant, a method of transporting it to every cell and a means of removing the products of respiration (carbon dioxide and water).

How gases enter and leave

Gaseous exchange takes place through vast numbers of microscopic holes in the surfaces of leaves and young stems called **stomata**. They are capable of opening and closing. Their structure will be considered on p. 106. On the surface of woody stems and roots there are different openings called **lenticels** (p. 193) which serve the same purpose.

Conduction within the plant

Inside the plant these openings lead to a series of spaces between the cells which form a continuous network all over the plant. The spaces are very large in the leaves, much smaller in other parts of the plant. The air spaces are lined with water and the oxygen in the air spaces dissolves in this and passes through the porous cell walls into the cytoplasm where the sugar is broken down into carbon dioxide and water with the liberation of the energy. The carbon dioxide passes out into the air spaces by a similar method.

The whole system works by diffusion; as the oxygen is used up by the cells a gradient develops between the cells and the air in the spaces, and similarly between the air in the spaces and the air outside the stomata and lenticels, so oxygen passes in. In the same way, as more carbon dioxide is given out by the cells a gradient occurs in the reverse direction and it passes out.

Aeration of roots

Most plants can aerate their roots by taking in the oxygen through the lenticels or through the surface of their root hairs (as their walls are very thin). They obtain this oxygen from the air spaces between the soil particles. But plants which have their roots in very wet places, such as ponds or marshes, are unable to do this. They are adapted to these water-logged conditions by having much larger air spaces which connect the stems with the roots, making diffusion from the upper parts much more efficient. The most usual adaptation is to have a hollow stem. Next time you are by a pond or marsh cut the stems of some of the plants which are growing there and see how many are hollow compared with a similar number of species of plants growing in normal soil. The problem of air transport is more difficult for trees and not many survive with their roots permanently in water. An exception is the mangrove tree of the tropics which sends up aerial roots above the surface and takes in oxygen that way.

15

Nutrition in green plants

Methods of nutrition

Before considering nutrition in green plants in some detail, it is useful to summarise the three main methods of nutrition used by living organisms as there is a close relationship between them.

1. Holophytic nutrition

This is the method used by green plants. It involves the building up of carbohydrates such as sugar and starch from simple chemical substances by a process called **photosynthesis**, and also the formation of fats and proteins by further processes. This is said to be an **autotrophic** or self-feeding method because in this way they *make their own food*. (Other organisms are dependent for their nutrition on those substances that green plants have produced; they are therefore said to be **heterotrophic** because they have no means of building up their own food from simple substances.)

2. Holozoic nutrition

This is the method characteristic of most animals. It occurs in three stages: a) **ingestion**—the taking in of the food at a particular place (in most animals the mouth); b) **digestion**—the processing of the complex food molecules by means of digestive enzymes until the products are soluble and have small enough molecules to be absorbed; c) **absorption**—the extraction of these products from the digestive system into the blood, from which it can be taken up by the cells and utilised. In the simpler animals, which have no blood system, the digested products diffuse directly into the surrounding cytoplasm.

3. Saprophytic nutrition

This is the method used by plants which have no chlorophyll, such as fungi (p. 28) and most bacteria (p. 133). Like all plants, saprophytes have no mouths, so there is *no* process of ingestion. Instead, enzymes are secreted by the cytoplasm through the cell wall on to the food, a process called **external digestion**. The soluble products are then absorbed and utilised.

You can demonstrate starch digestion by a saprophytic fungus by growing it on agar (p. 134) containing starch. The test for starch is to add some iodine dissolved in potassium iodide. Iodine solution turns starch blue-black. When the fungus mycelium has grown sufficiently, the agar is flooded with iodine; the agar will turn blue-black except in the immediate area where the fungus is growing. Here the agar is stained brown (the colour of the iodine) indicating that the starch has been broken down by the fungus (Fig. 15:1). Saprophytic fungi and bacteria thus feed on non-living organic matter and by their feeding process cause it to decay.

Parasites

In contrast to the saprophytes which feed on non-living material, there are **parasites** which are dependent on other *living* organisms for their food. Parasitic fungi have already been mentioned (p. 30). Animal parasites live either on their hosts (**ectoparasites**), e.g. fleas and lice, or inside them (**endoparasites**), e.g. tapeworms, often harming them in the process. Parasites may either feed directly on the living cytoplasm, e.g. the malarial parasite which feeds on red blood corpuscles (p. 51), or on the food of the host, e.g. gut parasites such as tapeworms (p. 25). The actual process of nutrition, however, is not basically different from either holozoic or saprophytic nutrition because the food is either ingested as in typical animals, or absorbed through the surface as in saprophytes. Thus it is better to think of parasitism as a different mode of life rather than a different method of nutrition.

Holophytic nutrition

Every green plant is composed of living protoplasm and the non-living substances that the protoplasm has produced. Many of these substances are complex, but all are synthesised from relatively simple chemicals which the plant finds in its immediate surroundings. Holophytic nutrition involves the taking in of these nutrients and their synthesis into more complex foods. There are two main stages in holophytic nutrition: photosynthesis, in which carbohydrates are synthesised and protein synthesis, when proteins are formed (p. 108).

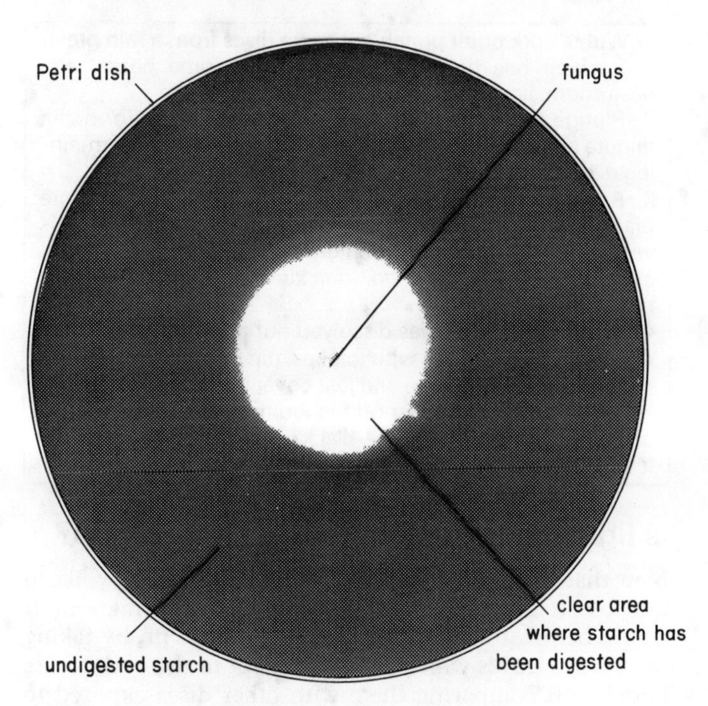

Fig. 15:1 Starch digestion by a saprophyte.

The substances which are synthesised are used for two main purposes: the growth of the plant, including the formation of new individuals (reproduction), and respiration, the means by which the plant obtains its energy for all its metabolic activities.

Photosynthesis

This is the basic process of plant nutrition. If green plants did not photosynthesise we could not exist, as all the food we eat, whether animal or vegetable, has its origins in this process. Because of the great importance of photosynthesis we will investigate it in some detail.

Our aim will be to build up, step by step, an understanding of photosynthesis by carrying out a number of experiments from which we can deduce certain facts about it, so producing as complete a picture as possible of the process.

We will start with the knowledge that when green plants photosynthesise carbohydrates are formed. Starch is a common carbohydrate found in plants so we can first see if it is present in leaves.

Testing for starch

> A potato is full of starch: pipette some iodine dissolved in potassium iodide on to its cut surface. Iodine turns starch blue-black. Look at the treated potato under a lens and notice how it is really speckled all over. Each speck is a solid starch grain.

Can starch be detected in green leaves?

It is not so easy to test for starch in a green leaf because the green colour prevents the starch from being seen even after treatment with iodine. But you can do it by first removing the chlorophyll with ethanol. Proceed as follows:

> 1. With a cork borer punch out some discs from a thin green leaf which has been in good light for some hours, e.g. nasturtium, lime or lilac.
> 2. Plunge the discs into a beaker of boiling water for half a minute and transfer them with forceps to a test tube containing a little ethanol.
> 3. Remove the bunsen from under the water bath before placing the test tube in the water. This is because ethanol is very inflammable and the contents of the tube might catch alight. The ethanol will boil quickly as its boiling point is below that of water.
> 4. When the ethanol has dissolved out all the green chlorophyll, remove the discs with forceps, dip them in water, place them in a clean test tube and just cover with iodine solution. After a few minutes, pour off the iodine, wash in cold water and note whether the blue-black colour is present in the leaves. If it is, we can conclude that starch is present.

Is light necessary for starch formation?

Now that you know how to test for starch in a leaf, you can find out if light energy is essential for its formation. It would be possible to adapt the last experiment, by taking discs from leaves which have been kept in the dark for 24 hours, and comparing them with other discs exposed to light for some time. Another way is to experiment on a single leaf and expose only part of it to light.

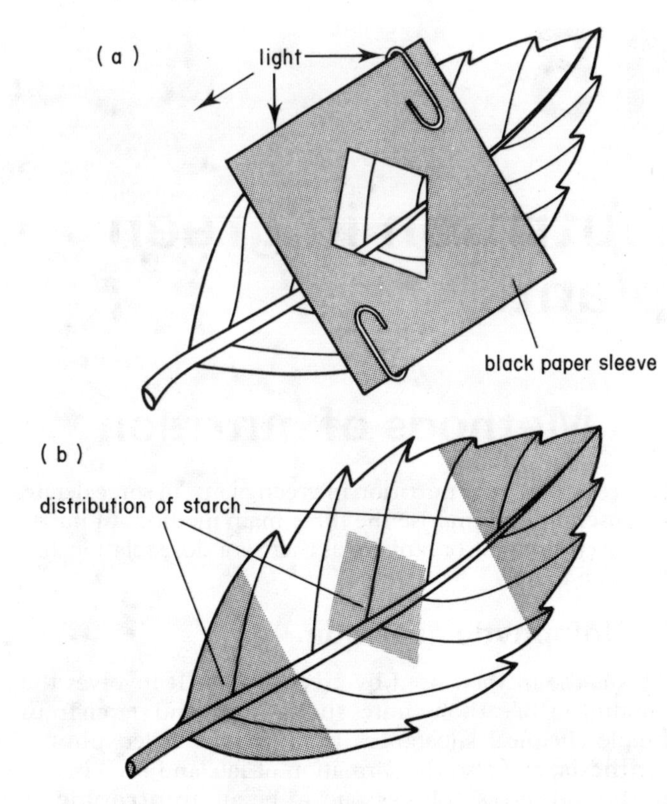

Fig. 15:2 a) Method of demonstrating the effect of light on starch formation in a leaf. b) The same leaf after testing for starch.

> 1. Make a sleeve of black paper as in Fig. 15:2. Attach it to the leaf of a tree or shrub where it will receive good light. Leave overnight and for as long as possible the next day.
> 2. Test for starch as in the previous experiment and note whether the distribution of starch corresponds to the part of the leaf which was exposed to the light.

What other factors could be involved in starch formation? As starch is found in green leaves exposed to light, it is possible that the chlorophyll which produces the greenness plays a part in its formation.

Is chlorophyll necessary for starch formation

To investigate this you can use a leaf which has its own built-in control. Certain plants, e.g. some forms of geranium, have **variegated leaves**, i.e. which are partly green and partly white or yellow (Fig. 15:3).

> 1. Remove a variegated leaf which has been exposed to light for some hours.
> 2. Draw the pattern to show the green and white parts of the leaf.
> 3. Extract the chlorophyll in ethanol as before and test for starch. Does the pattern of the starch correspond with the green part of the leaf?

From these experiments it can be shown that light energy and chlorophyll are both necessary for starch formation, but what is the starch made from? The starch molecule is composed of a variable number of units formed from glucose, each with the basic formula $C_6H_{12}O_6$, so

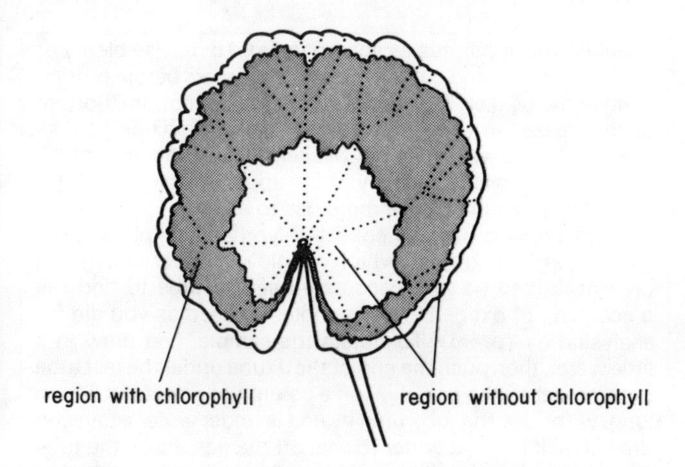

Fig. 15:3 A variegated leaf—used to determine whether chlorophyll is needed for starch formation.

obviously substances containing carbon, hydrogen and oxygen must be involved. A readily available carbon compound is carbon dioxide in the air. We know that in respiration carbon dioxide is given out by plants. Is it also used in the formation of starch when light and chlorophyll are present?

Is carbon dioxide taken in when green leaves are exposed to light?

You can find out if this occurs by carrying out an experiment involving bicarbonate indicator, the colour of which varies according to the concentration of carbon dioxide: no carbon dioxide—reddish purple; very little carbon dioxide, e.g. when the indicator is exposed to air—orange red; greater concentrations of carbon dioxide—yellow.

We could use this indicator to see if any changes occur in the carbon dioxide content of the atmosphere when green leaves are illuminated.

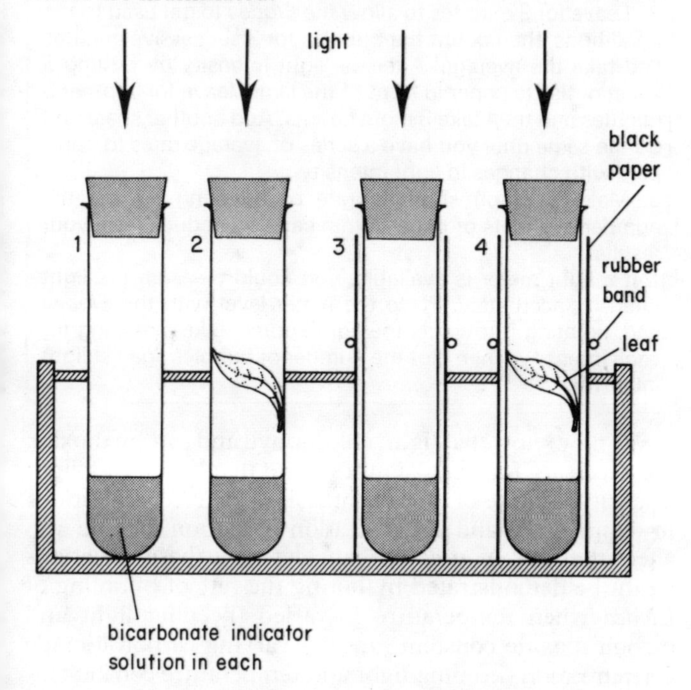

Fig. 15:4 Apparatus to determine if any changes occur in the carbon dioxide content of the atmosphere when green leaves are illuminated.

1. Wash out four bo'' g tubes thoroughly in tap water, rinse with distilled water a.d finally with a little bicarbonate indicator.
2. Put 2 cm³ of bicarbonate indicator in each and stopper at once (why?).
3. Wrap black paper around Nos. 3 and 4 to keep out the light.
4. Place similar green leaves which have just been removed from a healthy plant in sunlight in Nos. 2 and 4 in such a way that they do not slip into the indicator; do not put a leaf into Nos. 1 and 3.
5. Place all four tubes in a place where light from the window or bench lamp can illuminate them equally.
6. Shake each *gently* every five minutes and after half an hour compare the colours of the indicator in each by holding them against a white background. Compare tube 1 with 2, 3 with 4, 1 with 3 and 2 with 4. What can you deduce from each of these comparisons? What is your final conclusion of the effect of leaves in light and darkness on the carbon dioxide content of the atmosphere?

Green plants in the *dark* give off carbon dioxide as a result of *respiration*, but you should have discovered from the last experiment that when green leaves are illuminated the indicator changes to reddish purple. This suggests that not only is respiratory carbon dioxide being used up but some is also being taken in from the air. Is this carbon dioxide used in the making of starch or for some other purpose? We could try to find out by designing an experiment to see if carbon dioxide is essential for starch formation.

Is carbon dioxide necessary for starch formation?

1. Fit up the apparatus as shown in Fig. 15:5.
2. Put some caustic potash solution in one jar to absorb the carbon dioxide, and potassium bicarbonate solution in the other to ensure that a good supply of carbon dioxide is in the air above.
3. Put in each tube a nasturtium or other suitable leaf which has previously been kept in the dark to remove any starch. Absence of starch can be proved by taking out a disc from each leaf with a cork borer and testing as before.
4. Keep the leaves strongly illuminated for several hours, preferably all day, and then test for starch in each.

From your results deduce whether carbon dioxide is necessary for starch formation.

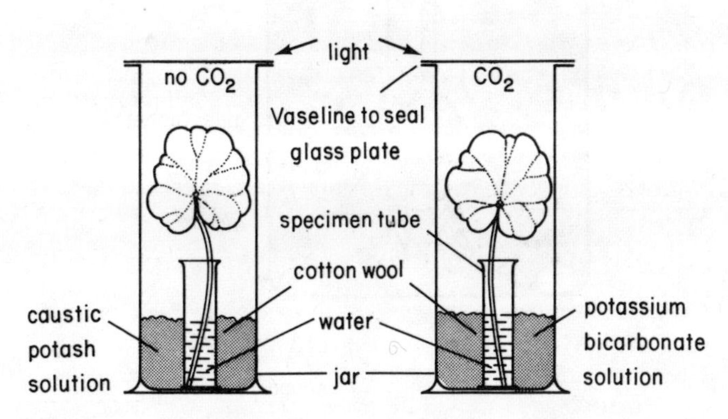

Fig. 15:5 Experiment to find out whether carbon dioxide is necessary for starch formation.

We described on p. 88 how scientists have used radioactive isotopes to 'label' certain atoms so that they could be traced from molecule to molecule. By a similar technique, using radioactive carbon in the carbon dioxide supplied to plants, it has been confirmed that it is the carbon atoms in *carbon dioxide* which are passed into the starch molecule.

From these investigations we should now be able to conclude that light, chlorophyll and carbon dioxide are all needed in the making of starch. The next question is whether any products are given off during starch formation. When considering respiration we saw that oxygen was used up and carbon dioxide given out. In photosynthesis we have seen that carbon dioxide is used up. Is oxygen put back into the air?

Is oxygen given off by green plants in the light?

It is not easy to investigate this on a land plant as it is already surrounded by air of which about one-fifth is oxygen, but by using a water plant any oxygen given off would appear as bubbles, as very little oxygen dissolves in water.

If you observe some Canadian pondweed, *Elodea*, in a beaker of water when it is in bright light, you will notice that bubbles of gas are given off from the cut ends of the stems. By collecting these in sufficient quantity you could test whether this gas is oxygen or not.

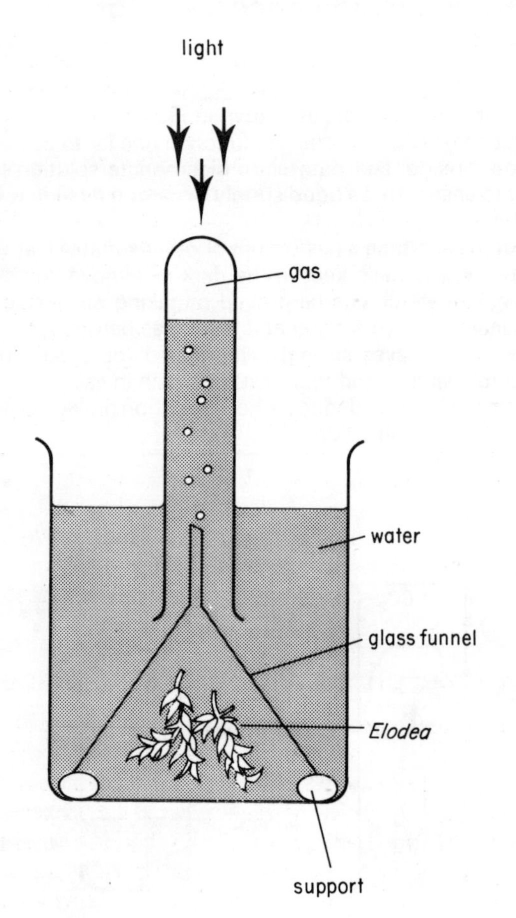

Fig. 15:6 Apparatus to determine whether oxygen is given off by green plants in the light.

Set up the apparatus as shown in Fig. 15:6. Use plenty of sprigs of *Elodea* and cut the ends with scissors before putting them in the beaker. The funnel should be kept off the bottom of the beaker to allow diffusion of carbon dioxide to take place from the water in the beaker. Use a bench lamp to illuminate the plants strongly. If the apparatus can be left for several days you will have more gas to analyse.

When enough gas has collected, you can simply test the gas with a glowing splint which will glow brighter if oxygen is present, but to be more accurate use a J tube to find the percentage of oxygen in the sample. Proceed as you did for analysing air (p. 96). To obtain the sample, first draw in a little water, then push the end of the J tube under the test tube until it reaches the gas, draw in a column of gas about 10 cm long, withdraw the tube until its end is under water again and draw in a little more water to seal off the gas. Place the tube under water in a sink for a minute and measure the column of gas accurately. Now proceed as before, first using potassium hydroxide to absorb any carbon dioxide and then alkaline pyrogallol to absorb the oxygen.

Calculate the percentage of oxygen in the gas. Is the percentage greater than that of air?

Does the intensity of light affect the rate of photosynthesis?

To find out, you could modify the last experiment by taking a single sprig of *Elodea*, counting the bubbles given out by the cut end, and then, by altering the light conditions, see if the speed of bubbling changes. Proceed as follows:

1. Fit up the apparatus as in Fig. 15:7. Shine a strong lamp on to the *Elodea* horizontally. The tube is immersed in a beaker of water to help keep the temperature constant. Keep the bench lamp the same distance from the *Elodea* during the experiment.
2. Leave for 3 minutes to allow the *Elodea* to get used to the conditions, then count the bubbles for 3 successive minutes and take the average. Alter the light intensity by putting a sheet of tissue paper in front of the lamp, leave for another 3 minutes and then take 3 more counts. Add another sheet and do the same until you have a series of average rates to compare with changes in light intensity.
3. Make a graph showing rate of bubbling against the number of sheets of paper. What can you deduce from your results?

If a light meter is available, you could measure the light intensity accurately. Place the meter level with the *Elodea* and, pointing it towards the light source, take a reading for each intensity. Then plot the number of bubbles against light intensity.

We now know that light, chlorophyll and carbon dioxide are necessary for photosynthesis and that by varying light intensity the rate of photosynthesis is affected. Variations in temperature and concentration of carbon dioxide also affect the rate. You could consider how these variations might be demonstrated by noting the rate of bubbling of *Elodea* when temperature is varied (keeping light and carbon dioxide constant) and by varying carbon dioxide concentration (keeping light and temperature constant).

It is of great economic importance to know the optimum conditions for photosynthesis for different crops in order to obtain maximum yield.

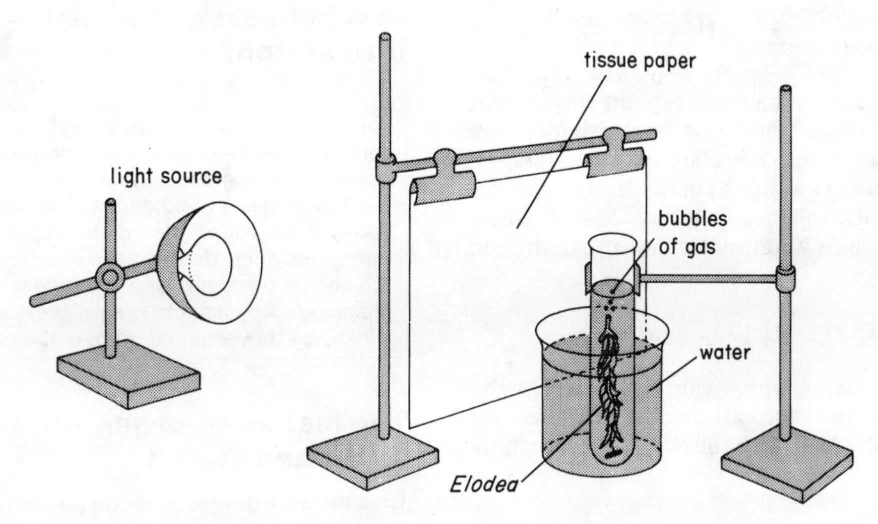

Fig. 15:7 Apparatus to determine the effect of light intensity on the rate of bubbling of *Elodea*.

We have assumed so far that starch is formed as a result of photosynthesis, but a molecule of starch consists of many glucose molecules connected together. It would therefore be a likely hypothesis that during photosynthesis glucose is formed first and then changed into starch.

Glucose belongs to a group of substances called reducing sugars. Their presence can be demonstrated using the following test.

Testing for reducing sugars

> Add 5 cm^3 of Benedict's solution to each of two tubes. Add a little powdered glucose to one and use the other as a control. Heat both to boiling point. Notice how the glucose causes the blue colour to change until finally a brick-red precipitate is formed. This precipitate still clings to the test tube after you have washed the contents down the sink. (It can be cleaned off with dilute hydrochloric acid.) The colour change occurs because the copper compound in the Benedict's solution is reduced to a copper compound which is red. Is there any change in the control tube?
>
> Now take another clean boiling tube, but this time add only a few grains of glucose to the Benedict's solution. Boil as before. Is there any change of colour?

When you test for reducing sugar in leaves you will be dealing with very small quantities of sugar, so remember that *any* change in colour is a good test for its presence. The colour varies with the quantity of sugar present.

Is reducing sugar present in leaves?

> Take three kinds of leaves which have been exposed to light for some hours: iris, geranium and lilac would be a good selection. Cut out four discs of iris with a cork borer and grind them up with a pinch of sand and a little water; filter and test the filtrate for reducing sugar with Benedict's solution. Repeat for the other two leaves and compare the results. Which leaves contained reducing sugar?

Your results will have shown you that some leaves do form sugar, but others apparently do not. A possible explanation of this result could be that in those which showed no reducing sugar, it was in fact formed, but it was converted into starch subsequently.

To investigate this possibility you could see if the plants which showed no reducing sugar in the last experiment are able to produce starch if kept in the dark and supplied with glucose.

Can leaves turn glucose into starch in the dark?

> Using plants which showed no reducing sugar in their leaves, keep them in the dark for two days. Punch out a number of discs. Test one to make sure that no starch is present.
>
> Take two Petri dishes, place 5% glucose solution in one and water in the other to act as a control. Float some discs on each with their undersurfaces in contact with the liquid. Keep in the dark for several days and then test for starch.

This experiment should clearly demonstrate that starch can be made from glucose sugar by the leaf and that this reaction does not need light. In fact, research has shown that glucose is the primary carbohydrate formed in photosynthesis, but in many plants it is immediately converted to starch. The conversion of sugar into starch is advantageous to the plant as sugar is soluble in water; if it became too concentrated, it would act osmotically and draw water into the cell. Starch by contrast is insoluble so there is no danger of this. However, as we have seen from our experiment, some plants do have some sugar in their leaves when they are exposed to light.

How does photosynthesis take place?

So far we have seen that light energy, chlorophyll and carbon dioxide are needed for photosynthesis to take place, and oxygen is released. Looking at the formula for glucose sugar, $C_6H_{12}O_6$, it can be seen that some hydrogen atoms have to come from somewhere. It has been shown by using radioactive isotopes that these come from water and that the oxygen given off also comes from the same source.

We can now express the process by the following equation:

$$CO_2 + H_2O \xrightarrow[\text{light energy}]{\text{chlorophyll}} C_6H_{12}O_6 + O_2.$$

But this simplified equation does not explain the various reactions which take place before the final products are formed. It is now known that there are two main stages. The first is the absorption of light energy by the chlorophyll, which changes it to chemical energy. The chemical energy is then used to split water molecules into hydrogen and oxygen:

$$2H_2O \rightarrow 4H + O_2.$$

During the second stage, which can take place in the dark, this hydrogen reduces the carbon dioxide molecules to form a simple carbohydrate with the generalised formula (CH_2O):

$$CO_2 + 4H \rightarrow (CH_2O) + H_2O.$$

This reaction is not a simple one and is catalysed by a series of enzymes. The final product is glucose sugar, $6(CH_2O)$, more correctly expressed as $C_6H_{12}O_6$. Hence photosynthesis, without including the intermediate reactions, can be represented by the following balanced equation:

$$6CO_2 + 12H_2O \xrightarrow[\text{light energy}]{\text{chlorophyll}} C_6H_{12}O_6 + 6H_2O + 6O_2.$$

(As all the oxygen released is known to come from the water, it is necessary to include 12 molecules of water in the equation to account for the 6 molecules of oxygen.)

Finally, when the glucose is turned into starch, the reaction involves the removal of water. The reaction is catalysed by the enzyme starch phosphorylase:

$$nC_6H_{12}O_6 \xrightarrow{\text{starch phosphorylase}} (C_6H_{10}O_5)_n + nH_2O$$

where n varies according to the type of starch.

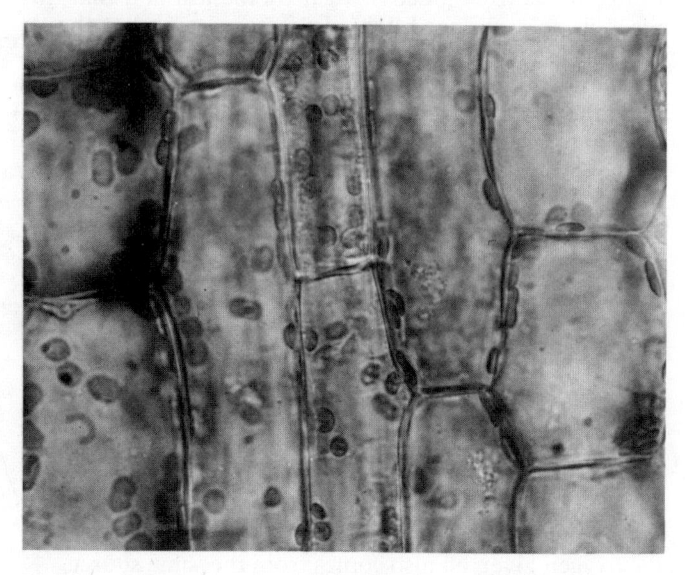

Fig. 15:8 High power photomicrograph of cells of an *Elodea* leaf showing chloroplasts.

In what part of the leaf is sugar turned into starch?

Take a very thin leaf which has been exposed to the light for some time and examine it in a drop of water under the high power of the microscope. A leaf from near the tip of the water plant *Elodea* is a good choice; the green chloroplasts which you will see round the edges of cells may be in a state of movement (Fig. 15:8).

Now lift the cover slip and add a drop of iodine before replacing. Any starch grains present will go blue-black. Where are they situated, all over the cell or in special places?

The leaf as an organ for manufacturing sugar and starch

Like all well-designed factories there must be good access for the raw materials, in this case carbon dioxide, water and light; good transport within it to bring the products together; and good means of removal of the final products, carbohydrates and oxygen.

Entry and transport of the raw materials

Stomata

Carbon dioxide enters the leaf through the stomata, which are usually situated on the underside. They are capable of opening and closing, and their movements are primarily affected by the intensity of the light and, to a lesser extent, by the evaporation of water. As a general rule it can be said that they open by day and close at night. The basic principle of the opening is that the two **guard cells** which surround each stoma act as osmotic systems and when the sugar concentration within them is high, water is drawn in and they enlarge. Because of special thickenings in their walls the guard cells can expand in certain directions only.

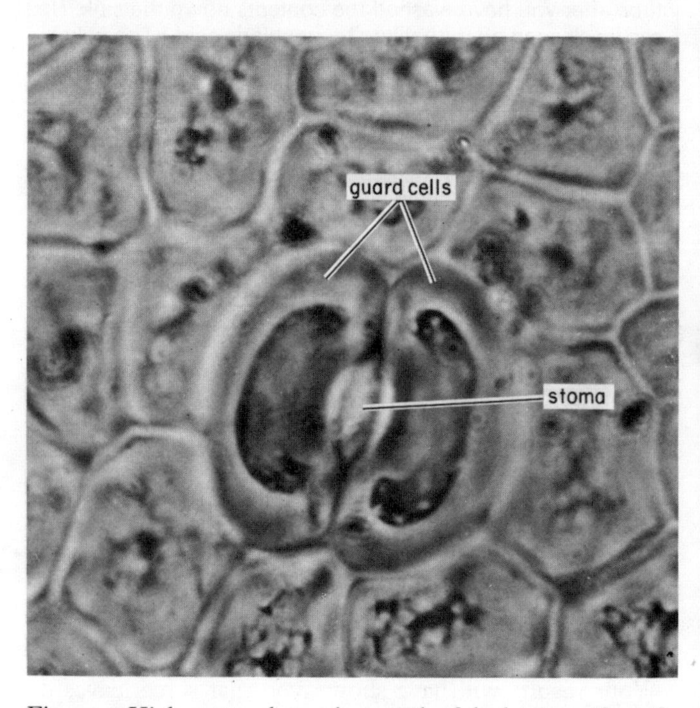

Fig. 15:9 High power photomicrograph of the lower surface of a leaf showing stomata.

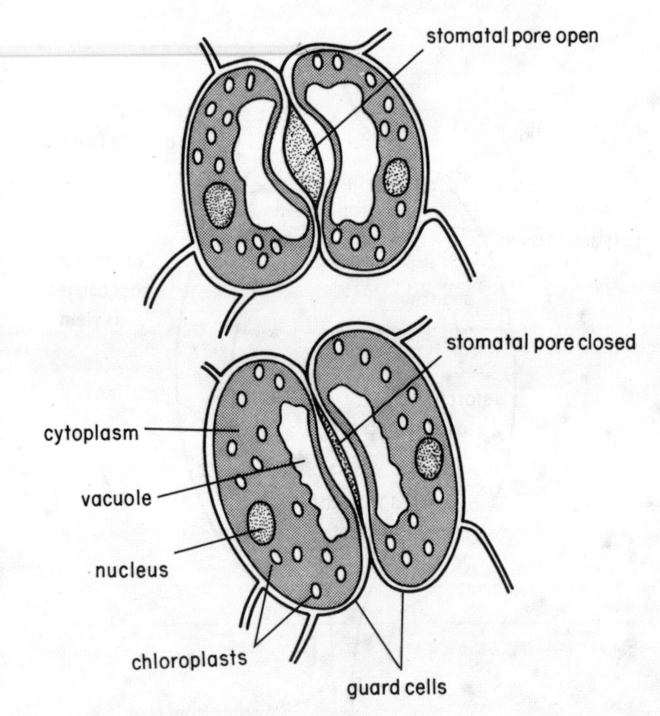

Fig. 15:10 Diagram of a stoma in surface view: (top) open, (bottom) closed.

As a result they move apart making the pore between them larger (Figs 15:9 and 10).

To understand stomatal movement it is important to realise that the guard cells of a leaf are the only epidermal cells to have chloroplasts. Hence, when illuminated, they form sugar while their neighbours do not. Unlike the mesophyll cells, the sugar made in the guard cells is not turned into starch until night time. So during the day, the sugar solution in the guard cells becomes stronger, water is drawn in, and the stoma opens. At night, when the sugar turns into starch, the osmotic strength of the sap in the guard cells becomes reduced and water is lost to neighbouring cells and the stoma closes. This is not the complete explanation regarding stomatal movement. There are still some factors involved which are not yet fully understood.

The passage of carbon dioxide from the atmosphere to the cells

Entry of carbon dioxide through the stomata takes place by diffusion. The process is very efficient because of the large surface area of the leaf and the vast number of stomata present. Once inside, the carbon dioxide passes to the mesophyll cells by diffusion within the intricate system of air spaces between the cells. It then dissolves in the water lining the air spaces, diffuses through the cell walls into the cytoplasm and so reaches the chloroplasts.

The transport of water

The water brought from the roots by the xylem is quickly spread to all parts of the leaf by a series of branching veins.

Hold up various leaves against the light and examine the network of veins with a hand lens (Fig. 15:11).

How light reaches the chloroplasts

The efficiency of this process depends largely on two factors: the surface area of the leaf relative to the volume, and its position with respect to the light source. A large surface area allows more light to reach the chloroplasts and if the leaf is at right angles to the light, more will penetrate.

Plants have various methods of presenting as much leaf surface as possible to light. A geranium, for example, if placed on a window ledge reacts to light coming from the side by bending the petioles so that the leaf surface is roughly at right angles to the incident rays. Also, in woods, you find that many plants growing in shade have larger leaves than those of the same species in the open and that many kinds arrange their leaves to give a minimum of overlap.

Internally you will also find adaptations. Turn to Fig. 5:13 and see how the chloroplasts are arranged in the cells in relation to the light. Where are they most abundant? Remember that light gradually becomes absorbed as it penetrates the tissues, so a very thin leaf will be more efficient than a thick one.

Removal of the products of photosynthesis

The oxygen diffuses out through the system of air spaces and through the stomata. The sugar formed on the surface of the chloroplasts is turned into starch grains and accumulates in the leaf during the day. At night the reverse process takes place and the sugar in solution passes into the phloem for translocation to other parts of the plant. You will now understand why in some of the experiments involving the formation of starch the leaves were 'destarched' by keeping them in the dark first.

Thus the structure of leaves is admirably adapted to bring together the carbon dioxide, water and light as efficiently as possible and to remove the oxygen and the carbohydrates when they are produced.

What happens to the carbohydrates formed in photosynthesis?

Carbohydrates are used up in three main ways:

1. Respiration

Most cells receive sugar via the phloem; this can be broken down in respiration to liberate the energy the cells require.

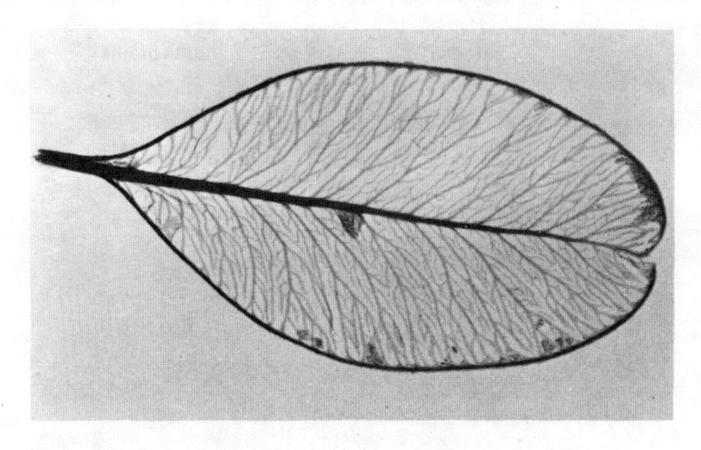

Fig. 15:11 A leaf 'skeleton' showing the arrangement of the veins.

2. Food storage

Some of the sugars may be stored for future use, either by the parent plant in special organs such as bulbs, tubers, corms or rhizomes, or for the next generation in seeds or fruits. The sugar is usually converted into starch for storage; occasionally it is stored as cane sugar, as in the roots of sugar beet. In some seeds it may be changed into an oil and stored as such. These storage organs are of great value to man because of their high concentration of food. Many of our vegetables are really plant food storage organs: potato tubers are storage stems; carrots, turnips, parsnips and beetroot are storage roots; onions are mainly composed of storage leaves; and beans and peas are seeds with large food stores. When oil is stored instead of starch this may be extracted for many purposes, e.g. olive oil from the fruit of the olive, and ground nut oil, among others, for the manufacture of margarine.

Examine the starch grains from various storage organs.

> Take a scraping from the cut surface of a potato and mount in water with a little iodine. Draw the starch grains carefully. Repeat with scrapings from peas, rice and wheat mounted in the same way. What differences can you see?

These grains are made in the cells of the storage organ from the sugar which reaches them from the leaves. The synthesis takes place within certain bodies which are like chloroplasts except that they contain no chlorophyll. These **leucoplasts**, as they are called, build up the starch grains, layer by layer, until they break through the outer membrane of the leucoplast into the vacuole of the cell. Sometimes the layers in the starch grain can be seen; did you see them in the starch grains from the potato?

3. Growth

Sugar is also taken to the parts of the plant which are actively growing. Growth results from an increase in the number of cells. Sugars are used in the formation of cellulose (a complex carbohydrate found in cell walls) fats and in the synthesis of protein, a basic component of protoplasm. All these processes are summarised in Fig. 15:13.

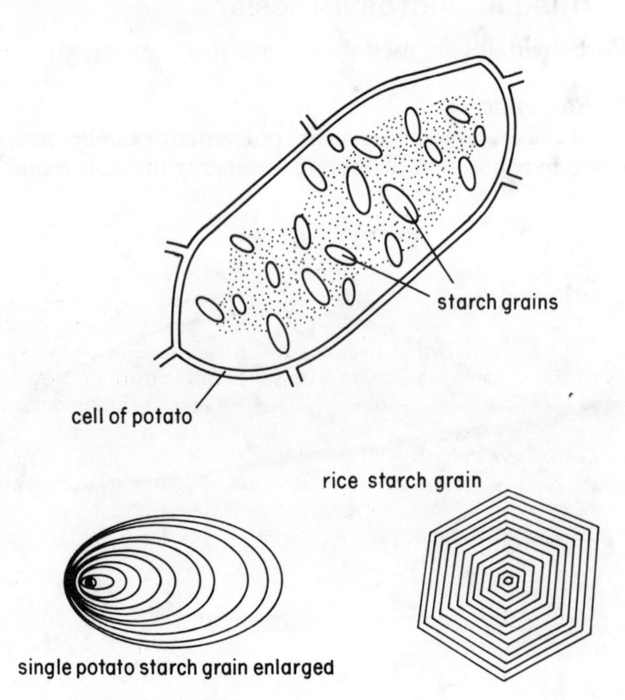

Fig. 15:12 Starch grains.

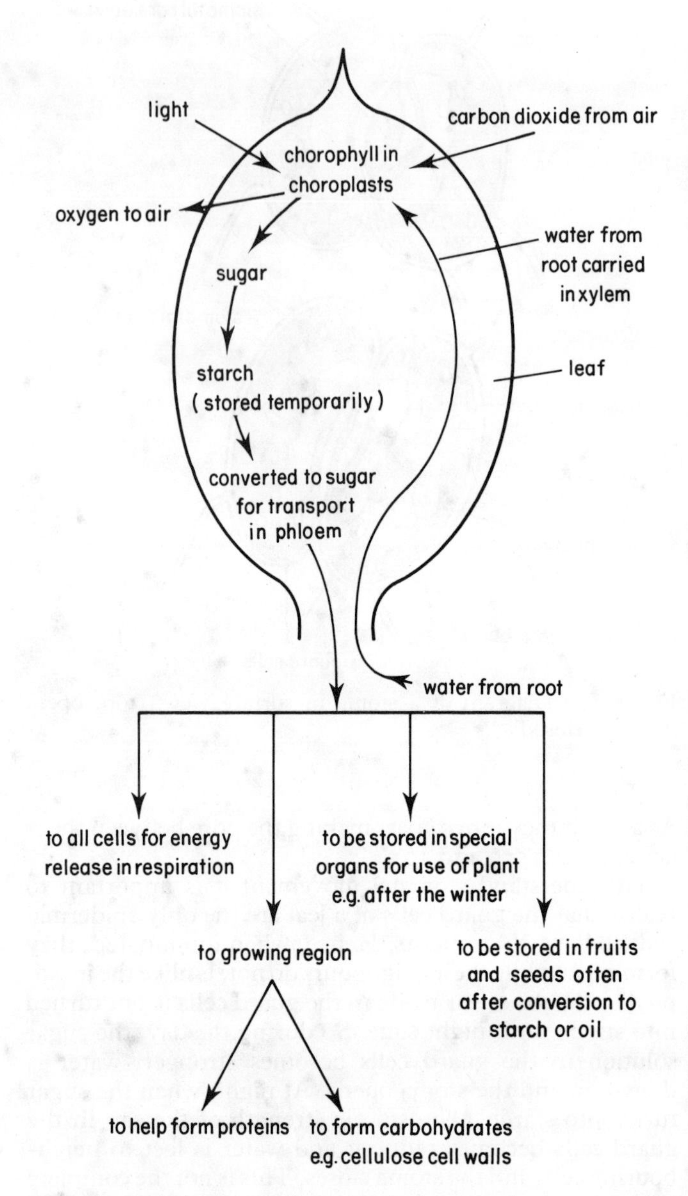

Fig. 15:13 Summary of the process of photosynthesis and the fate of the products.

Protein synthesis

Holophytic nutrition comprises two related processes, photosynthesis and protein synthesis. The former, as we have seen, leads to the formation of carbohydrates, the latter to protein.

All proteins contain carbon, hydrogen, oxygen and nitrogen, and frequently sulphur and phosphorus as well (p. 111). Proteins are synthesised from smaller building blocks known as **amino acids** by combining ammonia (formed from the reduction of nitrates) with certain complex chemicals derived from the breakdown of sugars formed by photosynthesis. Sulphates and phosphates are

TABLE 15:1. CONSTITUENTS OF VARIOUS CULTURE SOLUTIONS

Substance	*Complete Knop's solution*	*Lacking nitrogen*	*Lacking magnesium*	*Lacking iron*
Calcium nitrate	0·8	—	0·8	0·8
Potassium nitrate	0·2	—	0·2	0·2
Potassium dihydrogen phosphate	0·2	0·4	0·2	0·2
Magnesium sulphate	0·2	0·2	—	0·2
Calcium sulphate	—	0·6	0·2	—
Ferric phosphate	trace	trace	trace	—

Note: All constituents are in grammes/litre. The solutions should be made up with distilled water.

also used in the synthesis of some proteins. All these reactions are catalysed by enzymes (themselves proteins) and take place in the cell protoplasm.

Elements essential for healthy plant growth

All green plants need the elements carbon, hydrogen and oxygen for the formation of carbohydrates, proteins and fats. These elements are obtained by land plants from carbon dioxide in the air and water from the roots. In addition a number of other essential elements are needed for healthy growth. These are taken in through the roots, dissolved in water in the form of **mineral salts** (soil nutrients), e.g. nitrates. Of these elements, nitrogen, phosphorus and sulphur are needed for protein synthesis, iron and magnesium for chlorophyll formation, and potassium for cell division; calcium affects the permeability of cell membranes to solutions of salts.

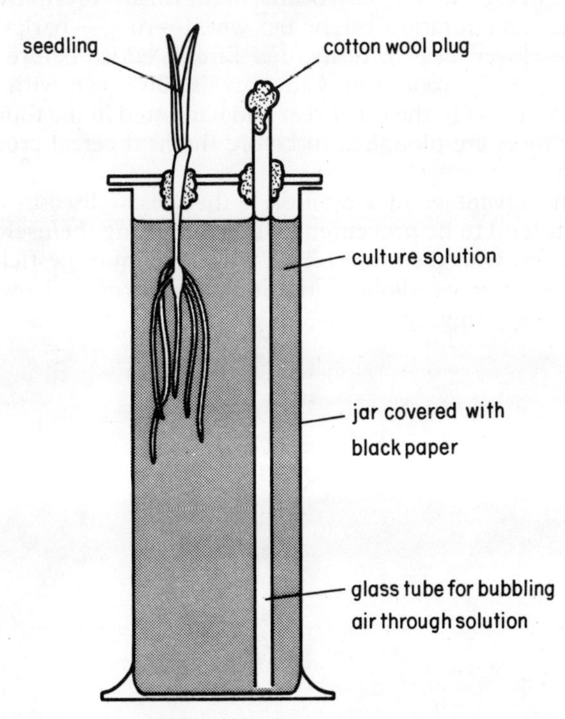

Fig. 15:14 Jar used in water-culture experiments.

In addition, there are others called **trace elements** because, although essential, they are only needed in minute quantities. They are used mainly for certain metabolic processes and include boron, manganese, cobalt and zinc. We can investigate the effects of some of the more important elements by depriving growing plants of some of them and seeing what happens. This can be done by using **water cultures**. This work was pioneered by the German scientists **Sachs** and **Knop** in the mid-19th century. They both devised culture solutions in which plants would grow healthily. Knop's solution is given in the second column of Table 15:1.

The basic principle is that several chemicals are used in making up the water cultures. By using different combinations of these salts, solutions can be produced which are deficient in *one* element. For example, to produce a nitrogen-deficient solution, both calcium nitrate and potassium nitrate are removed. This takes away calcium and potassium as well as nitrate. However, potassium is still present as potassium dihydrogen phosphate, so only the calcium must be put back. This is done in the form of calcium sulphate (see Table 15:1).

For example you could find what happens when nitrogen, magnesium and iron are missing by the following experiment:

1. Germinate some maize in moist sawdust. When the seedlings are several inches high, select five of equal height, carefully wash off the sawdust and remove with a scalpel the whole of the remaining fruit as this contains food reserves, including minerals.
2. Clean and sterilise 5 gas jars and set up each as shown in Fig. 15:14 transferring one seedling to each. One jar should contain the complete culture solution (the control), a second distilled water and the remainder the solutions deficient in nitrogen, magnesium and iron respectively. Surround the jars with black paper to prevent the growth of algae in the culture solutions.
3. Keep the plants in the light so that they can photosynthesise. Bubble air gently into the solutions from time to time to aerate the roots. Observe the growth of the plants over a period of several weeks, recording their height, size, leaf colour and growth of roots.

Obviously, you cannot come to definite conclusions using only one plant in each culture solution. Far larger numbers are really necessary to overcome possible errors due to variation between the individual plants used.

However, you should be able to establish in your experiment quite convincingly that the growth of plants is affected by the lack of certain elements.

How are minerals released from the soil?

When plants die their remains become incorporated in the soil as **humus**. This dead organic matter in the soil then decays due to the action of bacteria and fungi and as a result many essential nutrients such as nitrates are produced and returned to the soil.

Mineral salts may also be released from the underlying rocks by the action of soil acids and rain-water. The latter is also slightly acidic due to the carbon dioxide dissolved in it. The natural supply of mineral salts to the soil, however, is often greatly depleted by man through the growing of crops.

Natural and artificial fertilisers

When crops are regularly taken from the soil, the mineral salts locked up in the plants are not returned through the natural processes of death and decay. It follows, therefore, that crop yields will gradually diminish as more and more of these nutrients are removed, unless fertilisers are added.

In the famous Broadbalk Field at Rothamsted Experimental Research Station, Hertfordshire, wheat has been grown and harvested every year since 1843 without the addition of any fertiliser. Each year, for over 100 years, the yield has been measured, and during this time has decreased to about 45% of the original yield.

Farmyard manure is probably the best means of replenishing the soil but with the modern use of tractors, rather than horses, it is a scarce commodity. It takes time for the minerals to be released through bacterial action, but manure has the great advantage of preserving the soil structure through maintaining the level of humus. In a sandy soil the humus binds the sand particles together and helps to retain moisture; with a clay soil it tends to break up the fine particles, thus allowing water to drain away more easily.

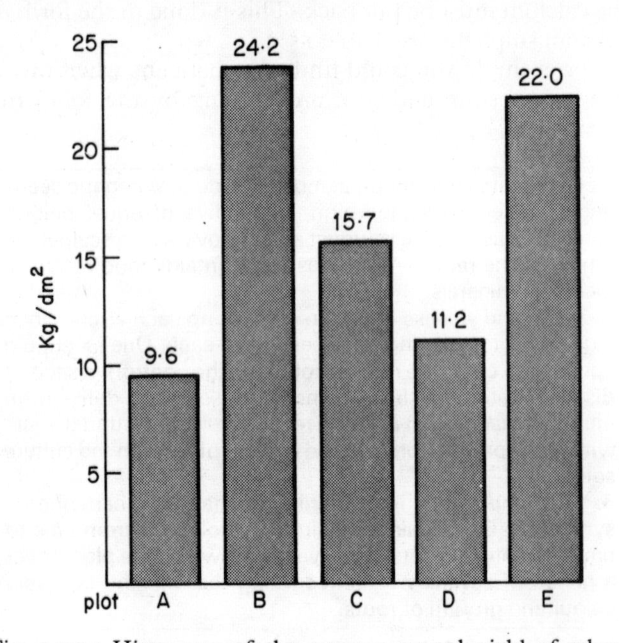

Fig. 15:15 Histogram of the average annual yield of wheat between 1852 and 1967 on five of the Broadbalk plots. A No treatment. B Farmyard manure. C Nitrogen only. D Potassium, phosphorus, sodium, magnesium (no nitrogen). E Potassium, phosphorus, sodium, magnesium, nitrogen.

Modern agriculture, with its intensive use of land, depends greatly on the use of artificial fertilisers. They are effective in promoting plant growth, but do nothing to maintain the soil structure. They are soluble and act quickly, but are easily washed out of the top soil by rain through a process called **leaching**. Therefore they have to be added just before the seeds are sown.

Many of these chemical fertilisers are used, but the commonest include ammonium sulphate, ammonium nitrate, sodium nitrate, 'superphosphate' (mainly calcium phosphate), and potassium sulphate.

The choice of fertiliser depends largely on the crops being grown and the type of soil. Cereals, for example, require more nitrogen than other types of crops while potatoes need more potassium. Knowledge about this has been obtained only by painstaking research. The Broadbalk Field at Rothamsted was divided into eighteen experimental strips in 1843. Wheat was sown in each and treated in different ways. The yields were compared each year. A summary of the main results is given in Fig. 15:15. They give convincing evidence that certain combinations of fertilisers do increase yields very considerably.

Crop rotation

In order to economise on fertilisers and to preserve the structure of the soil, it is normal practice in many places for farmers to sow crops in a particular rotation over a period of three or four years; but it is also common, now that intensive methods are used much more, for two crops to be harvested in one year. However, the basic method of rotation is to grow cereals in the first year, a root crop such as beet in the second, and crops producing a lot of nitrogen (p. 245), e.g. clover, peas, beans, in the third. Alternatively a four-year rotation might be: wheat—roots—barley or oats—clover, peas or beans. Fertiliser is added before the sowing of the root crop. Clover is usually sown with the barley or oats in the third year and harvested in the fourth; their roots are ploughed in before the next cereal crop is sown.

One advantage of a rotation is that pests, diseases and weeds tend to be prevented from establishing themselves; however, with the more recent development of pesticides and selective weedkillers, traditional patterns of growing crops are changing.

16

Food and diet

Why do we need food?

Food is a generalised term which includes all the materials taken in through the mouth which are necessary for the activity, growth and well-being of the body. We can divide these into the following categories:

1. Fuels to provide energy for movement and all the chemical processes going on in the body.
2. The raw materials for the growth of the body and the repair or replacement of tissues.
3. Substances concerned in regulating the chemical activities of the body.
4. Substances required for the health and protection of the body.

A good diet must not only include all these categories, but they should be in the right proportions for the person concerned. For man, these needs are met if the diet includes the following: carbohydrates, fats, proteins, mineral salts, vitamins and water. Our energy requirements are supplied by carbohydrates, fats and proteins; for growth and repair we need proteins, mineral salts and vitamins; for regulation and protection, mineral salts and vitamins are necessary, and water is needed for every process taking place in the body.

Carbohydrates

These provide energy when their molecules are broken down during respiration. They include starches and sugars and occur in large quantities in cereals, bread, potatoes and most root vegetables. They all have molecules composed of carbon, hydrogen and oxygen only, the hydrogen and oxygen being in the same proportion as in water, e.g. glucose sugar, $C_6H_{12}O_6$; sucrose, also called cane sugar, $C_{12}H_{22}O_{11}$; starch, $(C_6H_{10}O_5)_n$. The '*n*' outside the bracket may be any number between 300 and 1000, the figure depending on the type of starch. In other words a starch molecule is like a string of beads, each bead being a $C_6H_{10}O_5$ unit. When starch is turned into sugar by digestion, this molecule is broken down into its units and a water molecule is added to each. The reaction takes place in two stages, but can be summarised as follows:

$$(C_6H_{10}O_5)_n + nH_2O \xrightarrow{\text{enzymes}} nC_6H_{12}O_6.$$

Glycogen, a substance rather like starch, is an important fuel reserve in man; most of it is stored in the liver.

Fats

These include oils, a term used for fats which have a low melting point and hence are liquid at room temperatures.

Fats are also energy foods. Their molecules, like those of carbohydrates, consist of carbon, hydrogen and oxygen atoms only, but they differ from them in *not* having the hydrogen and oxygen atoms in the same proportion as those in water.

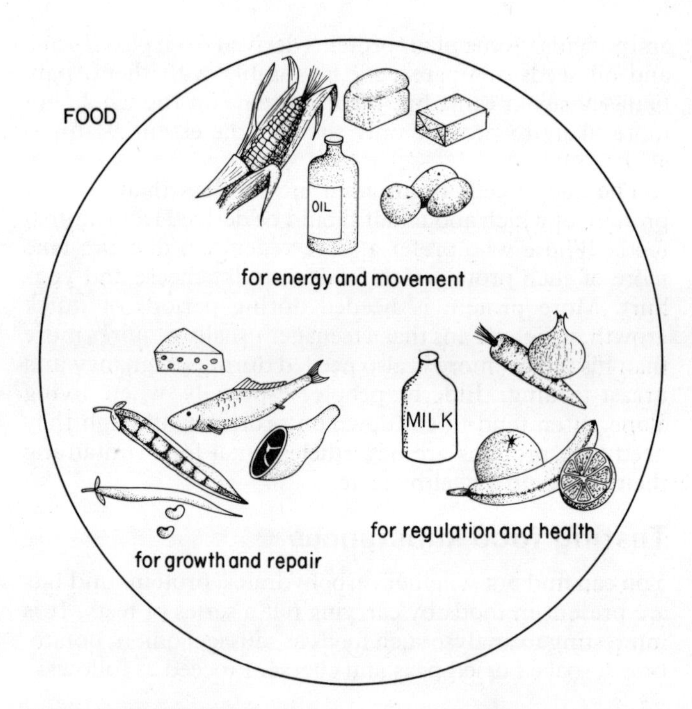

Fig. 16:1 Categories of food.

Fats are present in almost pure form in butter, margarine, olive oil and cooking fat, while meat, milk, egg yolk and nuts contain appreciable amounts.

Fat is easily stored in the body under the skin and around such organs as the kidneys. When stored under the skin, it acts as an insulating layer which helps to keep in the heat of the body: the blubber of whales is an extreme example of this. Stored fat also serves as a food reserve. As such, it is particularly important for hibernating mammals such as hedgehogs and dormice as they are dependent on fat to keep them alive during the winter when food is scarce.

Fat is also important to us as it acts as a vehicle for the intake of vitamins; many vitamins are only soluble in fat.

Proteins

These can also be used as a source of energy—indeed carnivores such as lions rely largely on them for this—but their unique and essential role is for the growth and repair of the body tissues. Without protein, nothing grows.

Proteins are extraordinarily complex chemical substances, containing atoms of carbon, hydrogen, oxygen and nitrogen and often sulphur and phosphorus as well. These huge molecules are built up from smaller units called amino acids which may be combined together in various ways; theoretically there is an almost infinite number of possible proteins that could be formed from them. Every animal species synthesises its own particular kinds of proteins from the amino acids it obtains from its food. Our own body requires about twenty amino acids; at least eight of these must be obtained from proteins in the diet—the remainder can be synthesised. Proteins derived from animal sources such as lean meat, fish, cheese, eggs and milk are extremely good sources of these eight essential amino acids. Some plant proteins derived from peas, beans and oil seeds compare very favourably with them, particularly soya beans, but plant proteins on the whole are more likely to provide only some of the essential amino acids.

The daily needs of an adult are not less than 70 g of protein, of which about half should be derived from animal foods. Those who prefer a more vegetarian diet can take more of such protein-rich foods as milk, cheese and yoghurt. More protein is needed during periods of quick growth, which means that a teenager usually requires more than his father; more is also needed during pregnancy and breast-feeding. Elderly people, especially when living alone, often tend to cut down on proteins; although they need less, proteins are nevertheless vital for maintaining their tissues in a healthy state.

Testing food substances

You can find out whether carbohydrates, proteins and fats are present in foods by carrying out a series of tests. It is interesting to analyse such foods as sausage, onion, potato, bread, soaked dried peas and cheese. Proceed as follows:

1. Grind up the particular food to be analysed with a pestle and mortar. Put a little of the product into a clean, dry test tube, add 2 cm^3 of ethanol and shake. If fat is present some of it will dissolve in the ethanol. Allow the remaining contents of the tube to settle, decant off some of the clear liquid into another test tube and add about 2 cm^3 of cold water. It will go milky if *fat* is present; fat is insoluble in water and so it comes out of solution as fine droplets, known as an **emulsion**.
2. Take the remainder of the ground-up food and mix it with water; some substances will dissolve, the rest will be suspended. Use this mixture as stock material, taking about 2 cm^3 of it for each of the following tests:

a) Add a few drops of iodine dissolved in a solution of potassium iodide. A blue-black colour indicates that *starch* is present.

b) Boil the sample for one minute with 2 cm^3 of Benedict's solution in a boiling tube; move the tube in and out of the flame carefully so the liquid does not spurt. A colour change from blue to green to orange and finally the formation of a brick red precipitate, indicates the presence of a *reducing sugar*, e.g. glucose. If only a slight change of colour occurs, reducing sugar *is* present, but only in very small amounts.

c) If *no* reducing sugar is present, boil a similar fresh sample for two minutes with three drops of dilute hydrochloric acid in a boiling tube, cool the tube under a tap, then add *small* quantities of solid sodium bicarbonate until the fizzing stops (this indicates that the acid is neutralised). Add 2 cm^3 of Benedict's solution and boil. A colour change as in b) indicates the presence of a *non-reducing sugar*, e.g. sucrose.

d) Add a few drops of Millon's reagent and heat gently; a pink or red colour indicates the presence of *protein*. This reagent is poisonous so it should be handled with care.

Compile a table showing which substances are present in the various food items analysed by the whole class.

Minerals

Mineral salts are essential because they provide many of the elements needed for growth, protection and the regulation of metabolic processes. Of these, elements such as sodium, potassium, calcium, phosphorus and iron are required in appreciable quantities.

Sodium, in the form of sodium chloride, is necessary to maintain the right osmotic balance of body fluids such as blood, hence enough has to be taken in each day to replace what is lost in sweat and urine.

Calcium is needed for muscle contraction and the formation of strong bones and teeth (along with phosphorus, magnesium and vitamin D). It is especially important during pregnancy as more is then required for the growth of the foetus. If insufficient calcium is present in the mother's blood at this time, the blood will extract some from the mother's bones and teeth which will suffer in consequence. A high calcium intake is also necessary when breast-feeding because calcium is an essential ingredient of milk.

Phosphorus is needed for bone and teeth formation, but in addition it is necessary for the adequate absorption of calcium from the gut. It is also an essential ingredient of ATP (p. 90), the substance used for transferring energy during the respiratory process.

Iron is essential for making **haemoglobin**, the red, oxygen-carrying pigment in the blood corpuscles. If iron supplies are inadequate **anaemia** results. Anaemia is a condition in which the haemoglobin content of the blood is below the normal for good health; the person becomes pale and lethargic and may have periods of breathlessness and dizziness.

Iodine is only needed in very small amounts, but it is essential because it is the active component of thyroxine, a growth-controlling substance formed in the thyroid gland.

In addition there are certain essential **trace elements**, so called because they are only needed in extremely small amounts; these include fluorine, copper, manganese and cobalt. Their importance can be illustrated by fluorine. In some places fluorine is present naturally in drinking water, but in others it is absent or insufficient. It has now been established by carefully controlled experiments that by adding fluorine to the water supply at a concentration of two parts per million, dental decay, especially in school children, is very greatly reduced. This is because the fluorine has a hardening effect on the enamel of the teeth. In Slough, for example, when the water supply was treated in this way the percentage of children with perfect teeth rose from 5 to 30%, and in South Shields dental decay was reduced by 44% compared with North Shields where no fluorine had been added to the water.

The main sources from which we obtain these minerals are given in Table 16:1.

Vitamins

If rats are fed on a diet of pure carbohydrate, protein, fat, mineral salts and water they will quickly die, but if a few cm^3 of milk are added daily they will thrive. This is because milk contains **vitamins**. Vitamins are organic substances which are found in a great variety of foods and are essential for life and health; they are required only in extremely small amounts.

The importance of vitamins for the health and protection of the body was realised long before their chemical composition was known, so for convenience they were called by letters of the alphabet. Today they are often referred to by their chemical names, e.g. vitamin C is ascorbic acid. Vitamins are important substances because

TABLE 16:1. MINERALS

Element	*Good sources of supply*	*Importance*
Calcium	Milk and cheese especially, also green vegetables and eggs.	For the development of strong bones and teeth. For muscle contraction. For clotting blood.
Phosphorus	Meat, fish, eggs. Wholemeal bread.	For the development of strong bones and teeth. For the absorption of calcium from the gut. For making ATP.
Iron	Liver, meat, eggs. Wholemeal bread. Water-cress, spinach and other green vegetables.	For making the haemoglobin in the red corpuscles of the blood.
Iodine	Sea foods, sea salt and common table salt.	For making thyroxine which is essential for growth.
Fluorine	Drinking water in some regions only.	Hardens the enamel of teeth and helps to prevent dental decay.
Sodium	Most foods.	For maintaining the osmotic balance of body fluids.
Potassium	Most foods.	For normal cell function and nerve action.

TABLE 16:2. VITAMINS

Vitamin	*Principal food sources*	*Functions*	*Deficiency effects*	*Other points*
Fat soluble				
A	Milk, butter, eggs, fish-liver oils; green vegetables, tomatoes, carrots. (These plants provide carotene from which vitamin A is synthesised.)	Keeps surface and lining tissues healthy. Essential for normal growth.	Infection of eyes, nose, throat and skin. Night blindness (inability to see in dim light). Slows down growth.	Can be stored in the body. Not destroyed by cooking.
D	Fish-liver oils, milk, butter, eggs. (Also formed in the skin by the action of sunlight.)	Concerned with the uptake of calcium and phosphorus.	Rickets, a disease found mainly in children, where bones fail to harden properly and consequently bend.	The vitamin is found in several forms. Rickets occurs mostly in children who receive little sunshine, e.g. in large cities.
E	Wholemeal wheat, eggs, butter.	Importance to man uncertain.	Sterility in some mammals.	Seldom deficient.
Water soluble				
Vitamin B complex:				
Thiamine (B_1)	Yeast, wholemeal bread, liver, peas and beans.	Essential for growth and general health.	Beri-beri, a nervous disease.	Beri-beri common in countries where polished rice is the main cereal in the diet.
Riboflavine (B_2)	Yeast, wholemeal bread, meat, milk, eggs, peas, green vegetables.	Keeps skin healthy.	Dermatitis (skin disease).	
Nicotinic acid	Yeast, liver, eggs, milk, cereals.	Concerned with health of skin, digestive and nervous systems.	Pellagra (a disease affecting the skin, alimentary canal and nervous system).	Pellagra common where maize is the main cereal in the diet; can be reduced through adding wheat flour.
B12	Liver.	Used in formation of red corpuscles.	Pernicious anaemia (deficiency of red blood corpuscles).	
C (Ascorbic acid)	Blackcurrants, oranges, lemons, grapefruit, tomatoes, fresh green vegetables.	Essential for health of skin.	Scurvy (bleeding from gums and under skin, weakness, poor healing quality of bones).	Easily destroyed by cooking.

many of them are involved with enzymes in vital metabolic processes.

Most vitamins are synthesised by plants (including bacteria), although mammals can partially synthesise vitamins A and D from substances of plant origin. We obtain vitamins either directly by eating plants, or indirectly from animals which have acquired them from plants.

Deficiency diseases

Diseases caused by lack of vitamins are called **deficiency diseases** to distinguish them from diseases caused by viruses or bacteria. Scurvy, beri-beri, pellagra and rickets are examples. Scurvy was a common disease among sailors in the days of long sea voyages, but **Sir Richard Hawkins**, as long ago as 1593, realised that the best treatment for scurvy was eating oranges or lemons, although it was not until the twentieth century that these were known to be effective because they contained vitamin C.

Table 16:2 summarises the more important facts about the various vitamins. Examine it carefully and note particularly the best sources of supply, because these should be present in a good diet. It is now a usual practice to enrich certain foods by adding vitamins; you can read about this on such things as breakfast cereal packets.

Gross vitamin deficiency is becoming unusual in countries where there is a high standard of living and the people have good mixed diets, but it has by no means been completely eradicated. For example, rickets is still prevalent in some northern industrial cities amongst populations who have immigrated from sunnier climates. Deficiency diseases are also still quite common amongst the very poor and especially in aged people who live alone. In many parts of the world, vitamin deficiency is a very serious form of malnutrition (p. 116).

Testing for vitamins

Most of the vitamins are very difficult to test for, but vitamin C (ascorbic acid) decolourises a blue dye called dichlorophenol-indophenol, DCPIP for short. This is because ascorbic acid is a powerful reducing agent. Try this test for yourself:

1. Place 1 cm^3 of freshly prepared DCPIP solution in a test tube. Take up 1 cm^3 of a 0·1% solution of pure ascorbic acid (made up from a tablet) in a syringe and add it drop by drop without shaking the solution until it is decolourised.
2. Use this test to see whether ascorbic acid is present in a variety of foods, provided they are not strongly coloured. Include both raw and cooked fruits and vegetables. Grind up each type of food with a pestle and mortar and shake with water. Any ascorbic acid should go into solution. Add this extract (filtered if necessary) to 1 cm^3 of DCPIP as before and note whether it becomes decolourised or not.
3. Summarise the results from the class by making a table showing which foods contain ascorbic acid. Did cooking the food make any difference?

Water

Water is an essential part of our diet for many reasons: digestion and absorption of food could not occur without it; it is necessary for transporting material within the body; all metabolic processes take place in a watery medium and water is a necessary raw material in many chemical reactions in cells.

Our bodies consist of up to 76% water by weight and, on average, a person loses about 3000 cm^3 of water every day as urine and through evaporation from the skin and lungs. This water must be replaced by what we drink and by the water in our food.

Roughage (dietary fibre)

This is the term used for the indigestible material taken in with the food; it consists largely of cellulose and lignin from plant foods. Although not strictly a nutrient, roughage is a necessary ingredient of food, because by adding bulk it stimulates the muscles of the gut wall to contract more vigorously, thus keeping the food on the move and preventing **constipation**. Constipation is the retention of the faeces within the gut for longer than normal, a condition which, if habitual, may lead to ill-health. Much 'processed' or 'purified' food lacks roughage, but fruits, salads, vegetables and wholemeal bread contain a good supply.

Fuel requirements

The amount of energy released from food can be estimated using the apparatus shown in Fig. 16:2. A known weight of the food is burnt in a current of oxygen and the heat given off is absorbed by a known volume of water. From the rise

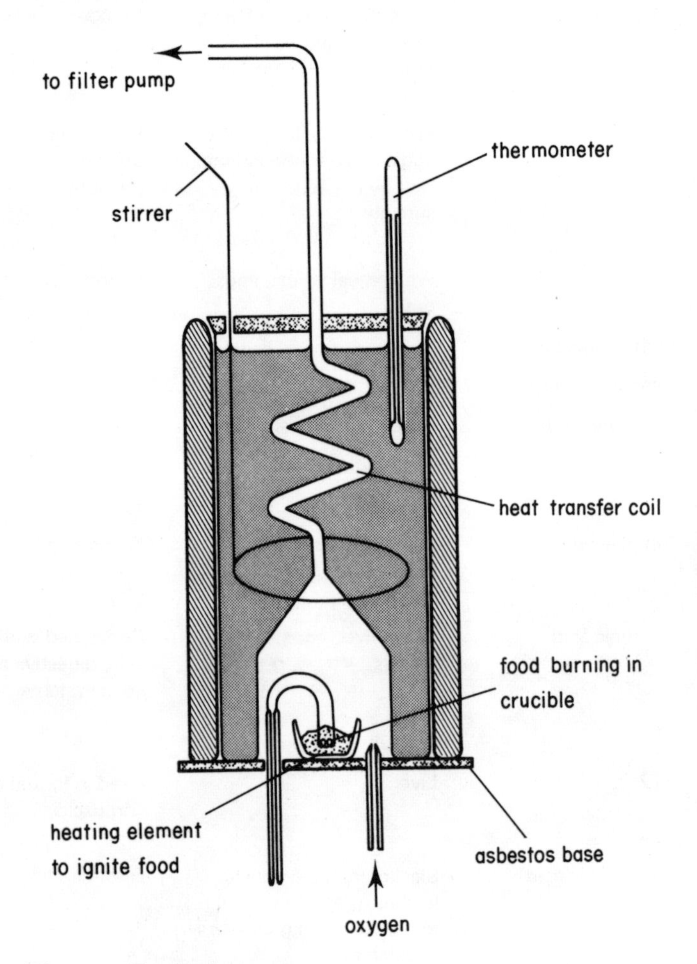

Fig. 16:2 Food calorimeter.

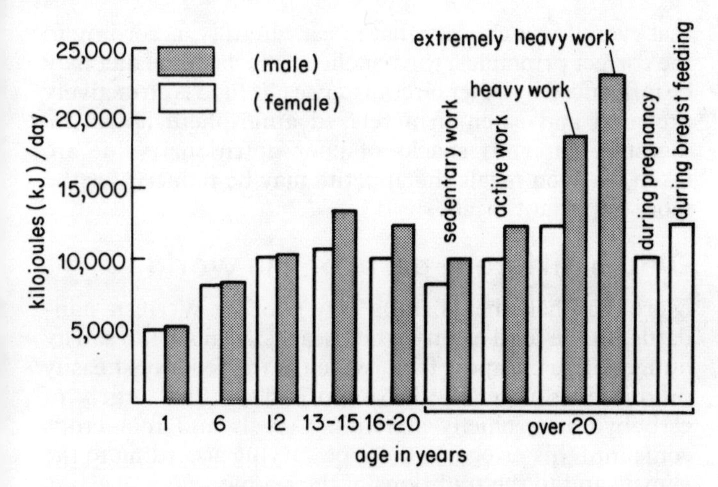

Fig. 16:3 Human energy requirements per day.

in temperature of the water, the energy released from the food can be calculated. Previously this energy value was expressed in kilocalories (1 kcal being the amount of energy required to raise 1 kg of water through 1°C); it is now expressed in kilojoules (kJ). The conversion is 1 kcal = 4·19 kJ.

It has been calculated that a man needs on average to eat enough food to provide him with about 7100 kJ of energy daily just to keep him alive. This is used for all the metabolic processes going on in the body and for keeping his temperature constant. The amount required over and above this figure varies considerably according to his activity (and hence his occupation) and his age. On average, women require about 10% less than men, but more is needed during pregnancy and when breast feeding a baby. Fig. 16:3 gives some approximate comparisons.

Why do you think more energy is needed during the period of adolescence (13–15)? What difference do you think climate would make to these figures? Consider the surface area of the body in terms of heat loss (and therefore more energy needed to replace it); how would the energy needs of men of the same age differ according to height and build?

All the energy a person needs is provided by the carbohydrates, fats and proteins that are eaten; if insufficient food is taken in, a person will lose weight because his reserve foods will be used up and even his tissues will be used as fuel for oxidation. If he eats too much he may put on weight. In Britain the average person eats about 9% more energy food than is necessary! People in different countries tend to obtain the energy they require from different foods according to availability and climate. Until recently, Eskimos, for example, ate far less carbohydrate and vastly more fat and protein than people in other parts of the world. Why was this?

Finding out the energy value of a meal

To do this you have to calculate the total weights of carbohydrates, fats and protein. Try this on a simple meal, e.g. fish and chips, or a picnic lunch of ham or cheese sandwiches.

TABLE 16:3. PERCENTAGE COMPOSITION BY WEIGHT AND THE ENERGY CONTENT OF VARIOUS FOODS

Food		*Percentage composition of*				*Energy (kJ/g or cm³)*
		Protein	*Carbohydrate*	*Fat*	*Water*	
MEAT AND POULTRY	Beef	23·5	—	20·4	54·8	12·0
	Roast mutton	25·0	—	22·6	50·9	13·1
	Bacon	9·9	—	67·4	18·8	27·9
	Pork	22·5	—	21·0	49·2	12·0
	Chicken	21·5	—	2·5	74·8	4·6
FISH	Cod	16·5	—	0·4	82·6	2·9
	Herring	19·5	—	7·1	72·5	6·1
	Salmon	22·0	—	12·8	64·6	8·8
	Trout	19·2	—	2·1	77·8	4·1
FATS	Butter	1·0	—	85·0	11·0	33·2
	Lard	—	—	100·0	—	38·8
DAIRY PRODUCE	Cheese	29·6	—	38·2	28·6	20·0
	Milk	3·3	5·0	4·0	87·0	3·0
	Eggs (boiled)	13·2	—	12·0	73·2	7·1
FRUIT	Apples	0·4	14·2	0·5	84·6	2·7
	Bananas	1·3	22·0	0·6	75·3	4·2
	Oranges	0·8	11·6	0·2	86·9	2·2
	Grapefruit	0·6	5·7	0·1	93·6	1·1
VEGETABLES	Potatoes	2·5	20·9	0·1	75·5	4·0
	Beans (dried)	22·5	59·6	1·8	12·6	14·8
	Peas (dried)	25·0	60·0	1·0	11·0	14·0
	Cabbage	1·6	5·6	0·3	91·5	1·3
	Lettuce	1·2	2·9	0·3	94·7	0·9
MISCELLANEOUS	Flour (white)	11·4	75·6	1·0	11·5	15·3
	Bread (white)	9·3	52·7	1·2	35·6	11·1
	Biscuits	9·0	70·0	10·0	11·0	18·0
	Honey	4·0	81·0	—	15·0	13·0
	Sugar (white)	—	100·0	—	—	17·1
	Chocolate	12·9	30·3	48·7	5·9	26·4

1. Weigh each item on the menu separately, e.g. the bread, butter and cheese from the sandwiches.
2. Calculate the weight of carbohydrate, fat and protein in each item by referring to Table 16:3.
3. Calculate the total weight of each class of food in the full meal and work out the energy values on the basis that: 1 g of carbohydrate releases 17·2 kJ of heat energy; 1 g of fat releases 38·5 kJ and 1 g of protein, 22·2 kJ.

Summary of the essential requirements of a good diet

1. It should contain enough energy food to meet the requirements of the person concerned.
2. It should contain enough protein to provide the right amino acids for the growth and repair of the body tissues. (These two requirements are met daily by eating about 70 g of protein, 100 g of fat and 500 g of carbohydrate.)
3. It should be varied to provide adequate salts and vitamins.
4. It should contain enough roughage to stimulate vigorous peristalsis.

A day's menu

Work out a good balanced menu for yourself to suit your own requirements. Refer to Table 16:3 and keep in mind the foods which provide the minerals (p. 112) and vitamins (p. 112). As a guide, here are some of the best foods arranged in categories; ideally, a good diet would contain daily something from each group.
1. Meat, fish, cheese, egg, beans, peas.
2. Butter, margarine.
3. Milk.
4. Potatoes, root vegetables, onions.
5. Cereals, bread.
6. Green vegetables (cooked or raw).
7. Tomatoes, oranges, or other fruit.

When considering diet it is also important to remember that even if you choose what to eat carefully, according to the correct principles, the beneficial effects of the diet may be influenced by other circumstances. If food is attractively prepared and eaten in a relaxed atmosphere it will be digested better. If snacks of poor nutritional value are taken between meals the appetite may be reduced for the more important meals.

Diet in different parts of the world

More than half the world is very poor by Western standards, so the food eaten is the cheapest available to satisfy hunger. The cheapest food, which is the food most easily produced, is vegetable in origin and consists largely of carbohydrates, chiefly starches. Cereals and root crops come into this category, the type varying according to the climate and to the traditions of the people.

In many parts of the world cereals and root crops are just about all that are eaten with the result that there is a gross deficiency of protein (especially of animal origin), vitamins and mineral salts; in other words it is not a *balanced* diet. When any of these three vital constituents are lacking a state of **malnutrition** exists. It is estimated that more than a quarter of the world population suffers in this way.

As a result of chronic malnutrition in children it is thought that some 300 million have grossly retarded physical growth, and through lack of vitamin A 300,000 children lose their sight every year. There are also millions of adults who exist on a diet of less than 8,400 kJ per day, barely sufficient to give them the energy to work (Fig. 16:3).

Vast numbers do not even get enough food for their minimum needs and are actually wasting away; this is **starvation**. The World Health Organisation makes the extremely conservative estimate that at least 60 million in Asia and Africa (more than the whole population of Britain) die of starvation every year due to droughts, failure of crops, poverty, wars and their attendant refugee problems; in some years vastly more.

Protein deficiency is common in India, Tropical Africa, the Nile basin, and parts of Central and South America, to

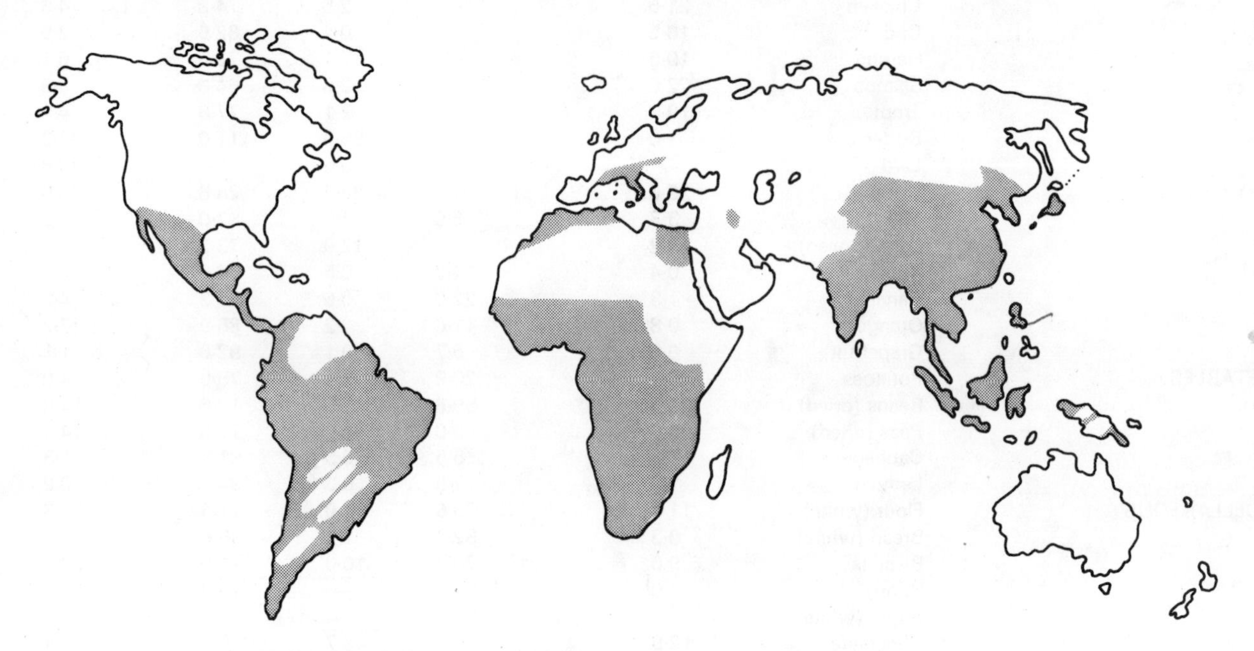

Fig. 16:4 Map showing regions where kwashiorkor occurs (shaded areas).

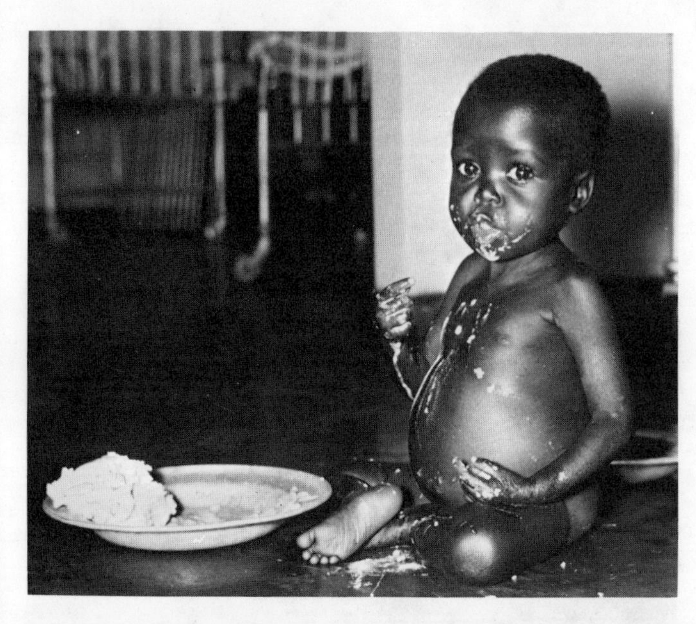

Fig. 16:5 Child suffering from kwashiorkor feeding on high protein diet in hospital.

name some of the more important regions. These are often areas of very dense population and deficiency may be due to a) dependence on crops which contain very little protein such as cassava and bananas, b) low production of domestic animals or fish, c) poverty. Protein is always expensive to buy, especially animal protein.

Protein deficiency in children is the cause of **kwashiorkor**, a horrible disease associated with stunted development, wasting of the muscles and gross swelling of the tissues due to their water-logged condition. The disease often develops in the child's second year, because when breast-fed he receives enough protein, but when another baby is born he is weaned to a diet consisting mainly of carbohydrate, and the symptoms begin to appear. In some parts of Tropical Africa up to 10% of the children suffer in this way.

This should be contrasted with the situation in many of the richer nations where over-eating causes more diseases than under-eating. As income increases, so the intake of animal protein becomes larger and excess energy food is laid down as fat. However, obesity (the state of being fat) is not always entirely due to excess feeding, as some people accumulate fat much more readily than others, but being sensible about diet is important because fatness brings great disadvantages. Fat people are less healthy and have a shorter life expectancy; the more weight there is to carry about, the more strain on the heart and the greater likelihood of high blood pressure and heart attacks. Fatness limits activity and good exercise is a basic need for health. Nobody likes being fat and very few need be. Twenty per cent of people in the United Kingdom are overweight, partly because they eat too much and partly because they eat the wrong kind of foods. Many older people become fat largely because they continue to eat the same quantity as when they were much more active. However, where there is sufficient determination and self-control, it is possible to prevent, or at least reduce, the condition. To slim, it is necessary to eat smaller amounts of high energy foods, especially those containing starches and sugars, but it is

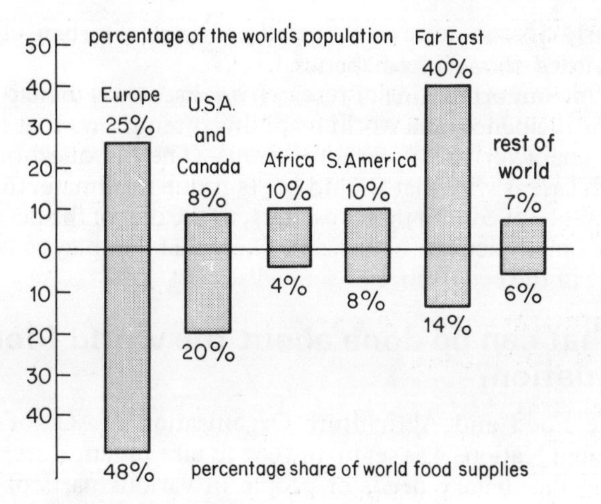

Fig. 16:6 How the world's food is shared.

essential to maintain a properly *balanced* diet. You could work out the effect of cutting out snacks such as sweets, biscuits, cakes, ice cream and potato crisps! (See also the section on p. 131.)

Intake of sugar alone is a very important factor in the diet of modern man. In the United Kingdom the average consumption of cane sugar has gone up from 6·75 kg to 54 kg per head per year since the early part of the last century due to more sugar beet being grown. This has not only contributed to obesity and its attendant ills, but has had various physiological effects which have led to an increase in such conditions as diabetes (p. 163), stomach ulcers and dental decay (p. 124). The following is an example of the type of evidence which suggests this association of disease with excess cane sugar:

a) Sugar consumption among Natal Indians is nine times higher than among Indians living in India, and the incidence of diabetes is more than ten times as high in the former.

b) In *rural* districts of Natal it has been estimated that Zulus consume 2·7 kg of sugar a year compared with 38·25 kg in *cities* such as Durban where Zulus eat a more Western-style diet. In a large hospital which serves the rural districts there were only ten cases of diabetes out of

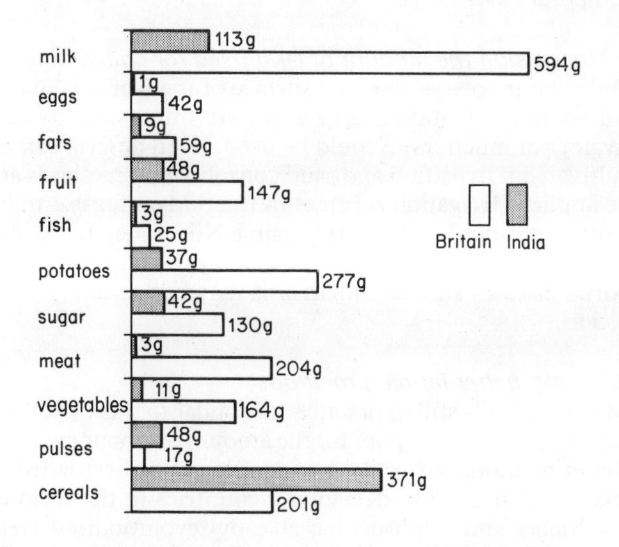

Fig. 16:7 Comparison of average Indian and British diets.

nearly 10,000 admissions, while a single Durban clinic recorded 1600 Zulu diabetics.

One important line of research on diseases is to map out their incidence on a world map; this often shows that they are characteristic of certain regions. The big question in each case is why that should be. Is it due to climate, to the presence of certain insect carriers, to the diet of the people, or to any other cause? It seems likely that diet plays a large part in the occurrence of some diseases.

What can be done about the world food situation?

The Food and Agriculture Organisation (FAO) of the United Nations was set up in 1962 to take practical steps to meet the dietary needs of people in various parts of the world. In addition many voluntary organisations have made very valuable contributions towards alleviating the problem. But basically the situation is a human one—do the peoples of the world really want the problem solved deeply enough and, if so, are they willing to make sufficient sacrifices to bring it about? The scientific know-how is available, but the cost of food production is escalating with the increasing cost of energy, so that the gap between rich and poor countries is widening. In spite of considerable aid from the more affluent countries, the poorer ones do not have the means to take the greatest advantage of what is available. It should be realised that when a government supplies aid to a developing country it is seldom a gift. Usually it is a loan involving conditions over the exports and imports of the country concerned. Only a more equal sharing of the world's finite resources will really solve the problem. The following are some of the helpful lines being pursued:

Fig. 16:8 Irrigating a sugar plantation, E. Africa.

1. Distributing food from places of plenty to places of need

It seems obvious that if there is a surplus in one place it should be sent to regions where it is needed. This applies particularly to products such as grain and dried milk which are more easily transported without loss of quality. Surplus cereals from the United States, Canada and Australia have been of great benefit in this respect, but this only provides temporary assistance.

2. Increasing the amount of land used for cultivation

Only about 10% of the land surface of the world is under cultivation. Obviously large areas are impossible to cultivate, but much more could be used. Even deserts can be cultivated if irrigation and sand consolidation methods are be applied. Irrigation schemes in the Sudan, for example, have been highly successful, using Nile water for wide-scale irrigation, but the cost is a major problem and water-borne diseases such as bilharziasis may be spread by such action.

3. Using better farming methods

Many methods still in practice are similar to those used for centuries. Yields are poor for the amount of labour exerted. Better methods are available if people can be persuaded to use them, and aid to developing countries in the form of machinery and fertilisers has already revolutionised yield in some parts. But advanced machinery is not necessarily

Fig. 16:9 Combine-harvesters at work.

the answer as this tends to produce unemployment; it is often better to use human labour to better effect. Mixed farms are more productive than those that specialise in one product, such as cereals, as in the former more animal food is produced and more manure becomes available to put back on the land for the growing of crops.

4. *Using new varieties of plant and animal species*

By crossing different strains and selecting useful varieties, new types of cereals, fruits, vegetables and domestic animals have been produced which have increased yields enormously and provided varieties suitable for growing or rearing under the very varied conditions found in different parts of the world. For example, by using new varieties of wheat in Pakistan between 1966–71 the annual yield increased from 4 to 8 million tonnes and in India during the same five years it rose from 12 to 20 million tonnes. Rice and maize varieties have been improved in a similar manner.

However, greater-yielding varieties require larger quantities of increasingly expensive fertilisers. Control of seed quality is also needed to ensure that the farmer gets the best available as yield may otherwise deteriorate. Research is also going on all the time to provide new disease- and pest-resistant varieties to replace older ones that have lost their resistance.

Similarly with animal breeding, varieties of sheep, pigs, cattle and poultry have been developed which, when given protein-rich diets, mature much earlier; hence more food for human consumption is produced in a given time. But again, protein-rich diets are expensive (see p. 244).

5. *Increasing supplies of protein*

a) *By increasing the fish supply.* It has been estimated that the world fish harvest could be increased by as much as 50%. Off some shores much more fishing could take place, but in the waters near highly developed countries over-fishing is already occurring and conservation measures will have to be taken to ensure a continued supply. Fish-farming is carried out successfully in some countries and could be further developed. An interesting experiment was carried out in Scotland whereby nutrients were added to the water of a sea-water loch. This caused the growth rate of plaice and flounders to increase significantly. This method could be successful in other localities where the water is more or less confined to a limited area.

Fig. 16:10 Fish farming in the Philippines.

b) *By using indigenous animals.* In some countries, notably in Africa, native animals can be used as a more profitable source of protein than the imported cattle. These animals, in contrast to cattle, have been in the country for millions of years and have become well adapted to the conditions there. They are largely immune to the local diseases, can extract more nourishment from the vegetation and need far less water. By a regular 'cropping' technique numbers can be kept at an optimum for the food available. One species of antelope, the eland, has been successfully domesticated and used for both milk and meat. In parts of East Africa, ranching of cattle is being practised on land also used by antelopes so that fuller use may be made of the vegetation as cattle and antelopes are selective feeders and do not compete unduly for the same food.

c) *By growing more leguminous crops with a high protein content.* Leguminous crops such as peas and beans need little nitrogenous fertiliser as they make their own with the help of bacteria (p. 245) using the nitrogen in the air. Some, such as the **soya bean**, *Glycine soja*, are very rich in protein and could be grown much more extensively in developing countries to improve the diet. Another species, the **winged bean**, *Psophocarpus tetragonolobus*, shows even greater promise as the green pods, leaves, seeds and tuberous roots are all edible and the protein content is extremely high, the seeds containing 34%! At present it is only grown locally in Papua New Guinea, but it is very suitable for growing in large parts of Central and South America, the Caribbean, Africa and West Asia where protein deficiency is high and the plant at present unknown.

d) *By producing artificial meat.* Synthetic meat is produced from soya beans and is already on the market; it is also being developed from a fungus. It looks like meat, costs far less, is easily transported and has a high nutritional value.

e) *By making protein-rich foodstuffs suitable for animal feed.* These can now be made from bacteria grown on methyl alcohol obtained from methane (North Sea gas) and coal. Pilot schemes have been very successful and full-scale production is now well under way which could produce hundreds of thousands of tonnes of the product every year.

6. *By enriching food*

Many basic foods have nutrients added to them to prevent malnutrition. Vitamins and calcium salts are added to bread and other vitamins to margarine. In developing countries dried skimmed milk is the commonest supplement. A biscuit made from wheat flour and ground nuts has been used on a large scale in India, Singapore and Fiji; its high protein content has greatly benefited the health of children in these places.

7. *By protecting food*

When food is stored it is often attacked by beetles, rodents, birds, moulds and bacteria. In parts of India, for example, this may reduce the food for human consumption by as much as 50%. Apart from the food actually eaten by pests, much of the remainder becomes contaminated and unfit to eat. It has been estimated that the food lost in the less developed countries after harvesting would keep 100 million people fully fed! So better methods of pest control,

Fig. 16:11 Rats do great damage to crops, especially grain. They not only eat it but contaminate it with their faeces.

storage and transportation could dramatically increase world supplies.

Finally it should be said that if those living in the affluent countries ate less meat this would release the grain used in producing that meat to countries which need it more (see also the pyramid of biomass, p. 244).

17

Feeding methods

Before food can be utilised by any animal it has to be obtained from some source outside the body and ingested, in most animals, through the mouth.

Methods of obtaining food

When we eat our food we use different tools for different substances. We would find knives unsuitable for peas and chopsticks for soup! In the same way animals feed on a great variety of food and many methods of capture and ingestion have been evolved to cope with particular problems.

Invertebrates obtain their food in many different ways. You will recall how an amoeba engulfs food particles by enclosing them by means of pseudopodia (p. 8), how a tapeworm absorbs digested food through its surface by diffusion (p. 25) and how the mouth parts of insects are adapted for biting, licking, sucking, or piercing and sucking (p. 48).

Another method of obtaining food, which is used by a great variety of animals, is known as **filter-feeding**. This method is typical of many animals which feed on very small organisms, or particles of organic matter in the water. Filter feeders are characteristically **sedentary** animals, i.e. they move very little or not at all; sponges, and bivalves such as mussels, are examples. The latter draw a current of water through their bodies by means of vast numbers of cilia and the food particles in the water are sieved off and used as food. You can observe this happening by placing *either* a live sea mussel in a small dish of sea water, *or* a freshwater mussel in fresh water and leaving it undisturbed until the two valves of the shell open, and the two siphons concerned with water flow are seen projecting between them (Fig. 17:1). If you then sprinkle some mud particles in the water you can see how they are drawn towards one of the siphons.

A single sea mussel draws at least 11·25 litres (2·5 gallons) of water through its body every day. Many mussel beds contain tens, sometimes hundreds, of thousands of individuals; so you can calculate the vast quantities of sea water that must be filtered by such mussel beds.

Although most filter-feeders are invertebrates, the whalebone whales also feed in this manner. These huge mammals feed largely on small shrimp-like crustaceans which occur in the plankton. When a whale feeds, it opens its jaws and takes in a great quantity of water containing planktonic organisms. It then nearly closes its jaws and, with the help of its tongue, forces the water out through a sieve of **baleen** or whalebone, which hangs from the edge of the upper jaw all round the mouth. It then swallows the plankton which has been sieved off (Fig. 17:2).

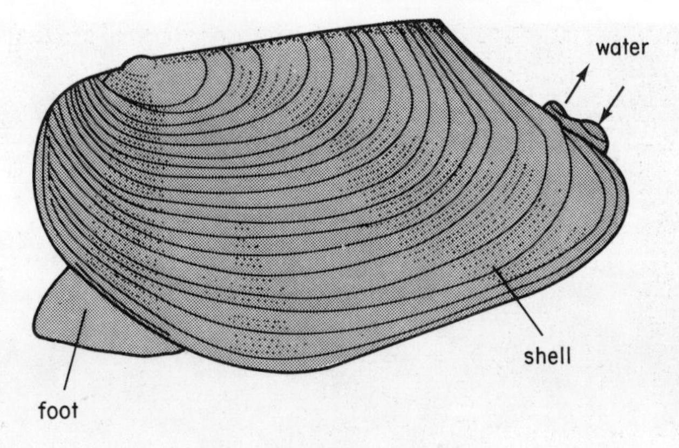

Fig. 17:1 Fresh-water mussel: a filter feeder.

Other vertebrates use many specialised methods of obtaining their food. We have seen how birds use beaks of many shapes and sizes according to the types of food eaten and the manner of obtaining it (p. 72). Mammals also vary greatly in their techniques. Consider first those which graze. If you observe sheep and cows grazing you will notice that a sheep uses its teeth to *bite* off the grass while a cow uses its tongue to *pull* it up. How would this difference determine the length of grass on which the two species could feed? What implications would this have for the farmer? The elephant also grazes, but it uses its trunk, which is an elongated nose, to grasp the vegetation and transfer it to the mouth. An elephant also uses its trunk when drinking: first it sucks water into its trunk, then it bends the trunk round until the end reaches its mouth so that it can squirt the water down its throat.

The carnivorous mammals use different techniques. Pack animals such as wild dogs and wolves run down their prey, using their teeth to bite it to death. They do not chew their food, but bolt huge pieces as quickly as possible before another member of the pack can rob them. Members of the cat family, which includes lions and leopards, leap on their prey and kill them with the help of their canine teeth; their rough tongues also enable them to rasp off the meat from the bones.

The anteater feeds almost exclusively on ants and termites and has to eat great numbers to supply its needs. Its long sticky tongue is an effective organ for catching them. You could think of many more examples of the specialised ways by which other vertebrates obtain their food.

We have already mentioned the importance of teeth in capturing prey; they are also used for preparing the food before it is swallowed. We will now study teeth in more detail.

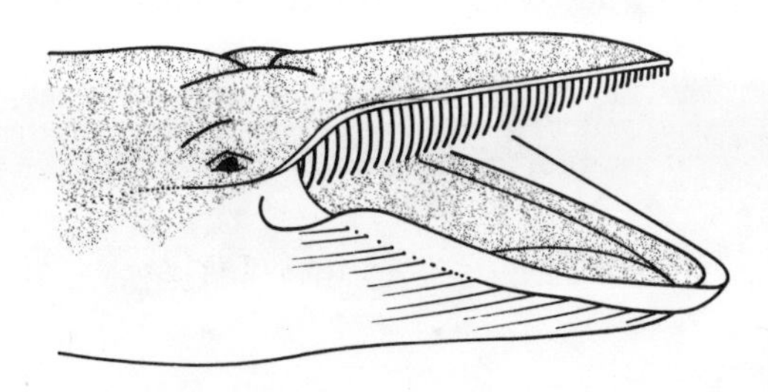

Fig. 17:2 Head of a fin whale showing fringe of baleen for sieving plankton.

Fig. 17:3 An elephant uses its trunk to grasp vegetation and put it into its mouth.

Teeth

How teeth have evolved

The teeth of all vertebrates are basically similar in structure to spines, called **dermal denticles**, found on the skin of some fish, e.g. dogfish, skates and rays (Fig. 17:4).

> Examine the dermal denticles on a dogfish. Stroke the back of the fish from the tail end forwards. The denticles give the skin a roughness resembling coarse sandpaper. Examine the mouth region. What happens to the denticles as the skin passes over the jaws?

In amphibians and reptiles the teeth have remained all alike, as in these animals they are only used for gripping the food. However, some snakes have certain teeth which are enlarged to form fangs. The snake uses these fangs to bite its prey and pass poison into it (Fig. 17:5).

Birds have no teeth as their function has been taken over by beaks, but mammals have teeth of various shapes and sizes which serve many different purposes. These range from the huge tusks of elephants, the largest known weighing over 100 kg, to the minute needle-sharp teeth of the shrews.

Structure of teeth

Examine various kinds of teeth such as those of a sheep or a dog, including one tooth which has been cut vertically.

Fig. 17:4 Section through dogfish skin showing dermal denticles.

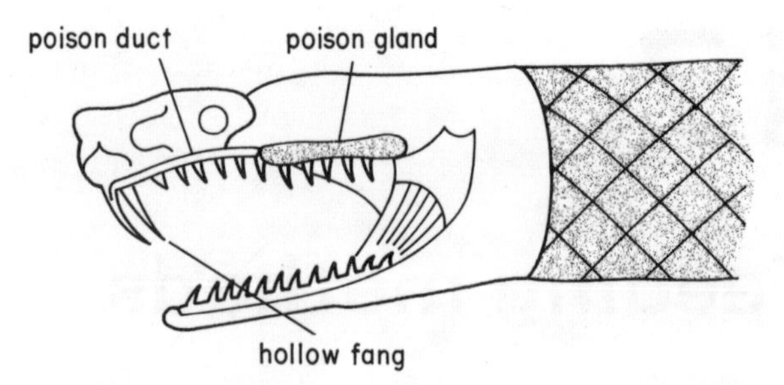

Fig. 17:5 An adder's fangs—specialised teeth for poisoning prey.

All mammalian teeth are embedded in sockets in the jaw (Fig. 17:6) and have the same essential structure. The main portion is composed of a bony material, the **dentine**, which is a living tissue. The part of the tooth which projects from the jaw is covered by a hard non-living secretion, the **enamel**. Where the tooth meets the gum, the enamel is replaced by cement which fixes the tooth firmly into the fibrous lining of the socket. Each tooth has one or more roots which are open at the base; these allow blood vessels and nerves to penetrate into the central part or **pulp cavity**. These roots almost close in most teeth, so preventing further growth, but in elephants' tusks, the incisors of rabbits and rodents, and the cheek teeth of herbivores, they remain open and allow growth to continue throughout life.

Human dentition

The term **dentition** is used to describe a complete set of teeth. We can find out something about our dentition by feeling our own teeth and looking at our neighbour's!

It is also possible to make a cast as dentists do when they are fitting false teeth or crowning a tooth. You do this by taking a piece of dental plaster big enough for all the teeth on one side of the mouth to bite upon. You then gently bite on it (*not* right through it) and when nearly set peel it off carefully. This gives you the impressions of your upper and lower teeth on the two sides of the cast.

Whatever method you use you will find that on one side of each jaw you should have two **incisors** in front. These

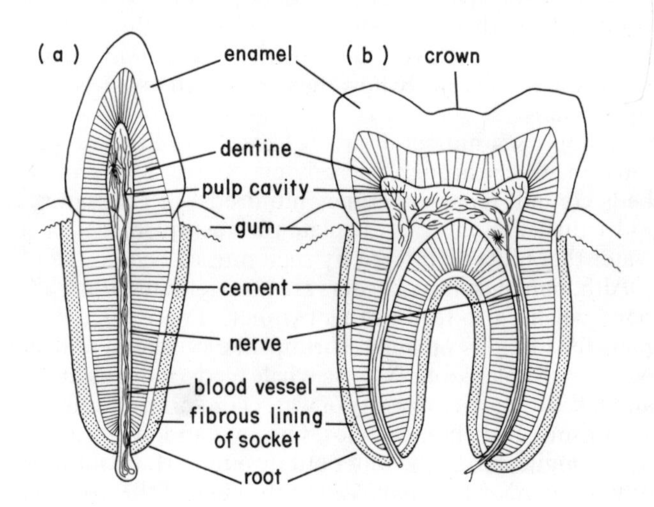

Fig. 17:6 Diagram of human teeth in longitudinal section: a) Incisor b) Molar.

have flat chisel-shaped edges. When biting an apple, for example, the top incisors slide down in front of the lower ones in a shearing action. Next comes a single **canine** which projects rather more than the incisors. This helps if you are trying to tear something tough. Behind the canines are the cheek teeth, the two **premolars** in front and the three **molars** behind. The last molar is called a wisdom tooth; you will probably not have any yet as they come much later than the others. Premolars and molars help to grind the food and their flattened surfaces come together neatly. If you put your finger between your molars and bite gently you will discover the great power they exert. The power increases the nearer the teeth are to the angle of the jaw; nutcrackers work on the same principle. You will also notice that when eating, your jaws are capable of sideways and forwards movements as well as up and down. This increases the efficiency of the various actions.

The difference between premolars and molars, apart from their position, is that during life there is only one set of molars, but there are two sets of premolars. The first set of teeth is called the **milk dentition**. At birth these teeth are already developing in the jaws, but they cannot be seen. The incisors are the first to grow through the gums and this happens when the baby is about six months old. The first premolars appear at about fifteen months, and at three years the full milk dentition, totalling 20 teeth, is functional. This consists of two incisors, one canine, and two premolars on one side of each jaw. At intervals after this the molars erupt (Fig. 17:7) and the milk teeth come out and are replaced by a second set. This complete set, totalling 32 teeth, is called the **permanent dentition** (Figs. 17:7 and 8). The details of this can be summarised by a dental formula, which represents the teeth in the upper and lower jaws on one side:

$$\underbrace{I\tfrac{2}{2}\quad C\tfrac{1}{1}\quad P\tfrac{2}{2}}_{\text{Present in milk dentition; replaced in permanent dentition.}}\quad M\tfrac{3}{3}\quad \text{i.e. 16 teeth on each side.}$$

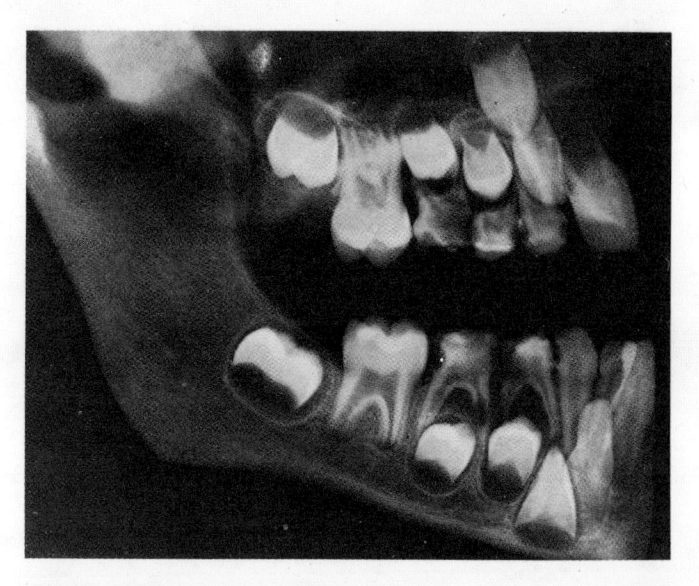

Fig. 17:7 X-ray photograph through the jaws of a 7-year old child showing permanent teeth embedded in the jaw below the milk teeth.

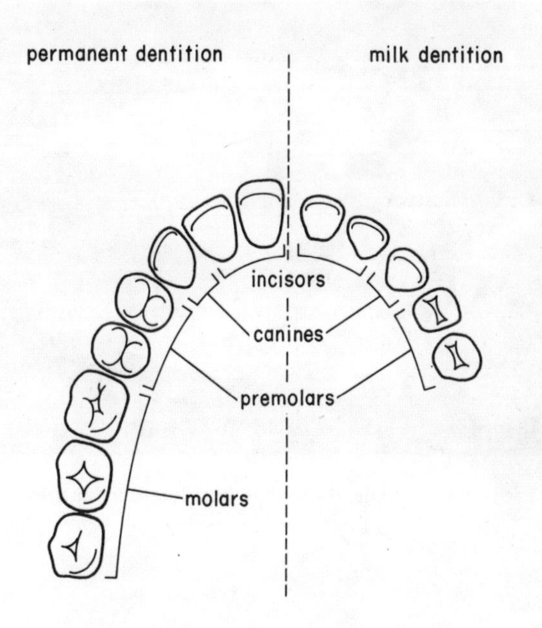

Fig. 17:8 Comparison of the permanent and milk dentition in man.

Other forms of dentition

Man is said to be an **omnivore** because he eats a great variety of food, both animal and vegetable. As we have seen, our dentition, consisting of all four kinds of teeth, is well adapted to cope with this wide range of foods. The **carnivores** which feed largely on meat, and the **herbivores** which feed on vegetation have characteristic dentitions which are especially suitable for these diets.

1. A carnivorous dentition

This is typical of lions, bears, otters, etc., but it is easier for you to study a cat or dog (Fig. 17:9).

A dog's skull has close-set incisors which are used for scraping off pieces of meat from the bone. They are also used for grooming the fur. Many species apparently derive pleasure from grooming each other.

The canines are the carnivore's chief weapons and are used for gripping the prey and may help in killing it. They are well adapted for these functions in being large, pointed in a backward direction and placed rather near the front of the mouth.

The cheek teeth are much more pointed than ours. The fourth upper premolars and the first lower molars are larger than the others and are called **carnassial** teeth. When carnassial teeth come together they act like shears to cut the flesh away; they are also used for cracking bones.

The two jaws have an up-and-down movement with little sideways play. This makes the action more effective, as it enables the incisors to meet exactly and the carnassial teeth to slide accurately against each other, to help the shearing action.

The dental formula for a dog is:

$$I\tfrac{3}{3}\quad C\tfrac{1}{1}\quad P\tfrac{4}{4}\quad M\tfrac{2}{3}.$$

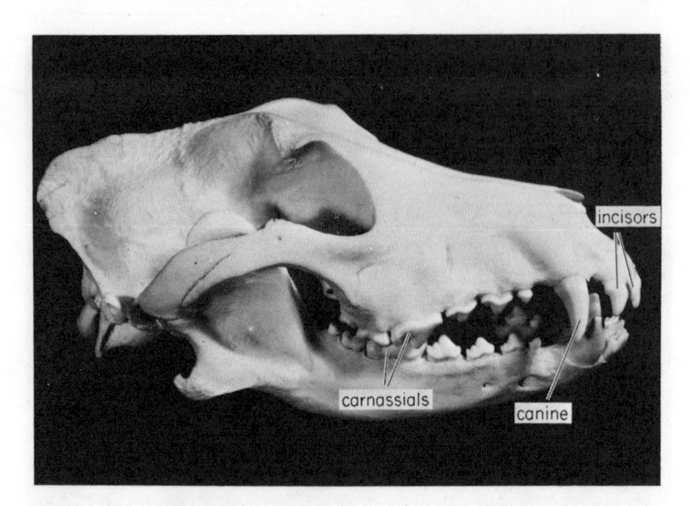

Fig. 17:9 Skull of dog showing a typical carnivorous dentition.

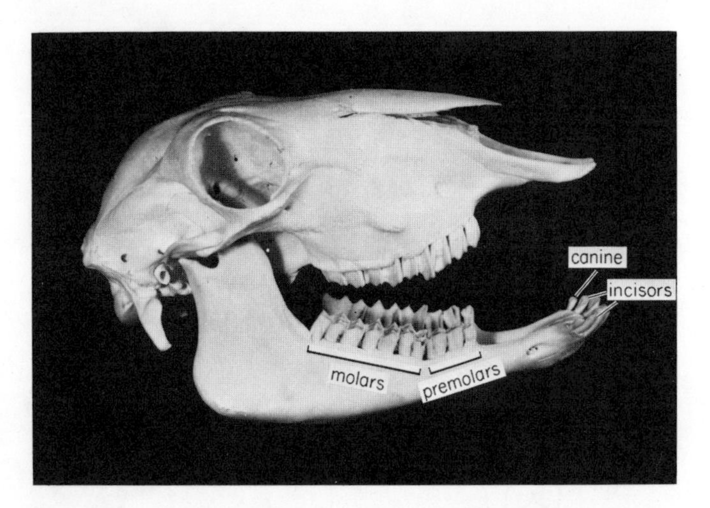

Fig. 17:10 Skull of sheep showing herbivorous dentition.

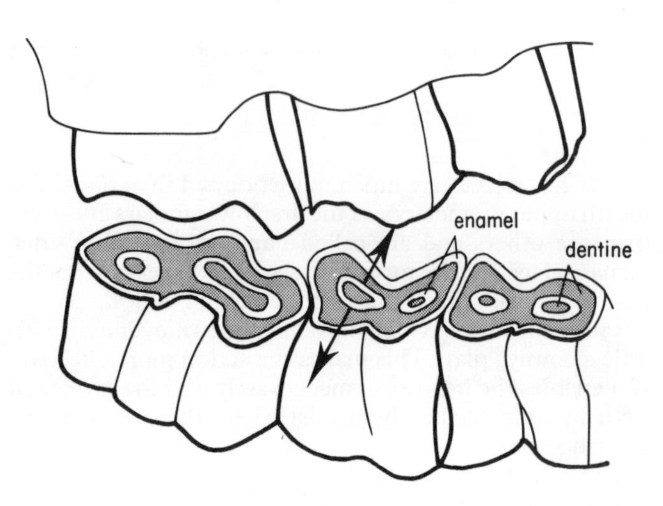

Fig. 17:11 Cheek teeth of sheep showing how the grinding surfaces fit together and move sideways.

2. A herbivorous dentition

Cattle, sheep, horses, antelopes and deer all have very typical herbivorous dentitions.

A herbivore's dentition contrasts greatly with that of a carnivore. If you look at a sheep's skull, for example, (Fig. 17:10) you will see that incisors and canines are present only in the lower jaw. In life these sharp teeth bite against a strong horny pad in the upper jaw; they are excellent for cutting tough grasses and other herbage. The large gap between the front and back teeth is the **diastema**; it allows the tongue to manipulate the grass.

All the cheek teeth are close together to form a large surface for grinding. When they first erupt they are more pointed, but quickly their crowns are worn away. As the enamel is harder than the dentine it wears more slowly and so projects to form hard ridges. Looked at from the side the upper ones have a W pattern and the lower ones an M, so that they fit exactly.

When you watch a herbivore chewing, the sideways movement of the jaw is very marked; this allows the ridged surfaces to slide on each other and pulverise the grass, a process of great importance for effective digestion.

The dental formula for a sheep is:

$$I\frac{0}{3}\ C\frac{0}{1}\ P\frac{3}{3}\ M\frac{3}{3}.$$

In other vegetarian animals such as voles, mice and rabbits the incisors have chisel edges which slide on each other; they are excellent for gnawing. As in sheep, a diastema is present, the cheek teeth are close together and their surfaces are flat and ridged for grinding. There are no canines. The dental formula for the rabbit is:

$$I\frac{2}{1}\ C\frac{0}{0}\ P\frac{3}{2}\ M\frac{3}{3}.$$

1. Examine various kinds of teeth such as those of a sheep or a dog, including one tooth which has been cut vertically.
2. Examine the skulls of such mammals as dogs, cats, sheep, rabbits, etc. according to availability. Note the adaptations of their teeth in relation to their diet as referred to in the text.

Care of the teeth

It has been said that man is unique in having three sets of dentition: milk, permanent and false! The need for false teeth and a lot of repair work could often be avoided if we treated our teeth better.

A report in 1979 stated: 'Each year $2\frac{1}{2}$ million rotten teeth were extracted from schoolchildren and 17,000 15-year-olds were fitted with false teeth. Nearly a third of children aged 5 had decay in 5 or more teeth and 3 out of 10 people over 16 were toothless'.

Dental decay is a major factor causing ill health. It starts when acid attacks the enamel, causing it to dissolve. The underlying dentine, being softer, then dissolves more quickly, allowing bacteria to enter. How is this acid formed? After a meal, food particles may lodge between the teeth or form a pulpy covering, **plaque**, especially near the base. This is particularly the case after one has eaten sugary and starchy foods. If this food is not removed, bacteria act on it and break it down into acids which then attack the teeth. If we eat such things as chocolates, sweets and ice lollies throughout the day we provide a continuous supply of sugar for the bacteria to act upon, so our teeth are liable to decay quickly. This is one very important reason why dental decay has increased so alarmingly in recent years. Eskimos and many Africans were proverbial for having almost perfect teeth, but today when many of them have changed from their traditional diets to those of a Western style with more sugary foods, the incidence of dental decay has greatly increased. In Britain dental decay is so serious that dentists have even been doing research into the possibility of sealing the grooves of newly-erupted teeth with a hard plastic, but this research is at present in a very experimental stage.

The fact that false teeth are poor substitutes for natural ones is one more good reason why it is very important to treat teeth with the greatest care. It is important to realise that bacteria which cause tooth decay are destructive when deprived of oxygen. So good dental care involves exposing as much of the mouth as possible to air. Brushing the teeth helps to do this as it not only removes the food particles and plaque on which the bacteria are feeding, but also deprives them of their protection against oxygen. Toothpaste is useful because it usually contains a mild alkali to neutralise the acid and a fine abrasive to clean the surface more effectively. Teeth should always be brushed in such a way that the gums are not pushed away from the teeth, as this exposes the weak spot below the enamel where the acids can more easily start their action.

Test the effectiveness of your particular brand of toothpaste for removing the bacterial film from teeth by carrying

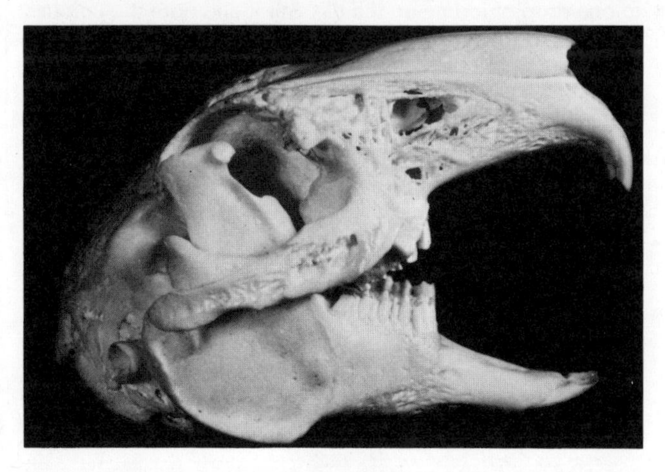

Fig. 17:12 Skull of rabbit—note the chisel-shaped incisors for gnawing.

out the following test. You will need a toothbrush and some toothpaste. The method is to use erythrocin, a dye which stains the bacterial film red.

> First rub some petroleum jelly over your lips to prevent them from being stained by the dye, then suck an erythrocin tablet. Examine your teeth in a mirror to see the extent of the staining on your teeth. Now brush your teeth thoroughly using the toothpaste and suck another tablet. The more effective the toothpaste, the less the teeth will stain the second time!

As well as brushing our teeth and avoiding eating sugary foods between meals, there are other ways of keeping teeth healthy. We have already seen that sufficient calcium should be present in the diet (p. 112), and that fluoridation of the water does much to harden the enamel and so keep the teeth healthy (p. 112). It is also important that we should have frequent, regular visits to a dentist for check-ups, as early treatment can often prevent the loss of a tooth. Babies and young children should be given hard foods to bite on as this helps to develop a good circulation to the teeth while they are growing, so aiding the conduction of the nutrients to them.

18

Making use of the food

The need for digestion

When we have eaten and swallowed our food it is still not inside our bodies in the strict sense of the word. We saw (p. 27) that earthworms were built up on a plan of a tube within a tube, the inner tube being the gut. The same applies to most of the higher animals. Before the food can be used it has to pass through the wall of the gut into the blood system which takes the food to the tissues which need it. Digestion is the process which makes absorption possible. Obviously solid foods have to be made soluble before they can pass through, but what about solutions? Do they need digesting too? It is not easy to experiment with a piece of living gut, but you could use instead a piece of cellulose Visking tubing which is rather similar to the gut regarding the substances which will pass through it. By filling this tubing with a solution of two foods, glucose sugar and starch, and placing the tubing in water, you could test the water to discover whether or not they are able to pass through the wall of the tubing. Proceed as follows:

1. Make up a solution of glucose and starch by adding 10 g of glucose to 100 cm³ of 5% starch solution.
2. Take a piece of Visking tubing a little longer than a boiling tube and knot one end after wetting it. Pour the solution into it and tie the end with cotton (Fig. 18:1). Rinse the outside of the tube with cold water to remove any solution which may have been spilt and place it in a boiling tube containing enough distilled water to surround the tube.
3. Test for the presence or absence of starch and glucose in the surrounding water immediately, and then after 5, 10 and 30 minutes. Do this a) by removing a sample with a syringe and testing for starch by placing a drop on a tile and adding iodine/potassium iodide solution, b) by boiling what remains in the syringe with Benedict's solution and testing for glucose (p. 105).

What do you conclude from your results, remembering that glucose has small molecules and starch very large ones?

Membranes such as the cellulose one used as a model gut, and those composed of living cells (like those lining the gut) allow some molecules through and not others. Digestion involves the breaking down of molecules until they are small enough to pass through. This is done in the gut by means of digestive enzymes, protein substances which act as catalysts and greatly speed up the reactions. We are able to produce many enzymes from special glands which act on the three main categories of complex food that we eat—carbohydrates, proteins and fats—and change them into soluble substances with small enough molecules to pass through the membranes of the gut. As a result, digestible carbohydrates are turned into simple **sugars**, fats into **fatty acids** and **glycerol** and proteins into **amino acids**.

Investigating a digestive process

Before describing how and where these reactions are carried out we will try out a digestive process for ourselves.

Is starch digested with the help of saliva?

When we eat starchy foods such as bread or cereals we mix them up in our mouth with saliva. Does the saliva contain an enzyme which helps to digest starch? When starch is digested, it is turned into a reducing sugar, so to find out if saliva is involved in this process, you would have to mix some starch with saliva and carry out tests to see if the starch was removed and reducing sugar appeared. But first you would have to be sure that no reducing sugar was present at the start. Proceed as follows:

1. Prepare a water bath by heating a beaker of water to 37°C (body temperature).
2. Rinse out your mouth thoroughly with water to remove any sugar. Take a mouthful of distilled water, swill it round for half a minute so as to mix it with the saliva and pass it into a boiling tube.
3. Test 2 cm³ of the saliva solution by boiling with 2 cm³ Benedict's solution to check that no sugar is present. (As a control also boil 2 cm³ Benedict's solution with water.)
4. Place 2 cm³ of saliva in a test tube and 2 cm³ of a 1% solution of pure starch into a second tube and place both in the water bath.
5. Put a series of drops of iodine/potassium iodide solution on to a white tile.
6. Mix the saliva and starch in one of the tubes, stir with a clean glass rod and immediately put a drop of the mixture on to one drop of iodine on the tile. Stir it and note the colour.
7. Clean the glass rod and repeat at $\frac{1}{2}$-minute intervals, noting any change of colour, and proceed until no further colour change takes place.
8. Finally add 2 cm³ of Benedict's solution to the remaining mixture and boil.

Is reducing sugar present? What is your final conclusion from this experiment?

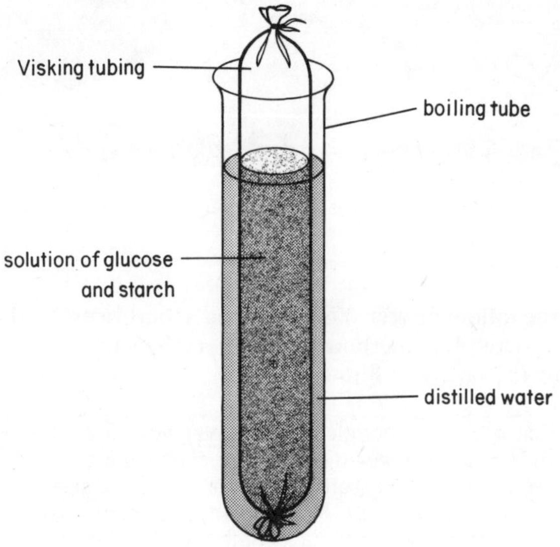

Fig. 18:1 Experiment to find out whether starch and glucose will pass through an artificial membrane.

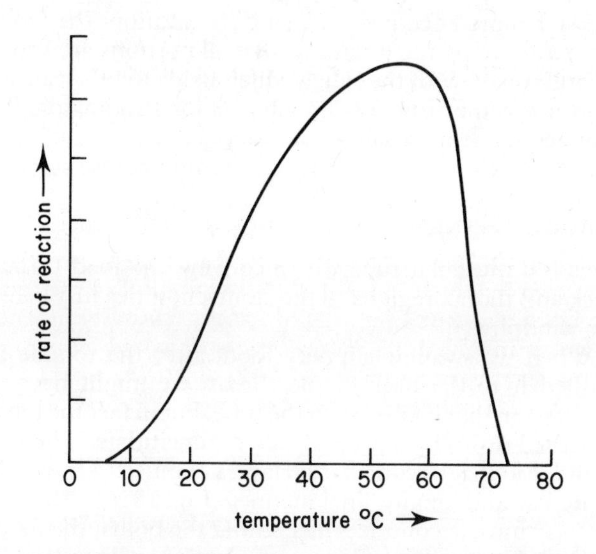

Fig. 18:2 Graph showing how the rate of enzyme action varies with the temperature.

What is the effect of temperature on the reaction?
You could find out the effect of temperature on this enzyme action by modifying the last experiment. This would involve taking equal quantities of starch and saliva and finding out how long it takes to turn the starch into reducing sugar when the mixture is kept at different temperatures. At the same time you could see the effect of boiling the saliva.

1. Take six test tubes and label them A—F.
2. Add 2 cm³ of 1% starch solution to A, B and C and 2 cm³ of saliva to D, E and F.
3. Boil the saliva in F.
4. Place A and D in a beaker of water kept at 20°C and the other four test tubes in a beaker of water kept at 37°C.
5. Allow a few minutes for the temperature to become stabilised, then add the contents of A to D, B to E and C to F, and take the time when each is mixed. Record the time taken for the mixtures to give no colour change with iodine on a tile.

What do you conclude from your results? What effect had the boiling on the saliva?

A similar kind of experiment can be used to determine the effects of acid and alkali on this enzyme action. Think out the experimental procedure you could use and the precautions you would take using dilute hydrochloric acid and sodium bicarbonate as the acid and alkali.

Digestive enzymes

We have now considered one enzyme action. In the gut there are many of these digestive catalysts which help to break down the carbohydrates, fats and proteins into soluble and absorbable products. They can be classified into three groups:

1. **Amylases** which act on carbohydrates and break them down into simple hexose sugars (those with six carbon atoms) such as glucose and fructose.
2. **Lipases** which act on fats and break them down into fatty acids and glycerol.
3. **Proteases** which act on proteins and turn them into various amino acids.

Many of these reactions involve several steps, each one being catalysed by an enzyme. For example, the enzyme that you investigated in saliva, **ptyalin**, converts starch to maltose sugar. In the small intestine another enzyme, **maltase**, converts the maltose to glucose:

$$\text{starch} \xrightarrow{\text{ptyalin}} \underset{C_{12}H_{22}O_{11}}{\text{maltose sugar}} \xrightarrow{\text{maltase}} \underset{C_6H_{12}O_6.}{\text{glucose}}$$

Note that the name of the sugar ends in *ose* and the enzyme which acts on it in *ase*.

Effect of pH and temperature on enzyme action

You should have discovered from your experiment on ptyalin that the enzyme reaction varied with temperature. Fig. 18:2 shows how the rate of the reaction varies and how the enzyme is inactivated if heated too much. From this graph determine the optimum and maximum temperatures for enzyme action and compare the optimum with your own body temperature.

Each enzyme also has its special requirements, some working best in an acid, others in an alkaline medium. As the pH (degree of acidity or alkalinity) alters considerably during the passage of food through the gut, different enzymes are needed in various parts which can work effectively under these particular conditions. Thus the gastric enzymes act in a strongly acid medium and those in the duodenum in a progressively less acid one.

Alimentary canal of man

We can now study the digestive system of man and see how

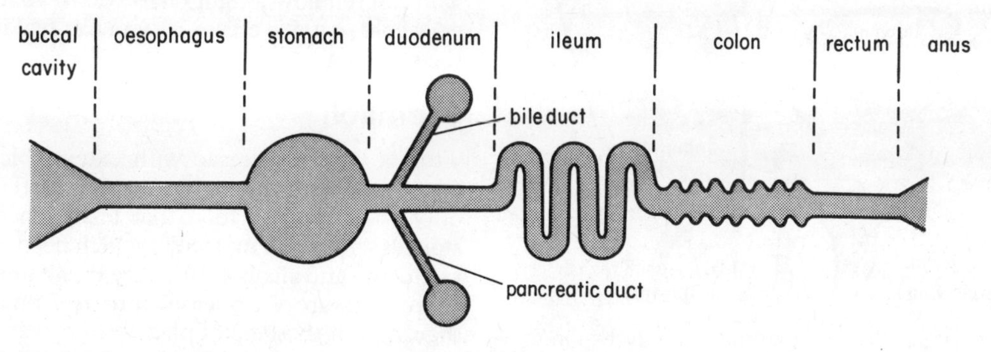

Fig. 18:3 Diagram showing how the alimentary canal is a much modified tube.

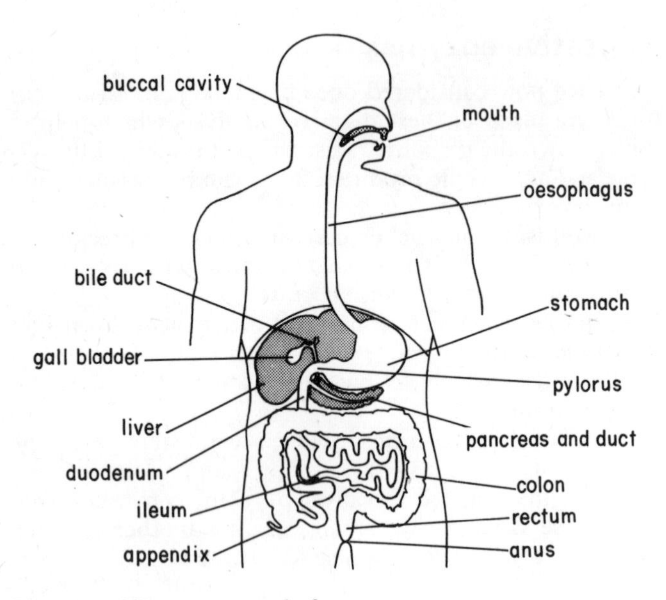

Fig. 18:4 Alimentary canal of man.

it is adapted to deal with food. It is best considered as a long tube open at both ends (mouth and anus) which is modified in different places to deal with the preparation, digestion and absorption of the food and the elimination of the remaining material or faeces. The tube is very long because digestion is a slow process and absorption is most effective where there is a large surface area in contact with the digested food.

Compare Fig. 18:3 with the more detailed diagram of the alimentary canal of man (Fig. 18:4).

Let us consider in more detail how each part of the system in man carries out its particular part in the process.

Buccal cavity

Here (Fig. 18:5) the food is broken up by the teeth, a process called **mastication**. This not only reduces the food to a convenient size for swallowing, but it also increases the surface area of the food so that enzymes can digest it more effectively. During mastication, the food is moved about by the tongue so that all portions are broken up and mixed with the saliva which helps to lubricate and bind it together into a ball or **bolus** for swallowing. The saliva comes from 3 pairs of salivary glands; it contains the enzyme ptyalin, which turns starch into maltose sugar.

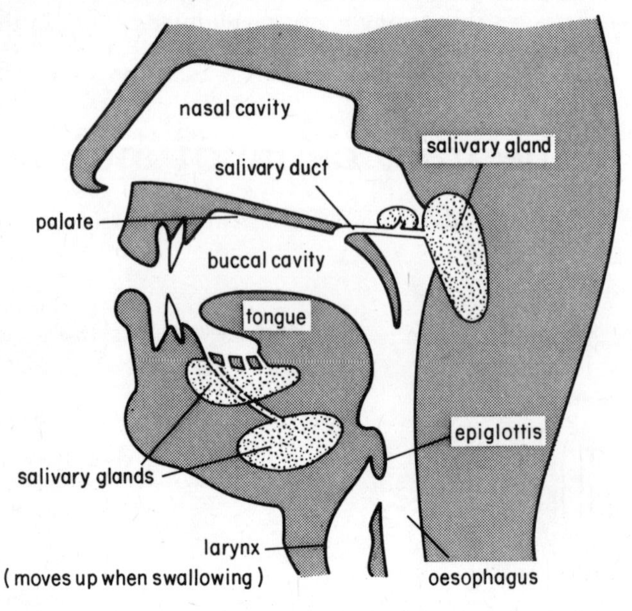

Fig. 18:5 Vertical section through the buccal cavity region of man.

Oesophagus

This is a muscular tube which conveys the food through neck and thorax regions to the stomach; it lies just behind the windpipe.

When we swallow slippery food, after the tongue has pushed it to the back of the throat, we might have the impression that it falls down the tube, but in fact the food is squeezed down by a process called **peristalsis**. The walls of the oesophagus contain two layers of muscle arranged in a circular and longitudinal manner (Fig. 18:6). When the circular muscle contracts just *behind* the bolus, the food is squeezed downwards; this occurs rhythmically, rather like squeezing a marble down a piece of rubber tubing with your fingers. When the longitudinal muscles contract *in front* of the bolus they cause the tube to widen, thus aiding the process. Peristalsis is so effective that even liquids can be swallowed when we are standing on our heads. Think what happens when a giraffe drinks; it is uphill all the way! You can tell how long it takes you to swallow if the food is very cold, e.g. ice cream, as you can feel it when it reaches the stomach.

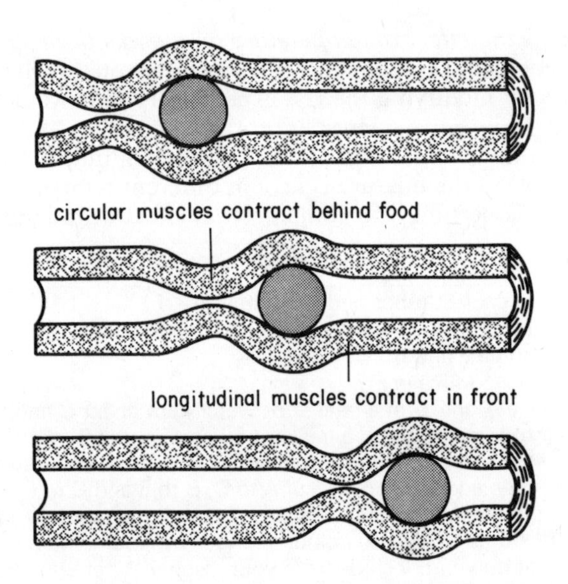

Fig. 18:6 Diagram showing the principle of peristalsis.

Stomach

This is a muscular sac with extendable walls primarily concerned with the early stages of protein digestion. The only absorption of food that takes place here is of substances with small molecules which need no digestion, such as glucose and alcohol. With the breakdown of tissue membranes, fats are also released into the stomach cavity, but no digestion of these takes place.

Both ends of the stomach are closed most of the time by

TABLE 18:1. SUMMARY OF DIGESTIVE PROCESSES (**enzymes** in bold type)

Region	*Name of secretion*	*Where produced*	*Contents of secretion*	*Action*
Buccal cavity	Saliva	3 pairs of salivary glands	(i) **Ptyalin** (salivary amylase) (ii) Water (iii) Mucin	Starch → maltose Lubricates food.
Stomach	Gastric juice	Glands in stomach wall	(i) **Pepsin** (ii) **Rennin** (iii) Hydrochloric acid	Proteins → intermediate products* Clots milk (caseinogen → casein). Provides acid medium for pepsin; kills bacteria.
Small intestine	Bile	Liver	Bile salts (no enzymes)	Emulsifies fats.
	Pancreatic juice	Pancreas	(i) **Trypsin** (ii) **Amylase** (iii) **Lipase**	Proteins → intermediate products* and some amino acids. Starch → maltose Fats → fatty acids and glycerol.
	Intestinal juice (succus entericus)	Duodenum	(i) **Sucrase** (ii) **Maltase** (iii) **Peptidases**	Sucrose → glucose and fructose. Maltose → glucose. Intermediate products formed from protein* → amino acids.

*Protein digestion occurs in stages whereby the molecule becomes progressively smaller. These intermediate products, e.g. peptones and polypeptides, are finally broken down into amino acids.

sphincter muscles; these contain fibres arranged in a circular manner which when contracted close the aperture. The most important of these is the **pylorus** which guards the entrance to the small intestine. Periodic waves of peristaltic action of the stomach wall (about every 20 seconds) cause the food to be thoroughly mixed up with the gastric juices. The food does not escape through the pylorus until it is in a semi-liquid state called **chyme**. Even when the stomach is empty muscular contractions may still occur; this may cause 'tummy rumblings' and sometimes a feeling of hunger.

Gastric juices are secreted on to the food from glands in the stomach wall, some producing hydrochloric acid and others, proteases. Of these, pepsin starts protein digestion and rennin, which is particularly important in babies, acts on the protein in milk (caseinogen) and changes it into casein, thus causing the milk to clot. The acid and pepsin do not digest the living tissues lining the stomach because a) the pepsin is secreted in an inactive form, pepsinogen, only becoming pepsin within the stomach cavity, b) there is a protective layer of mucus over the surface of both the stomach and subsequent regions of the gut. If for some reason this protective layer becomes less efficient, ulcers may result (p. 131).

The hydrochloric acid not only provides an acid medium for these enzyme actions, but it also kills bacteria and possibly parasite eggs which may enter with the food. In carnivores such as dogs and hyaenas the acid will dissolve the bones they eat.

The rate at which the stomach empties depends on the type of food being digested; a light meal in semi-liquid form may pass through the pylorus within an hour, but a heavy meal with a lot of protein and fat may remain in the stomach for three or four hours.

The small intestine

This is a long tube which varies in length according to its degree of contraction; in life it is probably about five metres long. Its two functions are to finish digestion and absorb the products. The first loop of the small intestine is the **duodenum**; it receives the bile and pancreatic ducts and is largely concerned with digestion. The very much longer hind region of the small intestine is the **ileum** which deals mainly with absorption.

When the pylorus opens, and the acid chyme from the stomach is squirted into the duodenum, three things happen:

1. The gall bladder in the liver contracts and the bile which is stored there passes down the bile duct and so into the duodenum.
2. The walls of the duodenum are stimulated to secrete into the blood a hormone (chemical messenger) called **secretin** (p. 163), which causes the pancreas to discharge pancreatic juice down the duct and so on to the food.
3. Glands in the wall of the duodenum discharge a digestive juice, **succus entericus**, on to the food. These three juices work together to bring about complete digestion. All are alkaline, and so they help to reduce the acidity of the chyme, thus providing a suitable medium for the enzymes to work in.

Bile is a green fluid which is secreted by the liver cells and stored in the gall bladder. It contains bile salts which give it the colour, and sodium bicarbonate which makes it alkaline. It contains no enzymes. The bile salts act like a detergent in breaking up any fat in the food into microscopic droplets, a process called **emulsification**. This process helps digestion as it increases the surface area of the fat for the enzymes to work on.

Pancreatic juice is secreted by the pancreas which is a

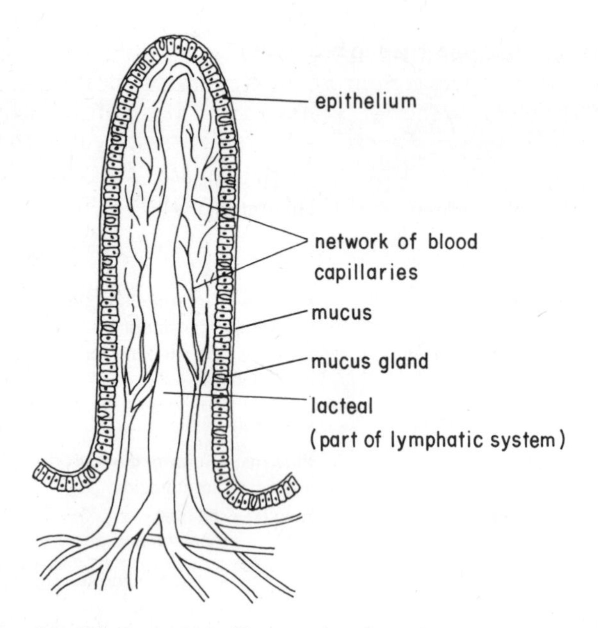

Fig. 18:7 Diagram of a villus much enlarged.

gland lying below the stomach; its duct joins the bile duct before entering the duodenum. It contains a protease, an amylase and a lipase.

Succus entericus, secreted by glands in the intestine wall, contains a number of enzymes which complete the digestion of carbohydrates into various hexose sugars and finishes the breakdown of proteins into soluble amino acids. A summary of the digestive processes is shown in Table 18:1.

Absorption of the products of digestion

Absorption of the end products of digestion occurs mainly in the ileum. The efficiency of the process is increased because:

1. The products pass along it slowly, perhaps taking as long as three or four hours.
2. Its walls are not only folded internally to form ridges, but its lining is thrown into millions of finger-like processes called **villi** which greatly increase the surface area for absorption.
3. The rhythmic contractions of the wall ensures that the fluid which is in contact with the absorptive surface is constantly changing.

Each villus (Fig. 18:7) contains a network of capillaries just below its surface so that hexose sugars and amino acids can quickly pass from the gut into the blood stream. These capillaries join up to form the large portal vein which takes these substances to the liver. The fatty acids and glycerol are absorbed mainly into the lymphatic system (p. 154), a branch of which—a **lacteal**—is present in each villus; these products eventually reach the blood system via a duct in the neck.

The large intestine

In man, this consists of the colon and rectum. By the time the contents of the gut reach the colon the main products of

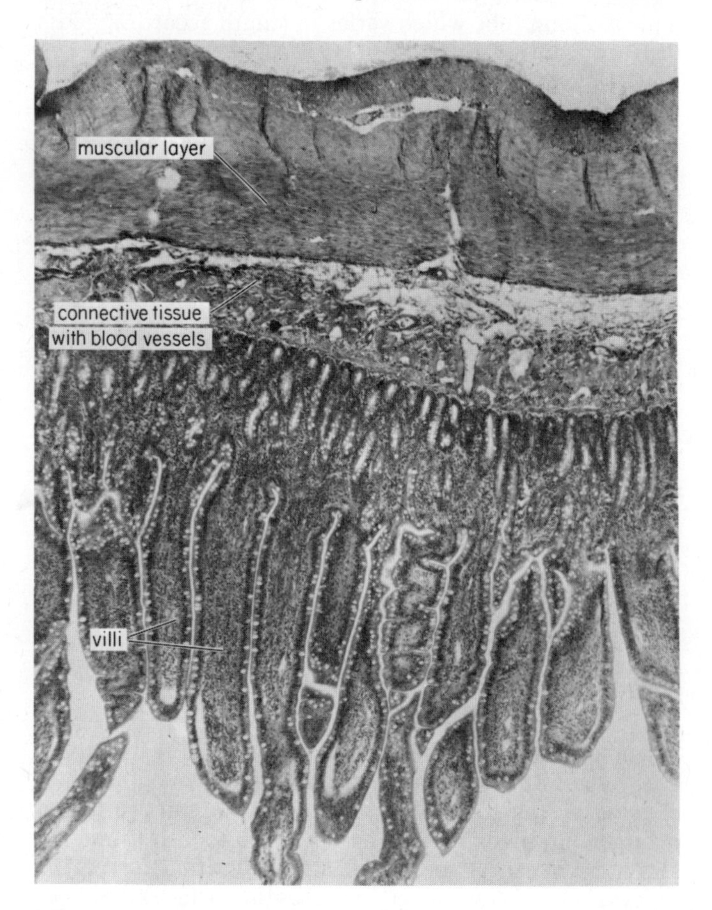

Fig. 18:8 Photomicrograph of a transverse section through the ileum of a cat.

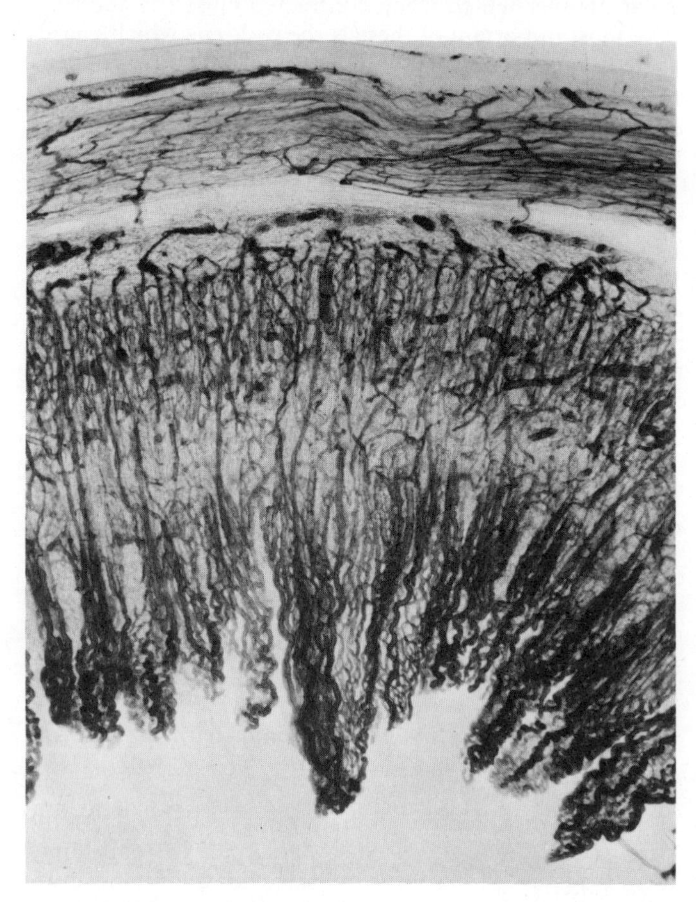

Fig. 18:9 Photomicrograph of a similar section with the blood vessels injected to show the circulation.

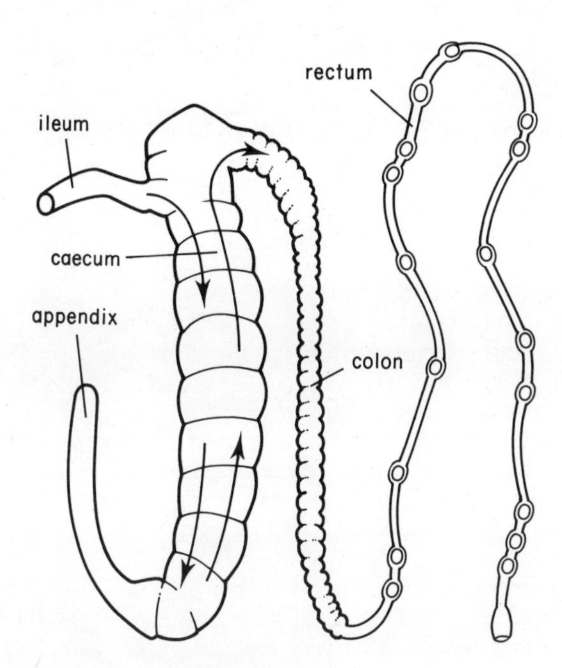

Fig. 18:10 Large intestine of rabbit.

digestion have been absorbed. The residue consists largely of water and solid material such as vegetable fibres and cellulose for the digestion of which there are no appropriate enzymes. In addition, bacteria are present in astronomical numbers feeding on the organic material. The chief function of the colon is to absorb most of the remaining water, leaving the solid material as faeces which pass into the rectum and are periodically eliminated through the anus.

In many mammals, especially herbivores such as rabbits, sheep and cattle, the large intestine also includes a large blind outgrowth, the **caecum** which ends in the **appendix** (Fig. 18:10). In man these are greatly reduced in size and have no digestive function.

The liver

As already indicated, hexose sugars are transported from the villi via the portal vein to the liver. An important function of the liver is to store this sugar temporarily in its cells in the form of solid **glycogen**, a more complex carbohydrate rather like starch.

The liver also regulates the amount of sugar that passes through it into the general circulation. Otherwise the level of sugar in the blood would rise steeply after the digestion of each meal. What osmotic effect would this have on the body cells? As sugar is constantly being removed from the blood by the body cells to replace what is used up in respiration, more sugar is needed to take its place. The liver supplies this need by converting some of its stored glycogen back into sugar and releasing it into the blood stream.

The amino acids cannot be stored in the body but travel round the general circulation and are selectively absorbed by the cells which need them for building up their proteins. Any surplus amino acids are broken down by the liver into a carbohydrate which is stored as glycogen and urea which is passed to the kidneys for excretion. This process is called **deamination**.

Fatty acids and glycerol eventually arrive in the liver via the lymphatic system and the blood stream. Here the fatty acids are broken down into smaller units. Some are converted by the liver cells into carbohydrates while others are reconverted into fats by cells in various parts of the body, especially those under the skin and in the connective tissues which surround the kidneys and other organs. This fat acts as a reserve of fuel which can be oxidised in respiration to liberate energy.

Additional functions of the liver include the secretion of bile (p. 129), the storage of iron for the manufacture of more haemoglobin, the formation of fibrinogen, a protein used in blood clotting, and the conversion of certain poisons, which may enter the bloodstream as a result of bacterial action in the colon, into harmless substances—a process called **detoxication**. By being a centre of metabolic activity the liver is also an important source of heat.

Health aspects of the alimentary canal

The human alimentary canal usually functions remarkably well considering how badly we treat it on occasions! Sometimes it rebels, and we either feel sick or have indigestion.

Vomiting is the body's method of ridding itself of unwanted or harmful substances from the stomach. The peristaltic movements of the stomach and oesophagus reverse their normal direction and the food is expelled. There are many causes of vomiting, but one of the most common is over-eating, especially when the food contains a high proportion of fat. Vomiting also occurs when we eat something very indigestible or poisonous.

When we feel 'bilious' or 'liverish', it is often the result of having eaten 'rich' meals over several days. The liver is unable to cope with the excessive fat and we get a feeling of nausea and sometimes a headache.

Indigestion is a general term used when there is difficulty in digesting food. Healthy people can usually avoid indigestion by: a) having simple, well-balanced meals, b) eating them in a leisurely manner, c) thoroughly masticating the food, and d) avoiding taking violent exercise soon afterwards. We can learn a lot from other mammals which after a meal have a good sleep!

A more serious form of indigestion is caused by stomach and duodenal ulcers. These conditions occur more often in people who may be described as hurried or worried. Thus ulcers occur more often in busy people who get into the habit of hurrying over meals and rushing from one activity to another without sufficient rests—such as doctors, school-masters, members of parliament, stock-brokers and business executives. Those who are able to relax, who are not continually tensed up, and who live at a slower pace, seldom get ulcers.

For good health it is necessary to empty the bowels regularly. If the food residues remain in the colon for too long, the bacteria present have more time to produce harmful substances which may be absorbed by the blood. Constipation can often be avoided by having plenty of roughage in the diet (p. 114).

The digestive process in herbivores

Because herbivores consume vegetable material they take in great quantities of cellulose, but like all mammals they

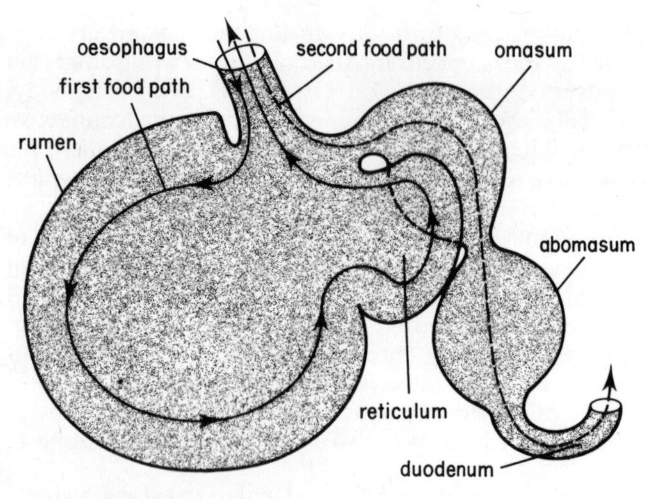

Fig. 18:11 Sheep's stomach.

are unable to digest it as no enzyme is present that will break it up. However, some of the bacteria living in various parts of the guts of herbivores are able to do this. In the rabbit, these bacteria occur in vast numbers in the large caecum, but in species which 'chew the cud' such as sheep and cattle, they largely occur in their specialised four-chambered stomachs.

In sheep and cattle, the herbage is first chewed and mixed with saliva. It is then swallowed and passed first to the **rumen** (Fig. 18:11) and then to the **reticulum**, the first two chambers of the stomach. In both these chambers astronomical numbers of bacteria start the breakdown of cellulose. This action continues when the food, or **cud**, is regurgitated into the mouth and subjected to further pulverisation by teeth. When the cud is swallowed once more it is directed towards the third chamber, the **omasum**, where it is churned up and some water is extracted from it. It then enters the fourth chamber, the **abomasum**, which is the part that corresponds to the stomach of other mammals. Here hydrochloric acid and proteases are secreted on to the food and the bacteria are killed by the acid. The products of cellulose digestion by the bacteria are mainly fatty acids which are absorbed by the blood and become the main source of energy for the sheep or cattle.

Some mammals, such as rabbits, have the strange habit of eating their own faeces. When the food passes through the gut the first time it is only partly digested and the faeces which are formed are moist; in wild rabbits these are voided during the day when below ground. These faeces are then eaten at once and further digestion of their contents takes place. During their second passage most of the water is extracted from them so these faeces when voided are dry; these are the pellets you commonly see in grassy places in districts where rabbits are common. Can you see any possible connection between this habit and the ability of rabbits to live in very dry places?

19

Bacteria and viruses

We discussed (p. 101) the different methods of nutrition used by living organisms and you will recall that the basic method is holophytic—the typical method employed by green plants. Other organisms, because they lack chlorophyll, are dependent for their food upon the complex substances which green plants have made. Animals typically use the holozoic method of nutrition, taking their food into their bodies and digesting it before it is absorbed. Other organisms, such as *Mucor* (p. 29) and many other fungi, feed on dead organic matter by the saprophytic method of secreting digestive fluids on to the food and absorbing the soluble products. By contrast, parasites such as mildews (p. 30), the malarial parasite (p. 51) and tapeworms (p. 25), feed on the living tissues of other organisms or on the food eaten by their hosts.

We are now going to study bacteria, the great majority of which feed either as saprophytes or parasites. Many of those which feed saprophytically are very beneficial and, like many fungi, they cause the decay of dead material and thus return essential nutrients to the soil. However, those which act as parasites attack living organisms, including man, and thereby cause disease.

Bacteria are so important to man that in this chapter we shall also study their general biology and some of the ways in which they affect our lives.

Bacteria

These cannot be seen with the naked eye, but the largest are visible under the medium power of a microscope. Most come into the size range 0·5 μm – 8·0 μm where 1 μm = 1 micrometre (0·001 mm). If the average size of a bacterium is 1 μm you could calculate how many there would be if they were placed in line on a 10 cm ruler.

With very few exceptions, bacteria have no chlorophyll and, like fungi, their surrounding wall is made of protein and fatty substances, not cellulose as in green plants. They are non-cellular and the majority have no nuclei, although nuclear material is scattered within the cytoplasm.

They are classified into three groups according to their shape (Fig. 19:1):

1. **Cocci** (sing. coccus), which are spherical. Sometimes they are grouped together in pairs (diplococci), others in groups of varying sizes (staphylococci) and others in strings (streptococci).
2. **Bacilli** (sing. bacillus) which are rod-shaped. Sometimes these have long projecting cytoplasmic threads called flagella which are used in movement.
3. **Spirilla** (sing. spirillum) which are twisted like a corkscrew, the number of twists varying from one to many. Some of these also have flagella at their ends.

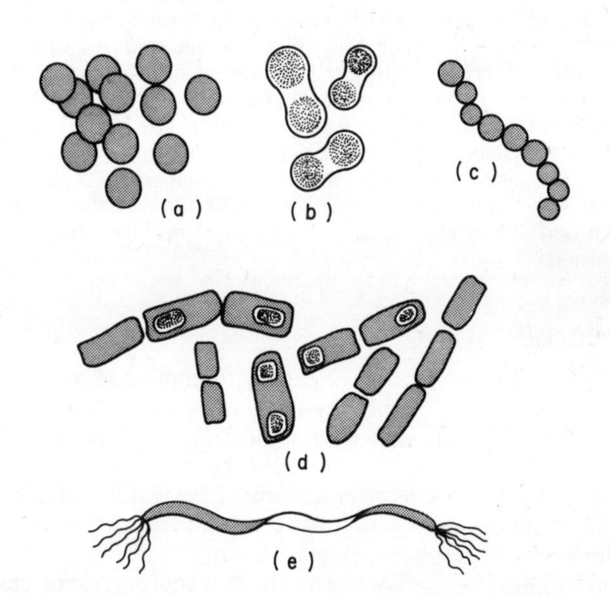

Fig. 19:1 Types of bacteria: a) Staphylococci. b) Diplococci. c) Streptococci. d) Bacilli (some with spores). e) Spirillum with flagella.

Although some bacteria have to have oxygen for the liberation of energy from food, others can do without it by respiring anaerobically and others, rather strangely, cannot exist if oxygen is present. Many species form spores within their bodies (Fig. 19:1d); these have hard resistant coats which allow them to remain alive for long periods under difficult conditions. When better conditions return the coat breaks down and active life is resumed.

Bacteria as saprophytes

If some chopped-up meat or a dead earthworm is left in a beaker of water, the water soon becomes cloudy, a scum forms on the top and it starts to smell. This is because saprophytic bacteria are causing the decay of the meat, the proteins being broken down into simpler nitrogenous compounds which are volatile and therefore smell. During this process of decay the bacteria absorb nutrients, grow and multiply very rapidly.

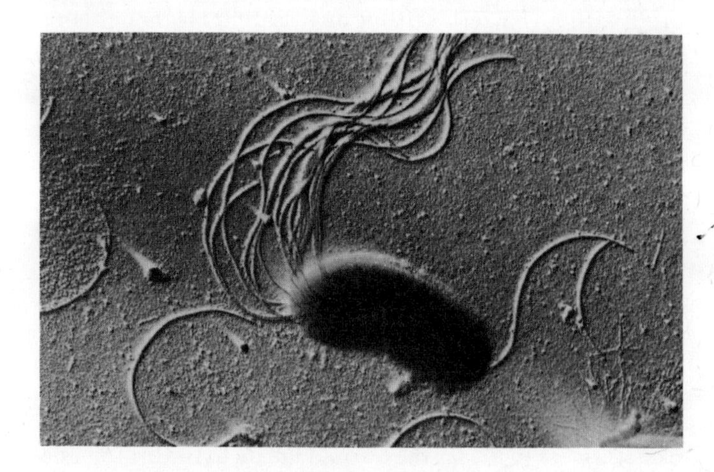

Fig. 19:2 Electronmicrograph of a bacillus with flagella.

Examine under the high power of the microscope a small drop of water in which something has decayed. If a microscope with an oil immersion lens is available, this would be better still. Focus very carefully and look out for bacteria of different shapes in a state of movement. The cocci joggle about in a haphazard way because they are being bombarded by molecules in the water (Brownian movement). Bacilli may move in a particular direction as a result of their flagella (not visible) while spirilla can move quite fast like animated springs.

Reproduction

This is a simple process of binary fission which can take place roughly every half-hour at room temperature. So, if there was only one bacterium to begin with, after 24 hours the number produced would have 14 noughts in it! This may be surprising, as after an hour there would only be 4, after $1\frac{1}{2}$ hours only 8 and after 2 hours 16. When large numbers of bacteria are produced in a fluid they tend to collect near the surface where there is more oxygen; here they produce a lot of mucilage and it is this which causes the scum on the top.

Where do bacteria come from?

In the Middle Ages it was believed that many living organisms arose spontaneously from non-living things. Thus it was thought that frogs were derived from the mud of ponds, maggots from bad meat and mice from dirty clothes left in a cupboard! As a result of more critical observations and experiments these conclusions were shown to be false, but the idea of **spontaneous generation** persisted for many of the smaller organisms and when bacteria were discovered it was widely believed that they arose from decaying material.

It was in 1861 that **Louis Pasteur** (1822–95), the famous French scientist, finally proved that spontaneous generation of bacteria did not occur and that food went bad because bacteria were carried to it via the air. The experiment he performed to prove this is a classical example of how a scientific experiment should be done using a control.

For this experiment he used a clear nutrient broth which, when exposed to the air, would normally have gone bad after a few days. First he had to kill off any bacteria already in the broth. This he could do quite easily by boiling it; but how could he then allow air to come in contact with the broth without bacteria settling on it as well? He thought out a most ingenious method for doing this. He put the broth in a flask with a long thin neck and then bent the neck into an S shape in a flame (Fig. 19:4). He boiled the broth to kill off any bacteria in it. Now, if the flask was cooled, he knew that the vapour would condense inside and due to the consequent reduction in pressure air would be drawn into the flask. So he reckoned that if the cooling was done slowly, the air would pass in but the bacteria, being heavier than air, would not pass beyond the bend in the neck. He prepared a number of flasks in this way and the experiment worked perfectly; however long the flasks were left, the broth did not go bad. But to make quite sure, he used a control experiment by cutting off the neck of one of the flasks so that air could enter *from above*. He reasoned that if bacteria were in the air they would fall on to the broth. The broth went bad within a few days!

Fig. 19:3 Louis Pasteur (1822–1895).

The success of these experiments depended on Pasteur's ability to kill off the bacteria which were in the broth to begin with by boiling, a process called **sterilisation**. Today heat is still used to sterilise apparatus and the media for culturing bacteria.

Growing bacterial cultures

Microbiologists today usually culture bacteria on nutrient **agar**. This is a jelly-like material extracted from seaweed to which various food substances are added, such as beef extract, to suit the requirements of the bacteria to be cul-

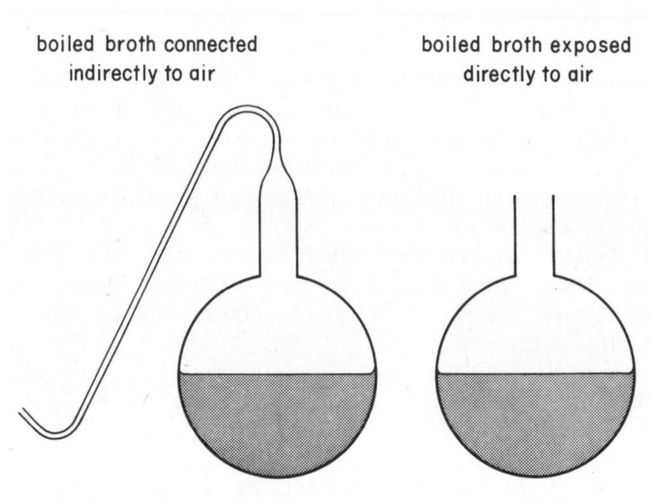

Fig. 19:4 Pasteur's flask.

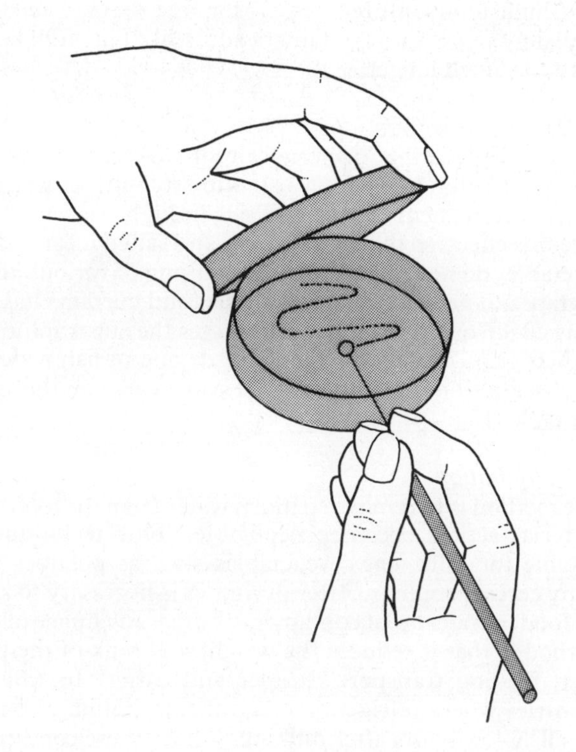

Fig. 19:5 Method of inoculating an agar plate using a nichrome loop.

tured. (Fungi may be cultured in the same way, a good medium being agar to which potato extract and glucose are added.)

The nutrient agar is dissolved in water, sterilised by boiling it under pressure in an autoclave (rather like a pressure-cooker) and then poured into sterilised Petri dishes where it sets on cooling to form a **plate**. The lid of the Petri dish is loose fitting and so allows air to enter, but bacteria are unable to do so—why is this? Tablets are commercially available which contain the ingredients for various types of medium and provide a quick alternative method for making up nutrient agar.

To culture bacteria a loop of nichrome wire is first held in a flame until red hot to sterilise it, and when cooled, dipped into the medium to be tested, e.g. a drop of milk. The lid of the Petri dish is then raised slightly and the loop stroked gently over the surface of the agar (Fig. 19:5); the lid is then quickly replaced. The Petri dish is placed in an incubator at 37°C for 48 hours. By this time any bacteria will have multiplied sufficiently to be visible as streaks where the loop touched the surface.

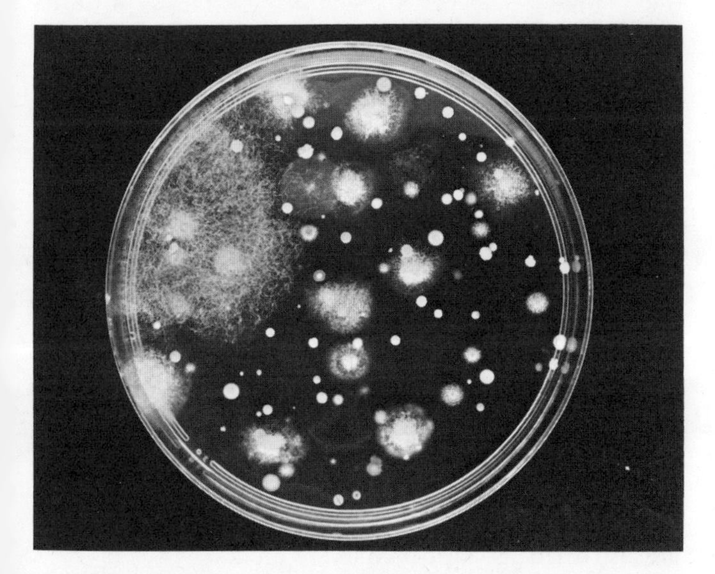

Fig. 19:6 Cultures growing on an agar plate after exposure to air for 30 seconds. The smaller colonies are bacteria, the larger, fungi.

Working in groups, you can now use this method to find out how widespread bacteria are. Each group will need eight sterilised Petri dishes containing sterilised nutrient agar. Label them 1–8 with a chinagraph pencil and treat them as follows:

1. To test for bacteria in the air: take the lid off and expose for 15 minutes.
2. To test soil: shake up a little soil in distilled water; sterilise the loop in a flame and when cool transfer a drop on to the agar, making a regular pattern (why?).
3. To test tap water: transfer a drop as in 2.
4. To test milk: as in 2.
5. To test your skin: make some fingerprints on the surface of the agar by touching it lightly.
6. To test your breath: hold the plate near your mouth and cough into it several times.
7. Inoculate from any other source, e.g. a scraping from your teeth or from under your finger nails.
8. Leave untouched to act as a control.

Place all dishes in an incubator at 35–37°C and examine them after about 48 hours.

If the plates were sterilised successfully No. 8 should have remained clear. Did you find bacterial colonies in all the others or only some? In some, you may have grown fungus colonies as well; they may be recognised by their hyphae (p. 29). If there is any particular colony you wish to grow as a pure culture you can do this by sub-culturing. To do this you touch the colony with the sterilised loop and stroke it over the surface of sterilised agar in another dish. If a pure culture is not obtained first time the process can be repeated.

Most of the bacteria grown will have been saprophytes which cause decay, but some could be disease-causing, i.e. **pathogenic**. Hence it is necessary to destroy the cultures when they are finished with by putting the Petri dishes, lids and all, into a bowl of strong disinfectant such as Lysol.

Let us now consider some of the important implications of the experiments you have carried out, first of all those concerned with decay.

Bacteria and food

You probably established that air contains bacteria, so if food is left exposed to the air bacteria will settle on it and cause it to decay. If the food is damp and the temperature warm it will decay all the more quickly. This is a great problem in the tropics and probably explains why traditionally so many curry powders, herbs and spices were used in preparing food, as they masked the disagreeable flavour of food which was going bad. Today we can prevent food from decaying in many ways, so allowing it to be stored for almost indefinite periods and transported all over the world. The more important methods are:

1. Refrigeration

This does not kill the bacteria, but the lower the temperature, the more it prevents them from multiplying and the slower their decaying action becomes.

Domestic refrigerators are usually kept at temperatures just above 0°C. This is low enough to reduce bacterial activity to an extremely low rate and at the same time to reduce the metabolic and transpiration rates of living food such as fruit and green vegetables so that they keep fresh for longer periods. Food does not remain fresh indefinitely at these temperatures and when it is taken out of the refrigerator bacterial activity will be resumed.

The use of a deep freeze is more efficient as the temperature is much lower and all metabolic activity is stopped, so food keeps indefinitely. By this method fresh meat, fish, fruit and vegetables can be 'put down' when supplies are plentiful and cheaper, and used when required.

In the food industry, refrigerated lorries and ships are used to overcome the problem of transporting perishable food over long distances.

The Antarctic continent acts like a giant refrigerator. You may have read that food left by Captain Scott during his famous expedition to the South Pole in 1912 was discovered many years afterwards still in perfect condition. It has been suggested that the Antarctic could become the world's storehouse for surplus food. Can you think of reasons for and against this idea?

Another example of natural refrigeration is when animals become embedded in ice. For example, a mammoth estimated at 39,000 years old was discovered in Siberia in such good condition that even its stomach contents could be identified.

2. Sterilisation by heat treatment

Several methods are used, all based on the facts established by Pasteur:

a) Canning. The food is subjected to high-temperature cooking, placed in sterilised tins, reheated under pressure and sealed when still hot. In this way bacteria in the food are killed and no others can enter. Bottling works on the same principle. These methods are very useful for meat, fish, fruit and vegetables.

If bacteria are not destroyed they give out carbon dioxide during their respiration and this causes the tin to bulge. The contents should on no account be eaten if this occurs as the bacteria will probably have produced toxins. Also, beware of tins which are badly rusted as these could be contaminated.

b) Pasteurisation of milk. You should have found that milk contained plenty of bacteria; this is not surprising, as milk is an excellent food for micro-organisms as well as us! Most of the bacteria it contains will merely cause it to go sour, but pathogenic forms may also be carried in milk and multiply quickly. Pasteurisation is a method used for killing off the majority of bacteria, including those disease-producing forms, without spoiling the flavour of the milk by boiling it. The usual method is to heat the milk to 72°C for 15 seconds, quickly cool it to 12°C and put it into sterilised bottles which are capped at once. Some decay bacteria may survive this treatment, so it will go sour, but not so quickly as untreated milk. Another method is to heat the milk to a very high temperature (135°C) for one second and immediately put it into containers and seal. The milk keeps fresh for much longer.

3. Osmotic methods

The principle is that bacteria cannot survive in an active state in a solution of high osmotic strength as water is drawn out of them. Sugar and salt in high concentration have this effect, so that honey, jams and salted meat, fish or vegetables do not go bad. The sun-drying of various kinds of grapes to produce raisins, sultanas and currants has the same effect, as the drying concentrates the sugar solutions in their cells. The salting and sun-drying of fish which is practised in the tropics on a large scale works on the same principle (Fig. 19:7).

4. Dehydration

The method is to remove so much water from the food that bacterial action becomes negligible. This technique is suitable for milk, eggs, vegetables such as potatoes and many cereal products. After drying, it is necessary to keep the food in waterproof containers. A great advantage of this method is that it reduces the weight and bulk of the product, making transport cheaper and easier. In tropical countries where refrigeration is difficult, liquid milk may 'go off' a few hours after milking. Under these conditions the use of dried milk is of the greatest importance.

Fig. 19:7 Fish being sun-dried, East Africa.

5. Chemical methods

Some foods are preserved by adding chemicals which kill bacteria but are considered to be harmless to man in the quantities used. This is the least satisfactory method as certain preservatives used in the past have been found later to be harmful. Strict regulations are now imposed to prevent the improper use of such preservatives. Substances still commonly used include benzoic acid and sulphur dioxide.

The smoking of fish is basically a chemical process as the smoke contains substances which are poisonous to bacteria and these become impregnated in the outer layers which are also dried and hardened in the process.

Pickling food in vinegar (which contains acetic acid) is another chemical method; it is successful because the acid kills the bacteria.

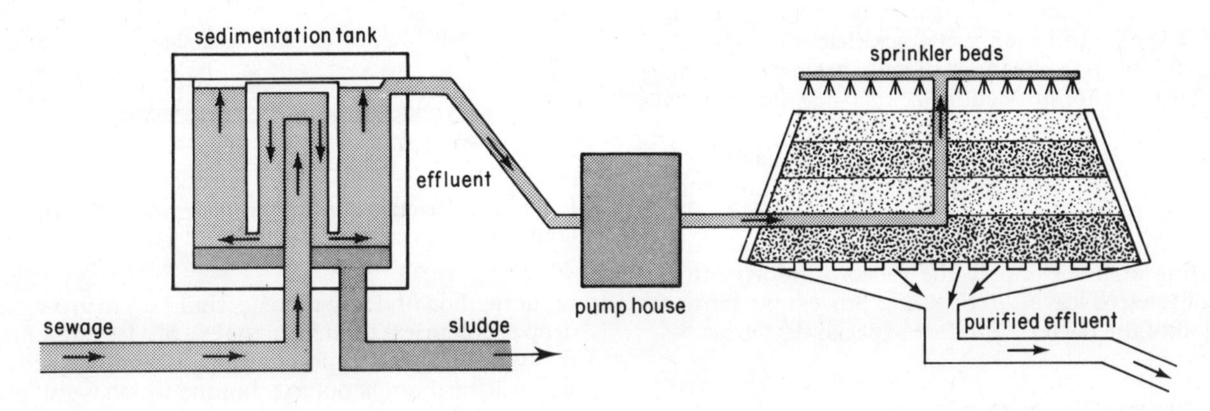

Fig. 19:8 Diagram showing the plan of a sewage works. (Not to scale.)

6. Irradiation

This modern and very effective method is to package and seal the food first and then irradiate it using radioactive cobalt which is powerful enough to kill all organisms which are present.

Bacteria and soil

Your experiment with a soil sample will have demonstrated that soil is teeming with bacteria. In fact, a pinch of soil will probably contain more bacteria than there are people in Britain. Most of these bacteria bring about the decay of organic matter in the soil. They are essential for life on our planet as they play a vital part in recycling the substances made by plants and animals so that they can be used once more (p. 245).

Bacteria and sewage

Fresh water, sea water, well water and even tap water all contain bacteria, hence any organic matter present decays just as it does in soil. This can be a big problem, for example, when large quantities of raw sewage are passed into lakes and rivers. In this case so much bacterial action takes place that nearly all the oxygen dissolved in the water is used up and other living things die in consequence. Another problem arises when large quantities of raw sewage are passed directly into the sea because it happens to be the cheapest method of disposal; this makes the water foul, and possibly dangerous to bathers, at many popular holiday resorts.

To avoid these problems, raw sewage can be treated beforehand so that the final products are harmless. One method often used in towns and cities is to utilise the action of bacteria under controlled conditions at a sewage works. The raw sewage from houses passes down concrete sewers to the sewage works (Fig. 19:8). Solid matter is first ground up, screened and then passed into a sedimentation tank from which the solids which form a sludge at the bottom are removed every few days. This can be processed and used as a valuable fertiliser. The effluent from the tank is then pumped to sprinkler beds where rotating arms spray it on to beds of stones or clinker through which the fluid percolates. Bacteria form a film over the clinker and as plenty of oxygen is present, they break down the organic matter very quickly into simpler substances which are safe to pass into the outlet to a river or the sea.

> Examine a drop of 'activated' sludge from a sewage works under the high power of the microscope. What types of bacteria can you recognise? What other organisms are present besides bacteria? Why do you think they are present?

Man's use of bacteria

In the dairying industry bacteria are used for a number of purposes. When cheese is made, bacteria are necessary for the formation of lactic acid from the sugar in milk; the lactic acid causes the milk solids to coagulate. In the formation of butter, the cream from which it is made is inoculated with certain bacterial strains which act on various substances present and so give the butter its characteristic flavour. Different strains of bacteria are also used in the making of yoghurt.

Bacteria are also essential for the formation of **silage**, an excellent fodder for dairy cattle. Silage is made from hay which is cut green and put in a silo. Bacteria ferment the sugar in the plants to form lactic acid and anaerobic conditions are produced which prevent the bacteria which cause decay from acting, hence the product retains much of the food value of the original plants and may be used in the winter when fresh grass is not available.

Fig. 19:9 A modern sewage works showing a sprinkler bed.

In the manufacture of vinegar, fruit juice is first fermented by yeast and then acetic acid bacteria (which are tolerant of low concentrations of acid) are used to change the alcohol present into acetic acid. It is the latter that gives the sharp taste to vinegar. Bacteria are also used for separating the fibres from the stems of the flax plants before they are processed as linen threads, an operation known as **retting**. In the leather industry bacteria, which are tolerant to low concentrations of acid, are also used during the tanning of hides and in the tobacco industry in the curing of tobacco leaves; in the latter process the fermenting action of the bacteria produces special flavours.

Bacteria as parasites

Bacteria and the spread of disease

So far we have been concerned mainly with the bacteria that can cause decay; now we come to those which cause disease.

Louis Pasteur, apart from his important work on decay, did much to increase our understanding of how bacteria cause diseases. His work on anthrax and hydrophobia (rabies) is thrilling to read about. However, it was a German doctor, **Robert Koch** (1843–1910), who first proved conclusively that a particular bacterium caused a certain disease. He also worked on anthrax, a disease which killed great numbers of domestic animals. When he examined the organs of animals which had died of anthrax he found them swarming with bacilli, and to prove the connection between them and the disease, he transferred some of them into a tiny cut he made at the base of the tail of a healthy mouse. The mouse soon contracted anthrax and died. But he wanted to know what happened to the bacilli inside the mouse; how did they multiply so fast? Could he possibly grow them outside the body of an animal and watch them under the microscope? He thought of an ingenious method of doing this; he would try to grow them in a drop of colourless fluid from the eye of a freshly-killed ox as this was a nutrient fluid of animal origin. To observe the bacteria, he thought out a technique which is still practised today known as the 'hanging drop' method (Fig. 19:11). Under the microscope, after hours of watching, he saw the minute rods dividing and growing, dividing and growing. Then, when the drop was crowded, he transferred a few to another drop, using a sterilised splinter of wood, and in this way found that he could sub-culture them time and time again. He now had cultures which had never been in the body of any animal. What would happen if he put some into a cut in another healthy mouse? He did this and the mouse died of anthrax! He had at last proved that this particular type of bacillus was specific in its action.

Fig. 19:10 Robert Koch (1843–1910).

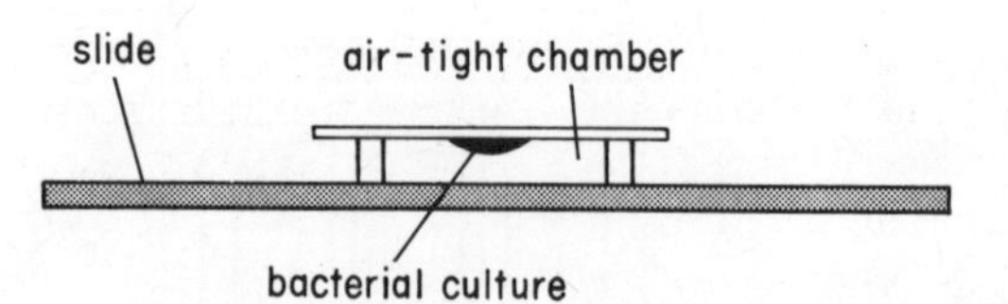

Fig. 19:11 Diagram illustrating the principle of the 'hanging drop' method of culturing bacteria.

When you grew bacteria on agar plates, you probably discovered that bacteria were present on the skin and in your breath when you coughed; if you tested material from your teeth or finger nails you will have found bacteria there too. Bacteria are also present in our nose and throat and in the gut. Most bacteria are harmless, some in the gut of herbivores help the digestion of cellulose (p. 132) and in man help the synthesis of vitamins of the B complex and vitamin K, however a few may cause disease. A disease which spreads from person to person is said to be **infectious**. How are bacteria spread? By studying the various methods of infection we can deduce sensible ways of reducing their effect.

1. Airborne infection

This occurs in droplets of saliva. When a person coughs or sneezes minute droplets are sprayed into the air which may then be breathed in by others. Epidemics spread quickly in this way, especially in crowded places such as classrooms, cinemas and on trains and buses. Good ventilation reduces the risk of infection. In the case of dangerous infectious diseases which are transferred in this way the patient is completely isolated to prevent the spread of the disease.

2. Contact infection

Diseases spread by direct or indirect contact with the body of an infected person are said to be **contagious**. They may be spread when such things as towels, hair brushes and clothes are shared by others. Impetigo, a very infectious disease of the skin which often occurs in schools, is quickly spread in this way. Not all contagious diseases are due to bacteria; some, such as ringworm and athlete's foot, are caused by fungi.

3. Food infection

Some conditions often described as 'food poisoning' are caused by contamination of food by bacterial spores. This may occur if food is handled by somebody with dirty hands in food shops, restaurants or kitchens. Diseases such as typhoid can be spread in this way by a **carrier**, i.e. somebody who harbours the bacteria but does not have the symptoms of the disease. Stored food can also become contaminated by rats or mice or, when displayed in shops, by flies. The packaging of food has done much to lessen some of these risks. Shops and factories in many countries are regularly inspected to see that legal standards of cleanliness are maintained.

4. Infection through drinking water

Cholera, typhoid and paratyphoid may be spread if drinking water becomes contaminated by sewage, as the organisms concerned occur in the faeces of infected persons. These diseases are a great scourge, especially in the tropics and in countries where standards of hygiene are low. Here, epidemics of cholera and typhoid may kill vast numbers of people. They also become a major hazard after disasters such as floods and earthquakes when the drinking water is liable to become contaminated. It should be noted that these two diseases may be air-borne and food-borne, as well as being transmitted through drinking water.

Every year more and more water is needed in towns and cities for domestic and industrial use and often it is in short supply. It may be drawn from lakes, reservoirs or rivers and must be purified before it is used for drinking. In some large cities the same water is used many times before it finally passes into the river and so to the sea; each time it has to undergo treatment.

Fig. 19:12 Joseph Lister (1827–1912).

Bacteriologists employed by water boards test mains water at frequent intervals to see that no harmful bacteria are present in sufficient numbers to cause disease; chlorine, which can sometimes be tasted in tap water, is added in sufficient quantities to destroy them.

Disinfectants

The use of chlorine in the water supply is an example of the use of chemicals to destroy micro-organisms. Such substances are called **disinfectants**. A Scottish doctor, **Joseph Lister** (1827–1912), was the first to use a disinfectant to kill bacteria during a surgical operation. Surgery in those days was extremely dangerous, because even if the patient survived the shock of the operation, there was a great likelihood of death soon after, due to the wounds becoming septic, i.e. **gangrenous**. Lister believed that sepsis was due to bacteria in the air getting into the wound and causing the tissues to decay just as the broth had done in Pasteur's flasks. In 1865 he put his theory to the test by carrying out his operations under a fine spray of carbolic, a very strong disinfectant. The results showed that he was right, the wounds did not become septic, but the carbolic was so strong that it also destroyed the tissues, and healing was slow. After further experiments using milder solutions it became accepted practice to use what were then called **antiseptics** in surgery.

Today, types of antiseptics are used which are much more efficient, as they do little damage to living tissues. When the skin is severely broken by cuts or grazes it is always wise, after washing the area, to apply one of these modern disinfectants and cover the wound to prevent further invasion of bacteria from the air.

In surgery today antiseptics are still used, but to a much more limited extent as now the emphasis is on **asepsis**—surgery carried out under bacteria-free conditions.

Surgical operations

The operating theatre is designed to exclude bacteria. The walls and floors are smooth and non-absorbent, so that they can be washed easily with disinfectant; the corners are rounded so that no dust can collect and there are no shelves for the same reason. The fittings are mainly chromium-plated or glass, and easily cleaned. The air supply is filtered so that no dust can enter and the temperature is regulated to prevent shock to the patient. The instruments and all the dressings are all heat sterilised under pressure beforehand.

There is still the risk of contamination by the surgeons, anaesthetists, nurses and the patient; they may all bring germs in with them. To minimise this risk hands and arms are washed thoroughly, sterilised gowns and gloves are put on and a gauze mask is used to cover the nose and mouth. The patient is also prepared for the operation, the area of skin around the point of incision being shaved of hair and treated with a mild antiseptic, and a minimum of sterilised clothing is worn. By taking these extreme precautions against bacterial infection, surgery has largely become free from the dangers of sepsis.

Bacteria and our health

Much is done by local authorities and others to provide us with air, food and water which is free from harmful micro-

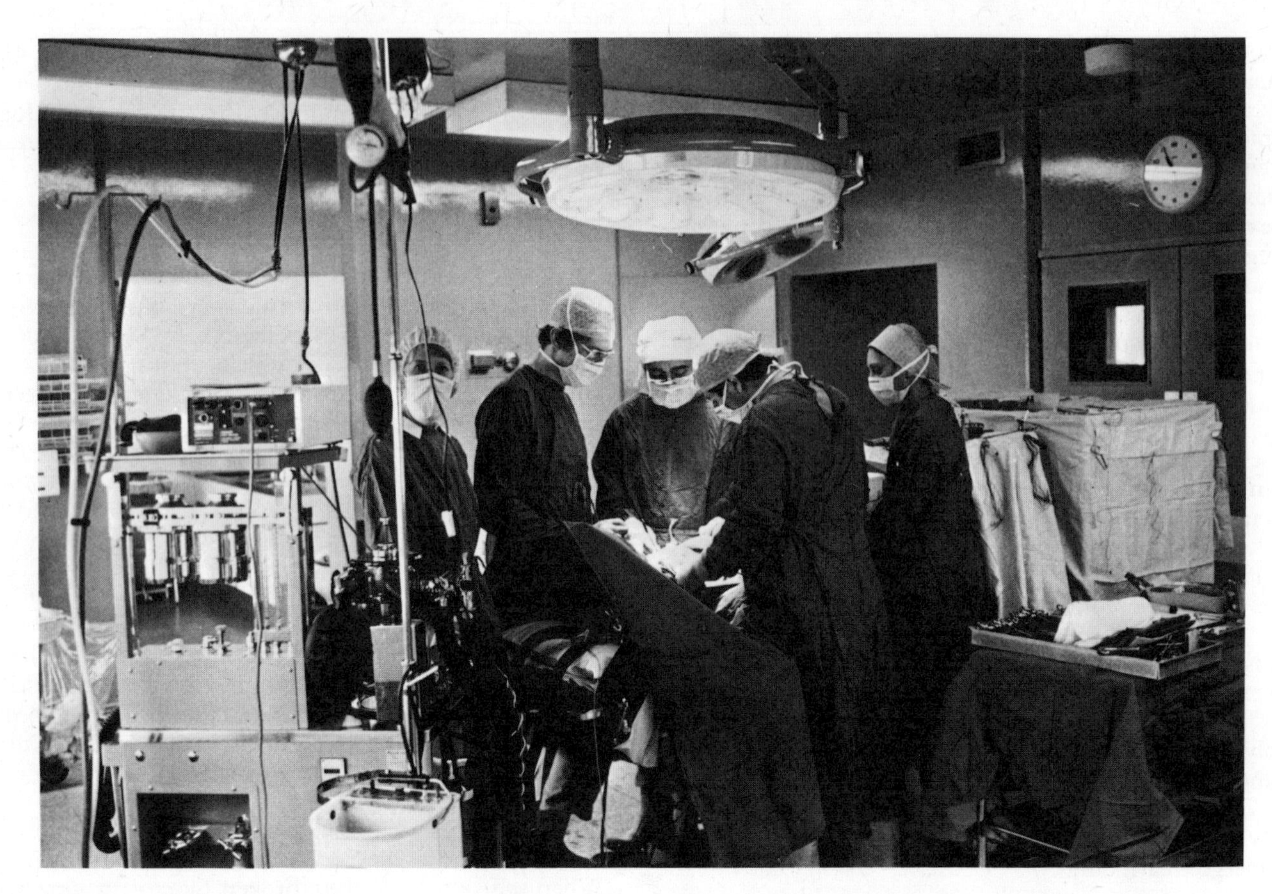

Fig. 19:13 A heart operation in progress.

organisms, but we still catch a lot of diseases which could be avoided if we were more careful over personal hygiene. Unfortunately we are not born with a set of instincts to keep the necessary health rules, so we need to acquire good habits of hygiene which often seem a bore until the reasons for them are understood. You should now be able to see the reasons for the following simple health rules:

1. Hands and nails should be washed before eating and preparing food.
2. Hands should be washed after using the toilet.
3. Hair should be washed and brushed regularly.
4. Baths or showers should be taken after games or other strenuous exercise.
5. Clothes, especially underclothes and socks, should be changed frequently.
6. Teeth should be cleaned regularly.
7. Brush, comb, towel and tooth-brush should not be borrowed or shared.
8. Food should be kept covered and pests such as flies, cockroaches, rats and mice should be destroyed.
9. Disinfectants should be used in lavatories and drains.

No doubt you will be able to think of some more.

Immunity against disease

When there is an epidemic, not everybody catches the disease. It is not just a matter of chance who gets it, because some people have more resistance to infection than others. Resistance to disease is a complex subject as many factors are involved, but one factor of particular importance is **immunity**.

Immunity is resistance to a particular infection; it may be present at birth or acquired during life by natural or artificial means.

1. Natural immunity

This occurs after we have recovered from some infection of bacterial or virus origin. If, for example, somebody becomes ill with diphtheria, the bacteria concerned produce poisonous substances called **toxins** which cause the symptoms of the disease. White blood corpuscles in the blood respond to these poisons by producing **antitoxins** (antibodies) which neutralise the effect of the toxins and so help recovery. These antitoxins prevent any recurrence of the disease as long as they remain in the blood. The duration of the immunity varies considerably according to the disease; it may be for life, e.g. measles, but it is more often for a much shorter period.

2. Artificial immunity

It is probable that during the 18th century 60 million people in various parts of the world died of **smallpox**, a disease caused by a virus (p. 143), but towards the end of the century an English doctor, **Edward Jenner** (1749–1823), discovered how immunity from the disease could be acquired artificially. During a terrible epidemic in England, Jenner noticed that people such as farmers and milkmaids who had previously contracted **cowpox** did not catch smallpox, while others all around were dying in hundreds. Cowpox is a mild virus disease caught from cattle which causes spots full of pus which are rather similar to those caused by smallpox. Jenner wondered what

Fig. 19:14 Edward Jenner (1749–1823).

would happen if he put some of the pus from a cowpox spot into a scratch made on the arm of a boy who was healthy. He tried this experiment and the boy caught cowpox! (This happened long before Pasteur's experiments.) Then came the crucial question: having had cowpox would the boy contract smallpox or not? There was only one way to find out; he would have to inoculate the boy in a similar way with pus from a patient with smallpox. It was a grave risk, but he did it. The next few days were an anxious time for all concerned, but all went well; the boy did not get smallpox! To make doubly sure Jenner repeated the inoculation several months later, but again there was no effect. We now call this process **vaccination**. (In 1980, owing to the recent practice of vaccinating children when only a few months old and making it illegal for travellers to enter countries where the disease still existed, the World Health Organisation was able to announce the complete eradication of smallpox from the world: a magnificent achievement!)

Nearly a hundred years later, Pasteur made rather similar experiments with anthrax, having probably got the idea from Jenner's work. In order to study the disease he cultured the anthrax in flasks of chicken broth. He found that if a little of this was injected into an animal it quickly died, but quite by accident he found that if a stale culture, 3 or 4 weeks old, was used the animals contracted some of the symptoms, but quickly recovered. Pasteur did many other experiments before he thought of injecting fresh culture into those animals which had previously been inoculated with the stale culture, using others which had not been so treated as a control. The experiment was a brilliant success; those animals which had not been previously treated all died, while the others kept perfectly healthy.

These experiments showed that immunity from both smallpox and anthrax was brought about by using a mild strain or a weakened form of the organism concerned. However, it has not been found possible to produce weakened forms of many other infections.

Today, two additional techniques are used:

a) A pure culture of the organism is grown and the toxin is separated from the living agent by filtration, and a carefully controlled toxin is inoculated into a person to cause the body to produce its own antitoxin. Sometimes, as in diphtheria, the toxin has to be reduced in potency by various treatments before it is inoculated.

When we are inoculated against some diseases a reaction is set up in the body, for example, the arm may become swollen and sore, or a mild fever may develop for a few hours.

b) The toxin is periodically inoculated into an animal such as a horse (why not a smaller animal?) which builds up large quantities of antitoxin in its blood. Blood serum from the horse can then be inoculated into a person.

Both methods are used for diphtheria; the first is the best for producing immunity while the second is used to counteract the action of the toxins in a person who actually has the disease.

Inoculation against diphtheria on a massive scale has reduced the danger of this disease enormously. In the 1930s nearly 3,000 children died of it every year in England alone; now it is extremely rare.

Chemotherapy

Since time immemorial, chemicals, often extracts from plants, have been used in the treatment of disease. Some were effective, others were not. However, most were concerned with the symptoms; they did not kill off the organisms themselves. But as soon as it was realised that many diseases were caused by micro-organisms the search started for chemicals which would kill the organisms, but not the cells of the body at the same time.

Paul Ehrlich (1854–1914) was the first person to succeed. After years of research on many chemical substances, he produced one called **salvarsan** which was the 606th substance he had tried! It was used in the treatment of syphilis (p. 172). Since then (1910) a number of substances have been discovered and tested and some are still used most effectively today, but many have now been replaced by **antibiotics**.

Antibiotics

These are chemical substances produced naturally by certain fungi and micro-organisms. When you grew bacteria on agar you probably grew some fungi as well. **Alexander Fleming** (1881–1955) did very much the same thing in 1929 when he was culturing staphylococci on agar. He noticed that one culture had somehow become contaminated by a mould fungus. This accident was of great significance as he noticed that there was a clear area around the mould where no bacteria were growing and he jumped to the correct conclusion that the mould was secreting a substance that prevented the growth of the bacteria. The mould was identified as *Penicillium notatum*, a fungus

Fig. 19:15 Alexander Fleming (1881–1955).

allied to the green mould you often see on oranges, and its secretion was called **penicillin**, the first of the antibiotics.

It was not until 1938 that **H. W. Florey** and **E. B. Chain** at Oxford continued the research on penicillin to investigate its possible use in medicine for destroying pathogenic bacteria. They knew that penicillin had great potentialities, so when the war came in 1939 they made an all-out effort to produce it on a large scale. To speed up the research Florey flew to America and persuaded a team of scientists in Illinois to collaborate in the project. Eventually a method was found to produce pure penicillin on a large scale and this 'miracle drug' was made available in time to save vast numbers of lives before the war ended.

Since then other antibiotics have been discovered, an important one being **streptomycin**. It has been very effective in reducing tuberculosis in the developed countries from a major cause of death to a disease of minor significance.

Unfortunately it has been found that some bacteria after many generations may no longer be affected by a particular antibiotic, i.e. they become **resistant** to it. It is therefore

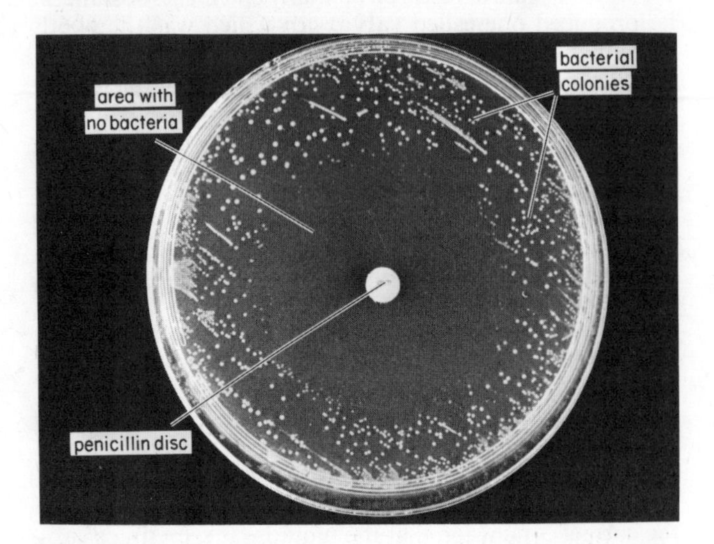

Fig. 19:16 Photograph of a plate of agar inoculated with bacteria showing the effect of penicillin on the bacterial colonies.

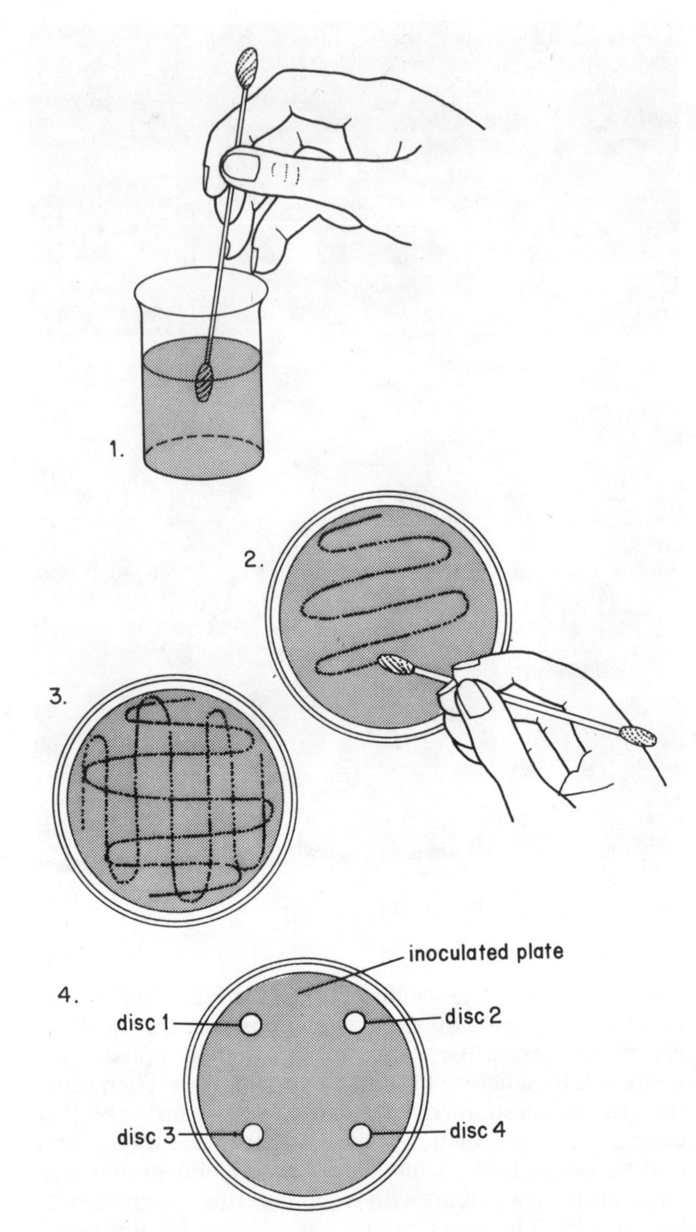

Fig. 19:17 Experiment to compare the effectiveness of two antibiotics in destroying bacteria. 1. Dip sterilised cotton bud in suspension of bacteria. 2. Streak the agar with it to form a regular pattern as above. 3. Turn the Petri dish through 90° and repeat the process. 4. The antibiotic-impregnated discs in place.

sometimes necessary when treating patients to find out which antibiotic is best to use. When, for example, a person is very ill with an infection of the lungs the doctor takes a sample of mucus from the back of the throat on a sterilised swab, places it in a sterilised tube and sends it to the pathology laboratory of a hospital. Here, agar plates are inoculated from the swab and incubated, and the resulting colonies of bacteria are tested with various antibiotics to see which kills them. The patient's doctor is informed and the correct treatment can then be given.

You can carry out a similar sort of investigation yourself using penicillin and streptomycin to see which is the more effective in destroying bacteria. Two species can be used, *Staphylococcus albus* and *Escherichia coli*.

For this experiment you will need two sterilised plates of nutrient agar, two sterilised cotton buds, a suspension of *S. albus* and another of *E. coli* and 8 discs which have been impregnated with the antibiotics (see details below).

Cover the surface of the first plate as evenly as possible with the suspension of *S. albus*. Do this by dipping the sterilised cotton bud in the suspension, and using the minimum amount of fluid, stroke the bud over the surface of the nutrient agar to form a series of loops (Fig. 19:17). In the same way, inoculate the second plate with the suspension of *E. coli*, using the second cotton bud. Label each plate.

To each plate add the following four discs, using sterilised forceps:

1. Penicillin of 5 unit strength
2. Penicillin of 10 unit strength
3. Streptomycin of 100 μg strength
4. Streptomycin of 25 μg strength

Fix the lid on each by putting adhesive tape round it and incubate the plates for two days at 37°C. A clear region round a disc will indicate the absence of bacteria and hence the effectiveness of the antibiotic.

What conclusions can you draw concerning 1) the action of each antibiotic on the two species of bacterium, 2) the effect of differences in concentration of the antibiotics?

It has recently been discovered that one in five of us has a type of bacterium growing on our skins which has remarkable properties. If you grow a culture of this species and soak up a minute amount of it on a disc of sterilised filter paper and transfer it, as you did with the penicillin disc, to a plate of *Staphylococcus*, a strange thing happens. When the plate is incubated and the colonies grow a clear patch forms round the skin colony. These bacteria are producing their own antibiotic! You could describe it as chemical warfare between different species of bacteria.

It has been found that hospital patients who have this species on their skin are far less likely to have an infection in their wounds following an operation than those who have not. Fortunately, this useful species can be transferred from one person's skin to that of another so it is possible to pass on this method of protection. So, some of us carry our antibiotic factory with us wherever we go!

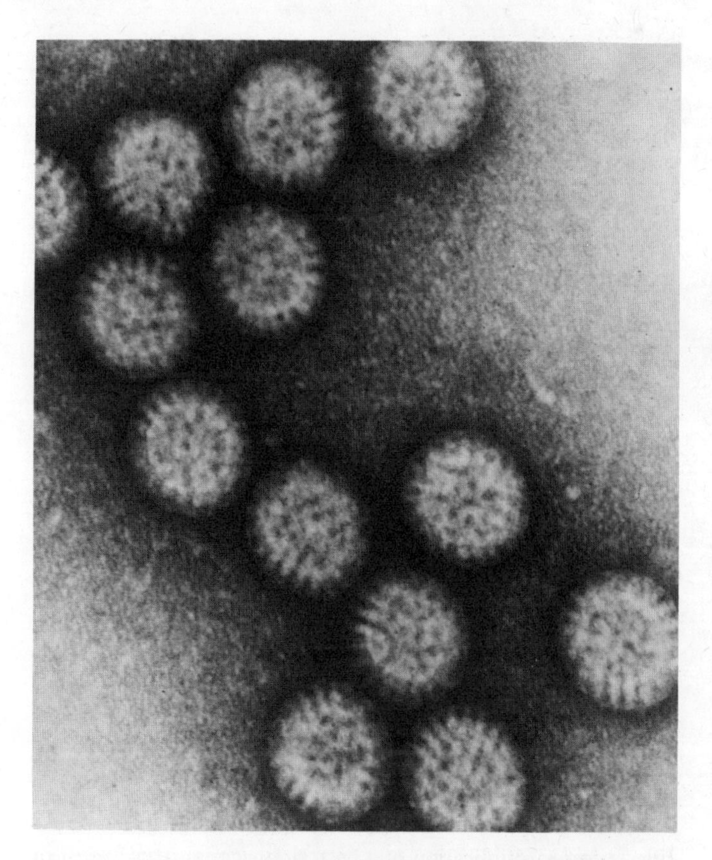

Fig. 19:18 Electronmicrograph of human rotavirus. × 200,000.

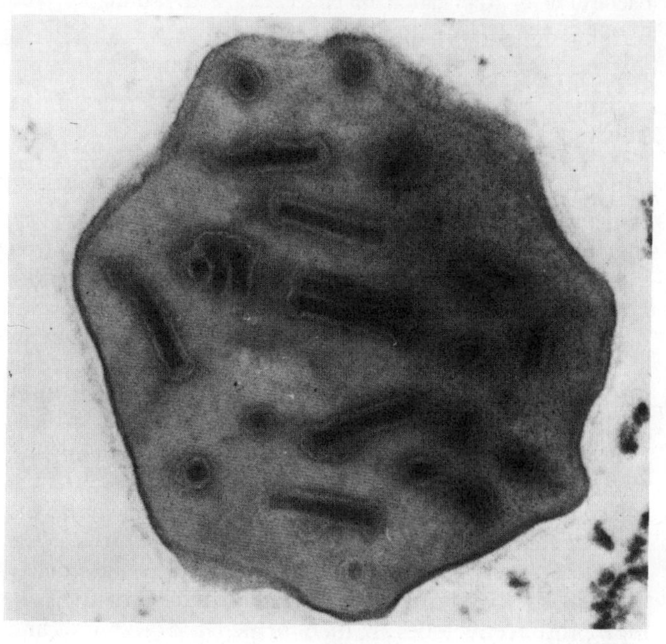

Fig. 19:19 Electronmicrograph of virus particles in the cell of an insect. The virus is used to control an insect pest of grassland in Africa. × 192,000.

Viruses

Viruses are incredibly small, some approximately 0·00001 mm in diameter, far smaller than the smallest bacteria. The largest can just be seen under the best light microscopes, but the majority have never been seen directly, only photographed under the electron-microscope. Some are rod-shaped, others are many-sided, often with a tail-like projection. The smaller viruses act like chemicals and can be crystallised. Viruses differ from typical living cells in having no nucleus, cytoplasm or surrounding membrane, the smaller ones at least consisting merely of a protein envelope enclosing a giant **nucleic acid** molecule. Nucleic acids occur in viruses in two distinct forms known as DNA and RNA. They are similar to the genetic material found in all plant and animal cells (p. 273).

Most viruses cause disease in plants and animals and many of the common infections of man are caused by them. We have already referred to smallpox (p. 140) and yellow fever (p. 52); others include influenza, the common cold, poliomyelitis, measles, German measles, chickenpox, glandular fever and mumps.

Viruses are inactive unless they penetrate living cells. Some invade the cells of bacteria, others plant cells and others, human and animal cells. They are often very specific in their action; in man, for example, one kind of virus will attack only cells in the skin (producing warts), another only

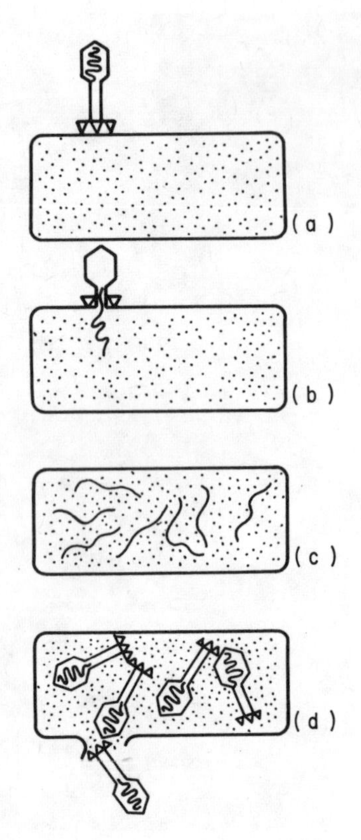

Fig. 19:20 Multiplication of a bacteriophage: a) Attachment of phage to wall of bacterium. b) DNA of phage passing into bacterium. c) Replication of phage DNA. d) Formation and escape of new phages.

certain nerve cells (polio) and another the cells of the salivary glands (mumps).

As viruses have no cytoplasm they produce no enzymes and hence are unable to carry out such functions as respiration, which are associated with living things. However, they have the power of multiplication. This multiplication is not the same as normal reproduction, as the virus does not grow or divide and cannot reproduce on its own, but only with the help of living protoplasm from a plant or animal cell. The way this happens has been discovered in viruses which attack bacteria; these are often called **bacteriophages** or just **phages**.

Phages become attached to the bacterium by their 'tail' (Fig. 19:20). The bacterial membrane dissolves and the DNA content of the phage passes into the cell leaving the protein coat behind. Inside the bacterium the virus DNA makes replicas of itself, using for the purpose the nucleic acids of the bacterium and other substances in the cell. New protein envelopes are then formed, the cell bursts and about 200 exact copies of the phage come out and infect other cells. Each cycle takes about 45 minutes. In this way the nucleic acids of the bacterial cells which are normally used for their own reproduction are 'commandeered' to make copies of the phages instead. In a similar way, if we catch influenza, the virus attacks *our* cells and makes use of the nucleic acids in these cells to produce more virus particles. These attack more cells and the cycle is repeated until vast numbers of cells are infected. Influenza is also extremely infectious and spreads on droplets of saliva from person to person. This is why it is important, for the sake of others, to stay at home if you catch the disease. You will know how influenza periodically spreads right across the world as a serious epidemic. It is estimated that during the epidemic of 1918–19 which followed the First World War half the population of the world caught influenza and over 20 million people died as a result. But influenza, like many virus diseases, is very variable; sometimes it is very virulent, at other times much milder. This may be because there are different strains of the virus, known as influenza A and B, and variations in a strain may come about by a mutation, a change in the composition of the DNA core of the virus.

Artificial immunity against virus diseases

We saw in the case of smallpox how a mild form occurred in cows called cowpox and immunity was produced by vaccination with cowpox virus. Much research has been done to find suitable vaccines for other virus infections, but one of the difficulties has been to grow the virus artificially. You cannot grow a virus on agar as it only multiplies within living cells, but it has been found possible to grow animal cells outside the body in bacteria-free nutrient material and to infect these cultures with a virus. An American scientist, **Dr Jonas Salk**, used this technique when working on **poliomyelitis**. He 'grew' the virus in a tissue culture of cells from a monkey's kidney and then inactivated the virus with formalin. This prevented further multiplication, but the factor was retained which caused the white blood cells of the person vaccinated to form its own antibodies. This vaccine gave a high degree of immunity.

Other treatments of virus diseases

Treatment of viruses by antibiotics is of use only with some of the largest viruses, such as the one causing **trachoma**, a very common cause of blindness in hot countries. What scientists are looking for is something which will prevent the virus from replicating its nucleic acid. A significant break-through occurred when it was found that when two different viruses invade the same cell one somehow interfered with the multiplication of the other. In 1957 **Isaacs** and **Lindemann** discovered that when this happened a

Fig. 19:21 De-capping of eggs before harvesting live virus in influenza vaccine production.

substance was made which they were able to isolate, which they called **interferon**. It does not directly destroy the other virus or prevent it from entering the cell, but it does interfere with its multiplication. Its great advantage is that it is naturally produced, is non-poisonous to body cells and affects a wide range of viruses. Research has continued on this very exciting substance and after 20 years methods have been devised for obtaining sufficient quantities for proper clinical trials to be carried out. However, it is yet to be seen how effective interferon will be for the treatment of diseases, although it holds out great promise.

Plant viruses

The first virus ever to be isolated was the cause of tobacco leaf mosaic, a disease of the tobacco plant in which the leaves have a mottled pattern of pale green and yellow where the cells have been attacked. In 1935 **Dr Wendell Stanley** made an extract from the diseased tobacco leaves and from it obtained crystals of the virus, which he then injected into healthy plants causing them to develop the disease. This virus is so stable that it can survive for many years in cured tobacco and people who handle it may spread the virus to other crops such as tomatoes which are susceptible to it.

You may have seen tulip flowers with attractive streaks of a different colour in the petals. This variegation is due to a virus attacking some of the cells and is of interest because it is shown in Dutch paintings going back to the 16th century, and thus represents the oldest plant virus known.

Many plant viruses are spread from plant to plant by aphids and their allies. When they feed on the sap of plants they plunge their mouthparts into the tissues and suck up the juices, and on going to another may transfer the virus. Plants attacked by a virus are usually weakened in consequence.

The study of viruses (virology) is a very exciting and fast-growing section of biology which has many implications for the future welfare of man. The connection between viruses and certain forms of cancer is currently being investigated, and also the production and application of new vaccines to combat other diseases. Research is also being carried out on the use of viruses in the control of insects and other pests.

20

Transporting materials within the animal

Every cell in the body has to have a continual supply of essential nutrients and oxygen in order to carry out its metabolic activities, and a means of removing the excretory substances which are formed within it. A brain cell, or a cell in the big toe, is a long way from the food which is digested in the gut and from the oxygen which enters the body through the lungs. Hence a transport system is essential for connecting up all the cells, wherever they may be, with the sources of supply of these vital substances. In the larger animals the **blood system** is the main transport system of the body.

Blood systems

The simplest way to see how a blood system works is to look at an animal which is fairly transparent. One of the best for this purpose is the water flea, *Daphnia*, but small specimens of the water louse, *Asellus*, or the water shrimp, *Gammarus*, are good substitutes.

Place a small *Daphnia* in a few drops of water on a cavity slide, lower a cover slip carefully and examine it under the medium power of a microscope. Locate the position of the heart (Fig. 20:1). Observe the movement of the heart carefully. Can you see any tubes coming out of the heart? Now turn to the high power and focus on the area near the heart and look out for any particles which are moving about. Do these particles flow along smoothly or in jerks? Do they move in a particular direction or in a haphazard manner?

These minute particles are blood corpuscles and their movement is due to the pumping action of the heart. In arthropods such as *Daphnia* the blood is not confined in tubes, but is pumped through a series of spaces which surrounds the organs. When blood returns to the heart it passes through openings in the heart wall which are controlled by valves. This type of circulatory system is called an **open system**. In the majority of animals, including ourselves, the system is said to be **closed** as the blood travels within tubes—the **arteries** and **veins**.

It is not easy to find out how the blood is flowing in our own blood system as we are not transparent like *Daphnia*, so it is not surprising that up to the 16th century people believed that blood flowed backwards and forwards like the tides.

In Europe, **William Harvey** (1578–1657) first demonstrated that blood actually circulated through arteries and veins in a particular direction. By many dissections and ingenious experiments he showed that the blood in arteries flowed away from the heart and that in veins towards it. He demonstrated the flow of blood in veins in this way. First he tied a band tightly round a man's upper arm which caused the veins in his arm to become prominent. He then placed one finger firmly on the middle of one vein and kept it there throughout the experiment. Then, with another finger, he squeezed out the blood from the part of the vein *above* this pressure point by stroking it firmly *towards* the upper arm. On releasing the pressure the blood did not flow back and fill the vein. When he repeated the action on the part of the vein *below* the pressure point, squeezing *down* the arm, on releasing the pressure this time the blood *did* flow back into the vein (Fig. 20:2). So he concluded that the blood in veins flowed towards the heart.

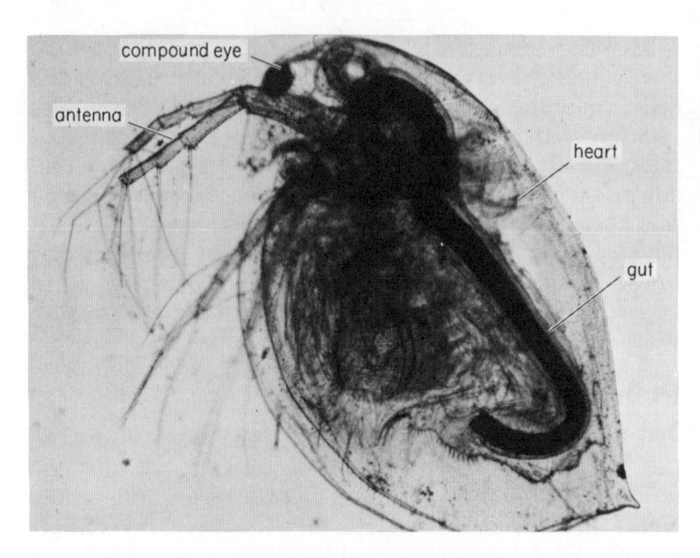

Fig. 20:1 A water flea (*Daphnia*) × 18.

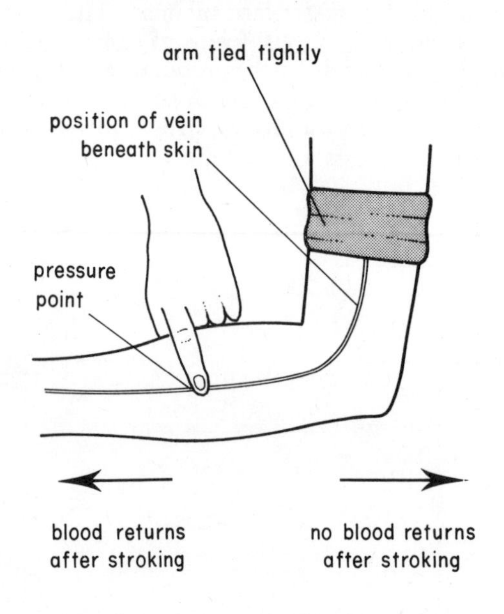

Fig. 20:2 Diagram showing the principle of Harvey's experiment. Pressure is exerted on the vein with one finger while with another finger the vein is stroked firmly up the arm and then down the arm.

Harvey was unable to demonstrate any connection between the smallest arteries and the smallest veins, but a few years later an Italian, **Marcello Malpighi** (1628–94), using a better quality lens, was able to see fine tubes connecting the ends of minute arteries with veins in the lung of a frog. These microscopic connecting tubes are called **capillaries**.

Blood circulation in man

The blood system consists of a heart which pumps the blood round, arteries which take blood away from the heart, veins which bring it back to the heart and capillaries which join the smallest branches of the arteries, the **arterioles**, to the finest branches of the veins, the **venules**.

How is it that every living cell in our bodies obtains the vital supplies that it needs? If the *same* blood went to all the organs, the substances it carried would soon be used up and the last organs to be reached would have none. For example, oxygen enters the blood at one place only—the lungs—but every organ needs oxygen, so in some way fresh blood has to reach each organ. This is brought about because each organ has a special circuit of its own consisting of an artery and vein and their branches, and the capillaries which join the arterioles with the venules. The only exception to this is the circulation to the lungs because *all* the blood goes through this circuit and becomes oxygenated, and on its return to the heart it is sent to all the other circuits and finally back to the heart again. You will see from Fig. 20:3 that because of this special circuit to the lungs the blood passes through the heart twice for each time it flows through any other part; this is known as a **double circulation**. Note that on one circuit the blood passes through the right side of the heart, but on the other circuit it passes through the left side.

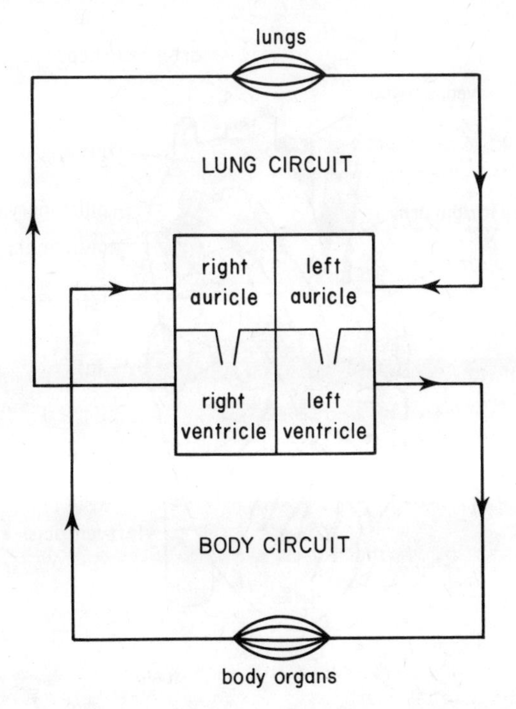

Fig. 20:3 Schematic diagram showing the principle of the double circulation.

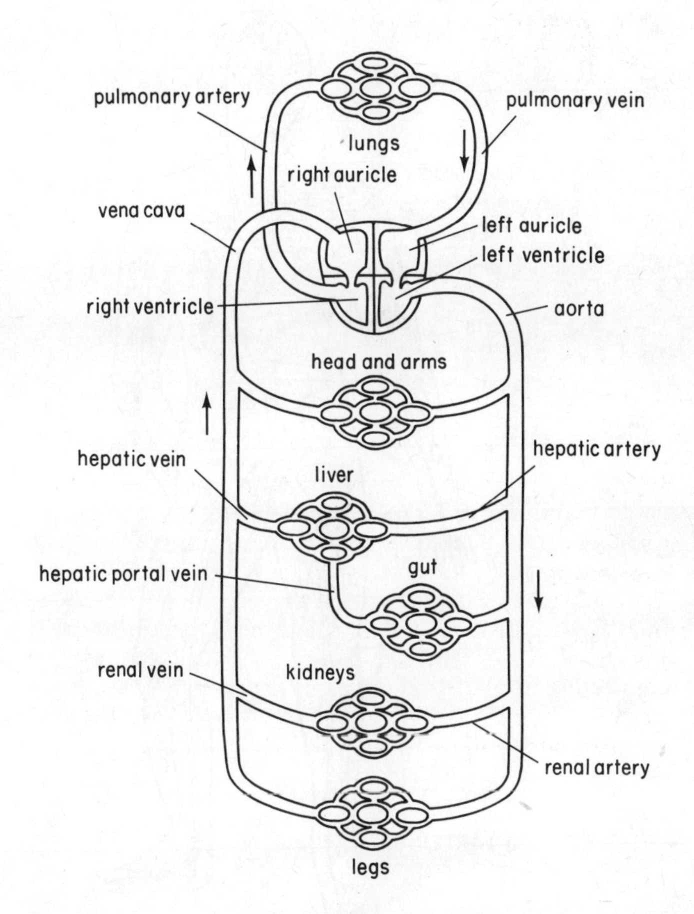

Fig. 20:4 The main blood circuits of the body (schematic).

Compare Fig. 20:4 with Fig. 20:5. The former shows the main circuits in highly diagrammatic form, the latter the natural positions of the main arteries and veins making up these circuits. In most circuits the artery and vein run alongside each other until they break up into finer branches. The main artery of the body, the **aorta**, runs in the mid-line alongside the main vein, the **vena cava**. The circuit concerned with the supply of blood to the gut is rather different from the others as the blood on leaving the gut does not return to the vena cava direct but passes first to the liver via the **hepatic portal vein**. This is the only vein in a mammal which has capillaries at both ends. Why is it functionally desirable that blood from the gut should go direct to the liver?

You should examine a dissection of a rat or rabbit and identify as many as possible of the main arteries and veins. We shall now study the component parts of the circulatory system in more detail.

The heart

The driving force of the circulation is, of course, the heart. Functionally the heart can be thought of as an organ composed of left and right sides, each side acting as a separate pump for the two circulations. Oxygenated blood from the lungs enters the left **auricle** (**atrium**), passes through a valve into the left **ventricle** and is expelled via the aorta to the head and body. On the other side of the heart, de-

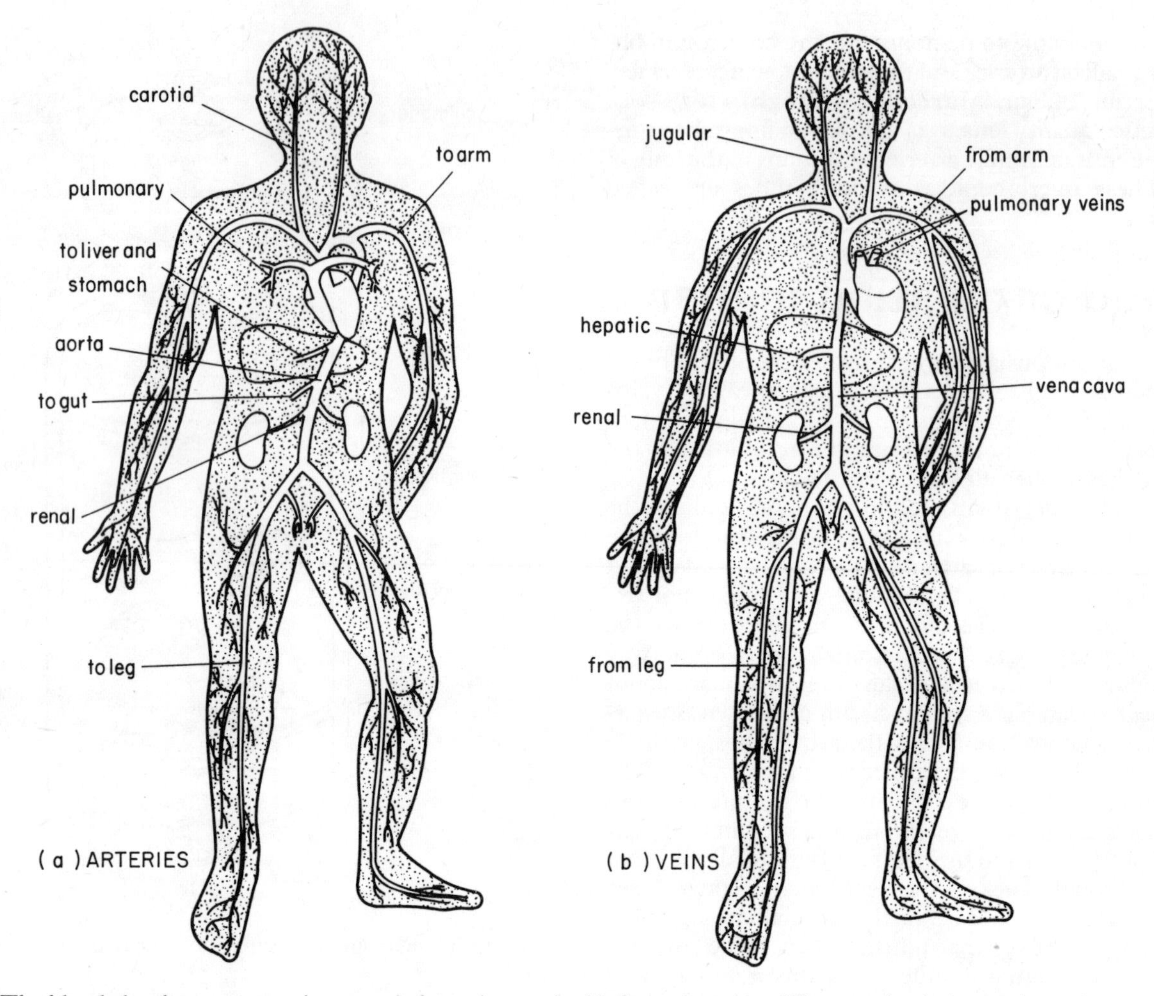

Fig. 20:5 The blood circulatory system in man: a) the main arteries b) the main veins. (The portal vein is not shown.)

oxygenated blood enters the right auricle, passes through a valve into the right ventricle and is then pumped through the pulmonary artery to the lungs.

> Examine the heart of a sheep, pig or cow. All are basically similar to the human heart. Study Fig. 20:6 and try to identify the various parts including the blood vessels.
>
> Also examine a heart which has been dissected to display the internal structures. Note that the valves between the auricles and ventricles are connected to pillar-like muscles in the walls of the ventricles by tough, non-elastic cords. These hold the valve flaps in position so preventing blood from flowing back into the auricles when the ventricles contract. If the aorta and pulmonary arteries have not been completely removed, you should be able to see inside them the semi-lunar valves which prevent blood from flowing back into the ventricles after the beat (Fig. 20:7). These valves work on the same principle as the valves in veins (Fig. 20:9). Compare the thickness of the walls in all four chambers of the heart. Can you think of any reasons for the differences?

The heart beat

This is a double action. First the two auricles contract, forcing blood into the ventricles, then the ventricles contract immediately afterwards, pumping the blood into the arteries. Between these two contractions, the valves between the auricles and ventricles close so that the blood is

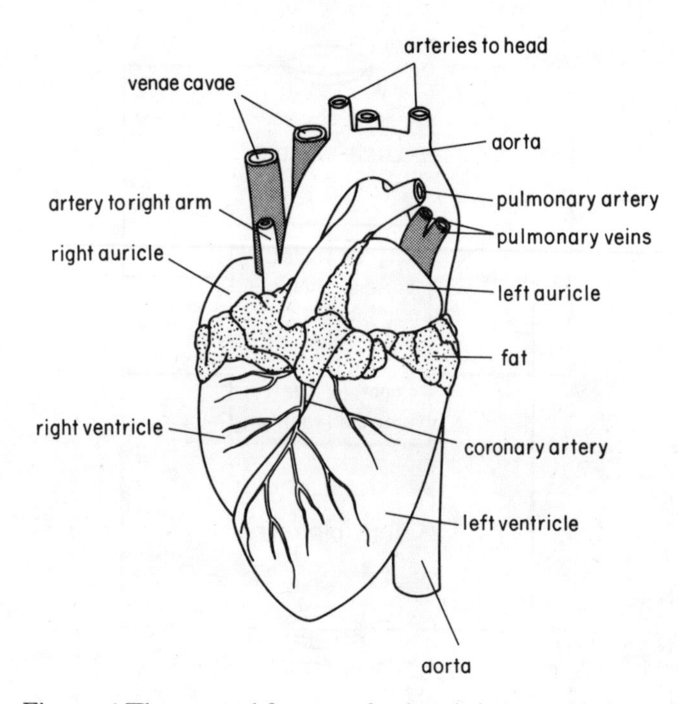

Fig. 20:6 The external features of a sheep's heart and its associated vessels, ventral view.

directed only into the aorta and pulmonary artery when the ventricle walls contract. At the same time as the ventricles contract the auricles relax and fill up again (Fig. 20:8).

The rate at which the heart beats in an adult averages 72 beats per minute. Find out your own rate by taking your pulse, or ask your partner to take it for you.

> The pulse in the wrist is the easiest one to take. Rest your left arm on the bench with the hand facing upwards, then place the tips of the first and second fingers lightly on the thumb side of the wrist. Count the number of beats in a 30 second period. Repeat two or three times and obtain an average figure. Calculate the number of beats per minute and compare your rate with the rest of the class. There will probably be considerable variation. How does the average rate of the class compare with the average adult rate?

You will know that your heart beats faster during and after exercise. You could find out what variation there is between members of your class in the way the heart reacts to exercise.

> In order to make the comparisons more exact, it is best for the class to take the same amount of vigorous exercise. As soon as the exercise is finished, take your pulse rate continuously over half-minute periods until it has returned to the original figure. You should plot a graph of your own pulse rate against time. Now compare the graphs of the whole class and find out:
>
> 1. What variation in recovery time is there between individuals? Do those with the lowest initial rates recover faster or not?
> 2. In how many cases does the rate fall below the original? If so, this is probably due to initial over-reaction followed by a corresponding adjustment.
> 3. Compare the results of those who train regularly for games, athletics or swimming with those who do not over-exert themselves. Can you draw any conclusions?

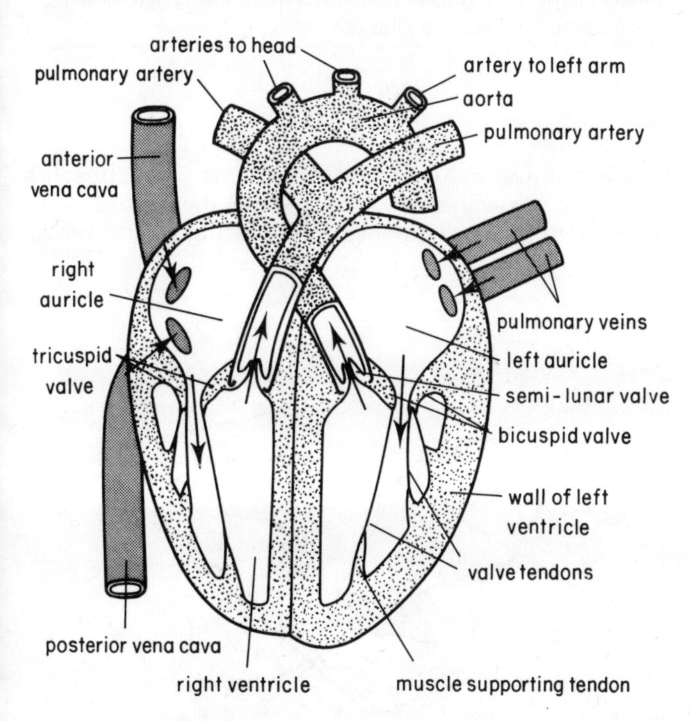

Fig. 20:7 Simplified diagram of a longitudinal section through the human heart. The arrows show the direction of blood flow.

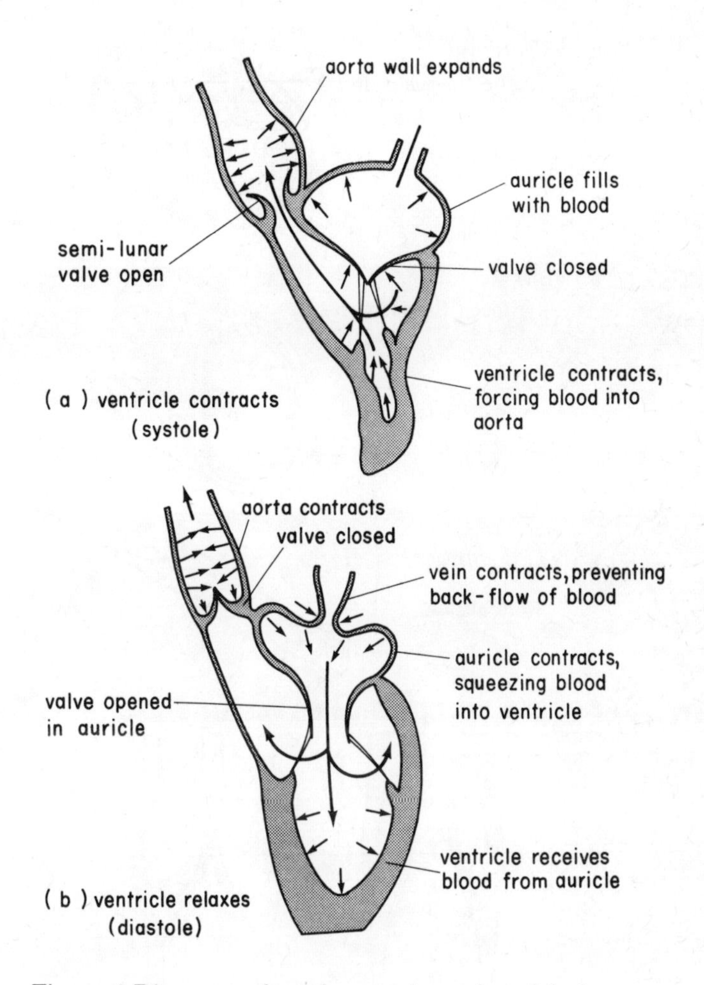

Fig. 20:8 Diagram to show the pumping action of the heart.

The mechanism by which the heart beat is regulated according to the overall needs of the body is controlled through nerves and chemical hormones (p. 162).

When we are active the body responds by pumping more blood to the muscles. This, together with an increased breathing rate, ensures that more oxygen and food reaches the muscles and that waste products are removed quickly.

Arteries

The system of arteries is rather like a tree which divides into smaller and smaller branches. When the heart beats it forces the blood into this system, but there is considerable resistance to its passage and the great arteries near the heart take most of the pressure. Their walls are consequently very strong but also elastic (Fig. 20:9a) so that they can dilate and take up the increased volume of the blood when the ventricles force more blood into them. Immediately afterwards, when the ventricles relax, the elastic walls of the great arteries return to their original diameter and force the blood along. This wave of expansion and contraction becomes less the further away the arteries are from the heart; this is because the walls of the more distant arteries are more muscular and less elastic. However, this wave can still be felt in the wrist as a pulse which, if counted, indicates the heart rate. You could, of course, take your pulse in many parts of the body if you could feel the appropriate artery easily. Try this for one of the leg arteries:

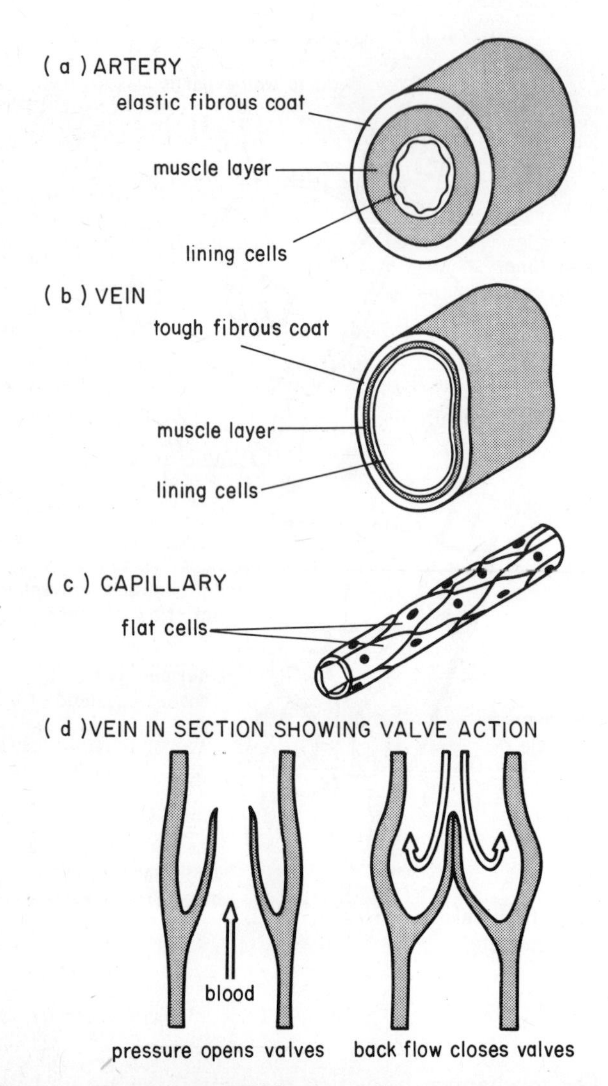

Fig. 20:9 The structure of blood vessels.

> Sit on a table with one leg dangling and the other resting on it so that the back of one knee rests on the knee of the other. After a time you will see and feel the leg which is on top give a series of small movements with each heart beat. If you do it for long you will reduce the blood flow to the leg and so develop 'pins and needles'.

The importance of the muscle layer in the *smaller* arteries and arterioles is that blood flow can be controlled according to their degree of contraction, thus some blood circuits can be given more or less blood depending on the demands of the body. You will see later how this affects the circulation to the skin (p. 202) and the gut (p. 162).

Veins

By the time the blood from the heart reaches the veins, the pressure is extremely low, a fact which accounts for their relatively thin walls. Veins depend largely upon movements in the surrounding muscles for squeezing the blood back to the heart. The larger veins contain valves to ensure that blood cannot travel in the wrong direction (Fig. 20:9d). Valves are especially important in veins where the blood flows against gravity, as in the arms and legs.

You can find out the position of the valves in your veins by examining your arm:

> Swing your arm round several times to fill the veins with blood, hold the arm vertically downwards and gently press your finger along a prominent vein—stroking it in the reverse direction to the blood flow, i.e. towards the hand. Can you see the swellings where you have pushed blood against the valves?

Some people, particularly those who are heavy and have to stand a lot, suffer from **varicose veins** in the legs. These are veins which have become swollen and coiled and they are caused by the valves failing to operate.

Capillaries

These microscopic vessels which join arterioles to venules are the *only* parts of the circulatory system where substances can enter or leave the blood (Fig. 20:9c). Their walls are so thin that some of the blood can pass out and bathe the surrounding cells and diffusion of dissolved substances can take place between blood and tissues.

Capillaries are present all over the body in every organ so that all living cells are near to some of them. It is the capillaries that cause the skin to be pink and lean meat to look red; they are too small to see individually, but they are so close together that they have this effect.

> 1. Put a drop of clove or cedar wood oil on to the colourless skin at the base of a finger nail; it will make the skin more transparent. Place your finger under a binocular microscope and shine a strong light on top of it; you should be able to see some of your own capillaries. Do they form a particular pattern?
> 2. Examine the external gills or the tail of a newt larva or a tadpole of a frog or clawed toad (*Xenopus*). To keep the animal still while you examine it under the microscope, anaesthetise it lightly by placing it in a 0.05% solution of MS222 (tricaine methane sulphonate). Note the pulsation due to the heart action. Replace the animal in fresh water as soon as you have finished your observations. What exchange of material is taking place in these capillaries?

Blood

This is a liquid tissue consisting of cells or **corpuscles** which float in a pale straw-coloured fluid, the **plasma**. You can separate these components by putting a few cm^3 of

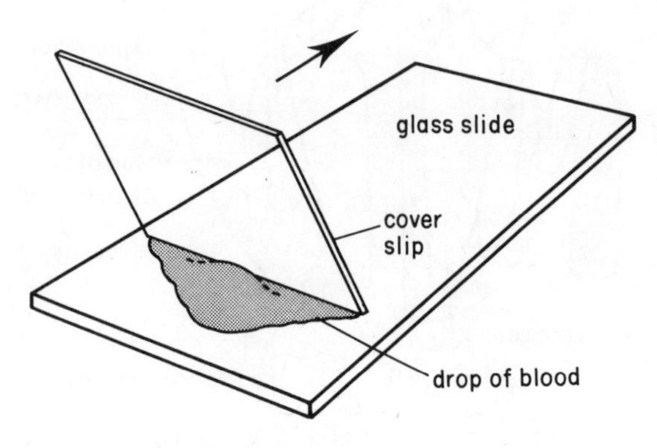

Fig. 20:10 Making a blood smear.

blood into a tube and centrifuging it. You can study these blood corpuscles by preparing and examining a microscope preparation, using your own blood if you wish. Precautions must be taken to ensure against infection as described.

1. Sterilise the back of a finger with some surgical spirit. Draw more blood into your finger by swinging your arm round, holding it downwards and wrapping a handkerchief tightly around the base of the finger.
2. Make a quick jab with a sterile lancet near the nail and put a drop of blood on to a very clean slide near one end. Never share a lancet with your neighbour.
3. Position a coverslip or slide *in front* of the blood drop (Fig. 20:10) and carefully push it away from the drop thus producing a thin smear. Allow the smear to dry. Sterilise your finger once more with surgical spirit.
4. Pipette some Leishman's stain on to the dried smear. Leave for five minutes, then carefully wash off excess stain with distilled water.
5. Gently wave the slide in the air to dry (or place it over a bench lamp).

Examine the preparation under the high power of the microscope; there is no need to use a coverslip.

Note the very numerous red corpuscles which have no nuclei, and the white corpuscles which may be easily distinguished because their nuclei will be stained blue. Compare their size with that of the red cells. Are their nuclei all alike?

Red corpuscles (erythrocytes)

Their function is to absorb, transport and release oxygen. They are produced continuously at the rate of ten million per second in the bone marrow of the ribs, sternum and long bones especially. Before being released into the blood they lose their nuclei and become flattened, bi-concave discs (Fig. 20:11). Each corpuscle consists of an elastic membrane which encloses cytoplasm and a high concentration of **haemoglobin**.

Haemoglobin, which is purplish red in colour, is a protein containing iron. Under conditions of high oxygen concentration one molecule of haemoglobin can combine loosely with four molecules of oxygen to form **oxyhaemoglobin** which is bright red. However, in conditions of low oxygen concentration the oxyhaemoglobin will give up its oxygen and revert to haemoglobin. Thus, when the blood passes through the tissues of the lung, oxygen is taken up, and when the blood reaches other tissues, oxygen is released.

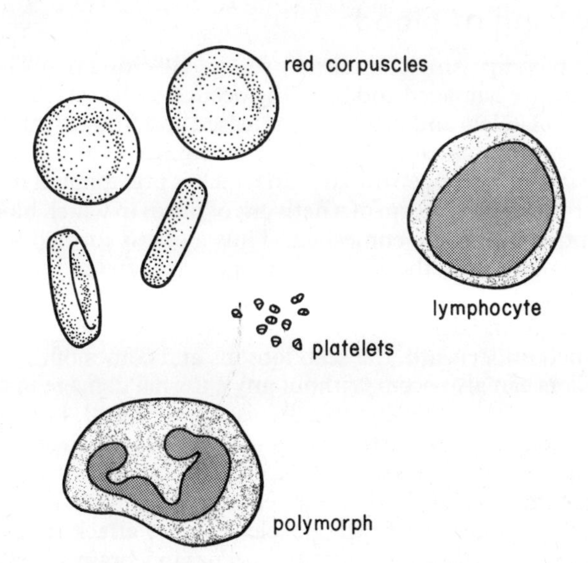

Fig. 20:11 Blood components.

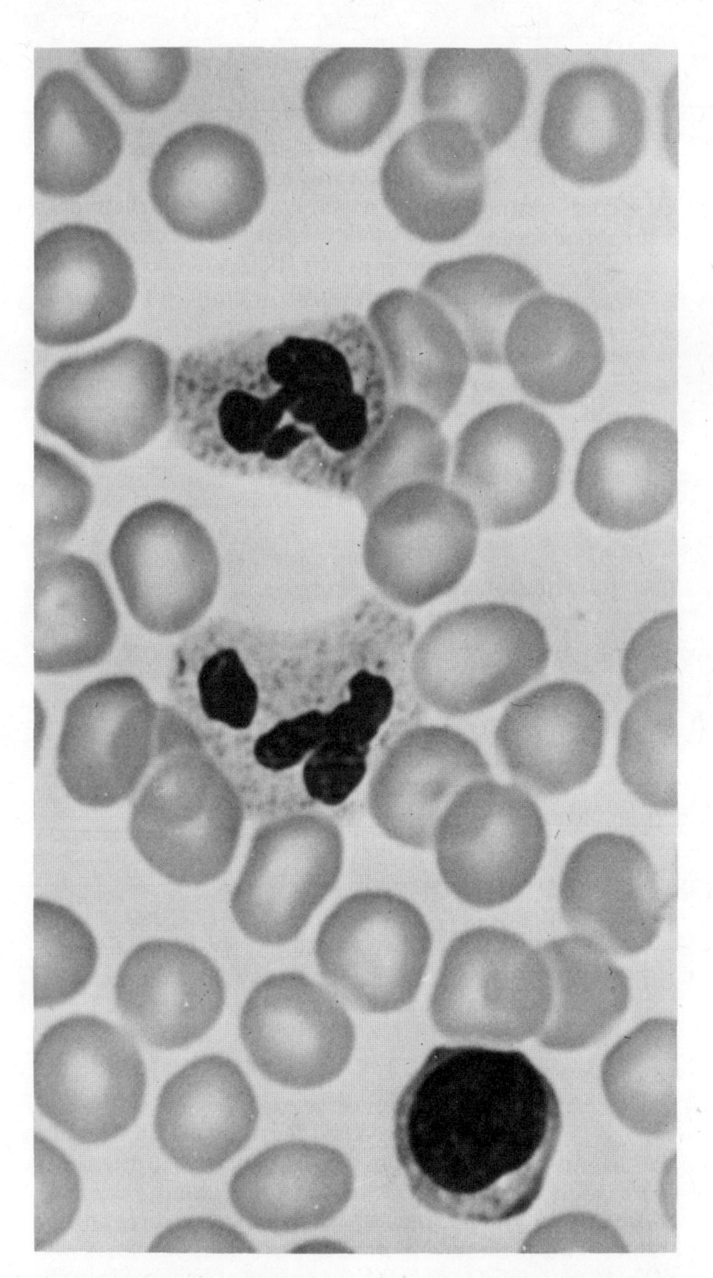
Fig. 20:12 High power photomicrograph of human blood showing a lymphocyte, two polymorphs and numerous red corpuscles.

LUNGS
oxygen concentration in alveoli higher than in the blood

haemoglobin (purplish red) + oxygen $\rightleftharpoons$ oxyhaemoglobin (bright red)

oxygen concentration in blood higher than in the tissues
TISSUES

The presence of haemoglobin in the blood allows 20–30 times more oxygen to be carried than if the oxygen was in simple solution.

In conditions where the amount of oxygen in the atmosphere is reduced, e.g. at high altitudes, the body responds by increasing the number of red corpuscles in the blood. This is one reason why climbers spend time acclimatising themselves at moderate altitudes before attempting the highest mountains. Without this training they would quickly become exhausted.

The red corpuscles become worn out after about four months and are broken down in the liver and spleen into bile salts and other substances which are used again to make more corpuscles.

Haemoglobin also combines with the poisonous gas carbon monoxide, which is a constituent of coal gas (not natural gas), car exhaust fumes and to a lesser extent in cigarette smoke (p. 99). When combination occurs, **carboxyhaemoglobin** is formed:

$$\text{carbon monoxide} + \underset{\text{(purplish red)}}{\text{haemoglobin}} \rightarrow \underset{\text{(cherry pink)}}{\text{carboxyhaemoglobin.}}$$

This reaction is irreversible, so it follows that if too much carbon monoxide is breathed in, there is insufficient haemoglobin left to combine with oxygen, and this condition quickly leads to death.

White corpuscles (leucocytes)

In human blood these are in the proportion of one white corpuscle to approximately 500 red. They are concerned with combating infection and disease. You will probably have seen in your blood preparation two types:

1. Polymorphs

These are formed in the bone marrow; they have an irregularly shaped nucleus. They can ingest bacteria in the same manner as amoeba, flowing round them and digesting them with enzymes. Polymorphs are able to squeeze through the walls of capillaries into the surrounding tissues, and if there is a wound, enormous numbers will migrate to the site to prevent the spread of bacteria into the body.

2. Lymphocytes

These have a large nucleus which nearly fills the cell. They attack bacteria by producing chemical substances, **antibodies**, which react with the surfaces of the bacteria and often cause them to clump together. The polymorphs can then ingest them. Disease-producing organisms (pathogens) liberate poisonous excretory substances or **toxins**. It is these which often cause fever. The lymphocytes produce other kinds of antibodies called **antitoxins** which neutralise the toxins, but this often takes time. Lymphocytes are found in great numbers in the lymphatic system (p. 154).

One of the big problems with organ transplants is that white blood corpuscles recognise, for example, a newly transplanted kidney as 'enemy' tissue and immediately invade and try to destroy it. To prevent this, the patient has to take various types of drugs to make his own defence mechanisms less effective. This, of course, will make him more susceptible to infection. Many of the deaths following the early heart transplants were due not so much to rejection of the 'foreign' heart but to secondary infection in other parts of the body, such as pneumonia.

Platelets

These are minute fragments budded off from certain large cells found in the bone marrow. They are concerned with the blood clotting process (see below).

Plasma

This is the fluid part of the blood; it contains a large number of substances in solution (see Table 20:1). The commonest mineral salt carried is sodium chloride. Some of the proteins cannot pass through the capillary walls as their molecules are too large, but sugars (chiefly glucose), amino acids and fatty substances can pass through by diffusion, provided there is a concentration gradient.

Transport of carbon dioxide

Most of the carbon dioxide is also carried in the plasma in the form of bicarbonate ions. When carbon dioxide is formed in the cells during respiration it builds up a concentration higher than that in the blood, and so the carbon dioxide passes along the diffusion gradient into the blood. Here it dissolves in the plasma and forms carbonic acid which then dissociates into hydrogen and bicarbonate ions:

$$\underset{\text{water}}{H_2O} + \underset{\text{carbon dioxide}}{CO_2} \rightarrow \underset{\text{carbonic acid}}{H_2CO_3}$$

$$\underset{\text{carbonic acid}}{H_2CO_3} \rightarrow \underset{\text{hydrogen ion}}{H^+} + \underset{\text{bicarbonate ion}}{HCO_3^-}.$$

When the blood reaches the lung capillaries the reverse action takes place and the carbon dioxide diffuses out of the blood into the alveoli, once more passing along the diffusion gradient:

$$HCO_3^- + H^+ \rightarrow H_2CO_3 \rightarrow H_2O + CO_2.$$

Clotting of blood

The blood plasma contains a protein, **fibrinogen**. When tissues are damaged and bleeding occurs, substances in the blood platelets and plasma are released and 'trigger' off the clotting mechanism. Basically, clotting involves the conversion of fibrinogen into **fibrin** which is precipitated from the plasma in the form of a network of fibres in which blood corpuscles become enmeshed. Thus a clot is formed over the wound and the bleeding is stopped. Later the clot hardens to form a scab which protects the underlying tissues from bacterial infection. When new skin has been formed underneath, the scab loosens and comes off.

Clots can also occur without any external damage to the tissues; this may be due to a diseased artery wall. Portions of the clot may be carried away into the blood stream and block a smaller vessel, so stopping the circulation to the organ concerned. If this happens in the coronary artery which supplies the heart muscle, a heart attack or **coronary thrombosis** results. If it occurs in a brain artery, it causes a stroke or **cerebral thrombosis**. A cerebral

TABLE 20:1. COMPOSITION OF THE BLOOD

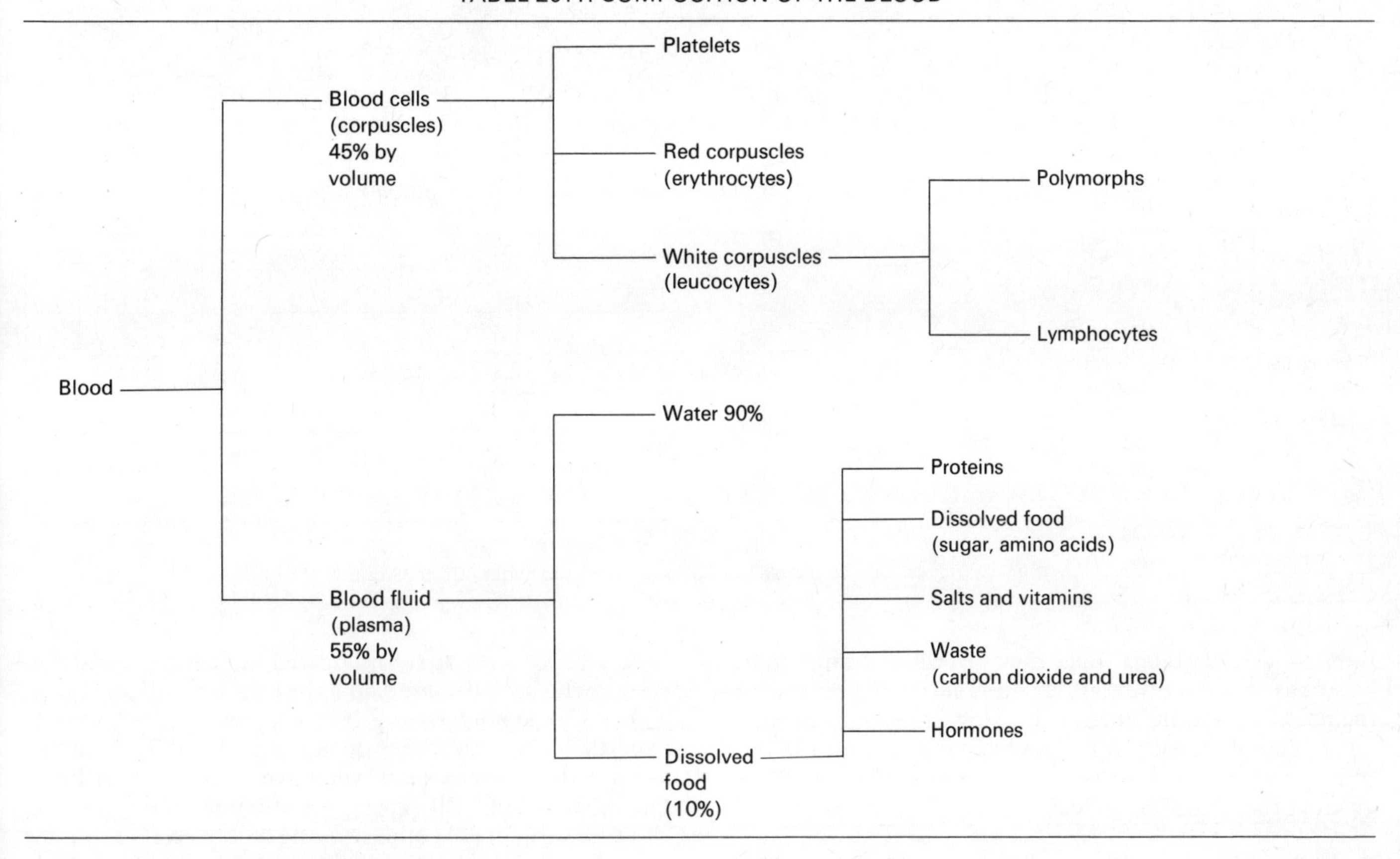

thrombosis can also be caused by the bursting of an artery in the brain. This is more usual in elderly people who have diseased arteries and a high blood pressure. A hereditary disease, **haemophilia**, is a condition where the blood is unable to clot properly, so wounds will not heal. In consequence a person with this disease may suffer a serious loss of blood through even a minor wound, or experience very severe pain and possible damage to the tissues through the pressure of blood collecting at the site of an internal injury, such as a bad bruise.

The composition of blood and its functions are summarised in Tables 20:1 and 20:2.

Blood groups

A person suffering from severe bleeding may have to have a blood transfusion, but first his blood group must be known. Mixing the wrong types of blood can be very dangerous and often fatal, as this causes the red corpuscles to clump together to form clots large enough to block a blood vessel. Blood types which do not mix are said to be **incompatible**. Two main types of protein called **antigens** occur on the surface of red blood cells. They are called A and B. People may have either of these, both, or none at all; so the four groups are known as A, B, AB and O.

TABLE 20:2. THE MAIN FUNCTIONS OF THE BLOOD

TRANSPORT	
1. Oxygen	Carried from lungs to tissues as oxyhaemoglobin in the red corpuscles.
2. Carbon dioxide	Taken from tissues to lungs as bicarbonate, mainly in the plasma.
3. Waste substances	Chiefly urea; formed in the liver and transported to the kidneys (p. 198).
4. Food	Most substances transported in solution in the plasma from the small intestine wall to the liver via the portal vein; then from the liver to the rest of the body. Most fatty substances enter the blood via the lymph (p. 154).
5. Hormones	Produced by ductless glands (p. 162); transported in the plasma.
6. Heat	Produced during respiration, particularly in muscles and in the liver. Distributed throughout the body, but regulated through the contraction or dilation of arterioles (p. 202).
DEFENCE AGAINST	a) White corpuscles (polymorphs) ingest bacteria.
HARMFUL ORGANISMS:	b) White corpuscles (lymphocytes) produce antibodies which neutralise harmful chemical substances.
	c) Substances in platelets and plasma cause blood to clot at a wound. Harmful bacteria are prevented from entering.

TABLE 20:3. BLOOD GROUP COMPATIBILITIES

		Antigens on donor's red blood cells			
		A (Group A)	B (Group B)	A and B Group AB)	NONE (Group O)
Antibodies in receiver's plasma	anti-B (Group A)	√	CLUMP	CLUMP	√
	anti-A (Group B)	CLUMP	√	CLUMP	√
	NONE (Group AB)	√	√	√	√
	anti-A and anti-B (Group O)	CLUMP	CLUMP	CLUMP	√

√ = will mix (compatible)

In Britain about 45–50% of people belong to Group O and about 40% to Group A

Similarly, the plasma may contain certain antibodies known as anti-A and anti-B. As these antibodies react with their corresponding antigen causing clumping, normal blood cannot contain, for example, antigen A on the red cells and its antibody, anti-A, in the plasma. The details are given in the following table:

Group	Antigen on red blood cell	Antibody in plasma
A	A	anti-B
B	B	anti-A
AB	A and B	NONE
O	NONE	anti-A and anti-B

Your blood group is determined by those of your parents according to Mendelian principles (p. 268).

The blood used for transfusion contains very little of the original plasma obtained from the donor, hence it is the corpuscles of the donor's blood that must be compatible with the plasma of the recipient. So, if Group A blood was transfused into a Group B recipient, the anti-A antibody in the receiver's plasma would immediately react with the A factor in the donor's red cells and cause clumping.

AB blood contains no antibodies and so will not clump the red cells of any group added to it; so people in this category are called **universal recipients**. Those of the O group are called **universal donors** as their red cells contain no antigens. This is summarised in Table 20:3.

The National Blood Transfusion Service is dependent upon blood donated voluntarily; it is stored in 'blood banks' ready for any emergency. The donors give a pint of blood each time. New donors between the ages of 18 and 65 years are always needed.

The rhesus factor

There is another antigen of red blood cells which is present in 85% of the people of Britain; this is known as the **rhesus factor**, as it was first discovered in rhesus monkeys. People who have this are said to be rhesus positive (Rh+). Those who do not have this factor are termed rhesus negative (Rh−). Normally they do not carry an antibody to this factor in their plasma. However, if Rh+ blood is transfused into the blood of a Rh− person, antibodies will be formed and these are capable of destroying Rh+ red cells. Under certain circumstances this is a potential hazard for babies.

If a Rh+ man marries a Rh− woman, some of the children are likely to be Rh+. At birth there is always some mixing of blood between the circulations of mother and baby and this may occasionally happen during pregnancy. So, if a child is Rh+ some of its blood will leak into its mother's circulation and cause antibodies to form in her blood. If the mother has more children, not all will necessarily be Rh+, but if they are, the amount of antibody in her blood often increases with each pregnancy, and in some instances the antibodies in her blood may pass into the baby's blood in sufficient quantities to produce very serious anaemia and even death. Fortunately these cases are infrequent, and when they do occur, the baby is given a complete transfusion soon after birth so that its blood is replaced by blood containing no antibodies to the rhesus factor. It is now possible for this transfusion to be carried out before birth. Another recently developed technique is for the mother to be given an injection shortly after the birth of her first child which prevents the Rh+ cells from stimulating the production of the harmful antibody.

The lymphatic system

If we lightly graze our skin a pale straw-coloured fluid oozes out; this is **lymph**. Much of it is derived from the fluid which seeps out through the blood capillary walls and bathes all our tissues. Lymph is the vital link between blood and tissues by which essential substances pass from blood to cells and excretory products from cells to blood (Fig. 20:13). It is similar in constitution to plasma, except that some of the plasma proteins are absent. Lymph con-

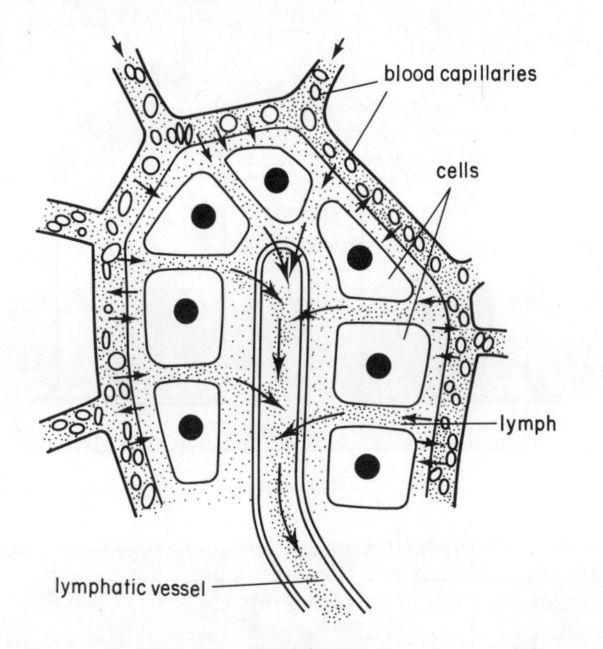

Fig. 20:13 Diagram showing the relationship between blood capillaries and the lymphatic system.

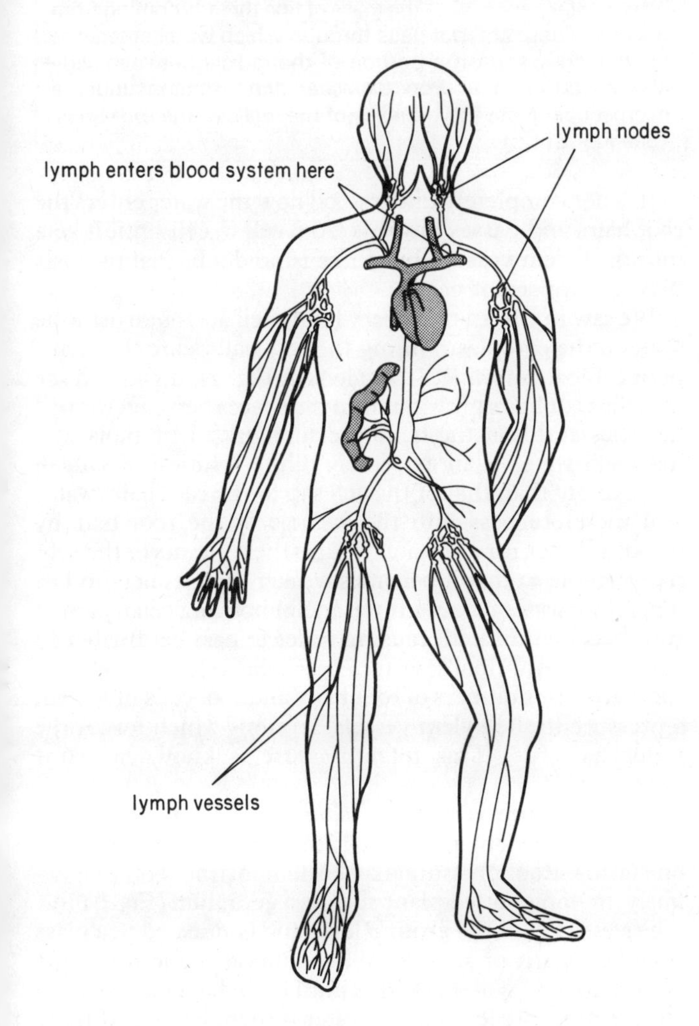

Fig. 20:14 The lymphatic system in man, much simplified.

tains no red blood cells, but considerably more lymphocytes than are present in blood.

Some of the lymph finds its way back into the blood stream, but most of it enters a network of blind-ending lymph capillaries. These lead to larger lymph vessels, rather like thin-walled veins. The surrounding muscles squeeze on these vessels and help to push the lymph along; one-way valves assist this process. At certain places the lymph passes into lymph nodes. These act as filters and poisonous substances are inactivated here. Lymphocytes are present in large numbers in these nodes and when we suffer from an infection the nodes swell up. For example, when we have a badly infected throat the glands on either side of the throat usually swell.

The tonsils which lie at the back of the throat, and the adenoids at the juncture between the back of the nose and the throat, are also made of lymphatic tissue. When we have tonsillitis the tonsils swell up in the process of combating the infection, causing them to become painful.

The lymph, having been 'purified' by the lymph nodes, finally enters the blood stream via two large veins on either side of the neck. The general distribution of the lymph channels and the main lymph nodes are shown in Fig. 20:14.

You will recall that the lymphatic system is also largely responsible for absorbing fatty substances from the intestine, the lacteals in the villi being part of this system.

Lymph, therefore, has both a circulatory and a defence function. It keeps the 'chemical environment' around the tissues constant. It is perhaps surprising that there is four to five times as much lymph in the body as there is blood.

21

Transporting materials within the plant

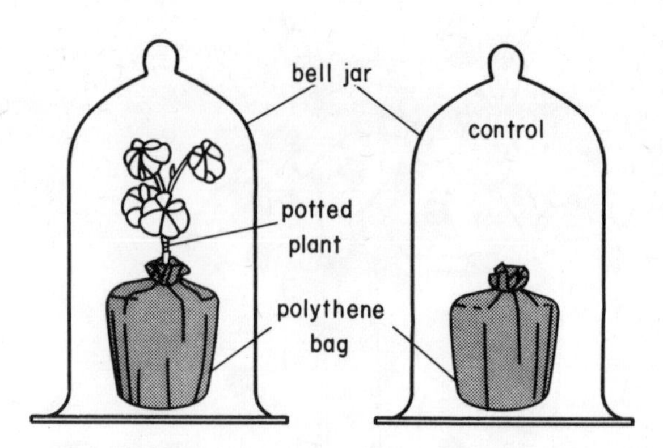

Fig. 21 : 1 Apparatus to determine whether water is given out by a green plant.

Is there anything in plants which corresponds to a blood system?

Every part of a plant needs water and if we grow plants in the house we have to water them regularly if they are to survive. It is therefore reasonable to suppose that there is a means of absorbing and transporting water from the roots to all the other parts. To test this hypothesis you can colour the water with a dye; if water is conducted the dye should stain the tissues in any region through which the water passes.

> Collect several small plants with their roots attached, wash away any soil which clings to them and place them in a beaker of 1% eosin so that the roots are covered by this red dye. Place in the same beaker some shoots from other plants which bear white flowers.
>
> After half an hour, take out one of the complete plants and cut it across at 10 mm intervals starting from the root end. Examine the cut ends with a hand lens and look out for traces of red.
>
> If you see any red colour, is the dye evenly distributed over the whole surface or is it confined to certain areas? How far has the dye travelled up the plant?
>
> Leave the other plants in the dye for twenty-four hours and then examine the stems, leaves and flowers. What deductions can you make from your observations regarding a possible transport system?

You may remember (Ch. 5) that in the root the xylem tissue was situated towards the centre while in the stem it was arranged in bundles near the outside. If the distribution of dye in your sections corresponds with that of the xylem it would confirm that this tissue in concerned with water conduction.

What happens to this water? Does it remain in the plant or is it given out again into the atmosphere? You could see if it passes out again by using the apparatus in Fig. 21 : 1, because if the atmosphere inside the bell jar became saturated with water vapour it would condense as drops on the cold walls of the jar.

> Set up the apparatus as in Fig. 21 : 1. Why should you water the plant first? Why should you enclose the soil in a polythene bag? Why should there be a control containing no plant?
>
> Leave the apparatus for a few hours, or overnight.
>
> If the walls of the bell jar become moist, test whether this is water using white anhydrous copper sulphate; it will turn blue in the presence of water.

From these experiments you should have deduced that a plant has a transport system for water. We now have to consider the mechanics of the process.

How is water absorbed?

> Examine some mustard seedlings which have been grown on wet filter paper. Note the mass of fine threads coming from the root. These are root hairs through which water enters the plant. Gently squash a portion of the radicle between slide and coverslip in a drop of water and examine under a microscope. Note the thinness of the walls of the root hairs. (See Fig. 21 : 2.)

It is not completely understood how the water enters the root hairs and passes inwards from cell to cell until it gets into the xylem vessels, but there is no doubt that osmosis plays an important part.

We saw in Ch. 12 that every living cell acts as an osmotic system, the cytoplasm lining the cell wall being the semi-permeable membrane. If you look at Fig. 21 : 2 you will see that the root hairs grow out into the spaces between the soil particles and that the hairs are surrounded by moisture. This soil water is an extremely dilute solution of salts—more dilute than that of the cell sap in the root hair; water will therefore pass into the vacuole of the root hair by osmosis. The entry of water dilutes the contents of the root hair vacuole so that it becomes weaker than its neighbour. Therefore water passes into the neighbouring cell which in turn becomes diluted, causing water to pass yet further in and so on until finally water enters the xylem vessels. As there are vast numbers of root hairs and root cells involved, a pressure in the xylem vessels develops which forces the water upwards. This total pressure is known as **root pressure**.

Root pressure is not the *main* cause of movement of water in the xylem as we shall see later, but it is certainly one factor. Root pressure may be demonstrated on a vigorously growing potted plant such as a geranium (Fig. 21 : 3). The stem is cut near ground level and connected to a glass tube by means of strong rubber tubing, the joints being bound tightly. Water is added until it can be seen above the rubber tube; the level of the water is then marked. If there is a pressure of water from below, the level will rise.

In spring and early summer, root pressure in trees can be quite considerable and if a branch is cut at this time, sap

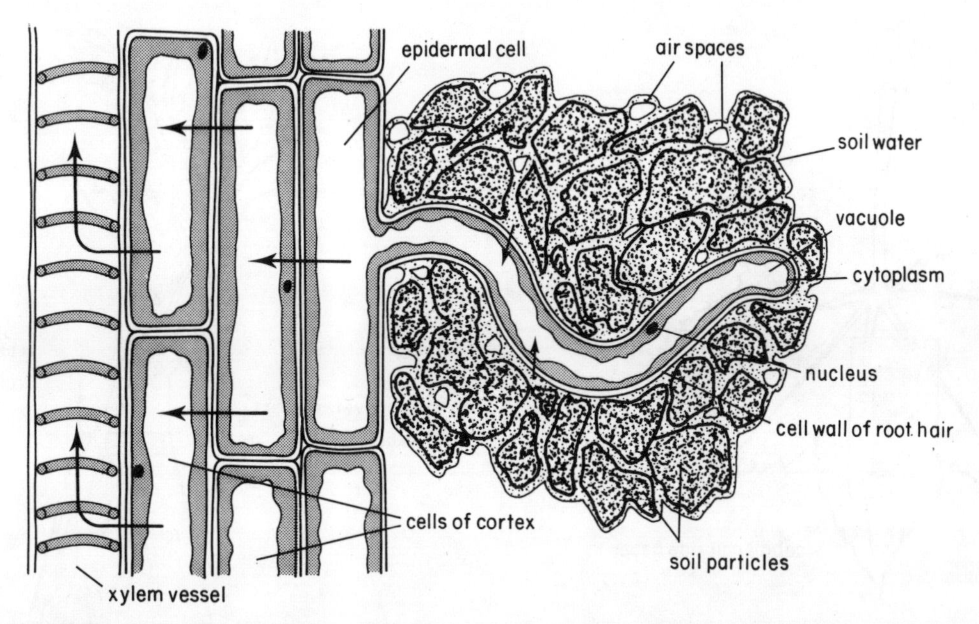

Fig. 21:2 Simplified diagram of part of a plant root in longitudinal section showing the relationship of the root hair to the soil water. Arrows indicate the movement of water.

oozes from the cut surface, but at other times the pressure is greatly reduced and plays little part in moving the water up the xylem.

How does the plant lose water?

We saw earlier that water evaporated from the surface of the shoot and it became visible as droplets when it condensed on the cold walls of the bell jar. This evaporation of water from plants is called **transpiration**. It occurs through the surface of the whole shoot, but some parts lose more water than others. You can investigate this by comparing the evaporation rate from the two surfaces of a leaf.

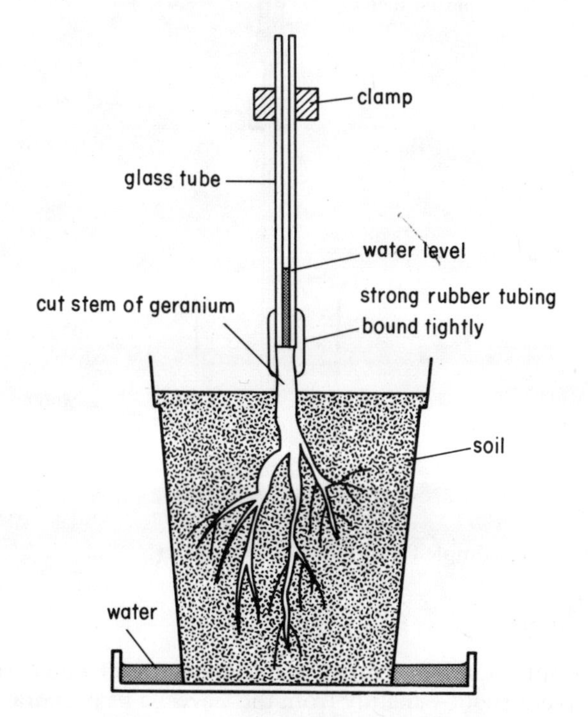

Fig. 21:3 Apparatus for demonstrating root pressure.

The principle of the experiment is to use small pieces of cobalt chloride paper on both sides of a leaf and see if one turns from blue (dry) to pink (damp) more quickly than the other.

> Use two rather different leaves such as sycamore and iris. Make sure that the pieces of cobalt chloride paper you use have been kept in a desiccator to remove all traces of moisture. Why should you handle them with forceps and not with your fingers?
>
> Place a piece of cobalt chloride paper on one of the leaves and cover it with adhesive tape, being careful to make an air-tight seal (Fig. 21:4). Turn the leaf over and repeat the procedure. Treat the second leaf in the same way.
>
> Which sides of the two leaves lose water most quickly? (The underside of the iris is the one where the mid-rib is most prominent.)

It is reasonable to suppose that this result could be related to the distribution of the stomata on the leaves. Are there more stomata on one surface than the other in the two specimens? You can find out in this way:

> Paint some nail varnish on to a part of each of the two surfaces of each leaf. When dry, carefully peel off each film and examine it under the microscope. The impression of each stoma will show up. Compare the number of stomata on the two surfaces of both leaves by counting all the impressions you can see under the high power field of the microscope. Choose five places for each surface and take the average, ignoring the stomata which are more than half out of the field of view.
>
> How do these results compare with those from your previous experiment?

From your observations you should have seen that leaves differ in the number and distribution of their stomata. In some leaves there may be as many as 20,000 stomata per cm^2 on the under surface and as many as 10,000 on the upper, but many leaves have no stomata at all on the upper surface.

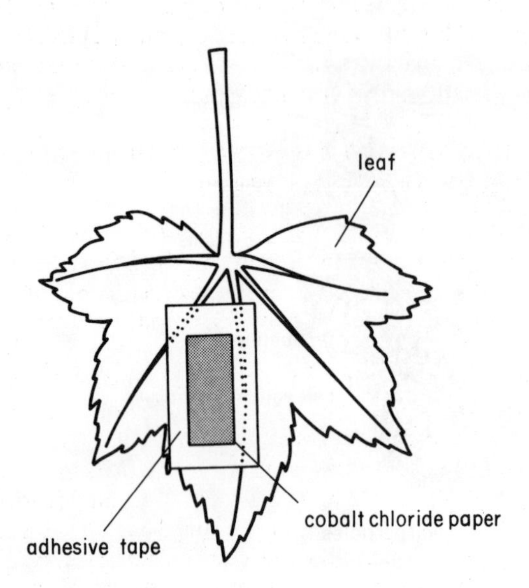

Fig. 21:4 Method of comparing the rates of transpiration from the two sides of leaf.

Leaves also differ in the thickness of their cuticle. This is a hard, waxy secretion which covers the outer walls of the epidermal cells. In leaves such as holly it is very thick and makes the leaf stiff; in flimsy leaves it is thin. Some water can evaporate through a very thin cuticle, but when thick, the cuticle is impervious to water. If there were no stomata, all water loss could be prevented by having a thick cuticle, but the plant has to have some means of taking in and giving out gases such as oxygen and carbon dioxide, hence stomata are needed and some water will automatically evaporate too. Transpiration is therefore an inevitable process in plants.

Plants living in dry or exposed situations are adapted by having stomata only on the lower surface of the leaf, often protected by hairs or situated in grooves or pits, while plants living in damp, shaded woodland need no such adaptations.

Transpiration has its dangers if there is insufficient water in the soil to replace what is lost. If you forget to water potted plants, the leaves soon wilt. This occurs when transpiration exceeds absorption, and the cells lose their turgidity. If wilting is prolonged, it may cause the death of some plants. Wild plants living in their natural situations are seldom seen to be wilted, although under extreme conditions this may happen. This suggests that for most plants the water absorbed is equal to the water transpired.

You could find out if this is true by using a **weight potometer**.

> Set up the apparatus as in Fig. 21:5.
>
> Measure the change in weight of the whole apparatus over a period of 24 hours. This will give you the weight of water lost to the atmosphere.
>
> Calculate the amount of water absorbed by the plant over the same period by noting the change in volume of water in the flask. Do this by finding out how much water needs to be added to bring the water level up to the original mark.
>
> As 1 cm^3 of water weighs 1 g, check if the weight loss in grammes equals the volume in cm^3 of water added.

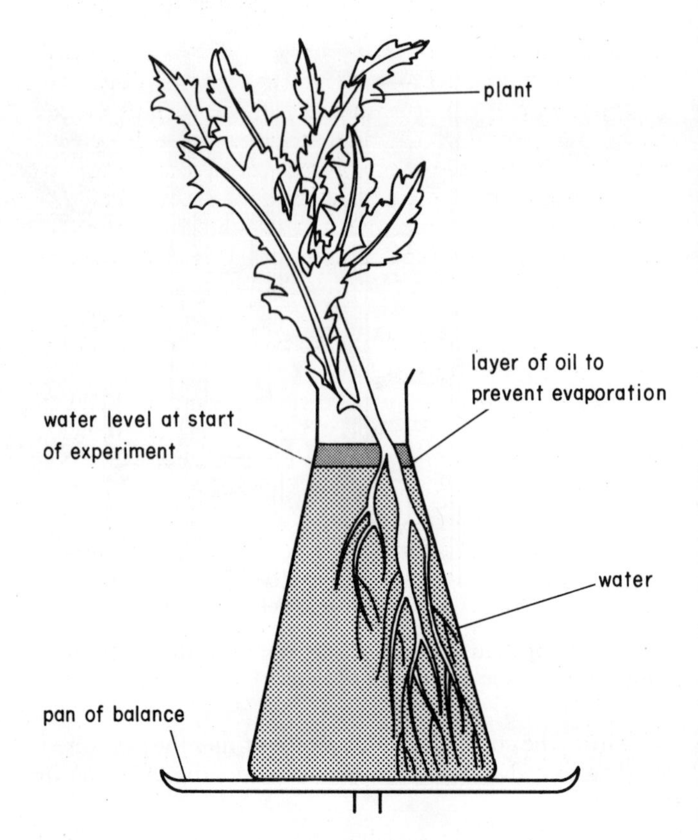

Fig. 21:5 A weight potometer.

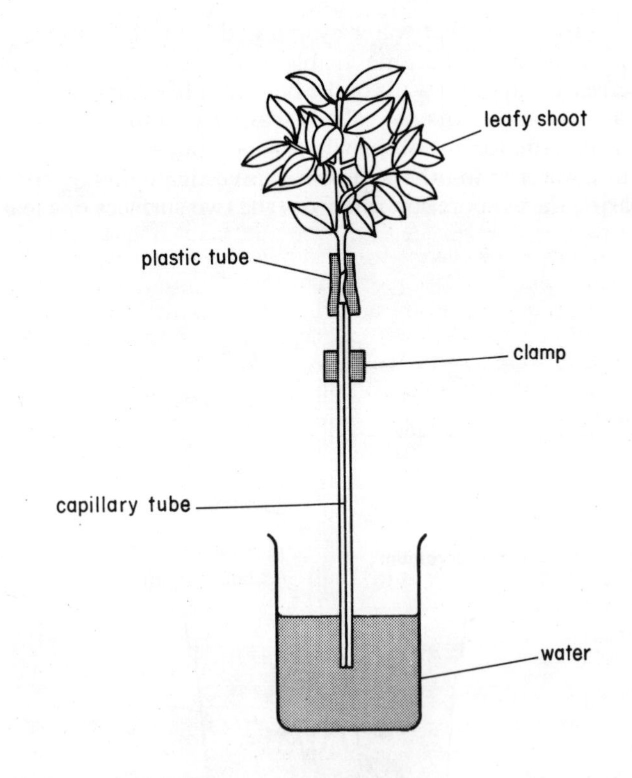

Fig. 21:6 A simple form of bubble potometer.

The rate of transpiration

Transpiration is essentially the evaporation of water from the plant shoot—mainly from the leaves. The evaporation actually takes place *inside* the leaves from the wet surfaces of the cells into the air spaces; it then diffuses out, mainly through the stomata. This diffusion process will depend on the conditions of the atmosphere outside, such as the humidity, temperature, the amount of wind and perhaps light. You can test the effect of these conditions on the rate of transpiration by using a **bubble potometer** (Fig. 21:6) which is more sensitive than the weight potometer and gives continuous readings. This potometer really measures the rate of *absorption*, but as shown in the last experiment, when plants are given plenty of water, absorption is, for all practical purposes, equal to transpiration, so this apparatus is used to measure transpiration.

The principle is that if a bubble of air is introduced into the capillary tube it will be drawn up by the water column as absorption of water takes place. The rate of movement of the bubble can be measured by taking the time for it to travel between two points marked on the capillary tube or on a scale fixed parallel to it. (An average of three readings is better than one—why?)

1. Cut a woody shoot round enough and with a diameter large enough to make an airtight joint with the plastic tube. Plunge it immediately into a jar of water to prevent air from entering the xylem vessels.
2. Assemble the apparatus under water to ensure that no air gets in for the same reason, taking care not to wet the leaves (why?).
3. Leave the apparatus for a few minutes to allow the plant to become adjusted to the external conditions.
4. Introduce the bubble by lifting the tube out of the beaker, blot the end with filter paper and replace. Take the time for the bubble to move between the two chosen points, then, with the tube under water, squeeze the plastic tube to expel the bubble, otherwise air will collect below the cut end of the shoot. Another bubble can then be introduced.
5. Using this apparatus, devise a means for comparing the rates of transpiration under different conditions. Give the shoot time to become adjusted to the change before taking readings. Here are some suggestions: an electric fan is a good, steady source of wind, and plastic hoods—one black and one transparent—could be used to cover the shoot loosely to show the effect of bright and dim light. What about the effect of temperature or humidity? Be critical of your methods to ensure that by altering *one* condition you do not alter another as well.

What effect on the rate of transpiration would you expect if you covered: a) the upper surface or b) the lower surface of each leaf of the shoot with petroleum jelly? Would any changes be dependent upon the species of leaf being used? What change in rate would you expect if you cut off half the leaves with scissors?

We can summarise the effect of external conditions on the rate of transpiration as follows:

1. An increase in temperature increases the rate because a) air can contain more water vapour at higher temperatures, so it increases the diffusion gradient between the air in the leaf and the air outside, b) it speeds up the process of evaporation.
2. An increase in the humidity of the air decreases the rate because the diffusion gradient is lessened, and if the air is saturated, transpiration stops altogether.
3. Wind increases the rate, because as soon as the air outside the leaf receives moisture from inside by diffusion, it is replaced by drier air, thus keeping the gradient higher.
4. Light. This has no *direct* effect, but light does cause stomata to open more widely (p. 107) and this may cause an increase in the rate of transpiration.

The mechanism by which the water travels through the plant

We have seen that there is a push from below due to root pressure on the columns of water in the xylem vessels, but this is seldom large and at some seasons it is nil. How does the water reach the top of a tree like a giant redwood 120 m high?

The principles involved can be demonstrated by using the apparatus in Fig. 21:7: a) is a leafy shoot immersed in water in a tube dipping into some immiscible fluid such as mercury; b) is a porous pot fitted up in the same way.

The wall of a porous pot is a mass of minute 'tubes' and when the water evaporates at the surface it causes a pull on the water columns and the mercury is drawn up. Similarly, when the leaves transpire there is a pulling effect on the continuous columns of water in the xylem vessels. The top ends of these vessels are surrounded by the leaf's mesophyll cells which contain sap, so the water is continuous from the xylem vessels to the walls of the mesophyll cells from which it evaporates into the air spaces causing the pull. The water column does not break because of its great tensile strength. This is a property of water you demonstrate every time you drink through a straw.

We now have a picture of the water-conducting system of a tree. Water is absorbed by osmosis from the soil by the root hairs and is passed into the xylem vessels which form a continuous system of tubes through root and stem into the leaves; here the water evaporates and passes into the atmos-

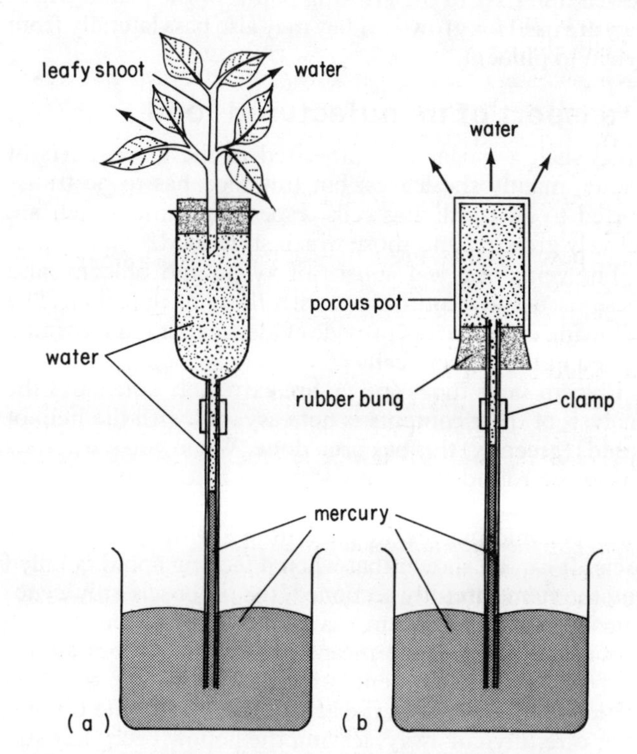

Fig. 21:7 Apparatus for investigating the effect on a column of water of a) transpiration b) evaporation.

phere. The evaporation creates the main pull from above, root pressure gives a variable and minor push from below. The result is a continuous column of moving water, the **transpiration stream.**

The amount of water passing through a plant is often considerable. For example, an oak tree can transpire as much as 900 litres of water per day (200 gallons). It follows therefore that areas of forest significantly affect the degree of saturation of the air above them, so that when air currents bring air which is already nearly saturated to a forest area, it becomes fully saturated and comes down as rain; this is why forest areas often have a higher rainfall than areas nearby.

Transport of mineral salts

You will recall that mineral salts are necessary for plant nutrition and that they are obtained from the soil in solution through the root hairs. The salts are in the form of electrically charged ions. Thus sodium chloride (NaCl) is in the form of Na^+ and Cl^-, and magnesium sulphate ($MgSO_4$) occurs as Mg^{++} and SO_4^{--}. They are *not* absorbed into the root hairs by the simple process of diffusion, as you would expect, for the cell sap of the root hairs contains a higher concentration of ions than in the soil water under normal conditions, and if diffusion did occur, the ions would pass from the plant to the soil. The detailed mechanism of absorption is not fully understood, but it must involve the use of energy by the cytoplasm, because if roots are deprived of oxygen their ability to absorb ions is reduced. It has been found that cytoplasm, when surrounded by solutions of ions, absorbs some ions more than others and in this sense cytoplasm can be said to be selective.

Once absorbed, the ions travel in the water in the xylem vessels and pass to the growing points of the plants where they are used for growth. They may also pass laterally from xylem to phloem.

Transport of manufactured food

Food such as sugar is synthesised in the green parts of plants, mainly the leaves, but this food has to be transported to all the living cells, especially those which are actively growing and those which store food.

The veins of a leaf consist of xylem and phloem, and these tissues are continuous with those of the stem. The following experiments provide evidence that food is transported in the phloem cells.

Phloem sieve tubes (p. 36) are extremely small and the analysis of their contents is not easy, but with the help of aphids (greenfly) this has been done. When you see aphids clustering round the young stems of roses or broad bean plants they are feeding on the plant juices. To obtain this juice an aphid pierces the plant tissues with its long proboscis. It can be shown that when a feeding aphid is killed and the stem carefully sectioned, the proboscis only penetrates as far as a phloem sieve tube. This proboscis also provides a ready-made means of obtaining the juice for analysis! The experiment can be done in this way. An aphid is killed while in the act of feeding and the body is then carefully cut away, leaving the hollow proboscis still inserted into the phloem (Fig. 21:8). It is found that because the contents of the phloem sieve tubes are under

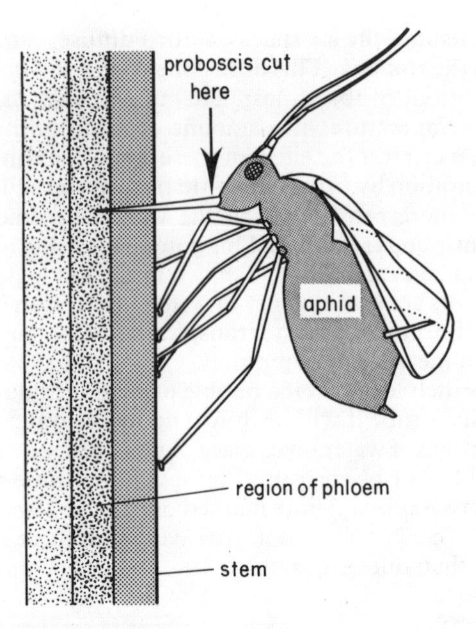

Fig. 21:8 Method of using the proboscis of an aphid for extracting fluid from phloem sieve tubes.

slight pressure the fluid slowly exudes from the cut end of the proboscis in the form of drops; these drops are then collected and analysed. The fluid is found to contain sugars and amino acids.

Not surprisingly, aphids absorb so much sugar from the phloem that they cannot assimilate all of it and it passes out of the anus as a sticky syrup called **honey-dew**. Leaves which have been attacked by aphids often feel sticky as a result.

Further experiments to illustrate the conduction of sugars by the phloem have been done by removing a ring of bark from a shoot to expose the wood. This in effect removes all tissues from the cambium outwards, including the phloem. After a few days, when the tissues above and below the ring were analysed it was shown that food had accumulated above the ring, but was not present below it. If left for some time, the stem increased in thickness immediately above the ring, but no growth occurred below it (Fig. 21:9). So any damage to the phloem all around the stem will prevent food from passing down to the roots and the tree will eventually die. This is a fact of great economic

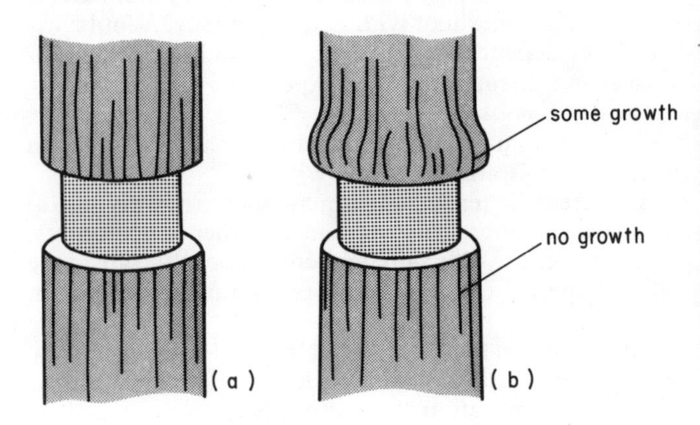

Fig. 21:9 Ringing experiment: a) ring of bark removed b) the same stem after some weeks.

Fig. 21:10 One of these photographs shows typical rabbit damage, the other damage by grey squirrels. Can you determine which is which?

importance because certain mammals gnaw the bark of trees to get at the food stored in the phloem, especially during hard winters when food is scarce. Voles do this to young saplings at ground level and rabbits can do much damage to older ones. Foresters find it economically worth while to enclose new plantations with wire netting to prevent rabbits from entering.

Foresters also encourage predators such as foxes, badgers, hawks and owls as they help to keep down the population of voles and rabbits. Grey squirrels too do great damage, particularly to beech and sycamore, and for this reason, in some parts it is impossible to grow these trees as a crop. When you next go into a wood look out for evidence of bark having been gnawed off saplings and trees. Note the species of tree, the position of the damage, whether the damage is recent or old, and the size of tooth marks if these are visible. From these observations you could find out which species had caused the damage. Also look out for the effect of such damage on the tree as a whole.

22

The endocrine system

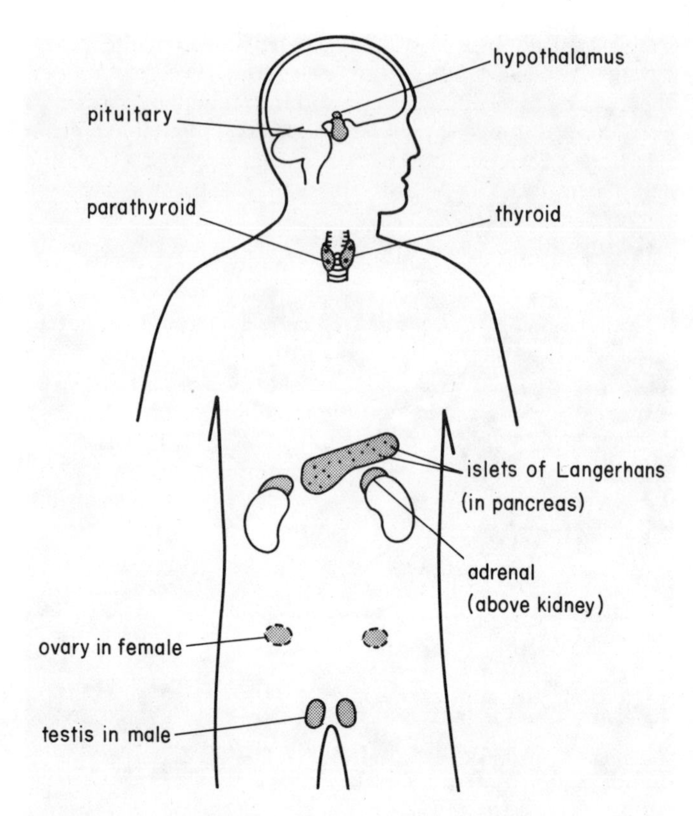

Fig. 22:1 The main glands of the endocrine system.

The term **endocrine** means 'internally secreting'. This system consists of all the glands of the body which do *not* pass their secretions down ducts, but pour them directly into the capillaries of the blood system which penetrate the gland; they are therefore called **ductless glands**. The names and positions of these glands are shown in Fig. 22:1.

Ductless glands secrete chemical substances called **hormones** which are carried by the blood to all parts and have a great influence on the body generally and, in some cases, on certain organs in particular, called **target organs**. It appears as if the hormone carries a coded message which only the cells of the target organ can decode and act upon. Hormones are often described as **chemical messengers** for this reason.

Hormones are not controlled consciously and are secreted by the glands in varying quantities according to the body's needs. The cells they influence respond at different rates so that hormones can be used both for slow reactions such as growth and high-speed emergency reactions such as those caused by fright.

Most ductless glands secrete several hormones which vary considerably in function. However, most functions may be grouped into four categories:

1. To co-ordinate the action of certain glands or organs.
2. To control metabolic processes.
3. To regulate growth.
4. To influence the development of sexual features, the process of reproduction and lactation.

As some glands secrete hormones which come into more than one category we will consider the hormones in the context of the glands that secrete them.

The adrenal glands

These are paired structures situated near the kidneys. They consist of two parts, each of which secretes quite different hormones. The outer part—the **cortex**—secretes several which are concerned with salt and water balance, maintenance of blood pressure and resistance to stress; the inner part—the **medulla**—secretes **adrenaline**, an important co-ordinating hormone.

Adrenaline is secreted when we become suddenly frightened. We know the feeling only too well: our heart starts thumping, we get a sinking feeling in our stomach, we break out into a sweat, our skin goes pale, our pupils dilate and it is even possible, sometimes, for our hair to stand on end. All these reactions are caused by adrenaline, and their combined effect, along with other internal reactions which we cannot feel, prepares the body for such actions as 'fight or flight'. In such an emergency the body suddenly needs a large supply of energy for the muscles and many of the fright symptoms we experience are connected with this action:

1. The heart beats faster and more strongly, so blood reaches the muscles more quickly.
2. The walls of many small arteries constrict; those in the skin cause paleness, those in the gut give us that sinking feeling. These actions combine to direct the blood to the parts which need it most, especially the muscles, the brain and the lungs.
3. The mouth opens, allowing more air to enter, and the breathing rate increases, so more oxygen passes into the blood and reaches the muscles, and more energy can be released.
4. More blood cells are released from the spleen, so more oxygen can be carried.
5. More glycogen in the liver is broken down into sugar, the level of blood sugar rises, and so more is available for energy release.
6. Sweating helps to reduce the temperature of the muscles.

In animals such as cats, the raising of the hair and the dilation of the pupils help to give the impression of greater size and fierceness; this may make the aggressor hesitate before attacking.

Adrenaline is also secreted when we are nervous or worried. It causes the rather unpleasant feeling we have when we are waiting for an interview, or for a match to begin, or when we are called to do something in front of a lot of people. It is also secreted when we have a bad dream and wake up with our heart thumping away and our skin sweating. Fortunately, the effects of this hormone do not last long as adrenaline is quickly destroyed after the emergency. However, adrenaline is not only secreted when there is an emergency; it is passed into the blood stream in small amounts all the time. But too much regular secretion of adrenaline over a long period, as may occur when people are living under constant stress, may cause the heart to become overworked.

The gut

The first hormone to be discovered was named **secretin**. This, like adrenaline, has a co-ordinating function. When food enters the duodenum from the stomach certain of the lining cells, as a result of coming into contact with the food, secrete this hormone into the blood stream. While circulating round the body the only organ it affects is the pancreas which is stimulated to secrete pancreatic juices, thus ensuring that the enzymes reach the food at the right time.

Another hormone, **gastrin**, causes acid to be secreted by cells in the stomach wall when food enters that organ and a further hormone **enterogastrone** stops acid production when the food has passed out of it.

The pancreas

We have seen that the pancreas secretes digestive enzymes down the pancreatic duct into the duodenum (p. 129) thus acting as a normal gland, but certain areas within the pancreas, the **islets of Langerhans**, behave as ductless glands and secrete the hormone **insulin**. This controls the level of sugar in the blood by causing excess sugar to be taken up by liver cells and converted into glycogen. Insulin is thus a metabolic hormone.

People who are unable to produce enough insulin suffer from **diabetes mellitus**. With this condition the blood sugar level rises above normal and as a result some sugar appears in the urine. Thus diabetes can be diagnosed by testing the urine for sugar with Benedict's solution.

Diabetes is a relatively common condition, but it varies in its severity. The normal treatment is to have regular injections of insulin prepared from the pancreas of animals. The quantity needed has to be accurately calculated according to the amount of excess sugar in the blood, and the diet is regulated to keep the sugar intake as constant as possible and at a low level.

The thyroid gland

This is situated in the neck just below the larynx; it consists of two lobes which lie on either side of the wind pipe and are connected together. It secretes the hormone **thyroxine**, a substance containing iodine. This explains the importance of having enough iodine in the diet to allow for the synthesis of the hormone (p. 112).

Thyroxine has two main functions: it influences the rate of certain basic metabolic reactions, especially the liberation of energy from glucose in respiration, and it controls the rate of growth and so affects development. If too little thyroxine is secreted during infancy, the child becomes a **cretin**. Cretinism is a condition where growth is stunted, sexual maturity is not reached and the person is mentally retarded. It can be cured through early treatment with thyroxine. Deficiency of thyroxine in adults results in **myxoedema**, a condition in which the person becomes sluggish, fat and slow-witted.

Over-activity of the thyroid is accompanied by a swelling of the gland so that the neck enlarges; the person becomes over-active, nervous and thin, a condition known as **thyrotoxicosis**. It can be treated surgically by the removal of some of the gland, or medically by the injection of a drug which counteracts the production of the hormone.

In amphibians, thyroxine is essential for metamorphosis, although growth continues without it. Tadpoles fail to turn into frogs when the thyroid gland is destroyed, but continue to grow into abnormally large tadpoles. If small concentrations of thyroxine are added to water containing frog tadpoles at the time when they are just beginning to form hind legs, they will metamorphose into frogs in a much shorter time than those kept as a control without such treatment. As a result, the young frogs are much smaller than normal as they have not had so much time to grow.

The parathyroid glands

These occur within the tissues of the thyroid and produce a metabolic hormone which controls the amount of calcium in blood and bones.

The gonads

The ovary and testis not only produce eggs and sperms but they also act as ductless glands secreting the sex hormones. The latter influence the development of sexual characteristics and, along with hormones from the pituitary, coordinate the reproductive cycles. We will consider their action more fully in Ch. 23.

The hypothalamus and pituitary gland

The pituitary and hypothalamus together exert a tremendous and vital influence on the body as they control the main glands of the endocrine system by regulating the quantity of each hormone in the blood. By controlling the balance between the hormones they allow the body processes to work efficiently.

The hypothalamus is the part of the brain which lies just above the pituitary gland. It forms an important link between the nervous and endocrine systems. Its main function, apart from secreting certain hormones, is to stimulate the pituitary to release hormones at appropriate times.

The pituitary gland is a small rounded gland connected by a stalk to the hypothalamus region at the base of the brain. It has two lobes, anterior and posterior. One of the hormones of the posterior lobe stimulates the uterine muscles to contract at childbirth (p. 170); another helps to control the water content of the body (p. 200). Both are made in the hypothalamus and stored and released when needed by the posterior lobe of the pituitary. However, it is the hormones of the anterior lobe that have the more widespread effect. They include the following:

1. A growth hormone which has a special influence on the long bones and those of the hands and feet. When people have too much of this hormone during development they become giants; if too little, they become dwarfs.
2. A group of **trophic** hormones, i.e. those which stimulate other ductless glands to secrete their hormones; they include one which stimulates the thyroid to form thyro-

xine, another which stimulates the adrenal glands to secrete cortisone and another that influences gonad development and the production of gametes.

Feed-back

Balancing the hormone levels in the blood by the hypothalamus and pituitary is achieved by a process called **feed-back**. There are many feed-back mechanisms, some highly complex, but we will illustrate the process by referring to a simple one involving the level of thyroxine in the blood.

The thyroid produces thyroxine. The amount it secretes varies according to the quantity of **thyroid-stimulating hormone (TSH)** secreted by the pituitary. The level of TSH is monitored by the hypothalamus and if it becomes too high, the hypothalamus causes the pituitary to reduce the secretion of TSH, thus lowering the level of thyroxine. This cycle is repeated whenever necessary and is an example of a feed-back mechanism. However, there is a further means of controlling the amount of thyroxine in the blood. If the level of thyroxine becomes too high this is also monitored by the hypothalamus which again causes the pituitary to release less TSH. So two feed-back mechanisms control the level of thyroxine, one by monitoring the TSH directly, the other by monitoring the thyroxine itself (Fig. 22:2).

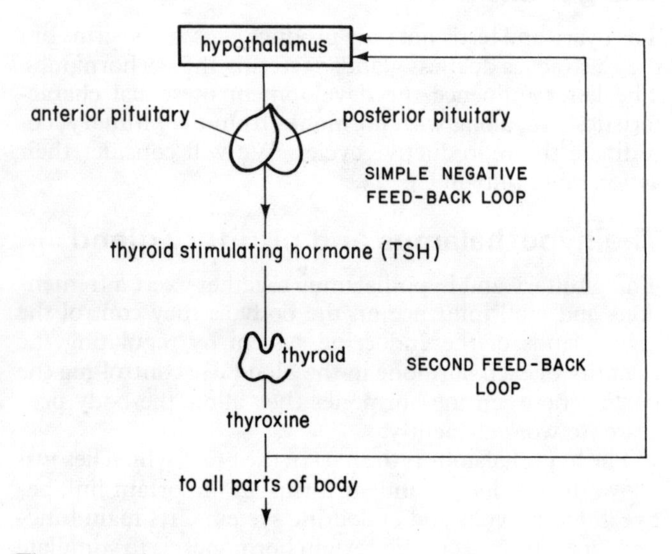

Fig. 22:2 The mechanism involved in controlling the level of thyroxine in the blood.

We will consider certain aspects of more complex feed-back mechanisms involving hormones when describing the ovarian and menstrual cycles concerned with reproduction (p. 171). However, feed-back mechanisms do not only control the level of hormones in the blood stream, they are used continuously and unconsciously to keep the whole body in a steady state, for example, by keeping the water content (p. 200) and the temperature (p. 202) of the body as constant as possible.

23

Human reproduction and development

Reproduction is a process which takes place in all living organisms. It is the ability to produce new individuals with the same general characteristics as the parents. Reproduction ensures that life is perpetuated.

Before discussing human reproduction we will summarise the principal methods of reproduction in organisms generally.

There are two main methods, **sexual** and **asexual**. When sexual reproduction occurs the new individuals are formed as a result of the fusion of two special nuclei from different cells usually derived from two parents; when asexual reproduction takes place the new individuals are not formed in this manner and only one parent is involved.

Asexual reproduction

There are many ways in which this is brought about, some of which we have already considered:

1. *Binary fission.* This occurs when organisms such as *Amoeba* and bacteria divide into two more or less equal parts. It is a quick and efficient method for simple organisms when conditions are favourable.
2. *Spore formation.* This occurs when an organism produces special structures called spores which are capable of growing into new organisms without the fusion of any nuclei. Many of the lower plants—for example, fungi, bacteria, mosses, liverworts and ferns—use this method; also some animals, such as the malarial parasite when it multiplies within the red blood corpuscles. Spores are usually produced in vast numbers, so theoretically a great many new individuals may result. In practice the majority seldom survive.
3. *Budding.* This occurs when a new individual gradually grows from the parent and eventually separates. We have seen this in *Hydra*, but sea anemones occasionally use the same method, the buds being formed internally. Yeast is a fungus which also produces new individuals by budding.
4. *Vegetative reproduction.* This term is used for a special type of budding much used by flowering plants; the buds grow out from the parent stem and form shoots which eventually separate into new plants. Propagation by runners, suckers, corms, bulbs, tubers and rhizomes are examples (p. 174).
5. *Parthenogenesis.* This method is unusual. It occurs when an egg cell remains unfertilized, but nevertheless develops into a new individual. It takes place in drone bees, aphids, water fleas (*Daphnia*) and the dandelion (p. 182).

Sexual reproduction

This results from the fusion of nuclei from separate sex cells called **gametes**, the fusion process being called **fertilization**. In all but the simplest organisms such as *Spirogyra*, the gametes are basically of two kinds: sperms which are derived from the male and ova (eggs) from the female. Sperms are motile and swim to the ovum to fertilize it. In flowering plants, however, the male gametes are formed in the pollen and are not motile. Fertilization may take place outside the body of the parent as in many fish and amphibians (**external fertilization**), or inside the female as in insects and mammals (**internal fertilization**). In animals the gametes are formed in special reproductive organs called **gonads**—testes in the male and ovaries in the female. When the gametes fuse a **zygote** is formed which grows by cell division into a new individual. (In some animals such as *Hydra*, earthworms and snails, both gonads are present in the same individual; they are said to be **hermaphrodite**.) The basic process of sexual reproduction can be summarised as follows:

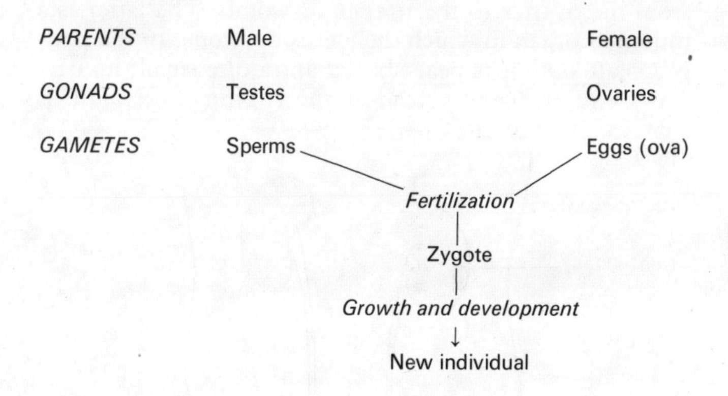

The stages of development vary somewhat according to the species so different terms are used for some of the stages. For example:

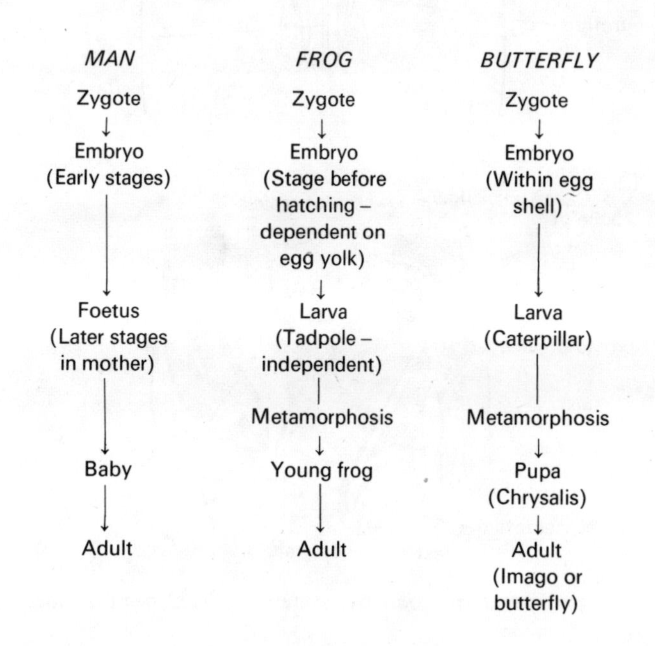

Human reproductive systems

In the male (Fig. 23:1) the primary reproductive organs are the two **testes** which produce the sperms. The testes lie in a special sac outside the abdominal cavity called the **scrotum**. In this position the temperature of the testes remains below that of the body, a condition which favours sperm production. The sperms are stored in the **epididymis**, an extremely long tube much coiled on itself which lies close to the testis. The epididymis leads to a long muscular tube, the **vas deferens** (pl. vasa deferentia). There are two of these, one from each testis, which unite at a point where they enter the **urethra**, just below the bladder. The urethra passes through the **penis**, the organ which transmits the sperms into the body of the female. The seminal vesicles and prostate gland secrete a fluid in which the sperms can be transferred during sexual intercourse.

In the female (Fig. 23:2), the two **ovaries** which produce the eggs are attached to the wall of the body cavity just below the kidneys. The two **fallopian tubes** carry the eggs from the ovaries to the **uterus** or womb. The latter is a muscular organ in which the foetus develops. In the non-pregnant state it is pear-shaped and quite small; its narrower end, the **cervix**, leads to the **vagina**. The vagina is muscular and connects the uterus with the outside world. It has in consequence two important functions: to receive the penis of the male during intercourse and to allow the baby to pass out of the body of the mother when it is ready to be born.

Sperm production, transmission and fertilization

Each testis consists of many greatly twisted, microscopic tubules bound together by connective tissue and is covered by a protective capsule to form a compact ovoid body (Fig. 23:3). The tubules produce the sperms by cell division (Figs. 23:4 and 5). Sperm production begins at puberty and continues throughout life. Puberty is the period of change from child to adult; it usually starts between the age of 11 and 13 years. Each sperm (Fig. 23:6), consists of a head, composed mainly of the nucleus, and a long cytoplasmic tail which enables it to swim towards, and possibly reach, an egg. The sperms collect in the epididymis and while there, they remain inactive.

Transmission of the sperms into the body of the female occurs during sexual intercourse. The penis is largely composed of spongy tissue and during intercourse it becomes stiff and erect due to the passage of blood into the spaces within this tissue; it can then be placed in the vagina of the female.

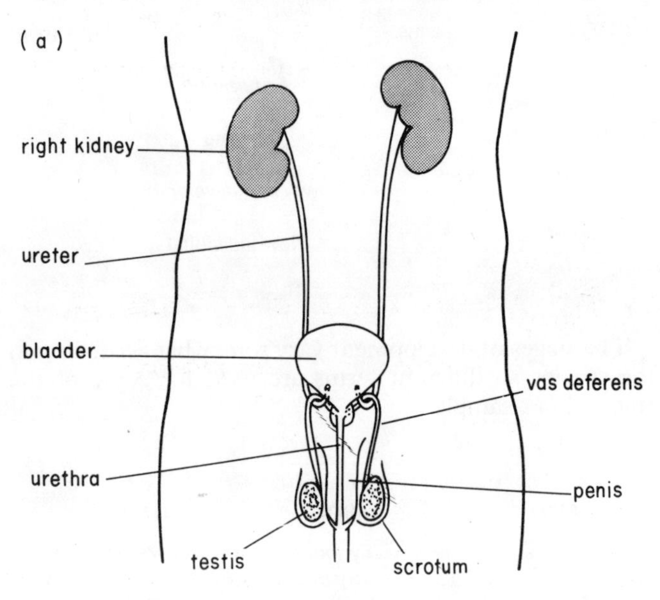

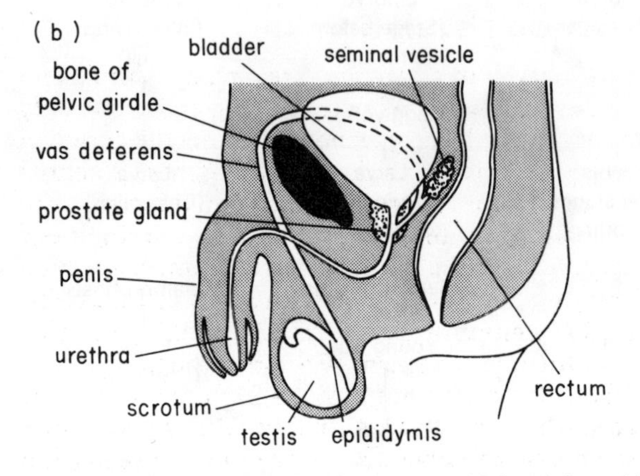

Fig. 23:1 Human reproductive system. Male: a) front b) side.

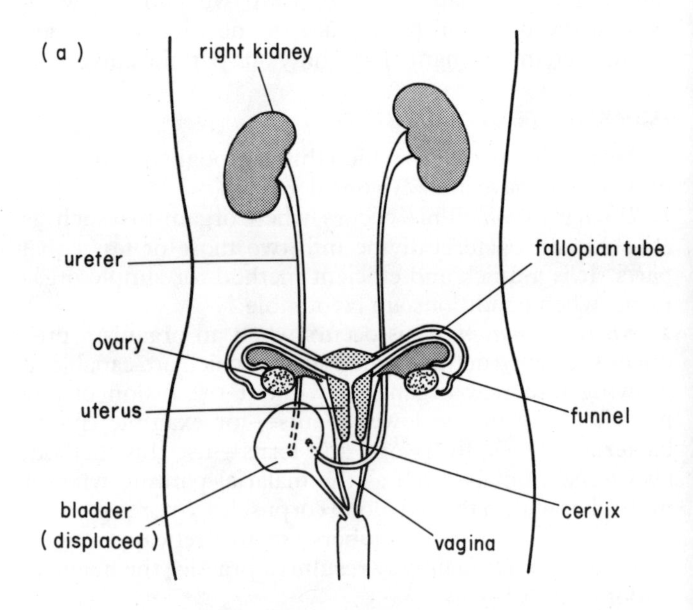

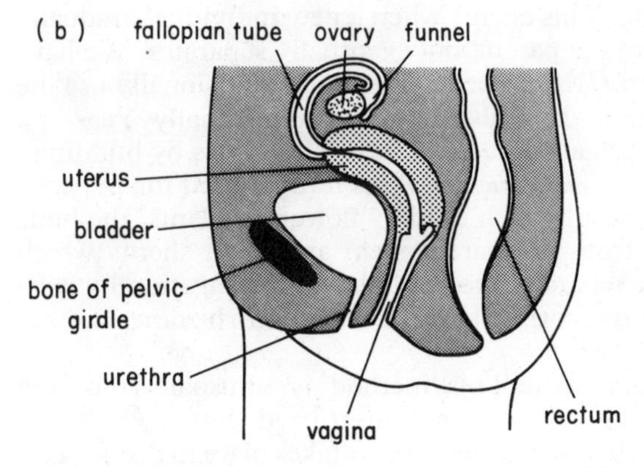

Fig. 23:2 Human reproductive system. Female: a) front b) side.

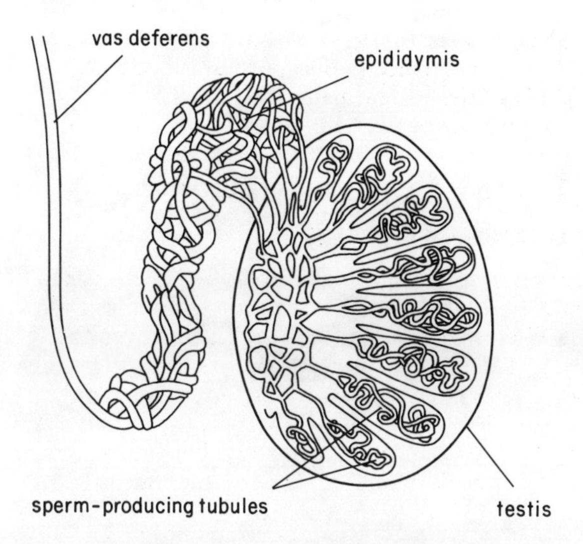

Fig. 23:3 Simplified diagram of a longitudinal section of a testis showing the arrangement of the tubules.

During intercourse, stimulation of sensory cells in the penis causes the walls of the epididymis and vas deferens to contract rhythmically and the sperms are passed into the urethra; here they mix with the fluid (**semen**) secreted by the seminal vesicles and prostate gland. This fluid, together with some 500 million sperms, is then ejaculated into the vagina; it supplies the sperms with nutrients and stimulates their swimming movements.

Some of this fluid, helped by movements of uterus and vagina, passes through the cervix into the uterus. Many of the sperms die, but a few thousand, by lashing their tails, may reach the fallopian tubes and a few hundred approach an egg. The sperms near the egg secrete digestive enzymes which enable a few of them to pass through the jelly surrounding the egg (Fig. 23:7), but only one penetrates the egg membrane which then, by a change in its composition,

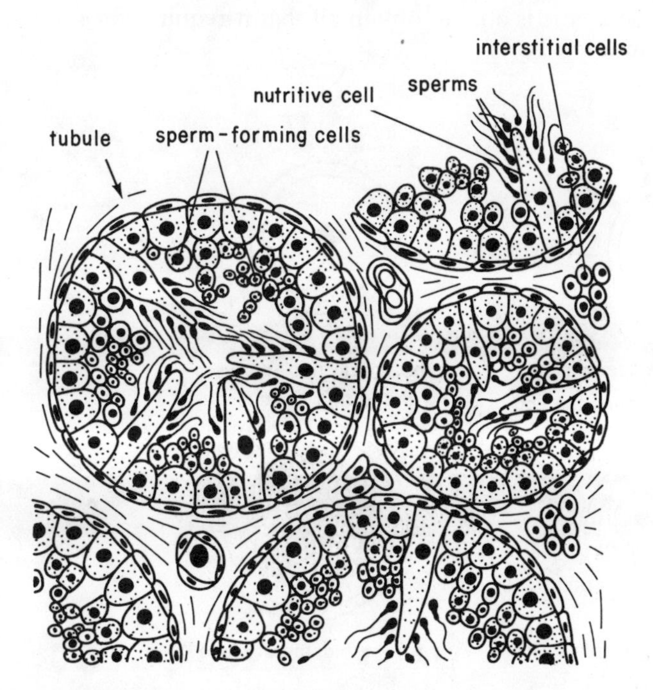

Fig. 23:4 Diagram of a section through testis tubules, much enlarged.

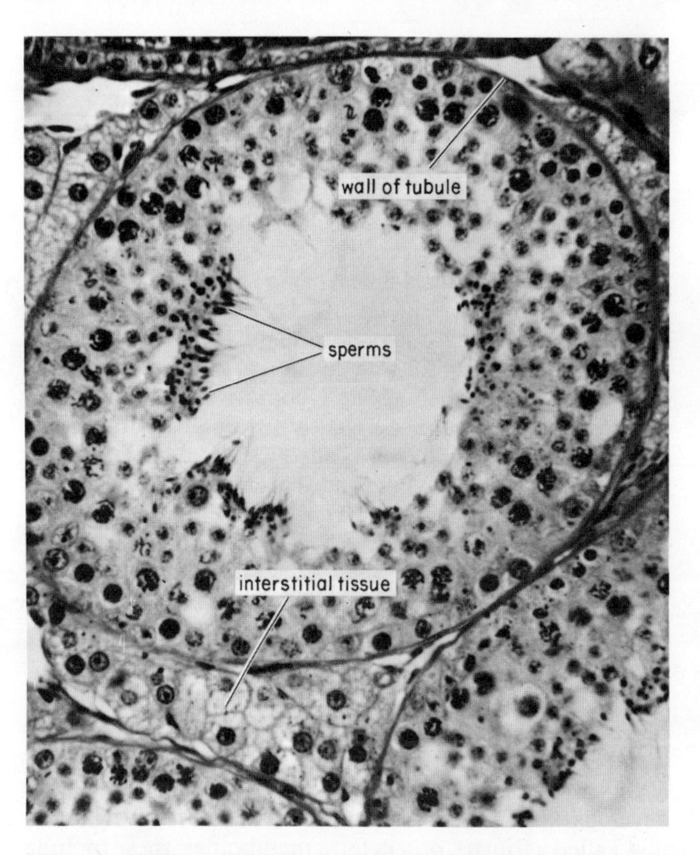

Fig. 23:5 Photomicrograph of a section through part of a testis.

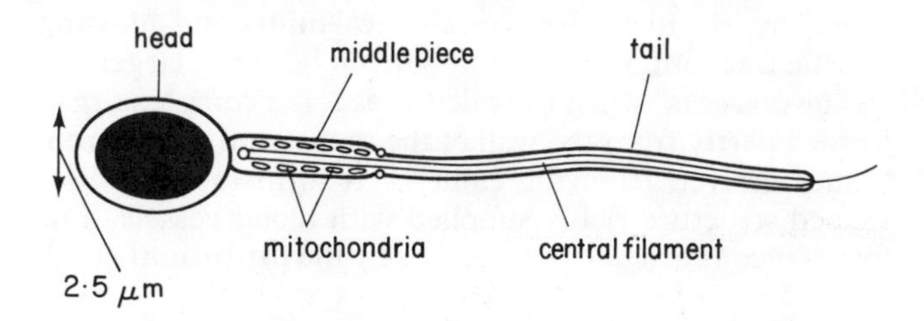

Fig. 23:6 Diagram of a human sperm.

prevents others from entering. Fertilization is accomplished when the two nuclei merge into one.

The ovary and ovulation

At birth the ovaries of a baby girl are already formed and contain many thousands of potential egg cells, but only about 500 of these will become mature during the period of reproductive life. It is not until puberty that these potential egg cells start to form mature eggs and from this time until the age of about 50 one or other ovary produces a single egg about every 28 days. Fig. 23:9 shows a diagram of a section of an ovary to show how the potential egg cell matures, surrounded by a group of actively dividing cells; in this way a **Graafian follicle** is formed. When ripe the follicle contains a fluid-filled cavity and projects from the wall of the ovary as a small bump. **Ovulation** then takes place, that is, the follicle bursts and liberates the egg into the **fallopian funnel** at the end of the fallopian tube. The egg is wafted along by the action of the ciliated cells which line the funnel and tube. The egg is still surrounded by a layer of cells derived from the follicle and is of a size just visible to the naked eye. It may not survive much longer than a day unless it is fertilized. Meanwhile, cells of the follicle from which the egg has been expelled divide rapidly to form a gland composed of yellowish cells, the **corpus luteum** (p. 171).

If it is fertilized the egg (now a zygote) starts dividing into 2, 4, 8 cells etc. until it becomes a hollow sphere of cells called a **blastocyst** (Fig. 23:7f). By this time it will have completed its passage down the tube and reached the uterus. Occasionally, during the first division of the zygote, the two cells which are formed separate completely and continue development quite independently; in this way **identical twins** are formed. They are called identical because they come from the same zygote and therefore have exactly the same hereditary characteristics. **Non-identical** twins occur when two eggs are produced and both are fertilized.

Implantation and development

The blastocyst, on reaching the uterus, sinks into its spongy, vascular walls, and becomes surrounded by maternal tissues, a process called **implantation**. From then onwards the developing embryo is dependent upon the blood of the mother for nutrients and oxygen.

Cell division continues rapidly and while some cells develop into the tissues and organs of the developing baby, now called a **foetus**, others form membranes: these include the **amnion** and those which help in the formation of the **placenta**. The amnion is a protective membrane which projects into the cavity of the uterus and encloses the foetus in a bath of fluid. The latter acts as a shock absorber, protecting the foetus from mechanical injury and allowing it some freedom of movement when it becomes larger.

The placenta, when fully developed, is a complex organ formed partly from the wall of the uterus and partly from tissues derived from the embryo. It forms a large disc-shaped structure richly supplied with blood vessels. The foetus is connected to the placenta by the **umbilical cord**.

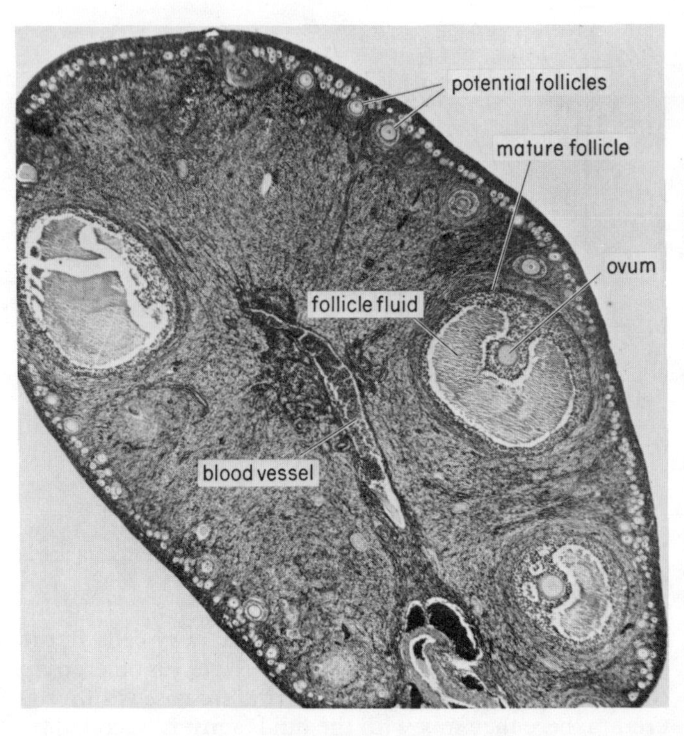

Fig. 23:8 Photomicrograph of a section through the ovary of a rabbit.

After about a month's development the foetus has its own blood circulation; this includes an artery which passes up the umbilical cord, breaks up into a mass of capillaries in the placenta and returns as a vein to the foetus. In the placenta the capillaries of the foetus become very closely associated with the blood vessels of the mother, but are never actually connected with them (Fig. 23:10). Even so, oxygen and soluble food material can easily and quickly pass from mother to foetus and carbon dioxide and nitrogenous waste can pass from foetus to mother. In this way the foetus is able to obtain all that it requires for growth until it is born.

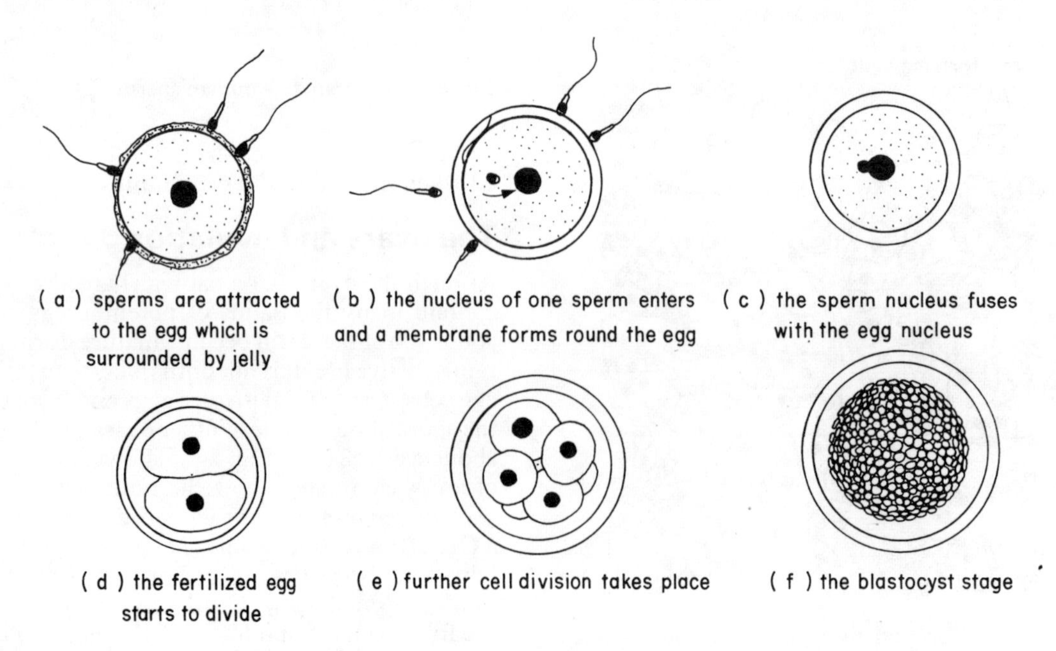

Fig. 23:7 Fertilization and early development of the egg.

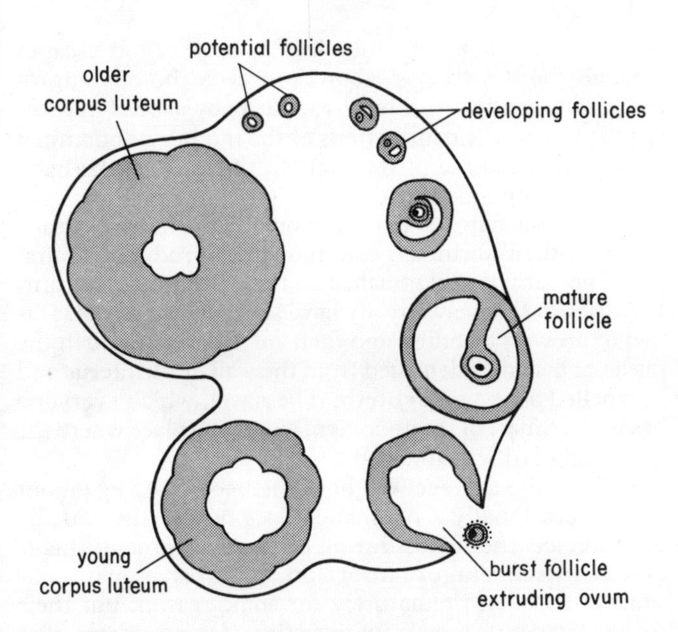

Fig. 23:9 Stages in the development of a follicle and corpus luteum in the ovary.

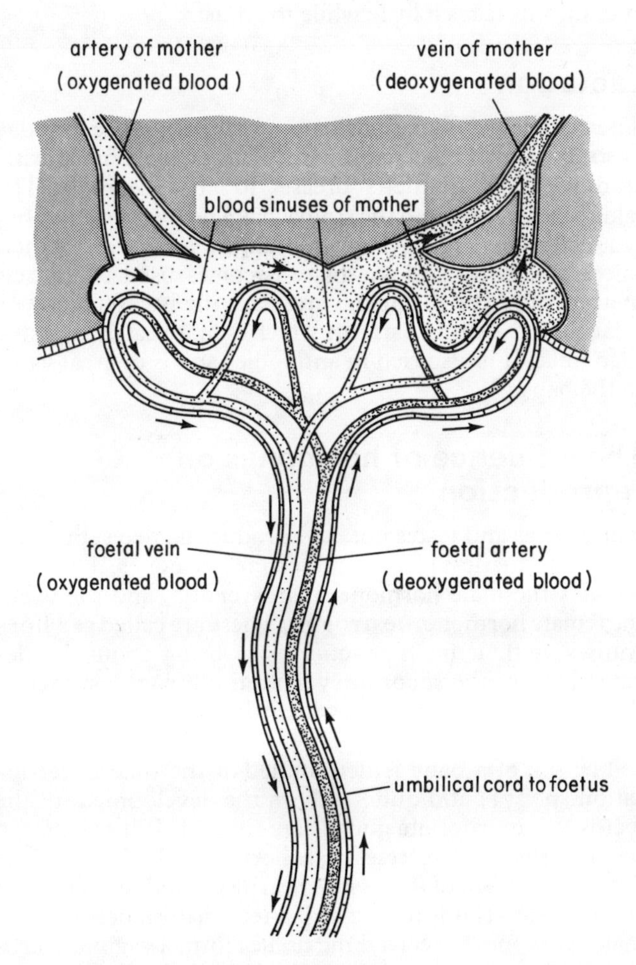

Fig. 23:10 Much simplified diagram showing how the circulations of mother and foetus associate in the placenta.

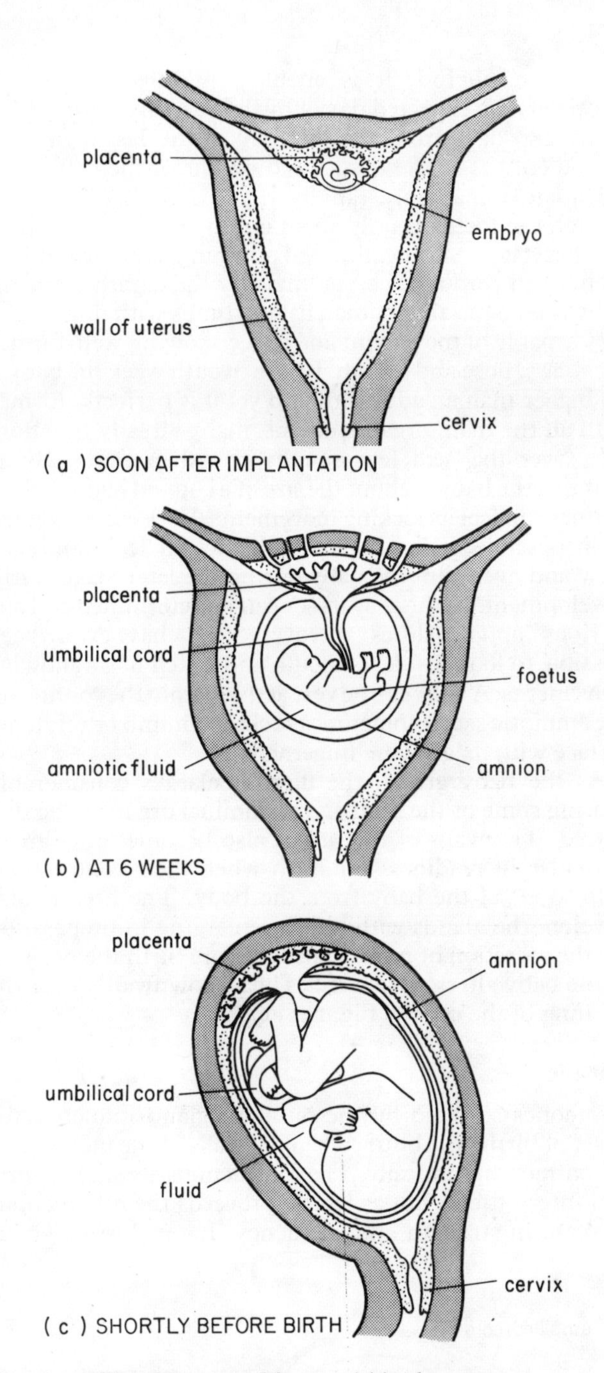

Fig. 23:11 Development of foetus within the uterus.

The substances that the foetus receives are selected to some extent by the placenta; most substances with large molecules are unable to pass through, so the foetus is protected from those in the mother's blood which might otherwise harm it. However, this protection is not perfect. Nicotine from tobacco causes rapid stimulation of the heart of the baby and in those mothers who smoke a lot during pregnancy, the carbon monoxide may adversely affect the baby's growth. Alcohol affects the growth and brain development of the foetus. The babies of alcoholic mothers are undersized and their intelligence greatly affected (p. 228). Other drugs also have their hazards. For example, in the early 1960s the taking of the drug thalidomide as a sedative during pregnancy tragically resulted in a large number of babies being born with deformities

The whole period from fertilization to birth is called the

gestation period; it is about 9 months (266 days). However, the expected day of birth is usually estimated as being 280 days from the first day of the last menstrual period (p. 171). The tissues and organs of the foetus differentiate remarkably rapidly. By the fourth week, although the foetus is only about 6 mm long, the rudimentary heart is already beating and pumping blood around the foetus and through the placenta. By the eighth week the baby is recognisably human having limbs with fingers and toes capable of movement and a face showing well-formed eyes, ears, nose and mouth. By the twelfth week the baby is no bigger than an adult's thumb yet it is perfectly formed with all the main organs present, many already functioning. Over the next few months growth is rapid. By $4\frac{1}{2}$ months, the baby is about the size of a cupped hand and the mother can feel its kicking movements. Very recently it has been possible, using highly sophisticated techniques, to view and even film the baby during the later stages of its development. In this way its sex can be determined and any obvious abnormalities discovered. It has even been possible to look inside its mouth and watch its swallowing movements. A baby can be very active within the confines of the amniotic sac; it sometimes sucks its thumb or scratches its face with its growing fingernails.

As the baby grows, the uterus enlarges considerably causing some of the mother's abdominal organs to be displaced. The walls of the uterus also become much more muscular in readiness for birth when their contractions help to expel the baby from the body. The breasts also develop, the glands within them enlarging in preparation for the secretion of milk. Shortly before birth the position of the baby alters so that it lies head-downwards near the opening of the cervix (Fig. 23:11).

Birth

Hormones secreted by the pituitary gland influence the onset of birth or **labour**, when the walls of the uterus start to contract rhythmically. The contractions are slight at first and intermittent, but as labour proceeds the contractions increase in strength and frequency. Later during labour the amnion breaks and the clear amniotic fluid escapes through the vagina. As labour proceeds the opening of the cervix gradually dilates as the baby's head presses against it, and the contractions of the mother's abdominal muscles assist those of the uterus to help force the baby towards the outside world.

At birth the baby takes its first breath, the lungs expanding from their shrunken condition; this produces its first cry. The baby is still attached to the placenta by the umbilical cord; this may already have started to constrict. It is tied to prevent bleeding and then cut. Soon after birth the placenta becomes detached from the wall of the uterus and is expelled as the **afterbirth**. The navel, which everyone has in the centre of the abdomen, marks the place where the umbilical cord was attached.

At birth the average weight of the baby is 3·2 kg (about 7 lb). Occasionally a pregnancy does not persist and the uterus evicts the foetus during early development; this is called a **miscarriage** or **abortion**. Some babies may occasionally be born prematurely for some reason, but their survival depends largely on their size; if they are over 2 kg they have a fair chance of survival, given specialised medical care.

For the baby, birth involves great changes. Previously it was dependent on its mother's circulation for food and oxygen, now it has to use its alimentary canal and lungs for the first time; previously its temperature was controlled by the mother, now it has to use its own heat-regulating system; this takes a little while to adjust.

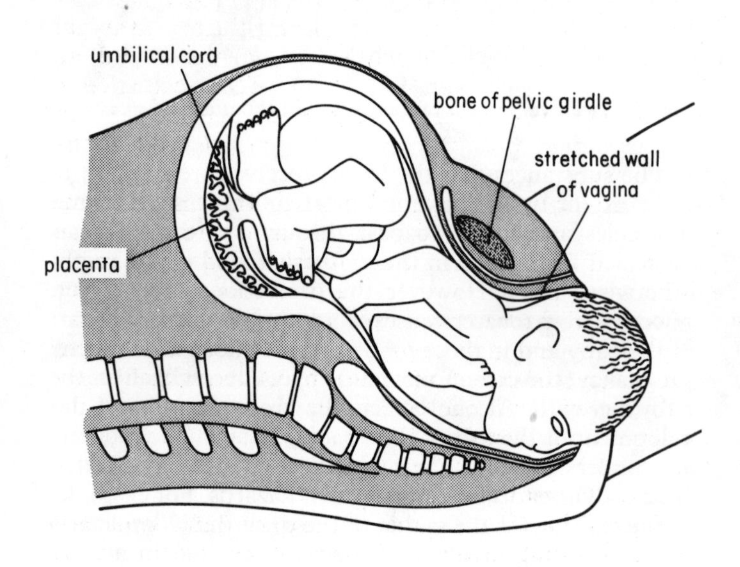

Fig. 23:12 Birth.

Lactation

The baby usually takes little nourishment for the first day or so although it has a natural instinct to suck. This sucking stimulates the mother's breasts to produce milk. The taking of the milk into the stomach is a new experience, but gradually the digestive glands adjust and secrete their juices efficiently and the baby becomes used to its new method of feeding. For the first days the breasts secrete a clear fluid called **colostrum** which is more easily digested than milk. It is also rich in antibodies and is of great value to the baby.

The influence of hormones on reproduction

The ovaries and testes not only produce gametes, but they also act as ductless glands and secrete hormones. The testis secretes the male hormone, **testosterone**, and the ovary the female hormone, **oestrogen**. These are called **sex hormones** as their main function is to bring about the development of the **secondary sexual characters**, that is, the characteristic features which distinguish males from females.

The sex of a baby is determined at the time of fertilization (p. 271) and quite early in the development of the foetus the appropriate gonads are formed. If it is going to be a boy the testis secretes male hormone which influences the development of the tissues nearby to produce penis and sperm ducts. If it is to be a girl, after a critical period, if no male hormone is received the tissues form a vagina, uterus and oviducts so by the time the baby is born it has the basic features of either a boy or girl. During childhood the

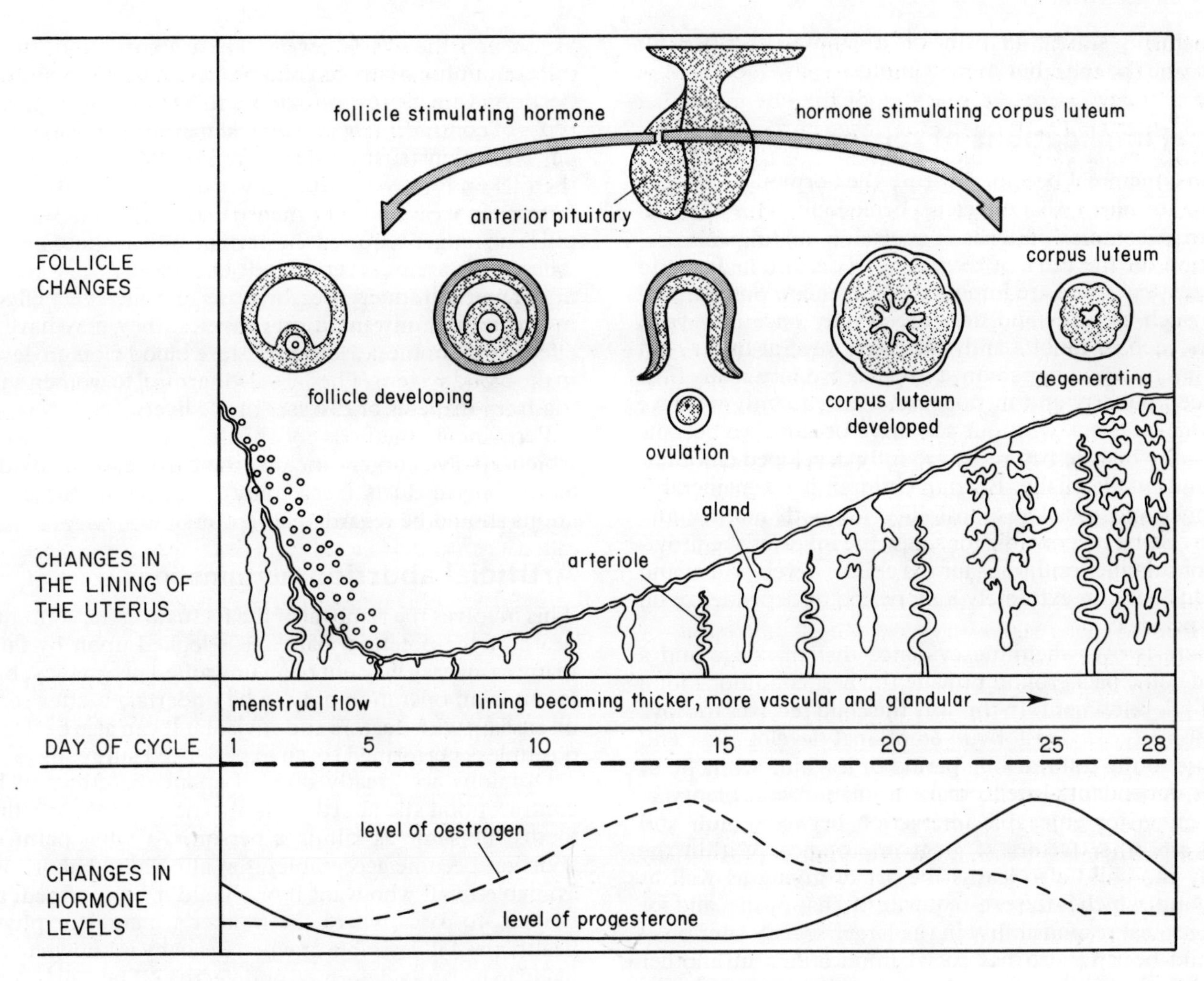

Fig. 23:13 Summary of the changes which occur during a human menstrual cycle of 28 days.

gonads remain more or less dormant and secrete very little sex hormone. But at the onset of puberty, as a result of stimulation from a hormone from the anterior pituitary, the gonads not only enlarge and produce gametes but they start secreting their sex hormones in much greater concentrations. This brings about further changes associated with puberty. Puberty starts in the male in the early teens, and in the female a year or so earlier.

In the male there is a rapid increase in height during puberty, hair grows on the face, chest and pubic region, the larynx enlarges and the voice deepens. The reproductive system itself also develops and becomes functional. In the female there is also rapid increase in height, the breasts develop, the pelvic (hip) girdle widens, the reproductive system becomes functional and **menstruation** starts.

The ovarian and menstrual cycles

Reproduction in the female is restricted to a period from puberty to **menopause** (about 50 years old). The latter is the time when the ovaries cease to produce any more eggs.

The **ovarian cycle** usually lasts 28 days and can roughly be divided into two halves. During the first 14 days, stimulation of the ovary by a pituitary hormone causes a Graafian follicle to become mature and an egg is shed. This is called **ovulation**. In the same period the ovary is stimulated to produce its own hormone, oestrogen. During the second fortnight, cells in the follicle from which the egg was shed quickly multiply to form a gland, the **corpus luteum**. This gland is stimulated by another hormone from the pituitary to secrete a further hormone, **progesterone**. So during the monthly cycle the levels of oestrogen and progesterone in the blood fluctuate. During the first fortnight there is a build-up of oestrogen, and in the second, an increase in progesterone (Fig. 23:13). This fluctuation of oestrogen and progesterone causes changes to occur in the lining of the uterus; these changes constitute the **menstrual cycle.**

By the time ovulation takes place the lining of the uterus is spongy and vascular and ready to receive an egg if it is fertilized. If the egg is *not* fertilized, the lining cells break away after about 14 days and there is some bleeding. This **menstrual flow** or **period** lasts four to five days. If, however, the egg *is* fertilized, the corpus luteum persists and continues to secrete progesterone which exerts a further influence on the growth of the uterus and the breasts. It also has the effect of suppressing further ovulation and causing the cessation of menstrual periods—usually the first sign that pregnancy has begun.

You will have realised that these ovarian and menstrual cycles are controlled by hormones whose concentrations change at different times of the cycles to promote the necessary effects. Their control is determined by feed-back mechanisms similar in principle to those described for thyroxine (p. 164), but more complex.

Social implications of reproduction

When a mammal becomes mature the hormones circulating in its blood also affect its behaviour. This is often shown as complex patterns of courtship, mating and preparation for the birth of the young. This also happens in humans, but there are important differences. Some mammals such as foxes and deer breed only once in a year, others such as rabbits and voles have several litters, but man has no breeding season, and can reproduce at any time between puberty and menopause. Consequently we have to come to terms with our sex drive because we become physically mature before we are fully developed emotionally and intellectually. Bearing children is a considerable personal and social responsibility for both mother and father. Being responsible for a new life involves the provision of suitable conditions for the child's development and humans have an extremely long period of dependence on their parents.

There is overwhelming evidence that marriage and a stable home background provide the best conditions for a child's development. In this way the child receives the love and security needed for full emotional development and can use to the full the long period of learning while he or she is dependent. Intellectual stimulation and plenty of time given for enjoyable interaction between adult and child are other factors of great importance. Within the family the child also learns the art of giving as well as receiving, which is the pre-requisite of a happy life, and for taking social responsibility in the larger society later on.

Child-bearing also has social implications in another way. A large family can be a happy and stable community, but frequently the children are deprived of the care, understanding and financial resources that they need, and many problems arise for both the children and their parents. Furthermore, with a marked population explosion in the world (p. 256), it is especially important to make sure that every pregnancy is a desired one.

Family planning

Responsible parenthood today demands that the family is properly planned. A happy, secure family is more likely to be created if the children are wanted in the first place. For this reason many people advocate some form of **contraception** (birth control) to prevent unwanted pregnancies, although others disapprove of this practice on religious grounds.

Various contraceptive methods are practised, some of the most common and reliable being those which aim to prevent sperms from reaching an egg during sexual intercourse. One of these is the **condom**. This is a sheath which encloses the penis and retains the ejaculated semen so that no sperms enter the vagina. A condom is made of thin rubber or plastic and is fitted on to the erect penis just before it is inserted into the vagina.

A second device is the **diaphragm**. This is a thin plastic cap which is inserted into the vagina so that it covers the cervix, thus preventing sperms from entering the uterus.

The **contraceptive pill** works on a different principle as it affects the ovarian cycle. There are different types of pill containing oestrogen and progesterone in various proportions which prevent ovulation (see Fig. 23:13). Most types of contraceptive pill are taken every day from the 5th day after menstruation starts until the 25th day. No pill is then taken for a week during which time the effect of the hormones wears off and menstruation occurs. When the pill is no longer required the normal cycle is again possible. These pills are of several kinds and should only be taken under medical supervision because, although very effective in preventing unwanted pregnancies, they may have side effects and, in rare cases, may cause blood clots to develop in the blood system. They are also harmful to women suffering from diabetes or diseases of the liver.

Permanent methods of birth control (sterilization) which involve surgery include the cutting of the oviducts or the sperm ducts (vasectomy). At present these operations should be regarded as irreversible.

Artificial abortion (termination)

This involves the removal of the foetus and the consequent death of the potential baby. It is looked upon by far too many as an easy way out of an unwanted pregnancy, but in fact it is an operation only to be undertaken after serious discussion with doctors and advisors. It can also be dangerous unless performed by an experienced surgeon.

Opinions are greatly divided about the ethics of both contraception and abortion. Is abortion killing? Is killing a foetus the same as killing a person? At what point does abortion become acceptable, if at all? Should abortion be available to all who want it or should it be confined, as in Britain, to cases where the mother's mental or physical health make it desirable or when the baby is known to be or likely to be badly deformed or defective? These are some of the questions that need to be considered. Whatever one's views, all would agree that prevention of a pregnancy is better than destroying a developing foetus.

The reality of the situation is that in 1979 in Britain, on average more than 350 legal abortions took place each day and, of these, 100 were on teenagers. What happens as a consequence of teenage pregnancies may be gauged by the 1978 figures: everyday, 23 16-year-olds became pregnant and of these four married before the baby was born, eight became single parents and eleven had abortions.

Diseases related to the reproductive system

These are termed **venereal** diseases (VD) or sexually transmitted diseases (STD). There are two of considerable importance, **gonorrhoea** and **syphilis**, both of which are caused by a bacterium and are spread by an infected person through sexual intercourse. Recent medical statistics show that the number of people contracting these diseases is increasing each year at an alarming rate and bringing untold misery. For example, there were 63,000 new cases of gonorrhoea in Britain alone during 1978, and of these, over 11,000 occurred amongst teenagers. This appears to be due to the great increase in promiscuity (casual sexual intercourse).

With gonorrhoea, some women do not, at first, show any obvious symptoms, so if promiscuous they can spread the infection. In the later stages of the disease the infection spreads through the reproductive system causing severe illness and sterility. A pregnant woman with the disease may also infect her child during labour and its sight may be endangered in consequence. In men the infection starts in the penis and may spread to the bladder and the kidneys with serious results.

With syphilis all parts of the body can be affected as the micro-organisms pass into the blood stream; infection of the foetus can also occur.

Both diseases have been treated effectively with antibiotics and when penicillin was discovered the incidence of these diseases dropped; however, today some strains of the organisms causing gonorrhoea have become resistant to penicillin and this has raised serious problems.

24

Reproduction in flowering plants

Flowering plants use both asexual and sexual methods of reproduction. The asexual method is also known as vegetative reproduction as it takes place by means of buds; sexual reproduction is brought about through the formation of flowers and the production of seeds.

Asexual or vegetative reproduction

Flowering plants reproduce vegetatively in many ways, but the principle involved is always the same; somewhere on the parent plant a bud develops and eventually becomes separated to form a new plant. At first the bud is always dependent upon food obtained from the parent, but later it develops foliage leaves of its own and so is able to photosynthesise for itself. Adventitious roots (i.e. those formed from a stem) also develop and these enable the young plant to absorb water and salts. Thus when separation from the parent occurs the new plant is completely self supporting. Different species form these reproductive buds from different organs. Some of these organs can be summarised as follows:

Modified stems

Rhizomes
These are stems which grow horizontally below ground level. They produce buds, some of which may grow out into side branches and come above the ground as shoots. Eventually the newly-formed rhizomes become separated from the parent by rotting and reproduction is finally effected. Examples include couch-grass, Michaelmas daisy, mint and iris (Fig. 24:1).

Tubers
These are the swollen ends of underground stems. Potatoes and Jerusalem artichokes can reproduce from these.

Runners
These are stems which grow along the surface of the ground and root at intervals. In the strawberry they arise from buds in the axils of the leaves and their terminal buds grow out to form new plants; adventitious roots develop at the same point. Other runners can arise similarly from the daughter plants. Eventually the connecting stems rot away and the daughter plants become independent.

Corms
These are short, erect underground stems. Crocus, gladiolus and montbretia are examples. Superficially they look like bulbs, but a true bulb contains fleshy leaves; a corm is solid.

Modified roots

Root tubers
These are swollen adventitious roots and are found in lesser celandine, dahlia and many orchids. New plants may grow from buds which arise near their point of origin with the stem (Fig. 24:1).

Modified buds

Bulbs
These are buds which have much stored food in some of their scale leaves. When they reproduce, new bulbs arise from buds in the axils of these scale leaves. Daffodils, bluebells and onions are familiar examples (Fig. 24:1).

Modified leaves

Some plants, such as *Bryophyllum* (Fig. 24:2), form buds from the edge of the leaves which quickly grow into tiny plants complete with adventitious roots. They drop off, the roots grow into the soil and new plants develop.

Examine a selection of these organs of vegetative reproduction in more detail and grow them to see how they reproduce.

Potato—a stem tuber. Examine a potato and note:
1. The 'eyes' which are the buds lying in the axils of curved scale leaves (or their scars).
2. The concave side of the scale leaves all point towards one end where the growing point is situated.
3. At the opposite end to the growing point is a scar where the tuber was attached to the parent plant.
4. The tough skin which is made of cork.
5. Many brown dots scattered over the skin. These are **lenticels**, small pores in the skin through which respiratory gases can diffuse.

As a longer term investigation you could place a small potato in damp earth against the glass side of a container so that you can see what happens when it grows, *or* grow one in light soil and after some weeks, when the foliage has grown well, dig it up very carefully so as not to brush off the new potatoes that have formed, and examine the whole plant.

You will find that some buds in the 'eyes' will form shoots; adventitious roots grow from their bases. The shoots bear scale leaves below the soil surface and from their axils grow rhizomes which swell up at their ends to form the 'new' potatoes. Above the soil surface foliage leaves are formed which produce the food which is translocated for storage in the new potatoes.

Fig. 24:1 Some organs of vegetative reproduction in flowering plants.

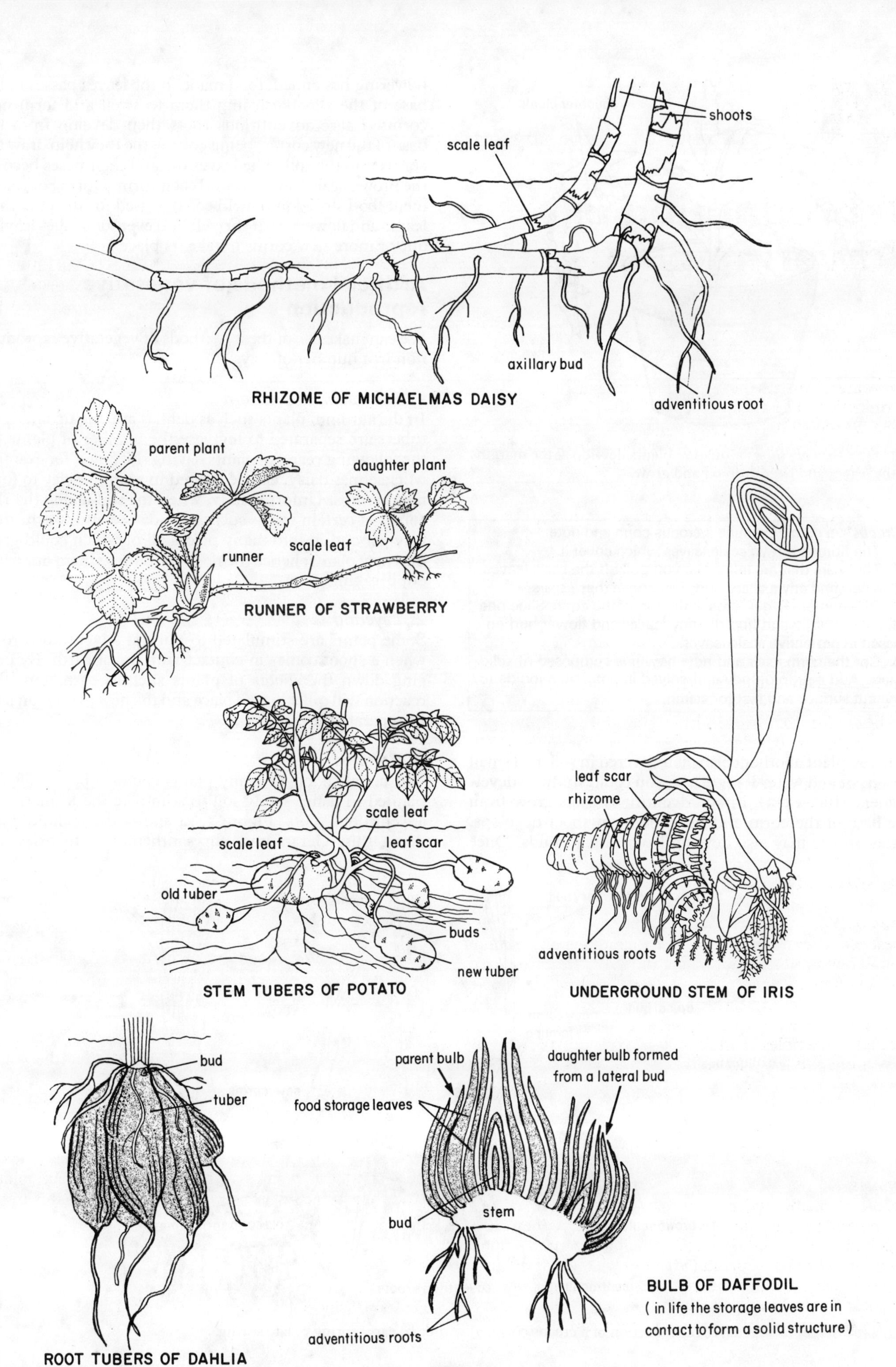
shoots
scale leaf
axillary bud
adventitious root
RHIZOME OF MICHAELMAS DAISY
parent plant
daughter plant
scale leaf
runner
RUNNER OF STRAWBERRY
leaf scar
rhizome
adventitious roots
UNDERGROUND STEM OF IRIS
scale leaf
scale leaf
leaf scar
old tuber
buds
new tuber
STEM TUBERS OF POTATO
bud
tuber
ROOT TUBERS OF DAHLIA
parent bulb
daughter bulb formed
from a lateral bud
food storage leaves
stem
bud
adventitious roots
BULB OF DAFFODIL
(in life the storage leaves are in
contact to form a solid structure)

Fig. 24:2 *Bryophyllum:* miniature plants develop at the margins of the leaves and later drop off and grow.

Crocus—a corm. Examine a crocus corm and note:
1. The fibrous brown scale leaves which cover it.
2. The scars that are left when you pull off the scale leaves. Can you find any axillary buds just above these scars?
3. One or more larger buds at the top of the corm. Slice one down vertically and find the tiny leaves and flower bud encased in protective scale leaves.
4. Cut the corm itself and note how it is composed of solid stem. Add a drop of iodine dissolved in potassium iodide to the cut surface and test for starch.

If you plant another corm, half covered in soil, in a small flowerpot and water it regularly, you could study its development (Fig. 24:3). First, adventitious roots grow from the base of the corm and then the main shoot develops. Other shoots may also develop from axillary buds. Once flowering has ended, food made in the leaves passes to the base of the shoots causing them to swell and form new corms. Large adventitious roots then develop from the base of the new corms; being contractile they help draw the corms into the soil. The leaves die and their bases become the brown scale leaves around each corm. During development food stored in the old corm is used for the growth of leaves and flowers, so it shrivels and eventually dies leaving one or more new corms to take its place.

Artificial methods of vegetative reproduction

We can make use of these methods of vegetative reproduction in a number of ways:

1. Separation or division

In the autumn, plants such as dahlias are dug up and their tubers are separated to increase the number of plants for the following year. All plants having rhizomes, for example Michaelmas daisy, can be treated in a similar way to form new clumps. Unfortunately, when the soil is dug the rhizomes of certain weeds such as bindweed and couch-grass may get cut up into many pieces all of which could grow into new plants; hence the need for pulling them out without breaking them.

2. Layering

Some plants are stimulated to put out adventitious roots when a shoot comes in contact with moist earth. By pegging down the shoots of plants such as carnations this reaction will quickly take place and the new plants can later be separated (Fig. 24:4).

3. Cuttings

The aerial shoots of many plants can also be cut off and planted in damp sandy soil to stimulate the formation of adventitious roots. Figure 24:4 shows the technique for doing this. Geraniums, chrysanthemums, fuchsias and

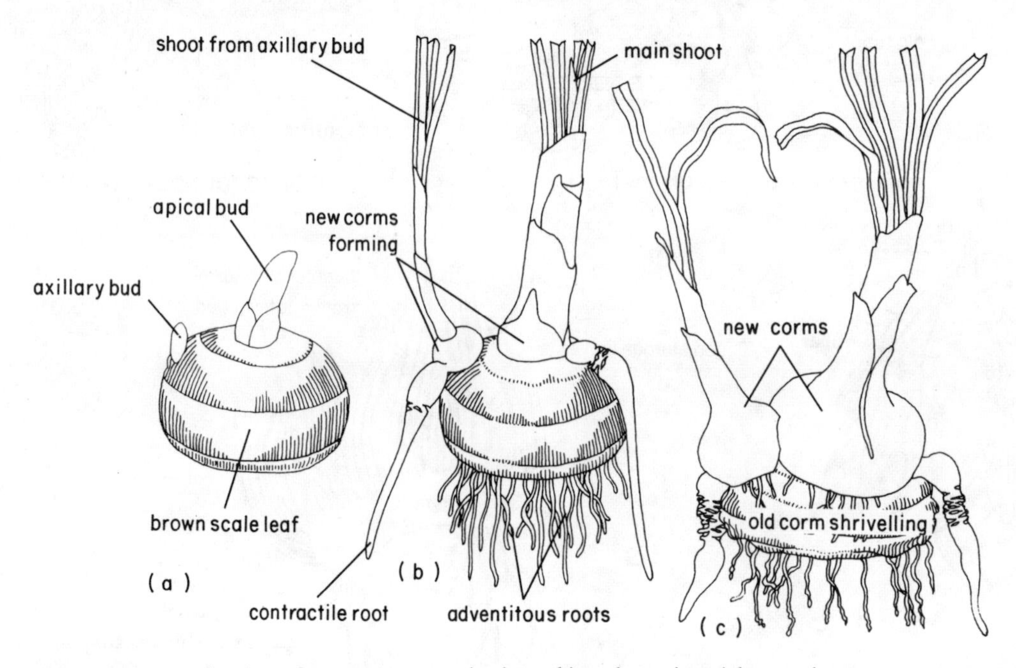

Fig. 24:3 Stages in vegetative reproduction of a crocus corm: a) winter b) early spring c) late spring.

many other plants are normally propagated in this way. Some plants which will not readily form roots on their cuttings may be helped to do so by treating the cut ends with a hormone powder; this stimulates their growth (p. 196).

4. Budding

This is the standard method of propagating roses. The principle is to take a bud from the rose to be propagated, called the **scion**, and insert it beneath the bark of the stem of a wild rose plant, called the **stock**. To be successful, the cambium tissue of each must be against that of the other. Why the cambium?

Budding should be done in the autumn after the scion has flowered. Using a sharp knife a leaf and its dormant axillary bud is cut out together with a thin slice of wood behind it (Fig. 24:5a). This wood is carefully removed to expose the cambium underlying the bud. A T-shaped cut is then made in the bark of the wild rose stock and the flaps peeled back to allow the scion to be inserted below it; it is then lightly bound in place. Finally the leaf is cut off leaving the bud and a short piece of petiole. In the spring it should grow rapidly and the stock shoot above the bud can be cut off.

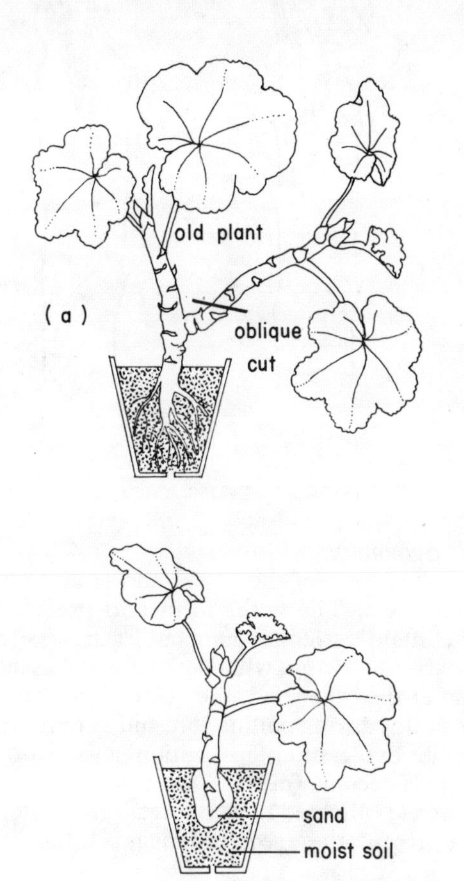

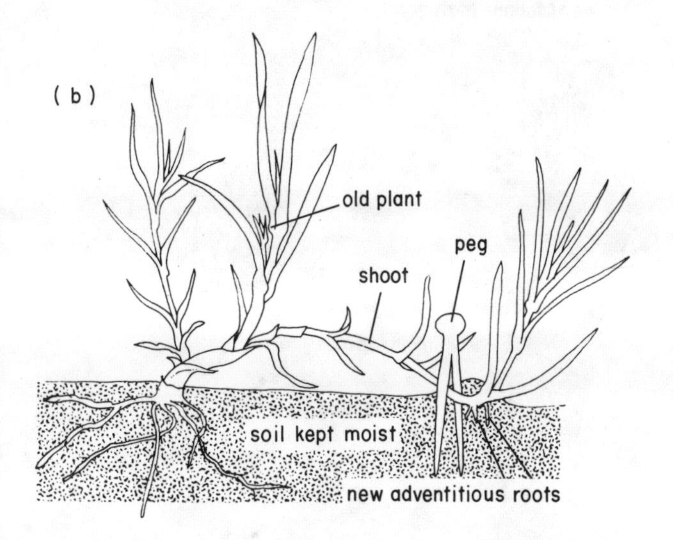

Fig. 24:4 Artificial methods of vegetative propagation: a) by taking cuttings (geranium) b) by layering (carnation).

5. Grafting

This method is much used for propagating fruit trees. In principle it is rather similar to budding but the scion used is a woody twig which is cut in such a way that its exposed cambium comes in close contact with the cambium of the stock on which it is grafted. The two are bound firmly together (Fig. 24:5b). Buds on the stock below the graft should be removed. Why do you think this should be done?

Perennation and food storage

Perennation is the ability to survive the difficult conditions of winter. Those plants which are able to do this are called **perennials**. The two requisites for survival are a means of protecting the delicate growing regions from cold, dry winds and frost, and the provision of enough food for growth in the spring before the new leaves take over this function.

Herbaceous perennials are those which have shoots which die down each winter, but also have special underground portions which survive. The buds on these are thus protected from severe frost. In the spring they form new aerial shoots. Rhizomes, tubers, bulbs and corms act in this way and are therefore perennating organs as well as organs of vegetative reproduction. You will have noticed that they also act as food stores; this enables growth to occur in the spring before the foliage leaves have been formed. This food store makes early flowering possible, a great advantage in the case of woodland plants as they can both flower and photosynthesise before the trees and shrubs above them come into leaf and cut off much of the light. Why do you think it is advantageous for flowering to occur during the well-lit period?

Woody perennials include trees and shrubs. These have permanent woody portions above ground. Many are **deciduous**, shedding their leaves in the autumn; others are **evergreen**, their leaves persisting through the winter as they have adaptations for reducing the rate of transpiration (p. 158). Woody perennials in cold climates protect their growing points by having winter buds, their scale leaves giving special protection from frost.

Some plants are unable to cope with winter conditions and survive only as seeds. These are the **annuals** which complete their life cycle within the year. **Biennials** take two years to complete their life cycle. During the first year they germinate and produce much leafy growth which forms the food; the latter is then stored in underground roots, usually causing them to swell up in consequence.

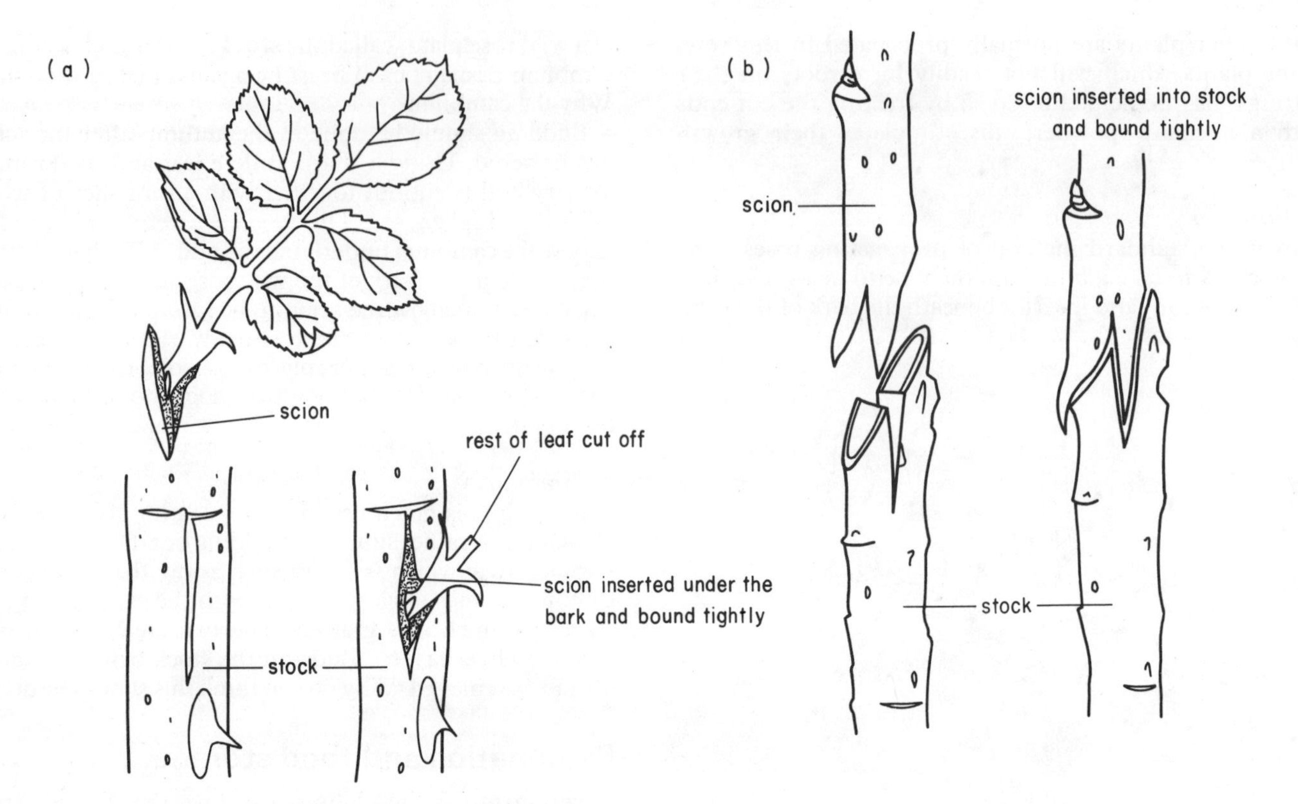

Fig. 24:5 a) Method of propagating a rose by budding. b) Method of grafting an apple twig.

During the winter the aerial shoots die, but the following spring the stored food is used in new growth with the formation of flowers, fruit and seeds before the plant dies. Carrots, parsnips and turnips are examples of biennials (Fig. 24:6).

Sexual reproduction in flowering plants

Structure of a typical flower

The flower is the part of the plant which is specially adapted to bring about sexual reproduction (Fig. 24:7). Flowers vary greatly in size and structure, but basically they all have a common plan. This consists of several **whorls** (rings) of modified leaves arising from the swollen end of the flower stalk called the **receptacle**. In most flowers four whorls are present:

1. The calyx

This is the outer whorl and is composed of **sepals** which protect the rest of the flower when in bud.

2. The corolla

This lies within the calyx and is composed of **petals**. In flowers which are insect-pollinated they are usually coloured to attract insects and may act as a platform for them to settle on.

3. The androecium

This consists of one or more whorls of **stamens** which produce the **pollen grains** in which the male gametes develop. Each stamen consists of a stalk or **filament** and two **anthers** at the top which produce the pollen.

4. The gynaecium

This is the centre of the flower and is made up of one or more **carpels**. The latter form and protect the **ovules** which contain the female gametes. Each carpel consists of a basal part, the **ovary**, which contains the ovules, a special portion at the other end, the **stigma**, which receives the pollen grains during pollination, and a connecting part, the **style**. The ovules after fertilization develop into **seeds** and the carpels become **fruits**.

In insect pollinated flowers, structures called **nectaries** may be present; these secrete a sugar solution called **nectar**

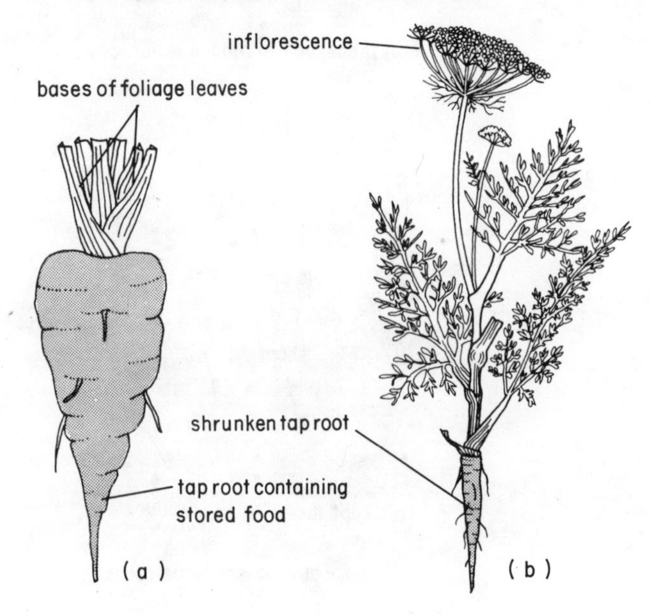

Fig. 24:6 Carrot—a biennial: a) at the end of the first year ($\times \frac{1}{2}$) b) forming flowers and fruits during the second year ($\times \frac{1}{6}$).

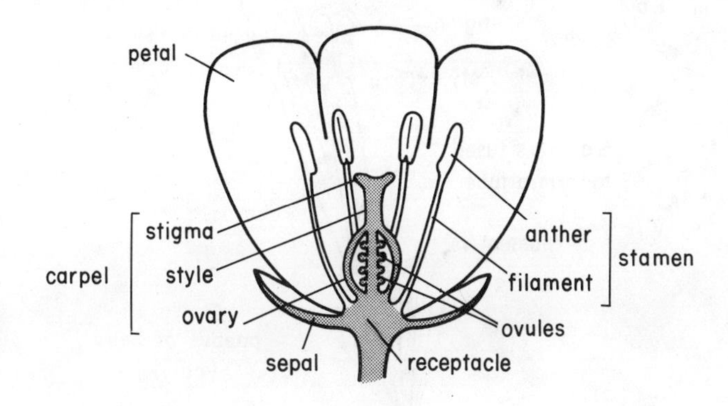

Fig. 24:7 The main parts of a flower.

to which insects are attracted. Nectaries often occur at the base of the carpels but in some flowers they may be modifications of the petals.

Variation in flower structure

In the course of evolution the basic plan of flower structure has become greatly modified, and if you consult an identification book you will see how flowering plants are classified into many families. When examining flowers you should notice the number of the parts; they are often in multiples of threes, fours or fives. The flowers may be regular, or irregular; some form open cups, others are tubular; some have their parts fused together, in others they are separate. Examine a selection to see how they vary; we will just look at three in detail.

Wallflower

This is a member of the family *Cruciferae* to which cabbages and cresses also belong (Fig. 24:8).

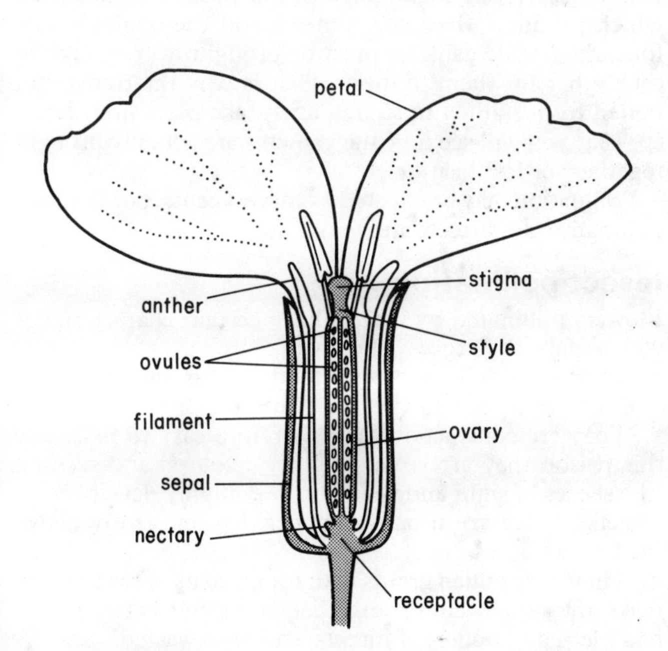

Fig. 24:8 Wallflower: half flower.

1. Examine the whole **inflorescence**, i.e. the whole group of flowers on a stem. Note that the buds or younger flowers are at the top, then come the flowers which are fully out and below these are the flowers which are withering or are completely over. The latter may have produced long fruits.
2. Examine a single flower. Note the calyx on the outside; it is composed of four sepals in opposite pairs. (Look also at a bud and see how the sepals enclose and protect the other parts of the flower.)
3. Remove the sepals with forceps to expose the corolla completely. Each petal has a flat part above which is coloured and acts as a platform on which insects can land, and a large claw which attaches the petal to the receptacle. The four petals are also arranged in opposite pairs and alternate with the sepals. Remove the petals to expose the androecium. Notice the position and relative lengths of the stamens—the two outer ones are shorter than the four in the middle.
4. Examine a stamen under a lens. Note the long filament and the two anthers at the end lying side by side and separated by a groove. Remove a stamen and examine its anthers under a microscope. Look for the pollen grains adhering to them.
5. Remove the remaining stamens to expose the gynaecium in the centre. Look for green, glistening bumps on the receptacle near the base of the gynaecium; these are the nectaries which secrete nectar on which bees and butterflies feed. How many are there?
6. Examine the gynaecium. The long cylindrical portion is the ovary; it tapers slightly to form a short style and ends in two stigmas. Cut the ovary transversely and look at the cut ends; it is divided into two compartments. Cut an ovary from another flower lengthwise between the stigma lobes and note the row of ovules which will develop into seeds if fertilization occurs. The structure of the gynaecium can be explained by the fact that it is composed of two carpels fused together, but the stigmas have remained separate.

Lupin

This is a member of the family *Fabaceae* (*Papilionaceae*) to which peas, beans and laburnum also belong (Fig. 24:9).

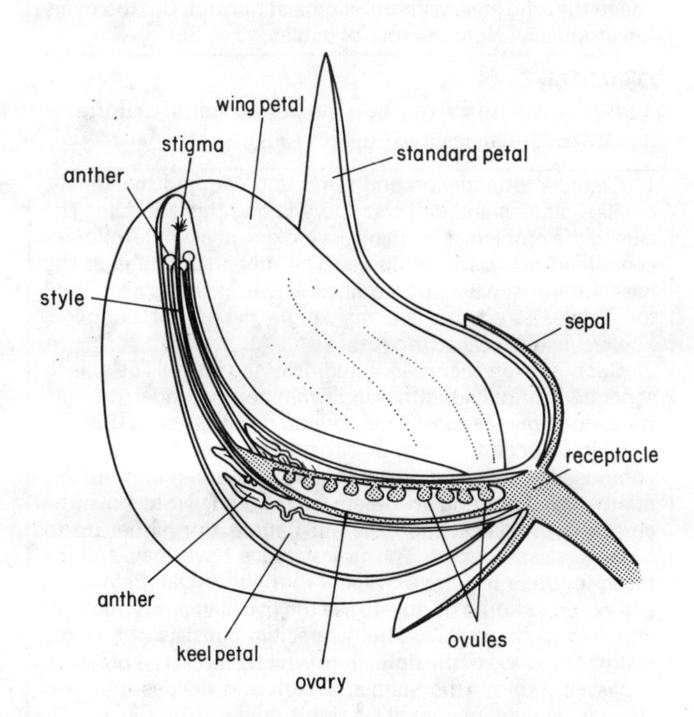

Fig. 24:9 Lupin: half flower.

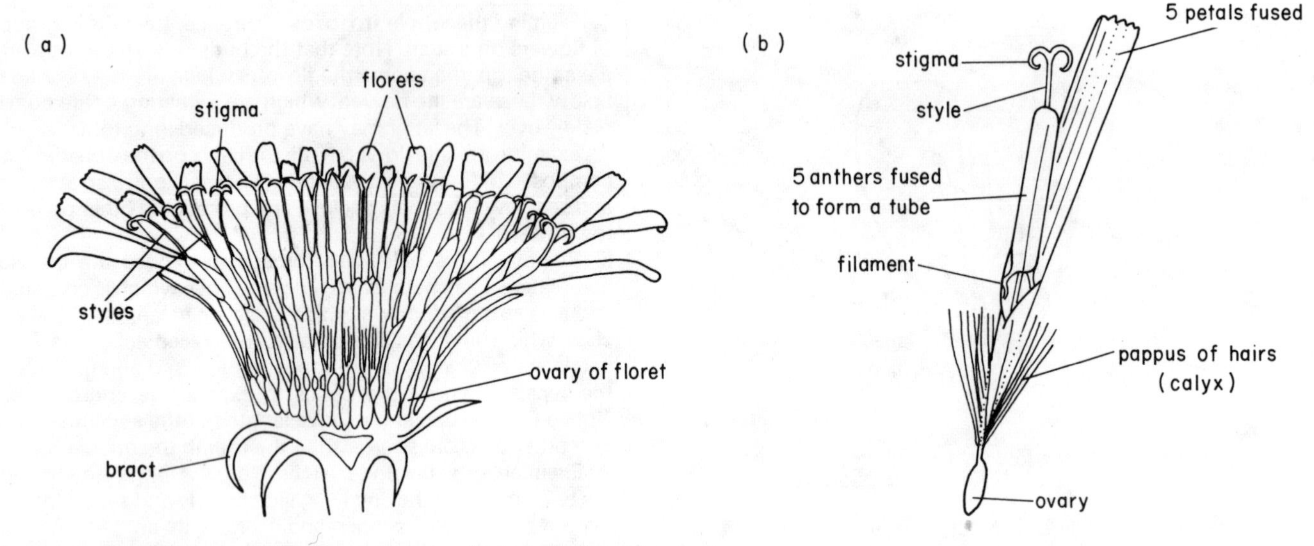

Fig. 24:10 Dandelion: a) half inflorescence (×2) b) single floret (×4).

1. Note that in this inflorescence, like that of the wallflower, the buds are at the top and the flowers which open first are the lowest ones.
2. Examine a single flower. In contrast to the wallflower it is irregular, the sepals are all fused together to form two lips and the petals are of varying shapes. The large petal which stands up at the back is called the **standard**; it makes the flower conspicuous and helps to attract insects. The two at the sides are the **wing** petals which act as a platform for insects, and in the centre are the two **keel** petals fused together to form a tube with a small hole at the end.
3. Remove the petals carefully and note how wing and keel petals fit together as in a ball and socket joint (p. 209). Inside the keel petals you will find both androecium and gynaecium. The former consists of ten stamens which are united by their filaments to form a sheath. Carefully cut this away and examine it. The gynaecium is now exposed in the middle; it consists of a single carpel, the ovary being pod-like and giving rise to the long style with the stigma at the end. Cut the ovary longitudinally. Note the row of ovules.

Dandelion

This is a member of the *Compositae* family along with hawkweeds, daisies and thistles (Fig. 24:10).

1. Examine a dandelion and see if you can find the sepals, petals, stamens and carpels. You will find this a puzzle. The clue to the problem is that you are looking at a whole inflorescence and not just one flower. The green structures at the base are *not* sepals, but modified leaves called **bracts**, and the yellow structures are *not* single petals but complete flowers, usually called **florets**.
2. Cut the inflorescence longitudinally and you will be able to separate the individual florets. Examine one under the low power of a microscope. There are no sepals as such, but the calyx is reduced to a ring or **pappus** of hairs. The corolla is composed of petals all fused together. You can find out the number by counting the teeth at the end. Note how the corolla is tubular at the base but flattens out higher up to become strap-shaped. The five stamens have their anthers fused together to form a cylinder round the style. Below the anthers you should be able to see the five filaments which are attached to the corolla. The gynoecium consists of a single ovary at the base of the floret from which a long style projects; it passes through the stamen sheath and divides into two stigmas. The ovary contains a single ovule.

In the dandelion the flower parts arise from a position *above* the ovary, so the ovary is said to be **inferior**. This is in contrast to the wallflower and lupin where the parts arise from below the base of the ovary; in these the ovary is said to be **superior**.

Some members of the *Compositae* such as the daisy have two kinds of florets, strap-shaped ones on the outside called **ray florets** and short tubular ones in the centre, **disc florets**. The ray florets have no stamens.

Pollination

Flowers have to be pollinated before their ovules can be fertilized and seeds produced. Pollination is the transference of pollen from the stamens of one flower to the stigmas of either the same flower—**self-pollination**—or more often another flower of the same species—**cross-pollination**.

Fertilization occurs later when male and female gametes fuse together. To make fertilization possible, the pollen which produces the male gametes and the ovules which form the female gametes must be brought near enough to each other for them to meet. That is why the transfer of pollen from stamen to stigma must take place first. Later (p. 184) you will see how the gametes are actually brought together for fertilization.

Pollination may be effected by many agents, but the chief pollinators are insects and wind.

Insect pollination

Flowers pollinated by insects have certain characteristics which make the process effective:

1. They provide insects with food: nectar and/or pollen. This causes insects to visit the flowers.
2. They are conspicuous and therefore easy to find. For this reason they are large, brightly coloured and scented (the senses of sight and smell are very highly developed in insects). They are usually supported prominently above the leaves.
3. They have pollen grains with rough coats. This tends to make the pollen clump together and cling better to the hairy legs and bodies of insects, so less is wasted.
4. Because each species has a characteristic scent, an

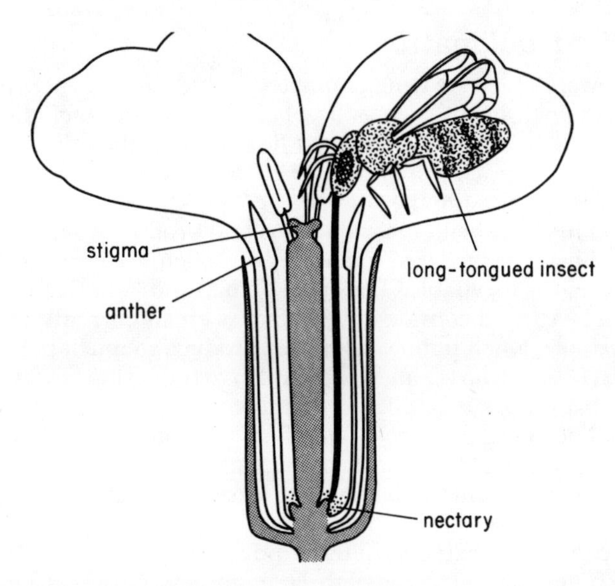

Fig. 24:11 Diagram showing how pollination is effected in a wallflower.

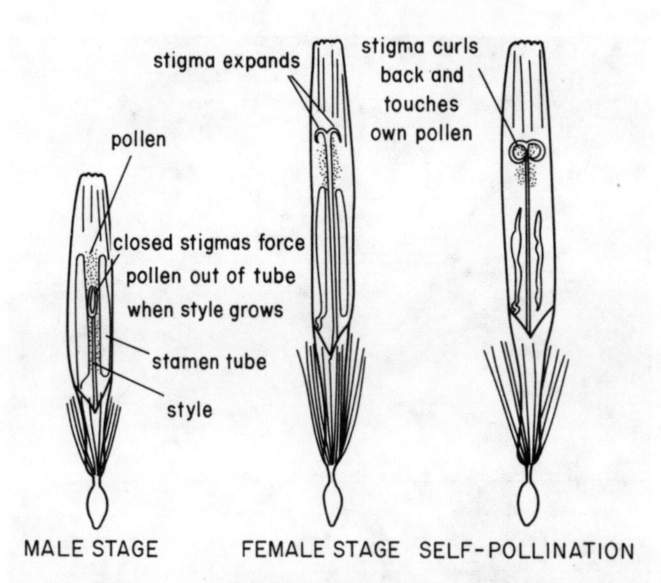

Fig. 24:13 Pollination mechanism in the dandelion.

insect, having once found nectar or pollen, is more likely to visit another flower of the same kind, so pollination is more probable.

5. The position of both stamens and stigmas are arranged so that the alighting insect is in the best position: (a) to collect pollen, (b) to deposit it on the stigma—a small target.

Let us see how these conditions are fulfilled in the flowers whose structures we have studied.

Wallflower

The flowers are brightly coloured and highly scented. The nectar at the base of the stamens can be reached mainly by insects with a long proboscis such as butterflies and bees. The stamens and stigmas are arranged at the top of the tube formed by the four petals. When the insect lands on the petals and puts its proboscis down the tube, its head rubs against the stamens and some pollen is brushed on to it. If it then visits an older flower, the style will have grown longer and the stigmas will project *beyond* the stamens, the best position to receive the pollen (Fig. 24:11).

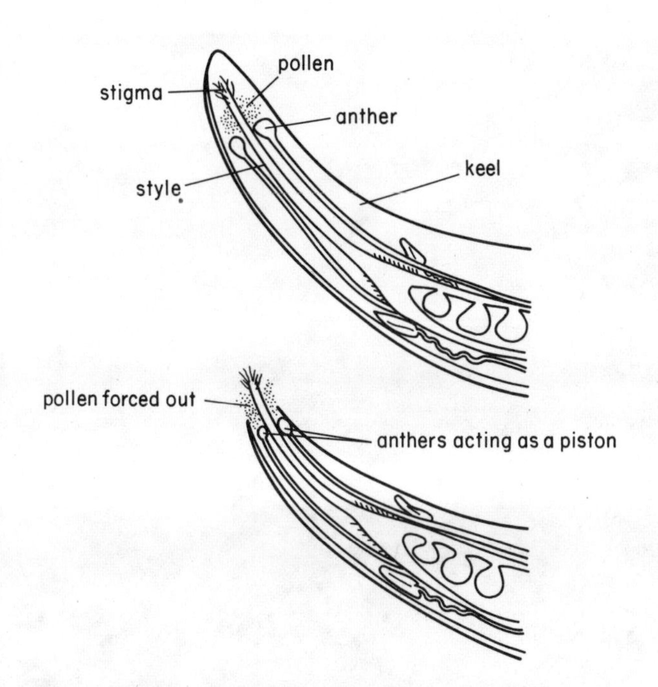

Fig. 24:12 Pollination in the lupin: (top) central part of flower cut open; (bottom) the pollen is pushed out of the end of the keel when the bee alights.

Lupin

The inflorescence is very conspicuous and, although no nectar is produced, a great deal of pollen is formed. The flowers are visited mainly by bees. In a young flower the stamens inside the keel produce a lot of pollen which collects at the end of the keel. When the bee alights on the wing petals its weight depresses them, and because of the ball and socket joint between them, the keel petals are also pulled down causing the stamens and stigma to act like a piston forcing the pollen out of the hole at the end of the keel on to the underside of the bee's body. (Examine a young flower yourself and pull the keel down and see how this happens.) As the flowers become older, the style gets longer, so that when a bee alights on an older flower it is the stigma that hits the underside of the bee and so receives pollen. Self-pollination is avoided because pollen will not develop on the stigma of the same flower (Fig. 24:12).

Dandelion

Individual florets are small, but the whole inflorescence is very conspicuous, highly scented and produces much nectar. The florets are pollinated mainly by bees and butterflies. The pollination mechanism is adapted for cross-pollination but self-pollination can occur if this fails.

> Take a dandelion which still has florets in bud in the centre. Select three florets as follows: one from the centre, one from further out and an old one from the outside. Examine all three under the low power of the microscope and note the differences in the position of the stigmas and in the areas where the pollen grains are clustered.

Fig. 24:14 Bee pollinating a dandelion.

The explanation of what you have seen is that in young florets the stamens shed their pollen into the stamen sheath, and as the style grows it acts like a piston and forces the pollen out of the top. At this stage the two stigmas are pressed closely together and cannot be self-pollinated. Later the stigmas expand and can receive pollen from another floret if an insect comes in contact with them. If this is not effected, then, later still, the stigmas curl round and touch the hairy style to which the floret's own pollen adheres and self-pollination may occur (Fig. 24:13).

This excellent method is used by all members of the *Compositae* but in the dandelion, in spite of the elaborate mechanisms, pollination has in the course of evolution become unnecessary as the seeds develop without any fertilization at all, a process called **parthenogenesis**. This has been proved by taking a dandelion before the flowers have opened and cutting off all the parts of the florets above the ovaries with a razor. In spite of this treatment the ovules still developed into seeds which are capable of germination.

As a project you could watch how other flowers are pollinated. Choose a sunny day. Note the types of insects which visit them and whether they are long- or short-tongued. Try to relate the shape of the flower and position of stamens and stigmas to the characteristics of the insect visitors. Interesting flowers to watch are those of fruit trees, antirrhinums, dead-nettles and poppies.

Can you explain the following?

1. a) Certain flowers give out their scent mainly in the evening; b) flowers which do this are usually white.
2. Bees visit certain flowers which appear red to us, yet they are unable to detect red as a colour.
3. Some plants may not form fruits well when they are grown in greenhouses. (Can you also think of a remedy for this?)

Wind-pollination

Flowers which are wind pollinated include grasses, nettles, plantains, shrubs such as hazel and many trees including oak and ash. Wind-pollinated flowers have the following characteristics, most of which help to increase the chances of pollen reaching the stigmas:

1. They are small, inconspicuous and produce no nectar.
2. They produce vast quantities of pollen grains as great wastage is inevitable. They are so small and light that they can be carried considerable distances on air currents. For example, when in flower, grasses produce so much pollen that people who are allergic to pollen may develop hayfever at that time, even if they live in towns.
3. They occur in a position where they are easily blown, often being grouped in hanging catkins, e.g., hazel (Fig. 24:15) or on stiff stems above the leaves, e.g., plantain (Fig. 24:16). Some, such as hazel, open before the leaves and so become more exposed to the wind.
4. They have stamens with large anthers borne on long filaments which hang or project well outside the flower. The anthers shake out their pollen when disturbed by the slightest breeze.
5. The stigmas are large and feathery so in exposed places have more chance of receiving the blown pollen.
6. They are more characteristic of species which are common and where many plants grow closely together.

Examine the flowers of one of the grasses and see to what extent the above characteristics are true for that species. Grass flowers are extremely small and grouped together in very compact inflorescences. Examine the inflorescence first and note how the stamens and stigmas project, and how large the anthers are. Now dissect out a single flower with a needle and look at it under the low power of the microscope. There are no sepals or petals to be seen and the stamens and gynaecium are partly protected by two green bracts (Fig. 24:18).

Fig. 24:15 Hazel: a wind pollinated shrub with catkins.

Fig. 24:16 Plantain: the flowers near the top of the inflorescence are in the female condition showing projecting stigmas, those below are in the male condition with projecting anthers.

Fig. 24:17 Rye grass (*Lolium*) in flower.

Avoiding self-pollination

In general, cross-pollination is better than self-pollination as it provides for greater variation within the species (Ch. 34). It is therefore not surprising that the majority of flowers have some device which makes self-pollination less likely or impossible. Here are some examples:

1. The stamens ripen before the stigmas, e.g. buttercup.
2. The stigmas ripen before the stamens, e.g. plantain.
3. The flowers are of more than one kind, the length of the style in each case being different, thus causing the position of the stigma to vary in relation to that of the stamens, e.g. primrose.
4. The flower has a stigma on which its own pollen will not develop, e.g. sweet pea.
5. Some flowers lack stamens, others on the same plant lack carpels, e.g. hazel.
6. All the flowers on some plants lack stamens, on other plants they all lack carpels, e.g. holly.

Anthers, ovules and fertilization

In order to understand how fertilization takes place we must now study the structure of a stamen and ovule and see how the gametes are formed.

Structure of the anther

Each stamen has two anthers; each anther is bi-lobed and within each lobe is a **pollen sac**. So in a transverse section through the anthers (Fig. 24:19) you see four pollen sacs. The pollen grains are formed inside the sacs by a process of cell division called meiosis (p. 269); this halves the number of chromosomes in preparation for fertilization. When ripe, a longitudinal split occurs between the two pollen sacs of each side and the walls curl back and expose the pollen (Fig. 24:20). The pollen may then be transferred by insects, wind or some other agent during the process of pollination.

Structure of the ovule

The number of ovules formed in the ovary portion of the carpel varies from one in the buttercup to many thousands in some orchids. Each ovule grows from a place inside the ovary called the **placenta**, the position varying according to the species. As the ovule grows it becomes surrounded by a protective coat except for a hole at one end called the **micropyle**. Within the ovule is the **embryo sac** in which various nuclei are formed by cell division, one of which enlarges and becomes the female gamete or egg cell.

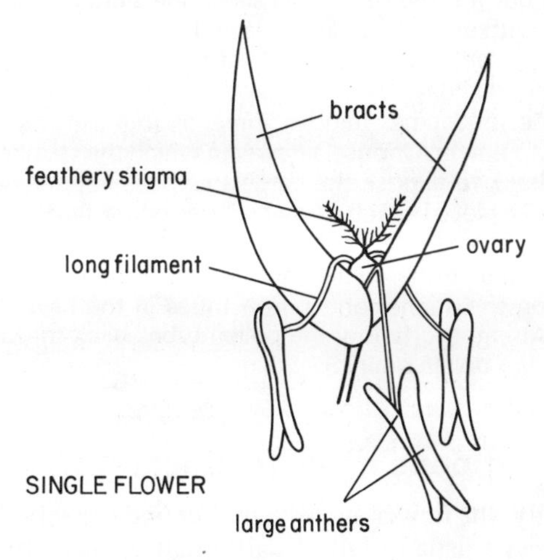

Fig. 24:18 Rye grass showing structure of single flower.

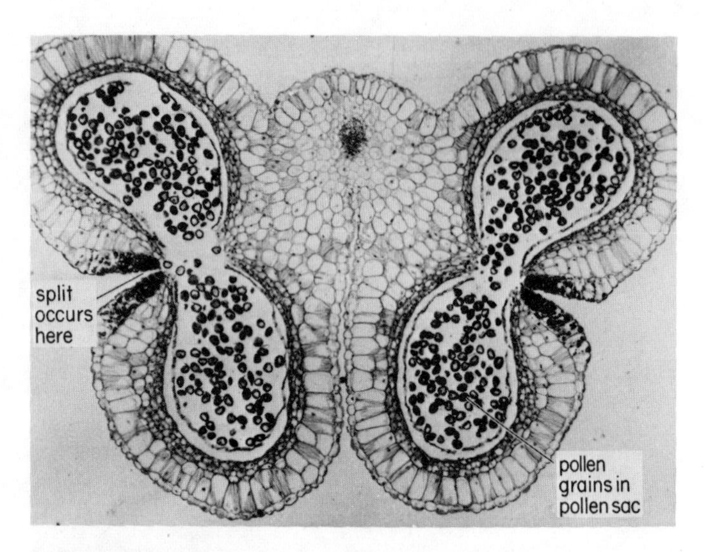

Fig. 24:19 Photomicrograph of a transverse section through the anthers of a stamen showing the four pollen sacs.

Pollen grain development and fertilization

A pollen grain, on reaching the stigma, absorbs sugar and water, swells up and starts to grow by putting out a hair-like tube—the **pollen tube**. This tube acts rather like the hypha of a fungus as it produces enzymes at its tip and dissolves the tissues of the style as it grows, feeding on the products. Eventually the pollen tube travels down the whole length of the style into the ovary, its direction of growth being influenced by a chemical which is secreted through the micropyle of an ovule (Fig. 24:21). During the early stages of pollen tube growth several nuclei are formed within it, two of which are male gametes. On reaching the ovule the end of the pollen tube bursts and one male gamete passes into the ovule and fuses with the female gamete. This is fertilization.

You can observe some aspects of this process as follows:

1. Add a little agar powder to a 10% sucrose solution, warm to dissolve it and then pour a little over the surface of several microscope slides; when cool the solution will set like a jelly. Scatter pollen from different flowers over the jelly; place each slide in a Petri dish with enough water just to cover the bottom but not the top of the slide (the slide can be raised slightly with modelling clay). Place the top on the Petri dish and keep at about 20°C. Examine after two or three hours and again the next day. Describe carefully what you see.
2. In the meantime, remove three or four carpels from a buttercup flower. Choose a flower in which the stamens have curled back to expose the carpels as pollination is likely to have taken place in such a flower. Place one or two carpels on a slide in some water, put another slide on top and squash firmly. Examine under the microscope and look for pollen grains on the stigma and pollen tubes in the cavity of the ovary. Attempt to follow the pollen tubes back through the style to the pollen grains.

What happens after fertilization?

Externally, the flower starts to wither and the sepals, petals and stamens usually fall off, although in some the sepals persist. At the same time the gynaecium enlarges con-

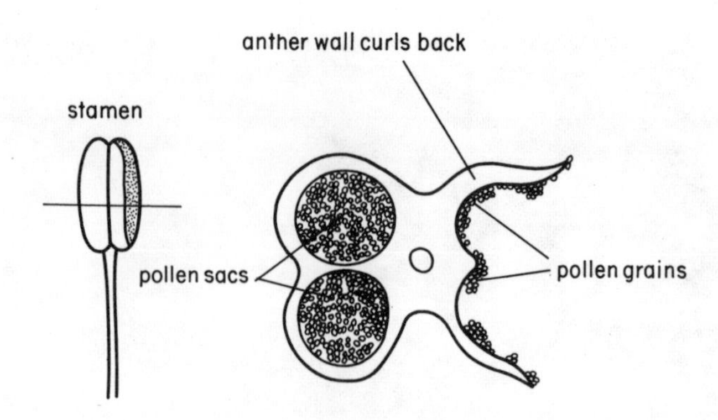

Fig. 24:20 Diagram showing how the anthers split and expose the pollen.

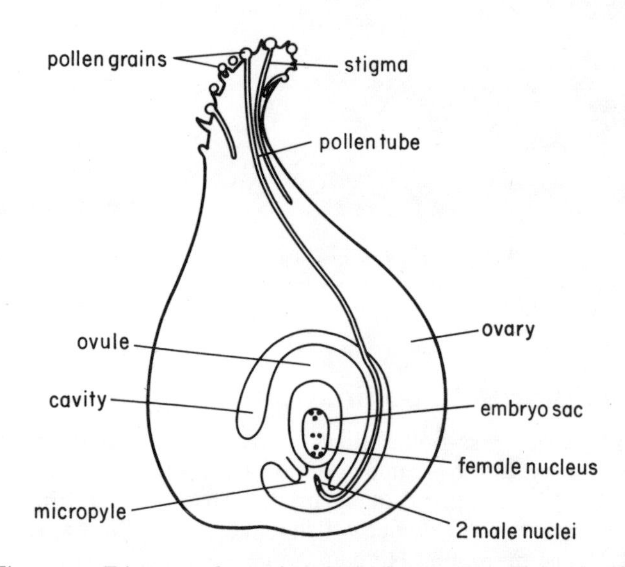

Fig. 24:21 Diagram of a section through a single carpel at fertilization stage.

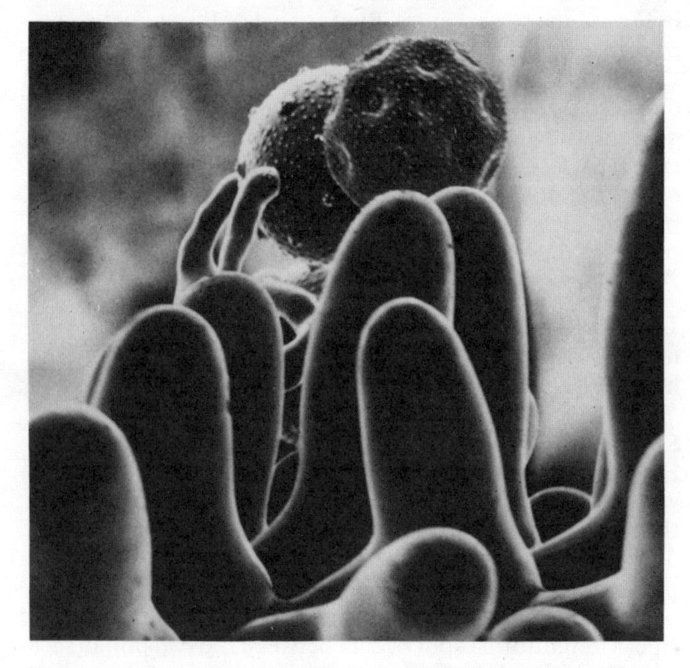

Fig. 24:22 Stereomicrograph of pollen grains on a stigma.

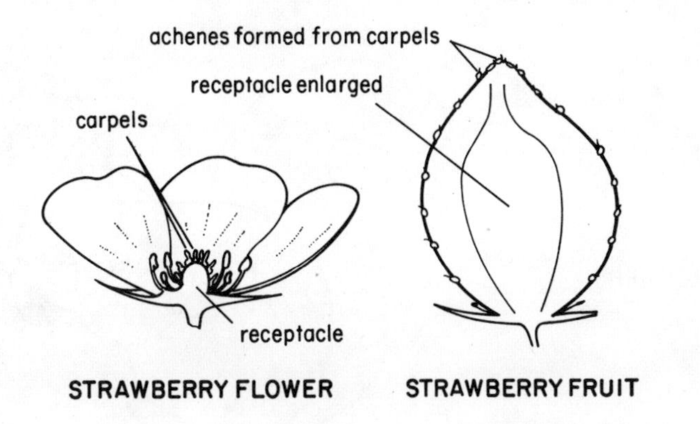

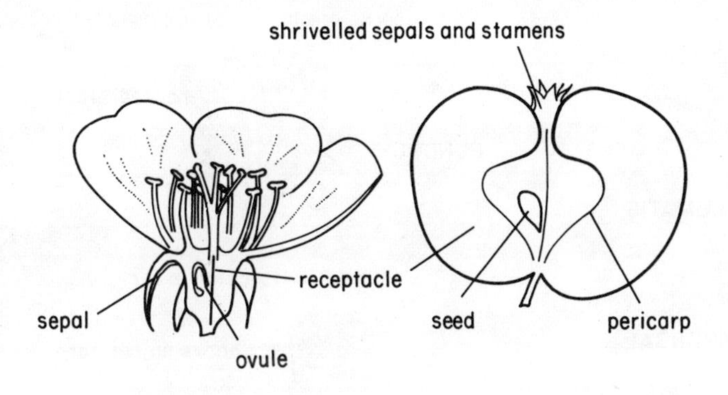

Fig. 24:23 Fruit formation.

Fig. 24:24 Selected fruits.

siderably. Inside the ovule, the zygote, formed by the fusion of the gametes, undergoes cell division to form an embryo plant consisting of a miniature shoot (**plumule**), a root (**radicle**) and one or two modified leaves (**cotyledons**). A food store is also laid down either round the embryo or within its cotyledons. This food comes from the green leaves of the plant. As the ovules develop into mature seeds their outer membranes become harder to form the **testa** and finally water is withdrawn from the tissues of the seed. It is now ready for dispersal. At the same time the ovary wall either dries up, or becomes fleshy, according to the species; it becomes the wall or **pericarp** of the fruit. Thus the carpel with its contained ovules becomes the fruit with its seeds.

Types of fruits

A true fruit is formed from the gynaecium *only*, but many common fruits such as apples and strawberries are called false fruits because they have developed from other parts of the flower in addition to the gynaecium. Thus the part of the apple you eat is the receptacle which has become fleshy and the core with its pips is the fruit with its seeds (Fig. 24:23). In the strawberry, the succulent part is the receptacle and the pips on the surface are the true fruits, each containing one seed (Fig. 24:23).

True fruits are said to be either **dry** or **succulent** according to whether the pericarp dries up or becomes fleshy. Most dry fruits are **dehiscent**, i.e. their walls split in various ways to let the seeds out; this aids seed dispersal. Others are **schizocarpic**, i.e. they break up into one-seeded portions. Other dry fruits are **indehiscent**, i.e.

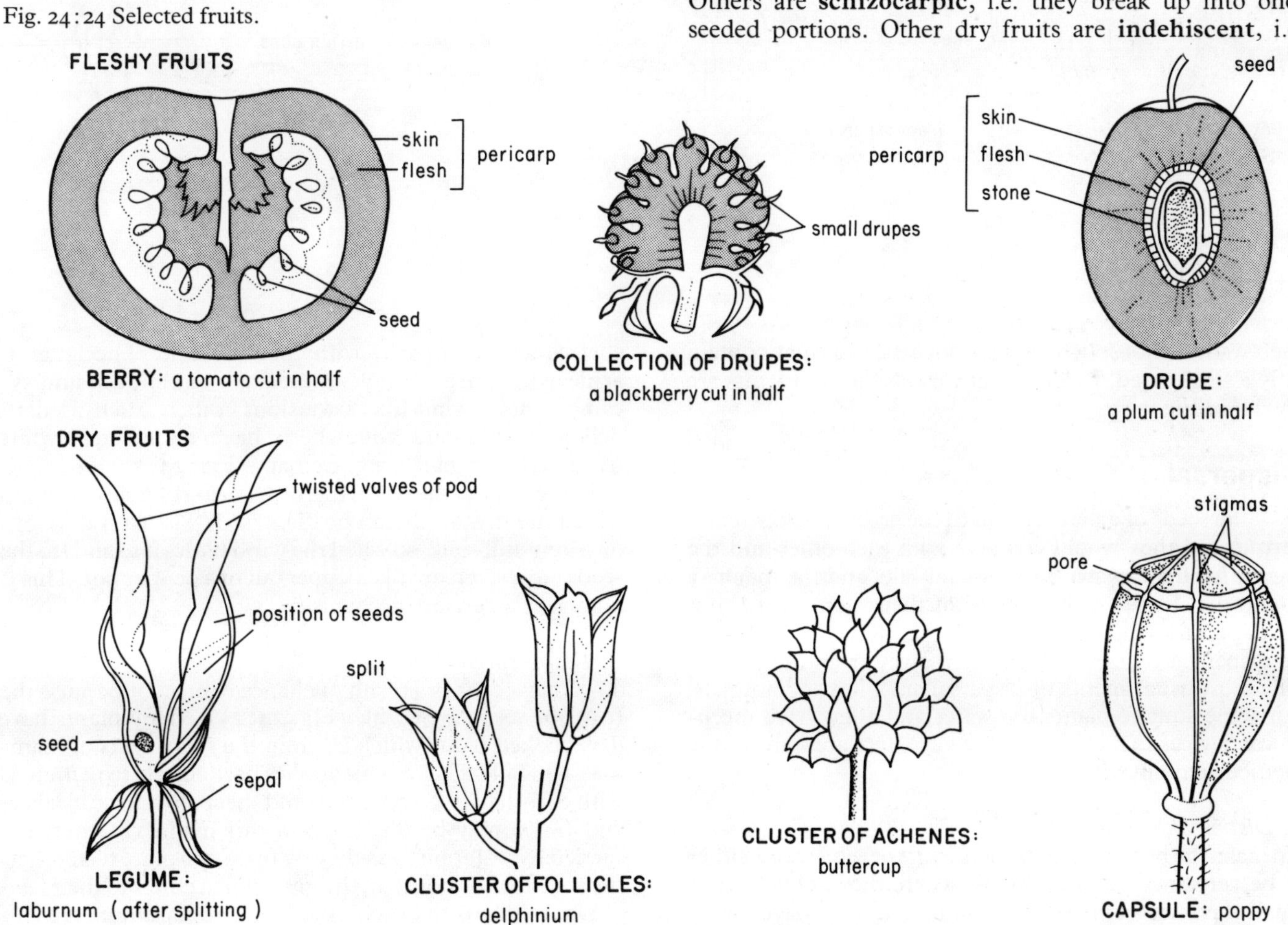

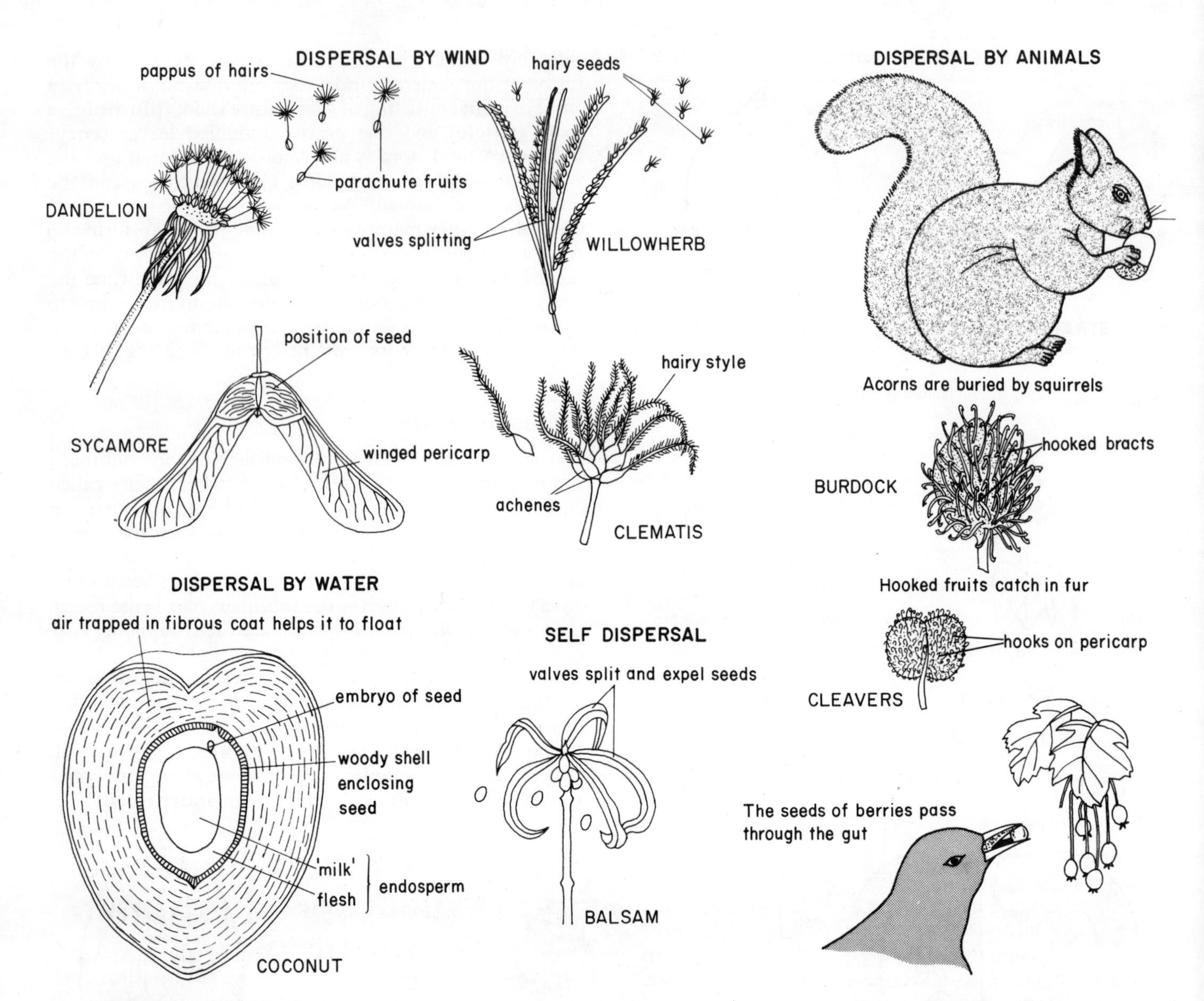

Fig. 24:25 Dispersal mechanisms.

their walls do not split as this is unnecessary as they contain only a single seed. Examples of various kinds of fruits are shown in Fig. 24:24.

Dispersal of fruits and seeds

If all the seeds of a plant fell on to the ground below it and germinated they would compete with each other and the parent for light, water and mineral salts and the majority would die. Plants cannot move of their own accord, but have special adaptations by which their fruits or seeds may be dispersed by some other agent away from the parent. The most usual agents of dispersal are wind and animals, but some aquatic plants use water and a few have mechanical devices to flick the seeds away. Here are some of the methods employed:

1. Dispersal by wind

To make the best use of wind the fruits or seeds need either to be very small, as in orchids, where they are almost as small as pollen grains and very numerous, or to have a large surface area compared with their volume. The latter is achieved in various ways: some such as elm, ash and sycamore have wing-like extensions, others such as dandelion, thistle and willowherb have parachutes, while others are very hairy, e.g. clematis (Fig. 24:25).

In the poppy, the ripe fruit forms a ring of holes, through which the tiny seeds can be dispersed. The fruit develops on a long stiff stalk, so when the wind blows it to and fro the seeds are shaken out like pepper out of a pepper pot. This is known as a **censer** mechanism.

2. Dispersal by animals

This may occur as a result of chance contact or because the fruits or seeds are edible (Fig. 24:25). Many plants have dry, hooked fruits which catch in the furry coats of mammals and fall off later. This applies particularly to mammals with curly hair, e.g. some dogs and sheep. Burdock, cleavers and *Geum* may be dispersed in this manner. Others are carried by water birds as they fly from one place to another, the seeds being lodged in the mud which clings to their feet.

Dry edible fruits such as acorns and hazelnuts are col-

TABLE 24:1. SIMPLIFIED CLASSIFICATION OF FRUITS

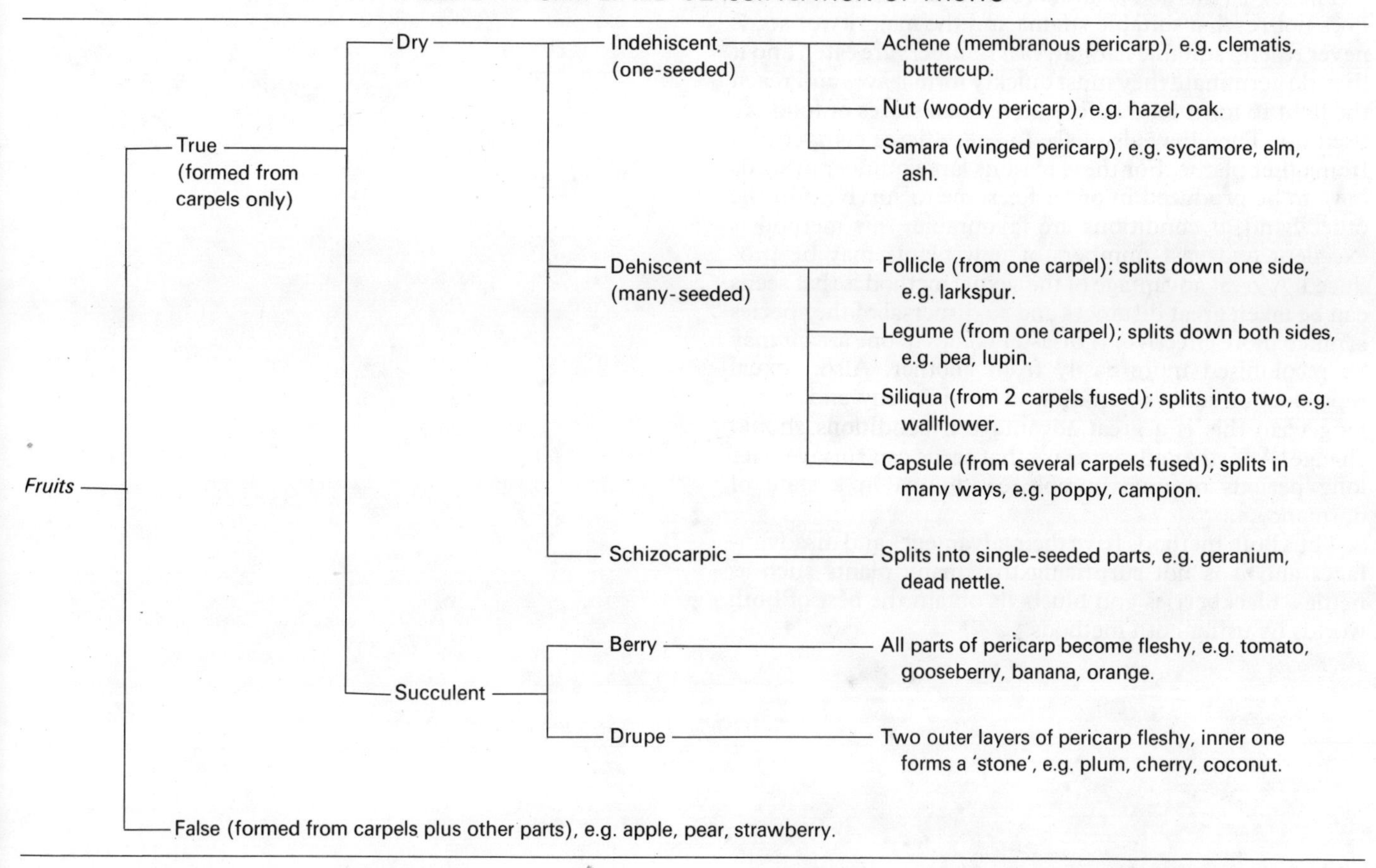

Fruits	True (formed from carpels only)	Dry	Indehiscent (one-seeded)	Achene (membranous pericarp), e.g. clematis, buttercup.
				Nut (woody pericarp), e.g. hazel, oak.
				Samara (winged pericarp), e.g. sycamore, elm, ash.
			Dehiscent (many-seeded)	Follicle (from one carpel); splits down one side, e.g. larkspur.
				Legume (from one carpel); splits down both sides, e.g. pea, lupin.
				Siliqua (from 2 carpels fused); splits into two, e.g. wallflower.
				Capsule (from several carpels fused); splits in many ways, e.g. poppy, campion.
			Schizocarpic	Splits into single-seeded parts, e.g. geranium, dead nettle.
		Succulent	Berry	All parts of pericarp become fleshy, e.g. tomato, gooseberry, banana, orange.
			Drupe	Two outer layers of pericarp fleshy, inner one forms a 'stone', e.g. plum, cherry, coconut.
	False (formed from carpels plus other parts), e.g. apple, pear, strawberry.			

lected by squirrels and jays and buried as a food store; if they are not found again they may germinate.

Succulent fruits may be dispersed as a result of being eaten by birds and mammals. They may be eaten whole and the seeds passed through the gut to be deposited with the faeces as in currants and holly, or they may be partially eaten and the seeds scattered as in plum, cherry and mistletoe. In the mistletoe the seed is sticky and adheres to the beak of the thrush which may press it into a crevice in the bark of a tree in order to wipe it off its beak. Here it germinates and the seedling becomes parasitic on the tree.

3. Dispersal by water

For this to be effective the fruit or seed must be buoyant enough to float. In coconuts this is achieved by air trapped in the outer covering (Fig. 24:25).

4. Self-dispersal mechanisms

Plants which flick their seeds out do so as a result of the uneven drying of their fruits. For example, when the pods of gorse, broom and lupin dry, tensions are set up in their walls which cause the pod to burst suddenly. When this happens the valves of the pod twist and the seeds are thrown out with considerable force; sometimes they travel several yards. Balsam also has an 'explosive' fruit (Fig. 24:25).

You might be interested to collect as many kinds of fruits as you can and classify them into their main groups (see Table 24:1) and note any adaptations that they have for dispersal.

Comparison of vegetative and sexual reproduction in flowering plants

We have seen how flowering plants may use two very different methods of reproduction, an asexual or vegetative method involving a process of budding and a sexual method resulting in the formation of seeds. Both methods have advantages and disadvantages.

The vegetative method is a relatively safe one; food is available from the parent plant for early development, the progeny can be quite large before becoming independent and so they can compete better with other plants in the neighbourhood for survival. This method results in the species growing in clumps or patches and in consequence competition between plants of the same species may increase, but this disadvantage may well be offset by the effect of crowding out plants of other species and the protection afforded to its own. These clumps are also very resistant to unfavourable external conditions and if the upper parts are destroyed they can form new aerial shoots from their underground buds to replace them. The chief disadvantage is that dispersal is limited to the immediate vicinity of the plant and the method is usually slow. There is also the fact that the daughter plants will inherit the exact characteristics of the parent, in other words, they will vary very little (Ch. 34). This may be an advantage in the short term, but not if the habitat changes.

The sexual method is far more hazardous as most pollen does not reach a suitable stigma and the majority of seeds never reach a suitable habitat; vast numbers are eaten and if they do germinate they must quickly form leaves and reach the light to form food before their small stores of food are used up. They may also have to suffer fierce competition from other plants. For these reasons large number of seeds have to be produced in order for some to survive. On the other hand, if conditions are favourable, this method is excellent and vast numbers of new plants may be produced. A great advantage of the sexual method is that seeds can be taken great distances and so dispersal of the species is much more effective. If disaster comes to one area it may be recolonised more easily from another. Also, sexual reproduction allows variation in the progeny and in the long term this is a great advantage if conditions should change. A further advantage is that seeds can survive over long periods of unfavourable conditions in a state of dormancy.

Thus both methods have their advantages and disadvantages and it is not surprising that many plants such as nettles, blackberries and bluebells obtain the best of both worlds by using both methods.

25

Growth

Growth is an inevitable consequence of reproduction. It is the permanent increase in size of an organism due to the formation of new protoplasm. All but the simplest organisms increase in complexity when they grow. For example, each one of us has grown from a single fertilized cell.

Growth is influenced by several factors. Heredity is important in determining the essential features of shape and size; for this reason tall parents, whose parents are also tall tend to have tall children (Ch. 34). However, within the general limits laid down by heredity, growth will also be affected by nutritional factors. If a person's diet is adequate for all needs, optimum growth should occur. It is probable that the increase in height of people in the more developed countries over the past 100 years has been due to their having had a more adequate diet. With many plants the nutritional factor is more easily seen, as those grown on poor soil with no nutrients added and with little water are much smaller than similar ones grown on rich soil with sufficient water. We have already seen (p. 163) that hormones greatly influence growth in people and are responsible for such extremes of growth as dwarfs (lack of hormone), and giants (excess of hormone). Growth hormones are also found in plants; they control the elongation of cells and determine the curvature of various plant organs and hence their shape (p. 196).

Mitosis

Growth in multicellular animals and plants involves cell division. When cells divide it is always the nucleus that plays the essential part. The nucleus divides first by a process called **mitosis** and the cytoplasm divides afterwards. Mitosis is basically similar in both animals and plants. The process involves thread-like structures in the nucleus called **chromosomes**. All but the most primitive species of plants and animals have a definite number of pairs of chromosomes in each of their nuclei and during

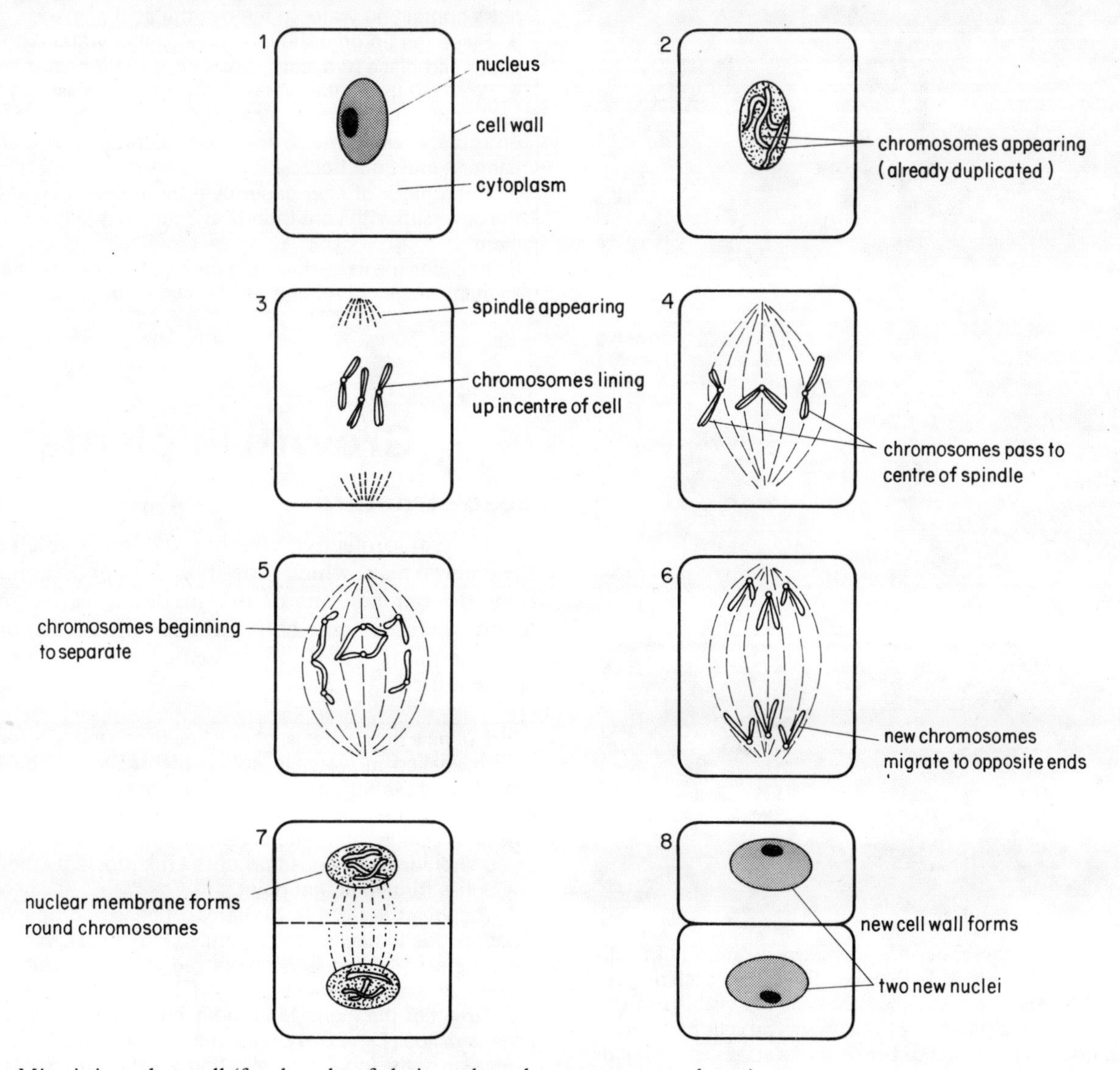

Fig. 25:1 Mitosis in a plant cell (for the sake of clarity only 3 chromosomes are shown).

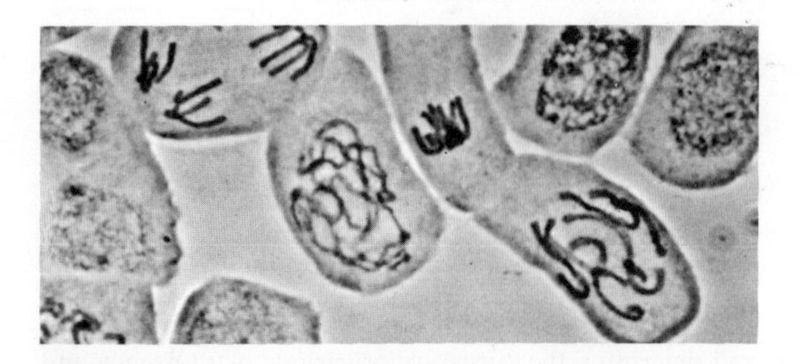

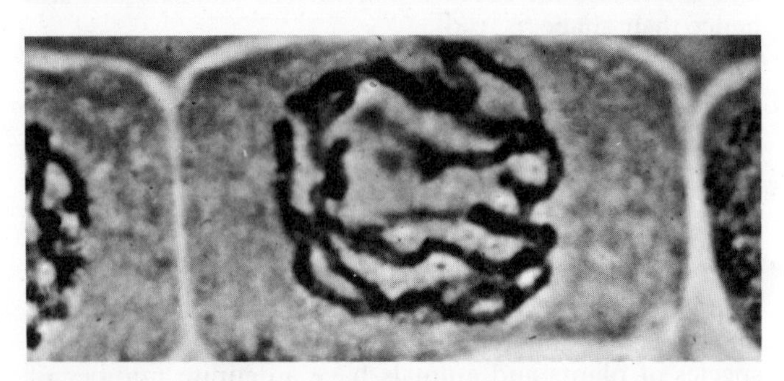

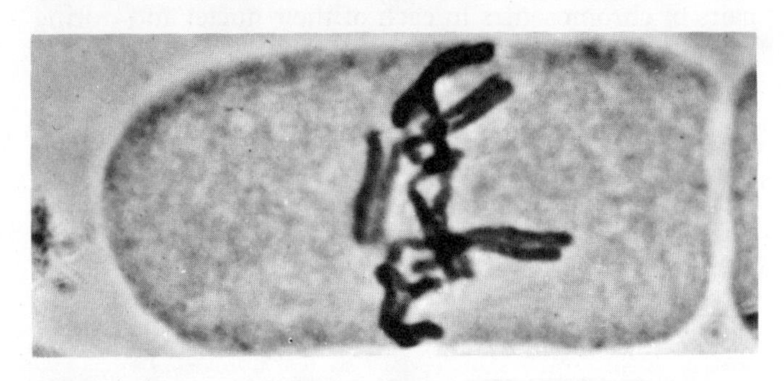

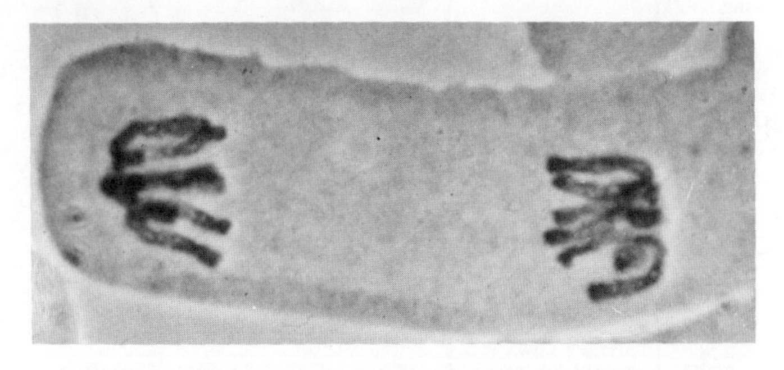

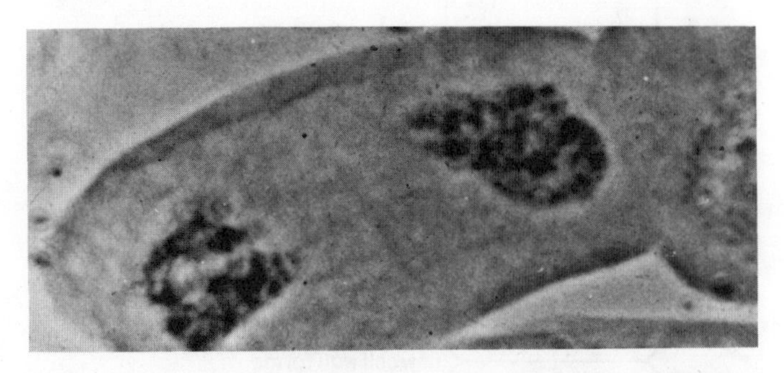

Fig. 25:2 Photomicrographs showing stages in mitosis in cells from the root tip of a crocus. In this species, each root cell nucleus has 6 chromosomes (3 pairs). The spindle is not visible in these preparations. The top photograph shows several cells in different stages of mitosis. The sequence below shows stages corresponding to 2, 4, 6, and 7 in Fig. 25:1.

mitosis this number is exactly maintained. Thus if a cell has 23 pairs, as in man, when it has divided into two, there will still be 23 pairs in each cell. This is accomplished by an exact duplication of each chromosome so that two new chromosomes are formed, one of which goes to one daughter cell and the other to the second. The significance of this exact duplication, known as **replication**, is discussed on p. 274.

Study the stages of mitosis in Fig. 25:1 to see the details of the process in plant cells. Note that when the nucleus has divided into two, a new cell wall is formed between them by the cytoplasm. In the division of animal cells the outer cell membrane constricts and complete separation into two cells occurs.

You can see some of the stages for yourself by looking at specially prepared longitudinal sections through the root tip of an onion (Fig. 5:5). You can also make a preparation yourself by separating the individual cells by a squash technique and then staining the nuclei:

1. Cut off the last 4 mm of an onion root. Place it in a test tube containing a little Molar hydrochloric acid at about 60°C for 5 minutes. This treatment will soften the tissues.
2. Decant the liquid carefully. Wash the root tip in a watch glass containing water to remove the acid.
3. Place the tip on a slide, remove excess water with a filter paper and place two small drops of acetic orcein stain on it.
4. Now gently tap the root with the end of a glass rod or scalpel handle until it is completely broken up. Place a coverslip on top and warm *gently* for a few seconds over a very low flame; it must not boil.
5. Put a piece of filter paper over the preparation and press the coverslip with your thumb, avoiding any sideways movement.
6. Examine the individual cells under the high power and find as many stages of mitosis as you can (Fig. 25:2).

Growth in plants

Seed structure

Seeds have a protective covering (the testa) which encloses the embryo plant which consists of a shoot (the plumule), a root (the radicle), one or two modified leaves (the cotyledons), and a food store which either surrounds the embryo and is called the **endosperm** or is stored in the cotyledons.

1. Examine broad beans, french beans and maize which have been soaked in water to soften them. Strictly speaking, a maize grain is a fruit containing one seed, the testa having fused with the pericarp. Compare their external features and note similarities and differences. Squeeze both beans gently; you should see the position of the micropyle by the bubble which is formed at that point.
2. Remove the testa from the two beans and open out the cotyledons to display the plumule (Fig. 25:3). Where is the food stored? What is it composed of? Carry out tests for starch, sugar and protein (p. 112).
3. Now cut the maize longitudinally, slightly to one side of the mid-line (Fig. 25:4). Note the size of the embryo and the position of the food store; also find out the nature of the food.

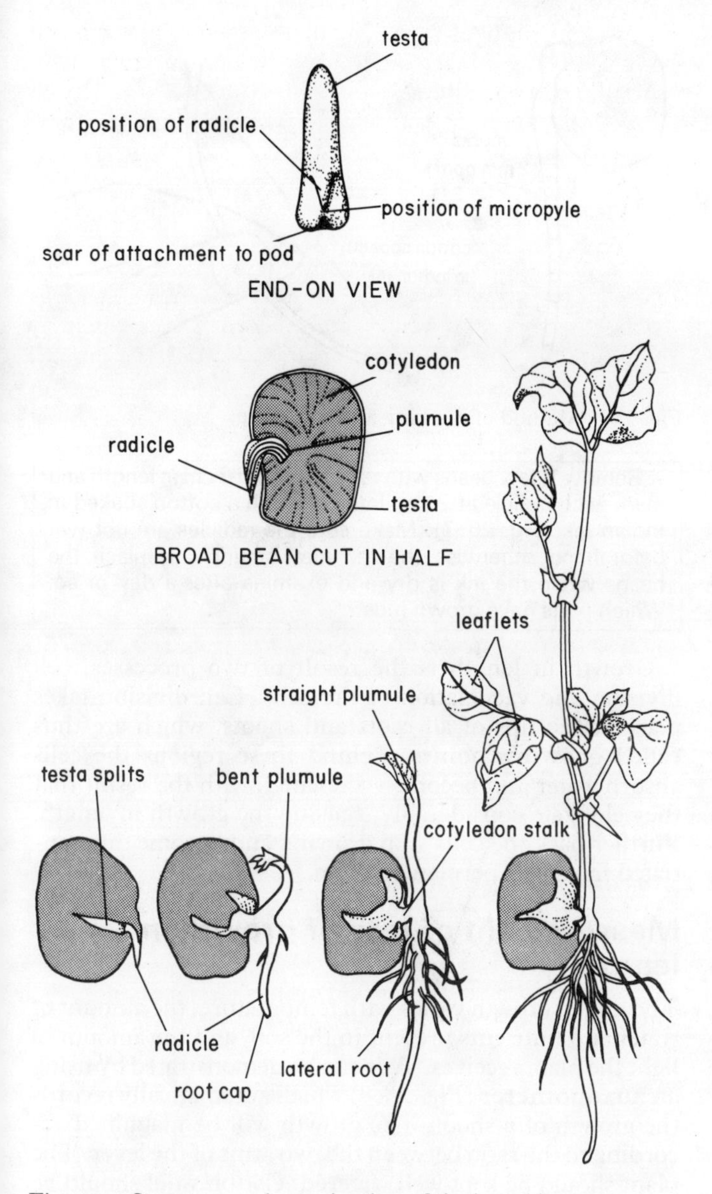

Fig. 25:3 Structure and germination of the broad bean.

Germination

Not all seeds germinate in the same way. In some, e.g., broad bean, the cotyledons remain in the seed below ground, so the seedling is said to be **hypogeal** (Fig. 25:3). In others, e.g. french beans, the cotyledons come above ground and turn green; this is known as **epigeal** germination (Fig. 25:4).

> You can compare the method of germination in broad beans, french beans and maize by placing one seed of each in a glass jar, kept in position by wet blotting paper as in Fig. 25:5. Over the next fortnight note how the plumule and radicle emerge and observe what happens to the cotyledons in each case.

Conditions necessary for germination

When you try to grow seeds in the garden they do not always come up as well as you had hoped; sometimes, they may not germinate at all. There are many possible reasons for this. First let us find out the importance of such factors as oxygen, water, temperature and light on germination.

> 1. Set up 5 large specimen tubes with cotton wool pressed into the bottom of each. Label them 1–5 (Fig. 25:6). Put 10 mustard seeds in each. Add enough water to soak the cotton wool in all tubes except tube 3. Remove oxygen from tube 4 by suspending a small tube of pyrogallol (an oxygen absorber) by means of a cotton thread and closing with a cork.
> 2. Put tubes 2, 3 and 4 in a dark box and 5 in the light. Keep all these at room temperature. Put 1 in a refrigerator (which is also, of course, dark).
> 3. Examine the tubes periodically. What are your conclusions concerning the factors necessary for germination?
>
> Comparisons of 1 and 2 will show the effect of temperature.
>
> Comparisons of 2 and 3 will show the effect of the absence of water.
>
> Comparisons of 2 and 4 will show the effect of the absence of oxygen.
>
> Comparisons of 2 and 5 will show the effect of the absence of light.

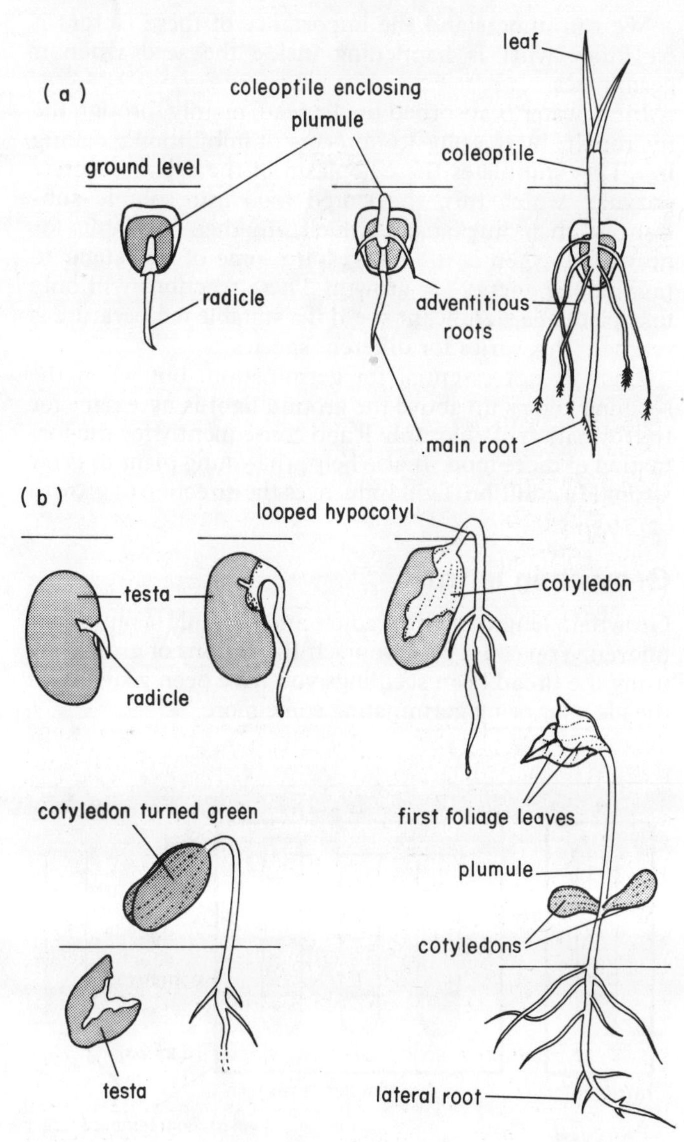

Fig. 25:4 Stages in the germination of a) maize (hypogeal) b) french bean (epigeal).

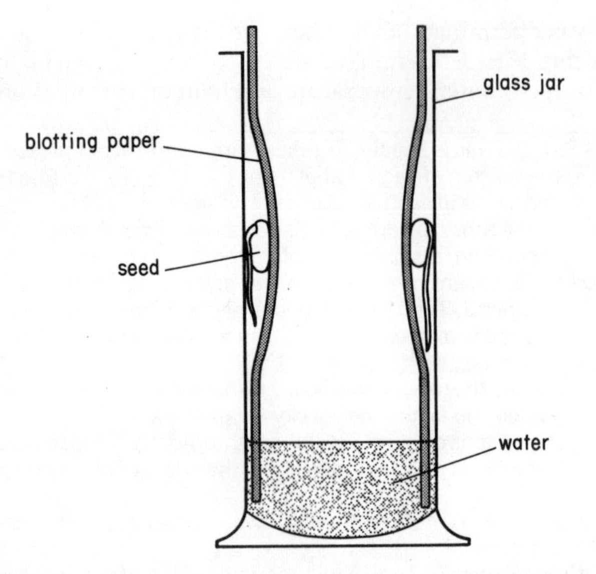

Fig. 25:5 Apparatus for observing the stages of seed germination.

We can understand the importance of these factors if we know what is happening inside the seed when it germinates.

First, water is absorbed by the seed, mainly through the micropyle, by the physical process of imbibition (soaking in). This stimulates the cytoplasm of the cells to secrete enzymes which turn the stored food into soluble substances, including sugar, which are then available for growth. Oxygen is used to respire some of this sugar to provide the energy for growth. These reactions will only take place at a significant speed if a suitable temperature is reached (this varies for different species).

Light is not essential for germination, but when the seedling comes up above the ground light is necessary for the formation of chlorophyll and consequently for the formation of more food. It also helps the young plant to grow strong. In addition, light influences the direction of growth (p. 195).

Growth in length

Growth in length of both radicle and plumule is rapid, but uneven. You can see for yourself the regions of growth by using the broad bean seedlings you have been growing in the glass jar or by germinating some more.

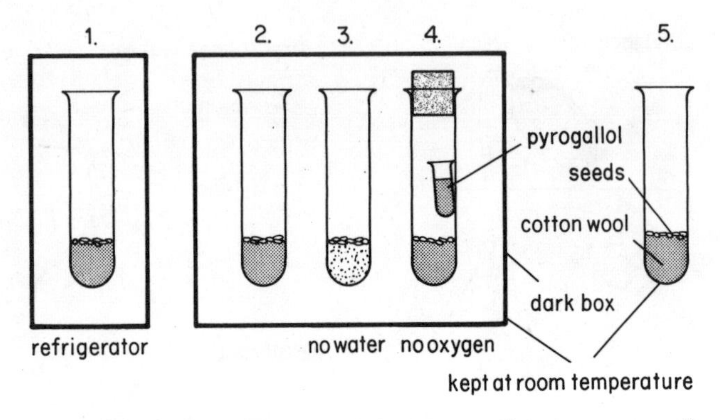

Fig. 25:6 Apparatus for determining the conditions necessary for germination.

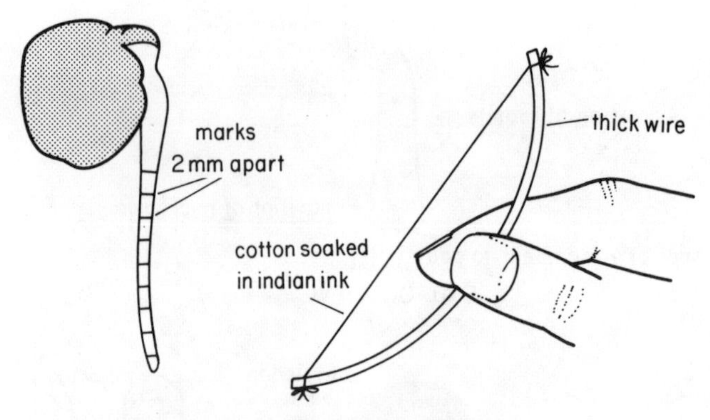

Fig. 25:7 Method of marking a bean radicle.

Remove a few beans with radicles about 3 cm in length and mark each radicle at 2 mm intervals with a cotton soaked in Indian ink (Fig. 25:7). Make sure the radicles are not wet beforehand otherwise the ink may smudge. Replace the beans when the ink is dry and examine after a day or so. Which parts have grown most?

Growth in length is the result of two processes, cell division and vacuolation of the cells. Cell division takes place at the tips of all roots and shoots, which are thus called **growing points**. Behind these regions the cells absorb water and become vacuolated, with the result that they elongate considerably, causing the growth in length. Further back the cells stop growing and become differentiated into their permanent form.

Measurement of rate of growth in length

The rate of growth varies with temperature, the amount of water and nutrients present in the soil, and the amount of light the plant receives. This can be demonstrated by using an **auxonometer** (Fig. 25:8) which automatically records the growth of a shoot. The growth will be magnified according to the ratio between the two arms of the lever. The plant should be kept well watered. Cotton wool should be used at the place where the cotton is fastened to prevent damage to the delicate tip. The cotton should be short so that if the humidity of the air varies any error due to changes in length of the cotton will be negligible. The compensatory weight should be just heavy enough to prevent the lever from pulling on the plant. The smoked disc is rotated by an electric motor at a speed of one revolution per hour. As the disc rotates, a bristle attached to the end of the lever makes a fine tracing on its smoked surface in the form of a continuous spiral. The distance between adjacent lines is proportional to the amount of growth that has taken place.

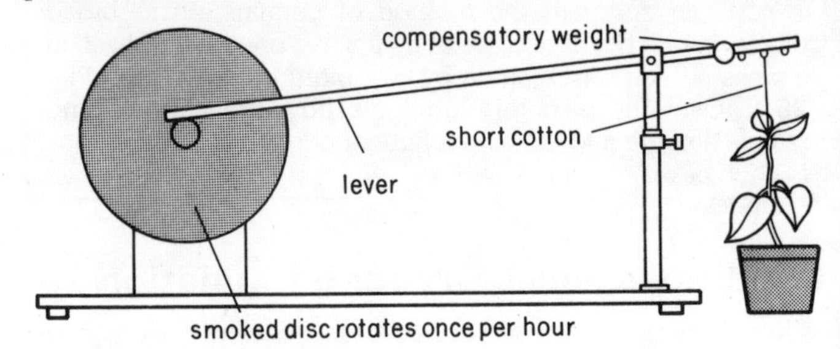

Fig. 25:8 Auxonometer.

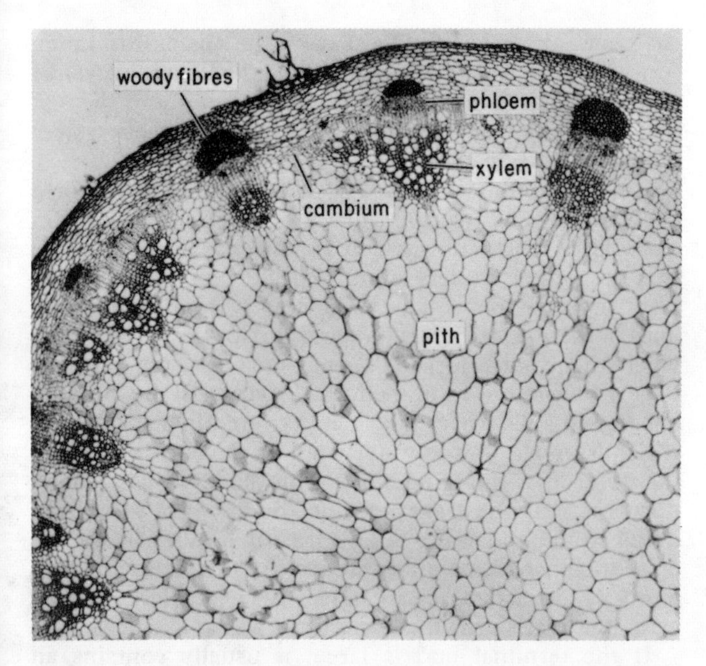

Fig. 25:9 Photomicrograph of a transverse section of a sunflower stem.

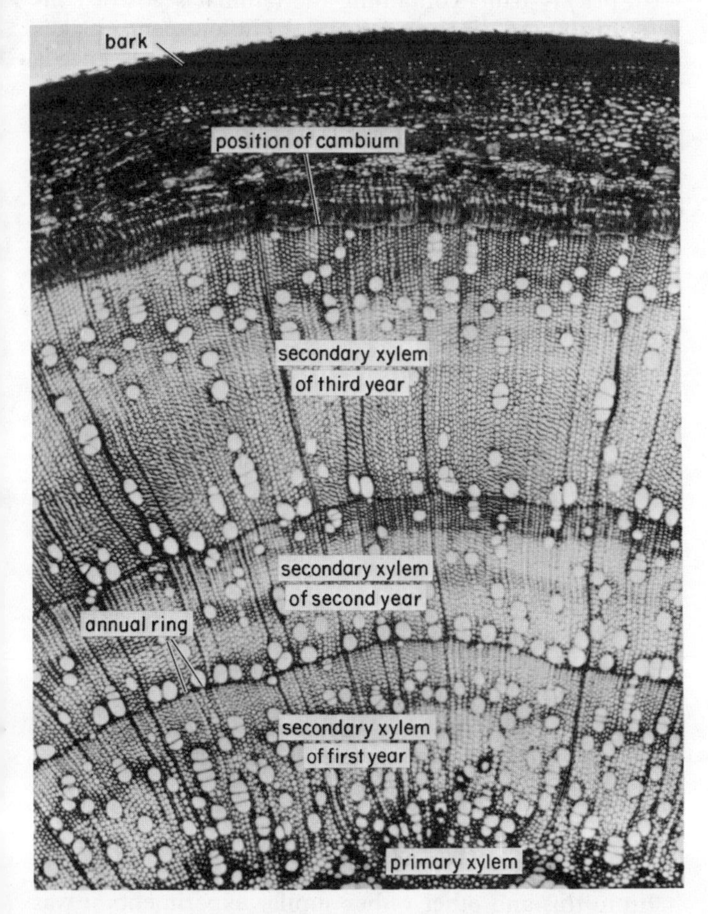

Fig. 25:10 Photomicrograph of a 3-year old stem of elm.

This apparatus, if examined after 24 hours, will indicate the effect of light and darkness on the growth rate if other factors such as temperature are kept constant. How could the apparatus be used to investigate the effect of temperature on growth?

Growth in thickness

As we have seen, growth in *length* occurs as a result of cell division and elongation at the tips of roots and shoots. Growth in thickness also results from cell division, but those concerned are the cambium cells which lie between the xylem and phloem. These cells divide in such a way that they form new xylem towards the inside and more phloem towards the outside. In very young stems the cambium is restricted to a position within the vascular bundles, but later it joins up to form a complete ring (Fig. 25:9). In woody plants, such as trees, this formation of **secondary** xylem and phloem continues throughout life, causing the trunk to get thicker and thicker. So a section through a 3-year-old stem (Fig. 25:10) shows 3 rings of secondary xylem making up the wood. New phloem is added each year, but this growth is not indicated by rings. The phloem lies just under the bark.

The rings in the xylem occur because growth is not regular. In the late summer and early autumn only small vessels and fibres are produced, in the late autumn and winter growth ceases and in the spring much larger vessels are formed. Thus the small autumn vessels are found next to the large spring vessels and their difference in texture is visible as a ring. As these rings occur annually the age of a particular branch or trunk can be determined by counting them.

Bark

As the stem thickens from within, the epidermis becomes stretched and eventually splits and peels off. Its protective function is taken over by a layer of cork formed from a ring of **cork cambium** which usually arises in the outer cortex. This layer of cork becomes thicker with repeated division of the cork cambium and becomes the **bark**. Cork is non-living and is impermeable to water and gases, but the growing stem still requires oxygen for respiration. This is obtained through **lenticels** which are pores in the bark, where the corky cells are loose and allow gases to diffuse between them (Fig. 25:11).

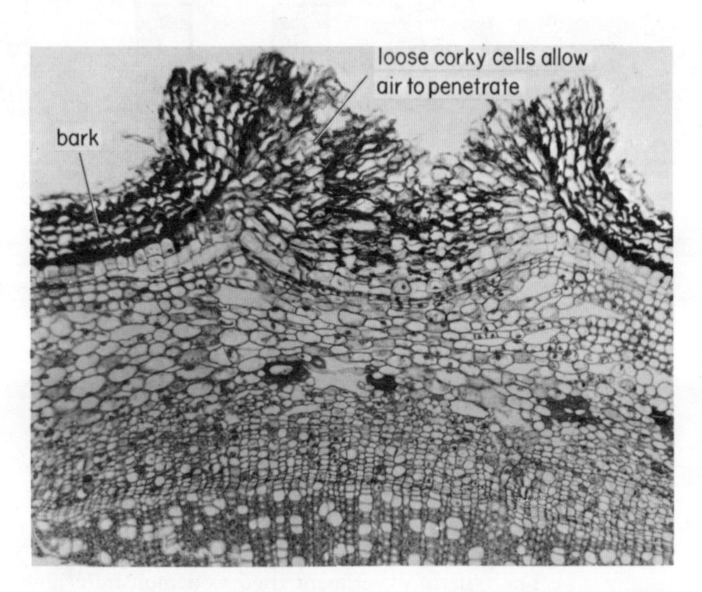

Fig. 25:11 Photomicrograph of a section through a lenticel of elder.

The winter condition

Trees are said to be **deciduous** when they shed their leaves as winter approaches. This is an adaptation to prevent more water being lost through transpiration than is absorbed through the roots, as at low temperatures water absorption is very slow even when the soil is saturated. The lenticels are also closed by a layer of cork to prevent further loss of water.

Before leaf-fall, any remaining food in the leaves is removed for storage and the chlorophyll breaks down into compounds which give the leaves their autumn colours. At the same time, a layer of loose cells, the **abscission layer** forms at the base of each leaf stalk and below it a layer of cork develops. This cuts off the water supply to the leaf which becomes dry and falls off leaving the corky layer which seals the wound.

If you examine winter twigs such as those of the horse chestnut (Fig. 25:12) you will see the leaf scars and within them a half ring of dots which are the ends of the vascular bundles. The leaf scars are in opposite pairs, each pair at right angles to those above and below it. Buds occur in the axils of these scars except (apparently) for those just below where the twig has forked. There were buds here, but they have grown out to form the forked stems. All the buds are encased in scale leaves which are covered in a sticky gum which is impervious to water, another device for preventing water loss. In spring most of the axillary buds remain dormant, but the large terminal bud always grows. In doing so the bud scales fall off leaving a number of scars close together called **girdle scars**. As the scales are shed annually the distance between two sets of girdle scars indicates one year's growth.

If the terminal bud is large, it usually contains an embryo inflorescence which will form one of the spikes of flowers which cover the tree in spring. Later the flowers form prickly fruits containing the familiar seeds or conkers. In autumn the fruits fall and the inflorescence stalk is shed leaving a **saddle scar**. It is when an inflorescence is formed that the buds below it grow out and the stem forks.

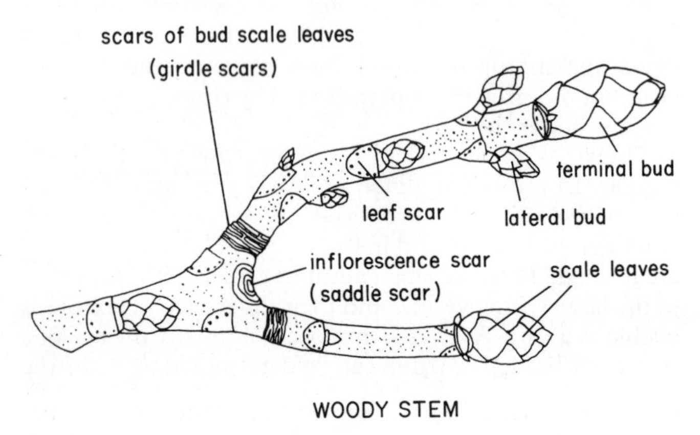

Fig. 25:12 Horsechestnut twig.

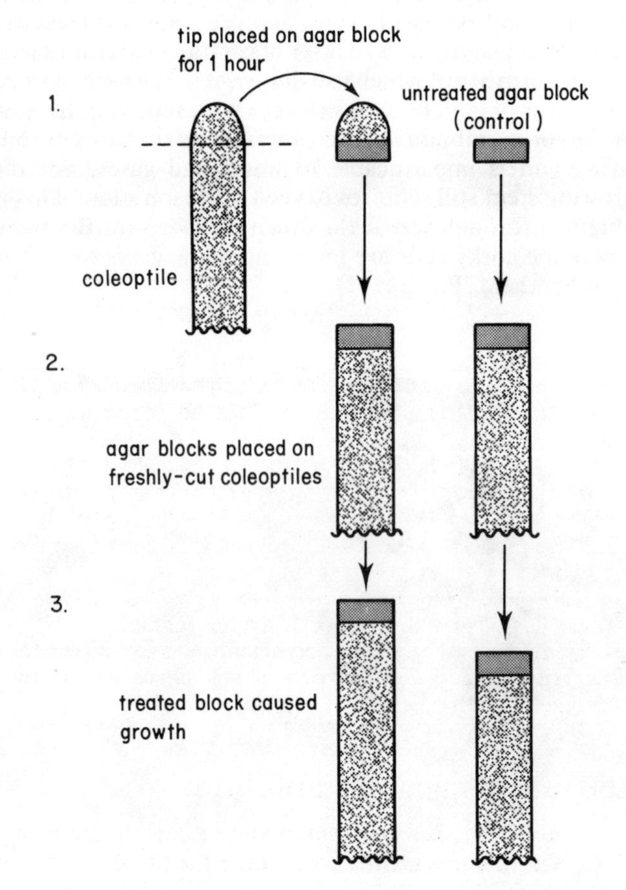

Fig. 25:13 The type of experiment used to demonstrate how auxin in the tip of the coleoptile (plumule sheath) controls growth.

1. Examine a horse chestnut twig. Determine if your specimen has previously flowered and, if so, how long ago. How old is the section nearest to the cut end? Confirm by counting the rings visible on the cut end.
2. Soak the terminal bud in methylated spirits to remove the stickiness. Remove the scales and find the fluffy mass inside which consists of tiny folded foliage leaves. Is there an embryo inflorescence in the centre?

Control of growth

We saw in Chapter 22 that growth in mammals was controlled by hormones secreted into the blood stream from the thyroid and pituitary glands. There are also comparable substances in plants. Consider this classical experiment:

A number of oat seedlings were grown in the dark and after a few days the tips of several coleoptiles were cut off and placed upright on tiny blocks of agar for one hour. Then further coleoptiles were decapitated and the agar blocks were carefully placed on the cut ends of half of them, while untreated agar blocks were placed on the others to act as controls. Within a matter of a few hours under warm conditions those with treated agar had grown considerably, while those with untreated agar had grown hardly at all (Fig. 25:13).

From this and other rather similar experiments it was concluded that the tip of the shoot produced a substance, soluble in water, which diffused backwards into the region of the stem behind the tip causing it to elongate. The agar, by absorbing this substance, acted in the same way as the tip when placed on the cut end.

This growth-regulating substance, first called **auxin**, was analysed and found to be indolyl acetic acid (IAA). It is now known to accelerate the elongation of cells. This is just one of the plant hormones.

What controls the direction of growth?

When you were growing seedlings, you will have noticed that whatever the position of the seed in the jar, the radicle grew downwards and the shoot upwards. Also, when plants are placed on a window ledge the shoots tend to bend towards the light, and when you grow runner beans in the garden they twine up the supports in a definite manner. Clearly the plant organ in each case is responding to some stimulus such as gravity or light or touch by a directional growth movement. Such responses are called **tropisms**.

Phototropism

This is the growth response of a plant organ to the stimulus of light; it is a directional bending in relation to the position of the light source.

> 1. Place about 10 oat grains in each of 3 small pots containing damp sawdust, planting them just under the surface. Let the seeds germinate in the dark by keeping them in an incubator at about 20°C for 4 or 5 days.
> 2. Make some small aluminium foil caps by moulding them on a matchstick. Place them over half the seedlings in one of the pots. The caps will keep the tips in darkness. Test their reaction to directional light by arranging them in boxes as shown in Fig. 25:14. Examine the seedlings after a few days. How has the direction of the light source affected the bending of the shoots? Which part of the shoot is most sensitive to light? In which region of the shoot has the bending occurred? Which set has grown the most?

This bending of a plant shoot towards the light source is called **positive phototropism**. The majority of *roots* show no reaction to light although a few such as mustard do bend away from a light source (they are said to be **negatively phototropic**). Leaves usually respond so that their surfaces are at right angles to the light. In woods, with the light coming mainly from above, the leaves of many trees such as beech grow in such a way that the lower ones find the gaps between those above them, so that they are exposed to the maximum light available. This arrangement is known as a **leaf mosaic**.

Geotropism

This is the response of a plant organ to the stimulus of gravity. As you have seen from growing seedlings, shoots grow upwards against gravity and roots grow downwards. How can you prove that this really is due to gravitational force? To do this you need to subject some plants to gravity and others to none. The latter would only be possible if the plants were grown in outer space where gravitational forces were not acting. Which way do you think the roots and shoots would grow if this were done?

It is not possible to eliminate the force of gravity under normal laboratory conditions, but it is possible to even out its effect by rotating the seedlings slowly and regularly so that all regions receive the same force. This can be demonstrated by using a **klinostat** (Fig. 25:15). An electric motor causes the cork disc to rotate about once every 15 minutes. The seedlings are pinned to the cork which is covered with soaking wet cotton wool. They are arranged so that the roots point outwards. A celluloid cover is placed on top to keep the air inside saturated with water and so prevent the seedlings from drying up. A similar apparatus should be set up which is *not* rotated, to act as a control. After about 2 days the roots in the control should all have grown downwards and those which were rotated should have gone on growing in the direction in which they were growing before.

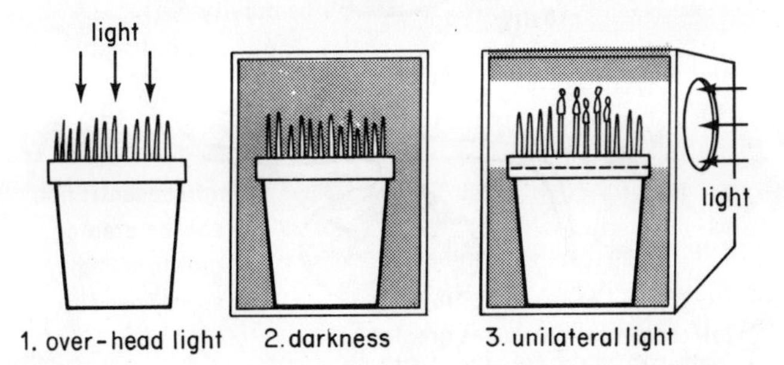

Fig. 25:14 Apparatus for determining the effect of light on the directional growth of seedlings.

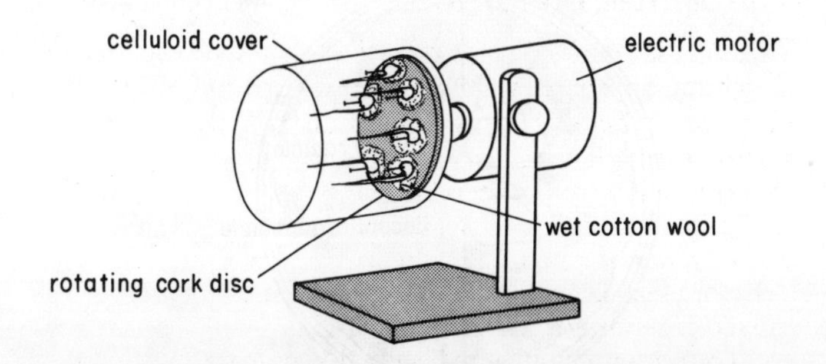

Fig. 25:15 Experiment demonstrating the effect of gravity on the directional growth of seedlings.

Which part of the root is sensitive?

> 1. Take two broad bean seedlings which have been grown in damp sawdust and which have straight radicles about 2 cm long.
> 2. Mark the radicles with Indian ink at regular intervals as you did in a previous experiment, and then cut off the tip of one of them.
> 3. Place them on wet cotton wool in a Petri dish as in Fig. 25:16.
> 4. Now place the dish in a dark place (why?) on its side in such a position that both radicles are horizontal.
> 5. Examine each day. Which part of the root is sensitive to the stimulus of gravity? Which part of the radicle causes the bending?

The mechanism of curvature

We should have seen from these experiments involving the effect of both light and gravity that it is the tip of the root or shoot which is sensitive to this stimulus, but it is the region behind it—the region of vacuolation—that responds. It is also a fact that when a root or shoot bends it is because the

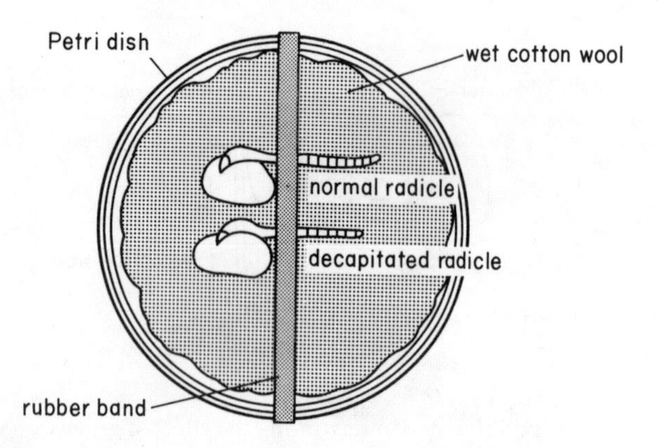

Fig. 25 : 16 Experiment to find out which part of a root is sensitive to gravity.

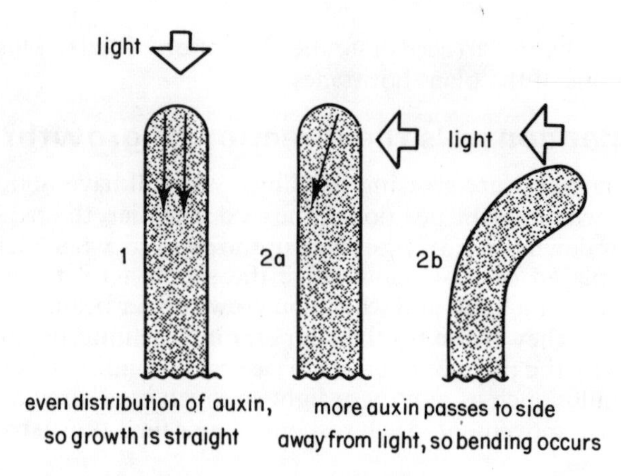

Fig. 25 : 18 The effect of light on the distribution of auxin.

cells on one side vacuolate more than on the other, thus extending them more. Could it be that in phototropic or geotropic bending more IAA reaches one side than the other?

This possibility was tested by using oat coleoptiles, as in the auxin experiment (p. 194), by cutting off their tips and replacing them at once eccentrically as in Fig. 25 : 17. After a few hours the coleoptiles had bent. It was also found that if the tips were placed on agar, so that the IAA could diffuse into it, and the agar block was put back eccentrically, the same thing happened. So IAA can cause bending if there is a greater concentration of it on one side of a plant organ. IAA is such a powerful growth regulator that one part in a million parts of water is enough to cause a growth curvature, but it has been shown that different organs respond differently to various concentrations. Consider the effect of gravity on root and shoot. When a seedling is placed horizontally (Fig. 25 : 19) the auxin tends to move towards the lower side of both root and shoot, but the root bends down

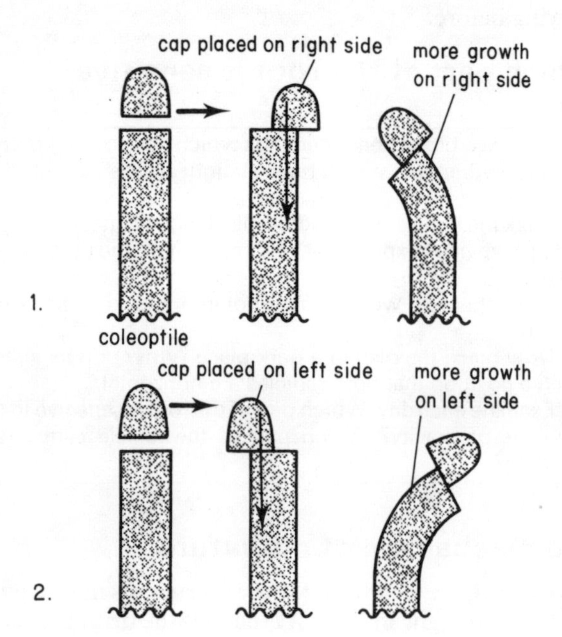

Fig. 25 : 17 The type of experiment used to demonstrate the effect of auxin on stem curvature. The vertical black arrows indicate the movement of auxin.

and the shoot up. Why should they not both bend in the same direction? It is now known that although the same substance is present in both root and shoot a higher concentration of IAA causes elongation in the stems, but in the root it is the lower concentration which causes the upper cells to elongate more, so the root bends down.

If too high a concentration of auxin is given, the plant will be killed. This is the principle behind the use of **selective weedkillers**. These are synthetic substances which act like plant hormones. When used in higher concentration different species of plants show varying tolerance to them, so they kill off some and not others. One substance known as **2.4-D** is widely used for spraying on lawns. It is absorbed through the surface of the leaves but affects monocotyledons such as grasses far less than dicotyledons such as dandelions and daisies which quickly die as a result. It is also used for spraying wheat fields to eliminate dicotyledonous weeds which would compete for soil nutrients. (See also Fig. 33 : 5.)

Hormones also control the formation of adventitious roots; hence the practice of dipping a cutting in a hormone powder before planting it. This has enabled people to propagate plants which otherwise do not readily form adventitious roots.

Hormones also affect leaf fall and the dropping of fruit.

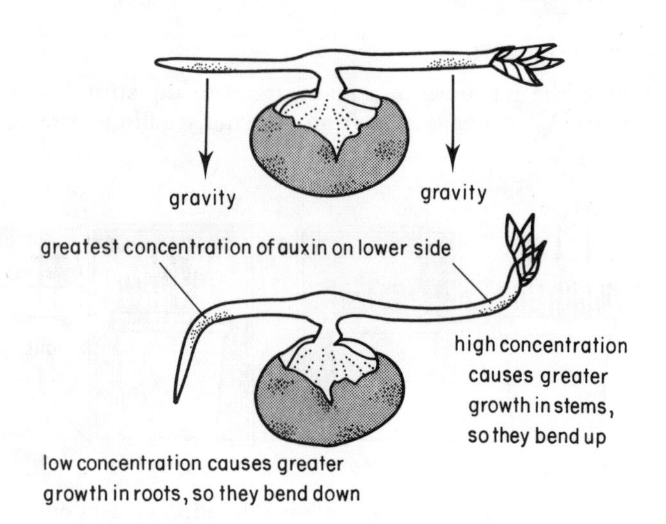

Fig. 25 : 19 The effect of gravity on root and shoot.

By spraying an apple crop with IAA it is possible to prevent the fruit from falling off prematurely.

Other tropisms

In this chapter we have concentrated attention on the effects of gravity and light on directional growth and the hormones which help to bring this about. These are not the only tropisms.

Haptotropism is the growth bending resulting from a touch stimulus. Many plants have tendrils which are used for support. If you stroke the inside of a tendril with a matchstick it will soon begin to curve. Other plants such as runner beans and bindweed have stems which are sensitive, and in consequence twine round any support they touch. **Hydrotropism** is shown by roots when they grow towards a source of water; this is easily seen along the banks of a river where the tree roots are often visible.

We have already described an example of **chemotropism** (p. 184) where the direction of growth of the pollen tube is controlled by a chemical secreted through the micropyle of the ovule.

All these tropisms are caused by greater growth on one side of a plant organ than on the other.

Comparison of growth in plants and animals

Similarities

1. Growth in both is brought about by mitotic division of cells.
2. Its rate and extent is modified by such factors as temperature and the availability of suitable nutrients, although the general pattern is determined by hereditary factors.
3. Hormones play an important part in controlling growth.

Differences

1. Growth is continuous throughout the life of the plant, but animals stop growing after a time, although cell division continues for repair and replacement.
2. Growth is restricted to special growing points in plants, but growth in animals occurs throughout their bodies.
3. As a result of 2., plants constantly change their shape, e.g. as buds put out new shoots, but animals keep roughly the same shape once their developmental stages have taken place.

26

Keeping the body in a steady state

Homeostasis

If you can imagine an organism living in a habitat where conditions never altered and where it was perfectly adapted to those conditions, there would be a perfect balance between that organism and its environment. In nature this never happens. Instead, we find many varieties of habitats and many changes of conditions within them, and in order for the organism to survive there must be many regulatory mechanisms which help to keep a state of balance when conditions change. The term used to describe this 'steady state' is **homeostasis**. In this chapter we shall examine some of the ways in which this balance is maintained.

Oxygen/carbon dioxide balance

One example of homeostasis we have already come across is the maintenance of the correct amount of oxygen within our bodies. When we take vigorous exercise much more oxygen is used by the muscles and so the amount of carbon dioxide in the blood increases. The brain responds to the higher level of carbon dioxide by increasing the rate at which nerve impulses are sent to the heart and respiratory muscles; this results in a faster heart beat and quicker and deeper breathing, and eventually the oxygen/carbon dioxide balance is restored.

Just as an organism has to maintain a balance with the *external* environment, so the living tissues have to maintain a balance with the body fluids which constitute its *internal* environment.

We have already seen (p. 86) how very sensitive blood corpuscles are to osmotic changes in the plasma. The problem of maintaining blood in a steady state is not simple because its composition is changing all the time. When blood passes through the capillaries it loses oxygen, sugar and other nutrients to the cells and at the same time gains carbon dioxide and other products of cell metabolism. The amount of water in the blood is also fluctuating. Let us now consider some of these problems in more detail.

The problems of water balance

The control of the water content of the body is known as **osmoregulation**. Such a mechanism is essential because if too much water is absorbed into the blood its osmotic strength would become less than that of the surrounding body tissues, and water would be drawn into them causing them to swell up. Conversely, if the blood became too concentrated, water would be drawn out of the tissues and they would shrink.

The problem, therefore, is to ensure that the water lost by the body is balanced by what is taken in. For an average person the figures would be something like this:

Water gained per day (cm^3)		*Water lost per day (cm^3)*	
water drunk	1600	water in urine	1650
water in food	900	water in sweat	650
water from tissue respiration	400	water in faeces	100
		water from lungs	500
	2900		2900

The control mechanisms ensure that if one of these factors alters—for example, when we drink a lot of water—there is a compensatory increase in the water lost in urine. Similarly, on a cold day there is less evaporation through sweating, but more urine is produced.

The problem of nitrogenous waste

When proteins are digested they are broken down into their constituent amino acids, which are then absorbed into the blood stream. Some are used by the cells in the normal processes of growth, i.e. they are built up into proteins again, but the body is incapable of storing excess amino acids. Each amino acid contains nitrogen as part of an amino group ($-NH_2$). For example, the simplest amino acid has the following formula:

```
H       H     O
 \      |    //
  N  —  C — C
 /      |    \
H       H     OH
```

or more simply $NH_2 . CH_2 . COOH$.

In the liver, amino acids not immediately required by the body are broken down through a process called **deamination**. The amino group is split off and converted into a soluble and relatively harmless substance, **urea**, which is then carried away by the blood; the organic acid which remains can be used as a source of energy in respiration. Other products of protein breakdown include ammonia and uric acid; their quantity varies in different vertebrates. All these substances are collectively called **nitrogenous waste**. These products, if allowed to build up too high a level, would quickly poison the body; their removal or **excretion** is therefore of vital importance. Now let us see how these two processes, excretion of nitrogenous waste and osmoregulation are carried out by the kidneys.

The urinary system

In Fig. 26:1 you will see that the blood flows into the kidneys via the two renal arteries. As these are short, large, and close to the dorsal aorta, does this suggest anything to you about the pressure of the blood entering the kidneys? Blood is removed from the kidneys via the large renal veins.

We can best think of the kidney as an organ which filters off some of the blood and then puts practically all of it back again, except for a small amount of fluid which passes down the two **ureters** as urine. The urine is temporarily stored in the **bladder**. There are two sets of circular sphincter

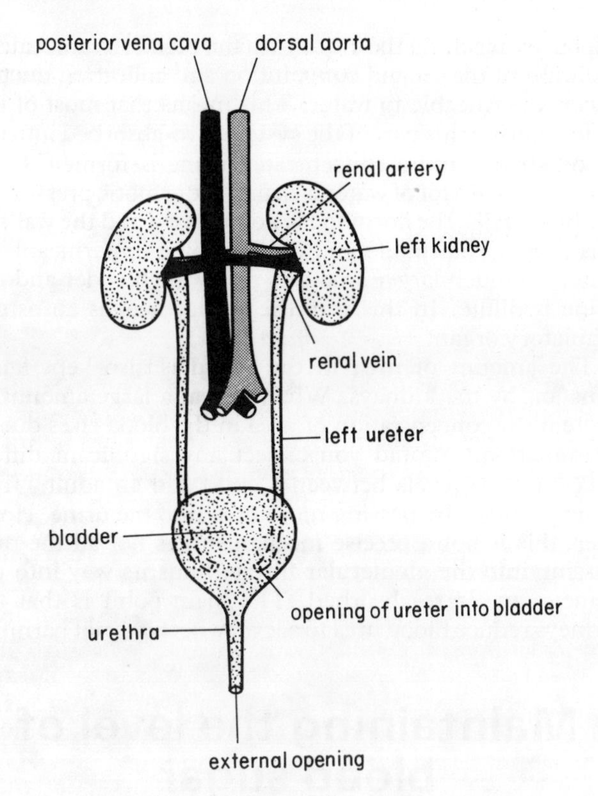

Fig. 26:1 Generalized diagram of the human urinary system.

muscles in the bladder. When the bladder is filling up both these muscles are constricted, so the exit is closed; however, as the pressure of the urine increases, the walls of the bladder are stretched and this triggers off an automatic reflex action which causes the upper sphincter to relax. But the lower sphincter, in contrast, is under the control of the will, and so urine can still be retained until this muscle is relaxed too. Control of urination is not possessed by very young children but is gradually learnt.

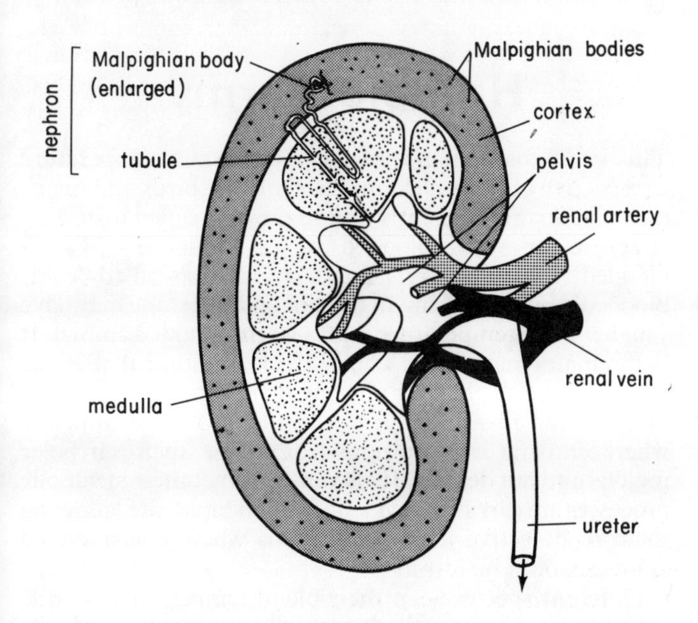

Fig. 26:2 Diagram of a human kidney cut in half longitudinally. A single much enlarged tubule is included to show its position.

The general arrangement of the urinary system can best be seen by studying a specially prepared dissection of a rabbit or rat.

The kidney

The structure of a kidney is best seen by cutting a fresh one longitudinally into two equal halves (Fig. 26:2): a lamb's kidney is suitable. You will notice that the kidney has an outer layer or **cortex** which is a darker red than the remainder because of the concentration of small blood vessels within it. An inner region, the **medulla**, is paler, and near the origin of the ureter and connected to it are some large spaces, the **pelvis**.

In the cortex are a large number of dark blobs, the **Malpighian bodies**. Each consists of a knot of arterioles, the **glomerulus** (Fig. 26:3) encased by a cup-like **Bowman's capsule** which is the blind end of a kidney tubule or **nephron**. One human kidney contains about a

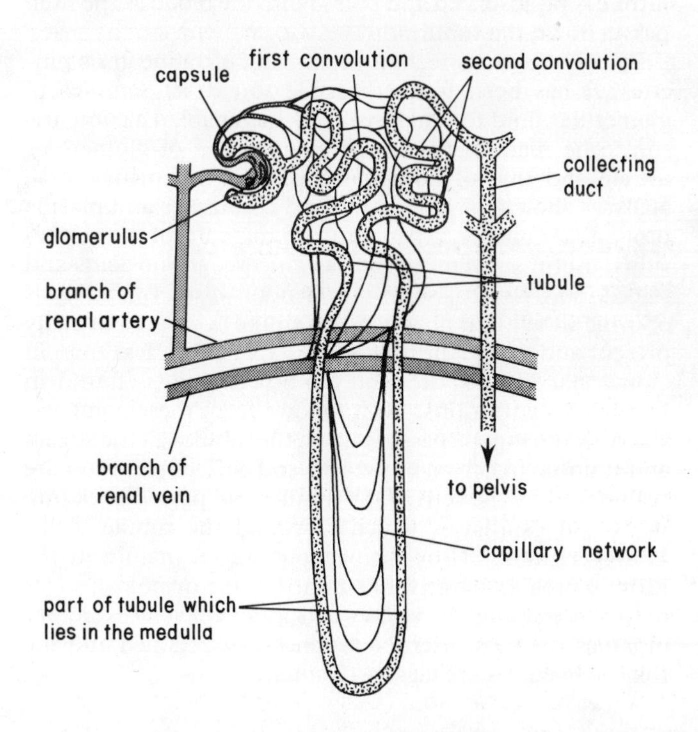

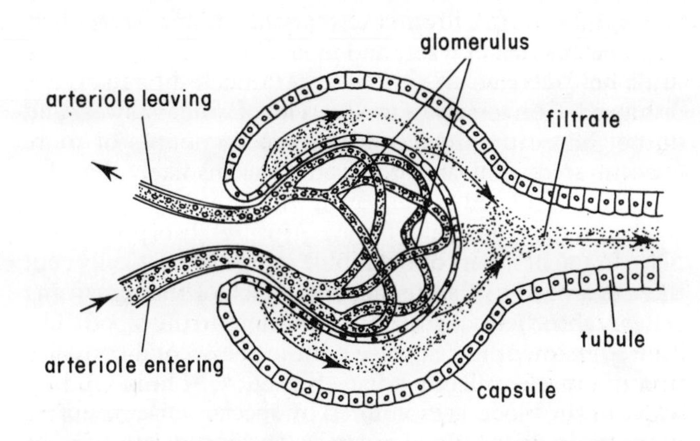

Fig. 26:3 (above) A single Malpighian body and nephron showing the blood supply. (below) A single Malpighian body enlarged.

million nephrons, each approximately 35 mm in length. Each nephron goes through many convolutions and loops deeply into the medulla. It eventually joins a collecting duct which opens into the pelvis.

What is the significance of all these glomeruli in the cortex? In Fig. 26:3 you will see that the diameter of the walls of the blood vessel entering the glomerulus is greater than that leaving it. This has the effect of building up the blood pressure within the glomerulus, causing fluid to be forced through the walls of its blood vessels into the capsule and so into the tubule. (The capsule has extremely thin walls which are in close proximity to the glomerulus.) The mechanism by which the fluid is forced out of the blood into the capsule is known as **pressure filtration**.

It has been calculated that every 24 hours an adult human filters off from the blood 170 litres of glomerular fluid—enough to fill the petrol tanks of five or six medium-sized cars! However, from this vast amount only 1.5 litres of urine are produced. In other words about 99% of the filtrate is re-absorbed and passed into the blood as the fluid passes down the tubules.

By the ingenious technique of using ultrafine glass pipettes, it has been found possible to extract samples of glomerular fluid for analysis. This has enabled a comparison to be made between the composition of glomerular filtrate and that of both blood plasma and urine. Such analyses show that although blood corpuscles and plasma proteins are unable to pass through the walls of the glomerulus, useful substances such as glucose, amino acids and salts do so, as well as some nitrogenous waste. But analysis of urine shows that no glucose or amino acids are normally present and the quantity of salts is very much less than in glomerular filtrate; urea, on the other hand, is found in higher concentrations. So it follows from these comparisons that during the passage down the tubules all the sugars and amino acids, most of the salts and 99% of the water are reabsorbed by the cells of the tubules and passed back into the blood capillaries which surround the tubule walls. However, most of the nitrogenous waste, mainly in the form of urea, is selectively retained in the urine.

In this manner the kidney acts as an efficient excretory organ as the most useful substances are retained and the toxic substances are largely eliminated.

The kidney as a homeostatic organ

The kidney is not only an excretory organ, it also *regulates* both the amount of water and urea in the blood. When we drink an excessive amount of water, much dilute urine is produced. Conversely, if we lose a lot of water as sweat and do not have anything to drink, small amounts of more concentrated urine are produced. It seems likely that this difference is related to the way in which the kidney tubules re-absorb the glomerular filtrate. Most re-absorption takes place in the first part of the tubule quite automatically, but there is a means of adjusting how much of the remaining water is absorbed according to conditions in the blood. The 'fine adjustment' takes place both in the second convolution and in the collecting duct (Fig. 26:3). The amount of water in the blood is monitored by special sense receptors in the brain. If the blood becomes too concentrated, i.e. its osmotic pressure is increased, these receptors stimulate the secretion of a hormone from the pituitary gland (p. 163) which, on reaching the kidney via the blood stream, causes the walls of the second convolution and collecting duct to become permeable to water. This means that most of the fluid entering this part of the system is re-absorbed into the blood stream and a concentrated urine is formed. Conversely, when a lot of water is drunk the osmotic pressure of the blood falls, the hormone is not released and the walls of this part of the tubule and duct become impermeable to water, so much larger amounts reach the bladder and the urine is dilute. In this way the kidney acts as an osmoregulatory organ.

The amount of urea in the blood is also kept fairly constant by the kidneys. When we eat a large amount of protein the concentration of urea in the blood rises due to deamination. (Would you expect any significant differences in urea levels between a child and an adult?) The kidney adjusts by passing more urea into the urine. However, this is not a precise mechanism, as not all the urea passing into the glomerular filtrate finds its way into the urine; some is re-absorbed. The main point is that the kidneys reduce blood urea to a level where it is not harmful.

Maintaining the level of blood sugar

Sugar, being an osmotically active substance, has to be carefully controlled, otherwise irreparable damage would be done to the brain and body tissues.

It is significant that the sugar absorbed from the gut is first taken to the liver and, if not immediately required, converted to glycogen as osmotically inert, insoluble granules. Although the input of sugar is irregular, the body requires a continuous supply for respiration. One of the functions of the liver is to ration out the sugar according to the body's needs. This involves the reconversion of glycogen to glucose. The process is reversible and is controlled by several hormones, one of which is **insulin** (p. 163).

Homoiothermy

This is the maintenance of a fairly constant temperature irrespective of the environment. Only the birds and mammals can do this efficiently and so they are called **homoiothermic** or warm-blooded in consequence. 'Warm-blooded' is not a very good term as a so-called 'cold-blooded' animal basking in the sun on a hot stone may have a higher body temperature than a warm-blooded animal. It is the maintenance of a constant temperature that is the important point. Homoiothermy brings many advantages to birds and mammals. It enables them to live in places where temperature conditions are extreme, such as in polar regions and hot deserts. It also ensures that their metabolic processes are carried out at a steady and rapid rate, allowing them to be active under conditions where cold-blooded animals would be torpid.

Different species keep their blood temperatures at different levels. Our so-called 'normal' temperature is 36·9°C (98·4°F), a hen's is 39·4°C (103°F), a blue-tit's 41·7°C (107°F), and a humming-bird's 43·3°C (110°F). Consider what differences these temperatures would make on their metabolic rate and what consequences this could have on their activities.

When we say our normal temperature is 36·9°C (98·4°F) we mean this is the *average* figure for the human population as a whole.

> Find out, through a class experiment, how much the 'normal' varies from person to person. Clinical thermometers used by doctors are the best, but if there are not enough to go round they should be washed in disinfectant before being used again. From the class results make a bar graph (histogram) recording how many in the class have temperatures at each 0.1°C reading. How does the class average compare with the population average? If it is different, can you suggest why?

You would also find it interesting to study the fluctuations in your own temperature by taking it at two-hour intervals throughout the day and plotting the results on a graph. You should start as soon as you wake up and note on the graph your activities just before you take each reading.

The temperatures of hibernating mammals such as bats, dormice and hedgehogs, fluctuate much more than ours. In winter they become almost cold-blooded, their body temperature dropping to a degree or so above their surroundings, but in summer their temperature is well above this and fluctuates very little. However, bats are peculiar because in daytime, when they go back to their roosts (caves, church towers, etc.), their temperature quickly drops; this means they have a daily cycle of temperature change as well as an annual one. Why is it advantageous for small, active creatures like bats to become cold when they are inactive, and how could this be disadvantageous?

Heat gain and loss

In order to maintain a constant temperature, the heat lost from the body to the external environment must be replaced by heat generated within the body. We have already seen that energy in the form of heat is released when food is broken down during respiration. It is this heat that warms

Fig. 26:4 A hibernating dormouse. Why is its curled-up condition significant?

our bodies. Some parts, such as the muscles and liver, release more heat than others, but even distribution is effected by the blood which transports the heat all over the body. Obviously more will be needed by those parts which are losing heat most quickly.

The body is continually losing heat: through the skin by radiation or conduction or by sweating, and to a lesser extent through breathing, urination and defaecation.

It is the skin which provides the greatest surface area of the body which is in contact with the environment, so we should expect to find the most important heat-regulating mechanisms there.

The skin

The skin is the largest organ of the body. You will see from Fig. 26:5 that the skin consists primarily of two layers, the **epidermis** and **dermis**. The epidermis also consists of several layers: a) The **Malpighian layer** composed of actively dividing cells which are constantly replacing the rest of the epidermis. b) The **granular layer**, consisting of living cells which are the products of the Malpighian layer. c) The **cornified layer** which is constantly being formed from below when the granular cells become horny and die. This dead layer is tough and protects the living cells from mechanical injury and fungal and bacterial attack. It also reduces the amount of water lost from the surface. On the soles of the feet and on the palms of the hands this layer may become very thick, especially in people who walk bare-footed or use their hands for heavy work, or for a particular activity such as guitar-playing. The outer portion of this layer is constantly being worn away; washing removes some of it, and the scurf on our heads which we may notice when we brush our hair is a sign that this is happening.

The **hair follicles** are pits lined by epidermal cells. The cells at the base, or **papilla**, are constantly multiplying, forming hair cells within the follicle. The constant adding of cells to the base of the hair causes it to grow, but the hair cells quickly become horny and die so that it is only the base of the hair that is living.

Opening into the follicles are **sebaceous glands** which secrete an oily substance which keeps the skin and hairs supple. Vigorous brushing of the hair stimulates the glands and this improves the condition of the hair—combing is no substitute. The fluid produced, known as **sebum**, also has antiseptic properties and in aquatic mammals especially it is very useful as a water repellant. For example, when an otter comes out of the water and shakes itself, its fur becomes dry very rapidly.

The **sweat glands** are present all over the body, but are especially numerous in the palms of the hands, the soles of the feet, under the arm pits and between the thighs. Each gland is a much coiled tube which lies deep in the dermis; it is connected to the surface by a spiral duct which opens on to the surface as a sweat pore. The cells of the sweat gland receive fluid from the surrounding blood capillaries and secrete it into the duct through which it passes to the surface of the skin. Sweat is a fluid which can be looked upon both as a secretion and an excretion; a secretion because it serves a useful purpose when it evaporates and cools the body; an excretion because it also contains some

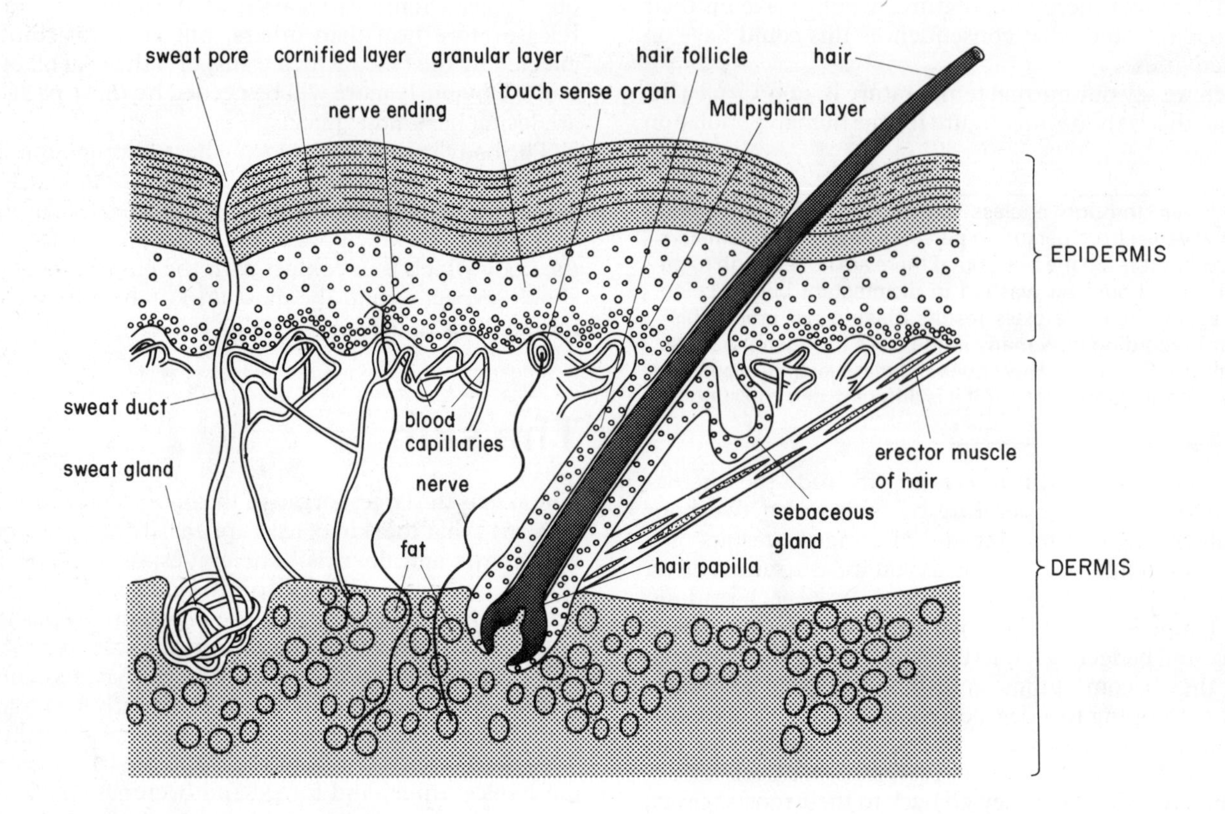

Fig. 26:5 Diagram of a section through human skin.

urea which is a harmful metabolic product. Sweat also contains some sodium chloride which is useful to the body; so people who sweat a great deal because of their occupations, or because they live in the tropics, need to take extra salt, either as tablets or with their food, to keep the salt and water balance of their body fluids correct.

The **dermis**, which is the inner layer of the skin, is made up of connective tissue and contains many blood capillaries, lymph vessels, sense organs which detect touch stimuli and temperature changes (p. 222), and elastic fibres which bring the skin back to shape when it is stretched or distorted. In the outer layers of the dermis there are cells containing the pigment **melanin** which protects the underlying tissues from the damaging effect of ultra violet light. When fair-skinned people become tanned by the sun it is due to an increase in melanin. Dark-skinned people who live in tropical countries have large quantities of melanin in their skins, a characteristic which is inherited. There are also muscles attached to the hair follicles. When contracted these muscles erect the hair and cause the skin to be covered in 'goose pimples'. Below the dermis there is usually a layer of stored fat, which acts as an insulator as well as a food store. In whales this layer is incredibly thick and constitutes the blubber from which whale oil is made.

Temperature regulation

This is brought about by a number of feed-back mechanisms. First let us consider how the skin reacts when the body becomes too hot. In the brain there are temperature-sensitive cells which act like a thermostat. If the blood becomes hotter than the 'normal' temperature, on reaching the brain it stimulates certain cells and nerve impulses are sent to the skin with two different results:

1. The walls of the arterioles in the dermis increase in diameter—this is called **vaso-dilation**. This allows more blood to flow into the capillaries near the surface of the skin so more heat is lost from the body by radiation. This is the reason why we go red when we are hot.
2. The rate of sweat production is increased and as the sweat evaporates it takes heat from the body (latent heat of vaporisation).

What happens when we become too cold? Our reaction is to shiver; this liberates heat when the muscles contract. Sometimes we stamp up and down and do some exercise; this has the same effect as shivering. But apart from the generation of more heat, the body responds by decreasing heat loss. There are two important methods:

1. The walls of some of the arterioles in the dermis contract, a process known as **vaso-constriction**. This has the effect of reducing the blood supply to the capillaries near the surface, so far less heat is lost through radiation, and as a result we go pale and sometimes blue with cold.
2. Sweat production is reduced to negligible amounts, so less heat is lost in consequence.

Temperature-reducing mechanisms are more effective if the surface area exposed to the atmosphere is increased and if the air is moving over the surface (or the surface itself moves). This is why African elephants flap their huge ears to help cool themselves; by so doing they increase their body surface by a third! You will have seen dogs panting with their tongues hanging out when they are hot. There are no sweat glands in the tongue; how do you think this helps them to cool down?

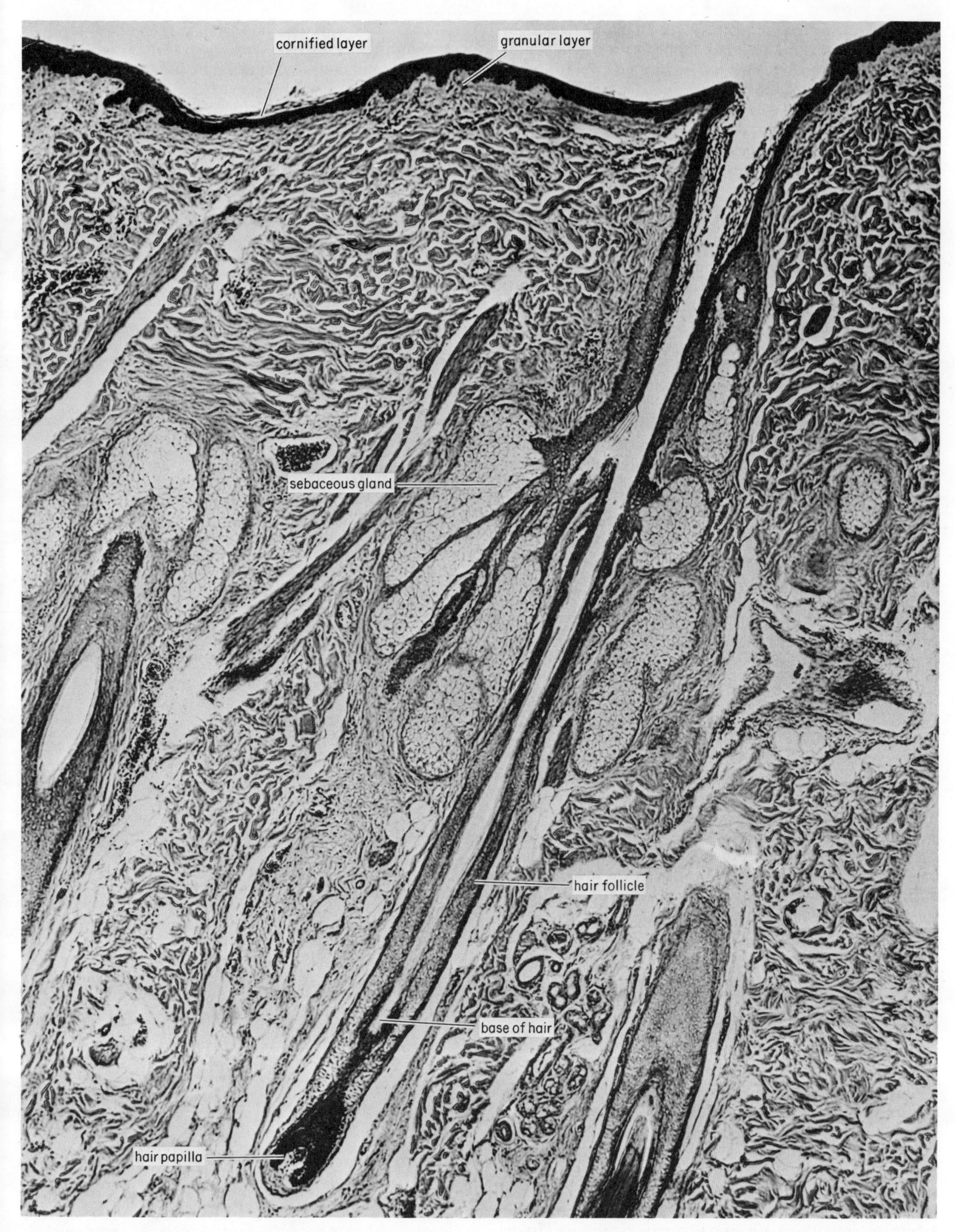

Fig. 26:6 Photomicrograph of a section through human scalp.

An interesting case of temperature regulation occurs in whales. The thickness of the blubber varies in different species, the more active ones having thinner layers of blubber. This is because the blubber is such a highly efficient insulator that during bursts of great activity so much heat is generated internally that the whale is in danger of cooking inside its jacket of fat unless the heat can be dissipated in some way. The problem is solved by the presence of a complex system of blood capillaries within the blubber. These dilate when the temperature of the blubber increases, so allowing a greater volume of blood to pass through and increasing the rate of heat dispersal. When a whale dies this mechanism cannot operate and after a time decomposition by bacteria raises the temperature of the body so much that when the whale is cut open the tissues are found to be blackened by the heat!

It follows from what we have learnt about the principles of maintaining a constant temperature that if it is necessary to conserve body heat because the habitat is cold, it is better for warm-blooded animals to have as little surface as possible compared with their volume. In other words, large animals would be more successful than small ones in a cold climate. Find out if this is true by making a list of the Arctic and Antarctic birds and mammals and see if you can find many small ones.

Another aspect of this is the fact that the smaller animals have a relatively large surface area and so they must eat a lot of energy food to make up for the heat lost. Consequently small animals such as shrews have to eat more than their own body-weight of food every day!

We referred briefly to insulation as a means of keeping in heat and mentioned the fat layer under the skin. In addition, the air around the skin is an excellent insulator if it is not moving. Fur and feathers serve this purpose admirably as they trap air close to the skin, and when a mammal or bird fluffs up its coat or feathers more air is trapped and the insulation is made more effective.

If you have understood these principles you should be able to work out the reasons behind the following facts:

1. Hedgehogs and dormice curl up when they hibernate, and you do the same when you get into a cold bed.
2. Eskimos are short, broad and fat people.
3. The Arctic hare and Arctic fox have much smaller ears than the brown hare and the common fox which occur further south.
4. A string vest under a cotton shirt keeps you warm in spite of the large holes in the vest.
5. A wind-cheater is used by mountain climbers to prevent sudden chilling.
6. Two thin pairs of socks are often warmer than one pair of thick ones.
7. Nylon shirts make you feel very hot in tropical countries, while cotton shirts are cooler.

To summarise: the skin is the main barrier between an organism and its environment. Its main functions in mammals are: a) to achieve homoiothermy, b) to protect the underlying tissues from mechanical injury, too much water loss, the effect of ultra violet light and invasion by bacteria and fungi, c) to act as an accessory excretory organ for eliminating some urea.

There are, of course, additional functions of the skin which are important. Vitamin D can be synthesised in the skin of mammals through the effect of ultra violet light (p. 114). The skin also makes an organism aware of its surroundings through its sense organs which detect touch, pain and temperature change. This function will be discussed in more detail in Chapter 28.

27

Support and movement

Support systems

All organisms need some support for their bodies and the larger they grow the greater this need becomes.

The method developed by most plants (p. 36) is through having rather rigid cell walls made of cellulose; the turgidity of these cells helps to give rigidity to the whole plant. This is aided by the presence of a tough cuticle on the outside and bundles of tough woody fibres and vessels inside. When woody elements are formed in dense masses, as in shrubs and trees, they play a major part in providing support. Although extremely effective as supporting structures, their rigidity prevents all forms of locomotion; however, plants have no need to move as they can feed without doing so.

Animals, by contrast, usually have to move to find their food and they have evolved methods of support which still allow freedom of movement. Most simple animals have little need for a support system as the majority are very small and many are aquatic, so the water buoys them up. Even some of the larger ones such as jelly fish can obtain enough support from the water to allow them to float. Do you think their shape would be significant in this respect?

Terrestrial animals such as earthworms gain support from the fluid inside the coelom. This fluid exerts pressure on the body wall, and so makes it fairly rigid. As a result the animal is able to move by means of rhythmic contractions of the body wall using the fluid 'skeleton' as a support (p. 27). Most land animals have evolved a more rigid system which not only gives much greater support but aids locomotion as well. The arthropods use an exoskeleton (p. 39) which gives excellent protection and support, but also has its disadvantages.

The vertebrates, on the other hand, have evolved an endoskeleton composed either of cartilage or cartilage and bone, which has the advantage of being able to grow internally without interruption.

The mammalian skeleton

The skeleton has three main functions:

1. To give support to the rest of the body and enable it to retain its shape.
2. To protect vital and easily damaged parts of the body from injury.
3. To help in locomotion—first, by having joints which give flexibility to the body, second, by providing a firm foundation for the attachment of muscles.

Before studying in more detail how the skeleton fulfils these functions we will first consider the tissues of which the skeleton is composed.

Cartilage and bone

Cartilage and bone, like all tissues, are composed of living cells and a non-living secretion or **matrix** which is produced by these cells. The matrix makes these tissues tough. Cartilage is more flexible than bone as the matrix is made of protein; bone also has a protein matrix, but in it calcium salts are deposited which give it much greater rigidity. You can see for yourself how efficient these salts are in giving rigidity to a bone in this way:

> Obtain a rib bone from the butcher and clean off any meat. Place it in a glass cylinder containing dilute hydrochloric acid and leave it for several weeks. The acid will gradually dissolve out the calcium salts but will not affect the protein matrix. Finally, wash the bone thoroughly and test its rigidity.

You will see from Fig. 27:1 that in bone tissue the cells are arranged in concentric circles around blood vessels, and the bony matrix laid down by these cells is in the form of a series of long cylinders. Cartilage, by contrast, contains no blood vessels within its matrix, although they are present in the surrounding sheath of tissue, so some diffusion of oxygen and nutrients can take place between the blood and the cartilage cells.

In the foetus the skeleton is first laid down as cartilage, but as development proceeds it is replaced by bone—a process known as **ossification**. This is a complex process involving a gradual removal of cartilage by cells from outside which invade it; other cells of a different kind then follow and lay down bone which replaces the cartilage which has been removed.

If you examine an X-ray of the limb bones of a child (Fig. 27:2) you will see a region between the head and shaft of each bone where cartilage is still present. These are the places where growth in length is still taking place. It is possible to estimate the age of the child from the size of these regions.

If a child's bone is fractured, it may bend and split on one side only, instead of breaking completely, because it is

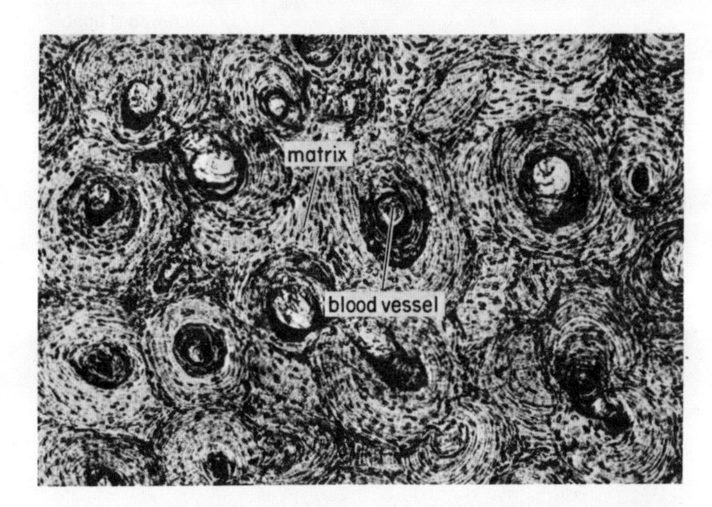

Fig. 27:1 Photomicrograph of a thin section of compact bone. The black dots arranged concentrically are the spaces in which the bone cells lie.

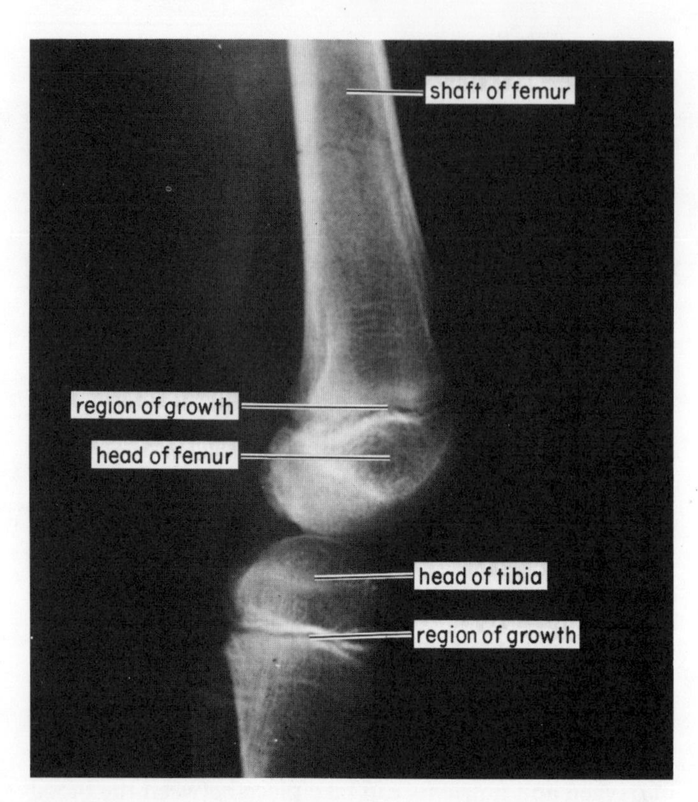

Fig. 27:2 X-ray photograph of the knee joint of a child aged 8. Note the growth regions where cartilage is still present.

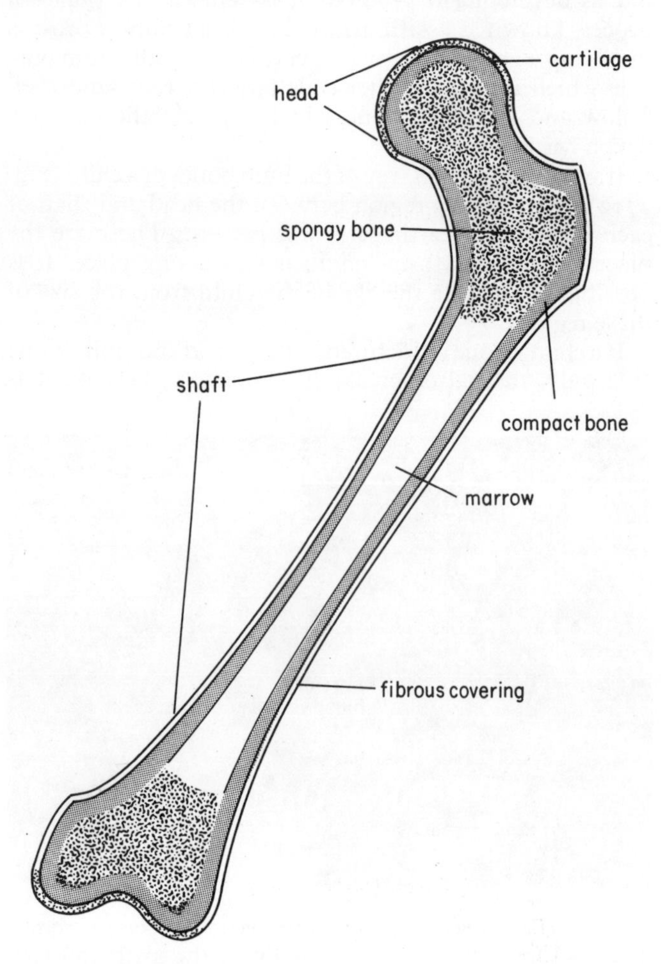

Fig. 27:3 Longitudinal section of a femur (diagrammatic).

incompletely ossified; this is known as a **greenstick fracture**.

Structure of a bone

Bones vary in shape, size and strength according to the functions they serve. A long bone, e.g. the femur (Fig. 27:3), has to be strong to give support to the body, but light enough to allow proper movement. This is achieved by having the hardest area of bone, called **compact** bone, on the outside to form a rigid cylinder, with looser tissue known as **spongy** bone further in. The hollow centre is filled with marrow where blood cells are manufactured and fat is stored. The ends are particularly strong and solid and are adapted to articulate with another bone at the joint (p. 209).

The general plan of the skeleton

All vertebrate skeletons are based on the same general plan. The skeletal plan for a mammal, when reduced to its simplest form, is shown in Fig. 27:4. Basically it consists of:

1. An **axial** part which runs down the length of the body; this consists of the skull, the vertebral column with its ribs, and the sternum or breast bone.
2. An **appendicular** part which consists of the girdles and limbs. The shoulder or **pectoral** girdle and the hip or **pelvic** girdle articulate with the fore and hind limbs respectively.

Carefully compare this simple diagram with that of the human skeleton (Fig. 27:5). The latter will give you details of the arrangement and the names of the major bones.

Examining the skeleton

It is not easy to obtain a human skeleton, although plastic models are sometimes available, but you should examine a mounted rabbit's skeleton to see the shape and position of the bones and try to deduce the main functions which they serve.

Also, if at all possible, examine and mount on card the bones of some of the small mammals. As explained (p. 82), these may be obtained either from owl pellets or from old milk bottles discarded at lay-bys etc. Small mammals enter these bottles through curiosity and fail to get out again, so complete skeletons may be obtained in this way. Wash the bones thoroughly and bleach by leaving them over-night in a dilute solution of hydrogen peroxide (take care not to get it on the hands as it may cause irritation). Now stick them on to cards arranged, as far as you can, in their correct positions. Although they are very small, you should be able to recognise the bones by comparing them with those of a mounted rabbit's skeleton.

The protective function of the skeleton

The skeleton, along with its attached muscles and ligaments, provides wonderful protection for the vital organs. The casualty wards of hospitals are full of people who are fortunate to be alive after serious road accidents, thanks to this protection, although there are limits to the shocks that bones can stand.

The skull is really a compact group of boxes fused together to form a single unit with the jaws attached. The

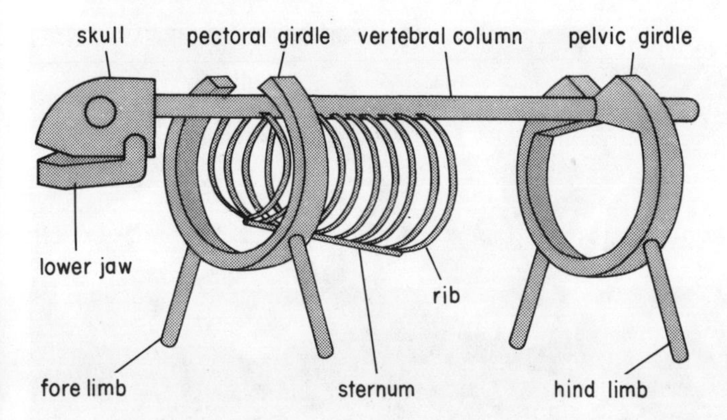

Fig. 27:4 Stylised diagram showing the general plan of the mammalian skeleton.

Fig. 27:5 Human skeleton.

largest box is the **cranium** which protects the brain and the smaller ones clustered round it are the **sense capsules** which protect the main sense organs—ears, nose and eyes. For obvious reasons, the eyes cannot be completely enclosed and so they are more vulnerable in consequence.

Similarly, the delicate spinal cord which leads from the brain is protected by the **vertebral column** which consists of a series of strong rings of bone firmly attached to each other by ligaments. Between the brain and spinal cord and their protective bony coverings are membranes enclosing a jacket of fluid—the **cerebro-spinal fluid**—which bathes these delicate organs and acts as a shock absorber. Compare this with the amniotic fluid which protects the foetus (p. 168).

The heart and lungs are well protected by a cage-like structure composed of the vertebral column, ribs and sternum, together with their attached muscles. The floor of this cage is formed by the diaphragm which separates the contents of the thorax from the abdominal organs. The

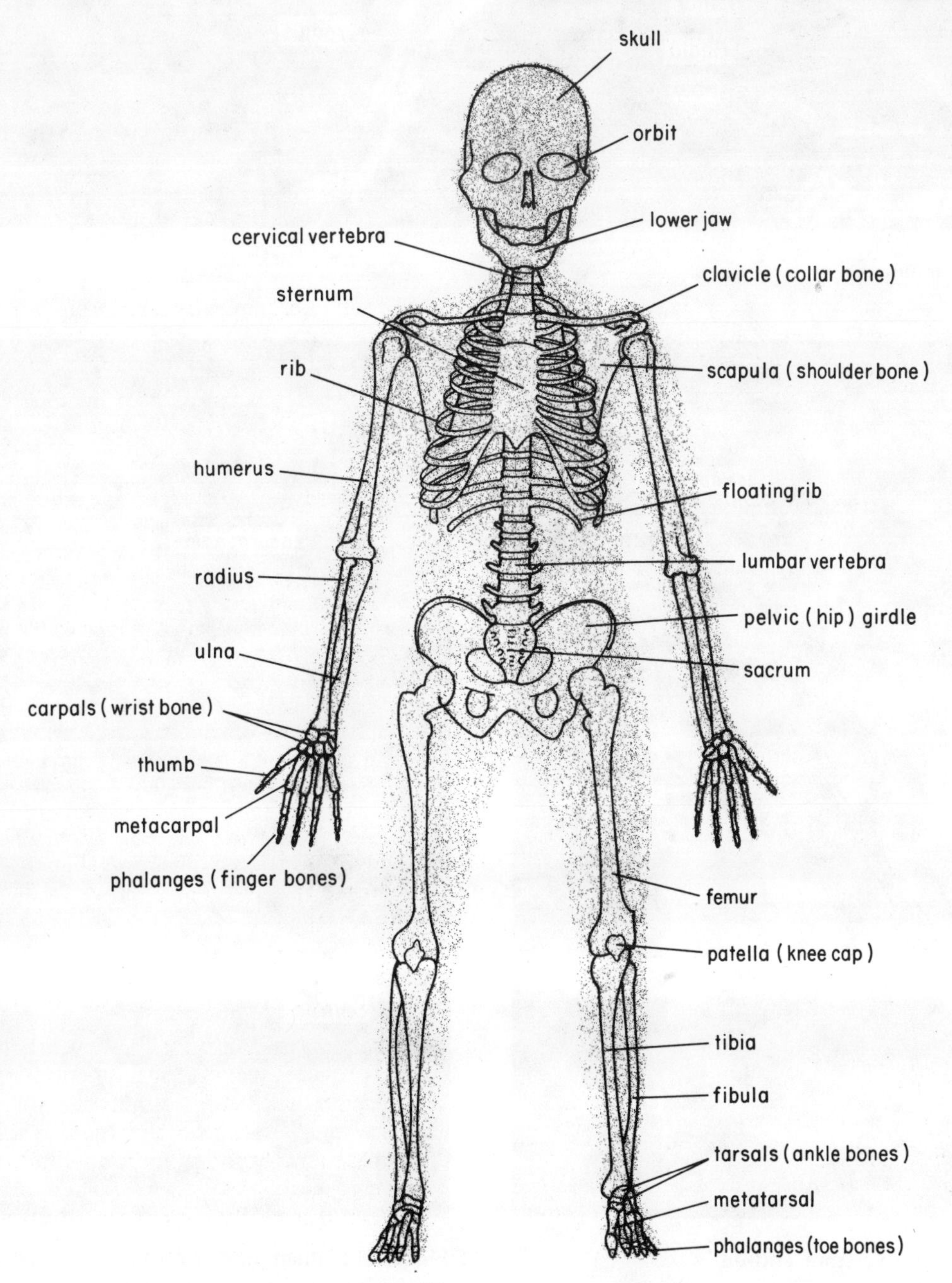

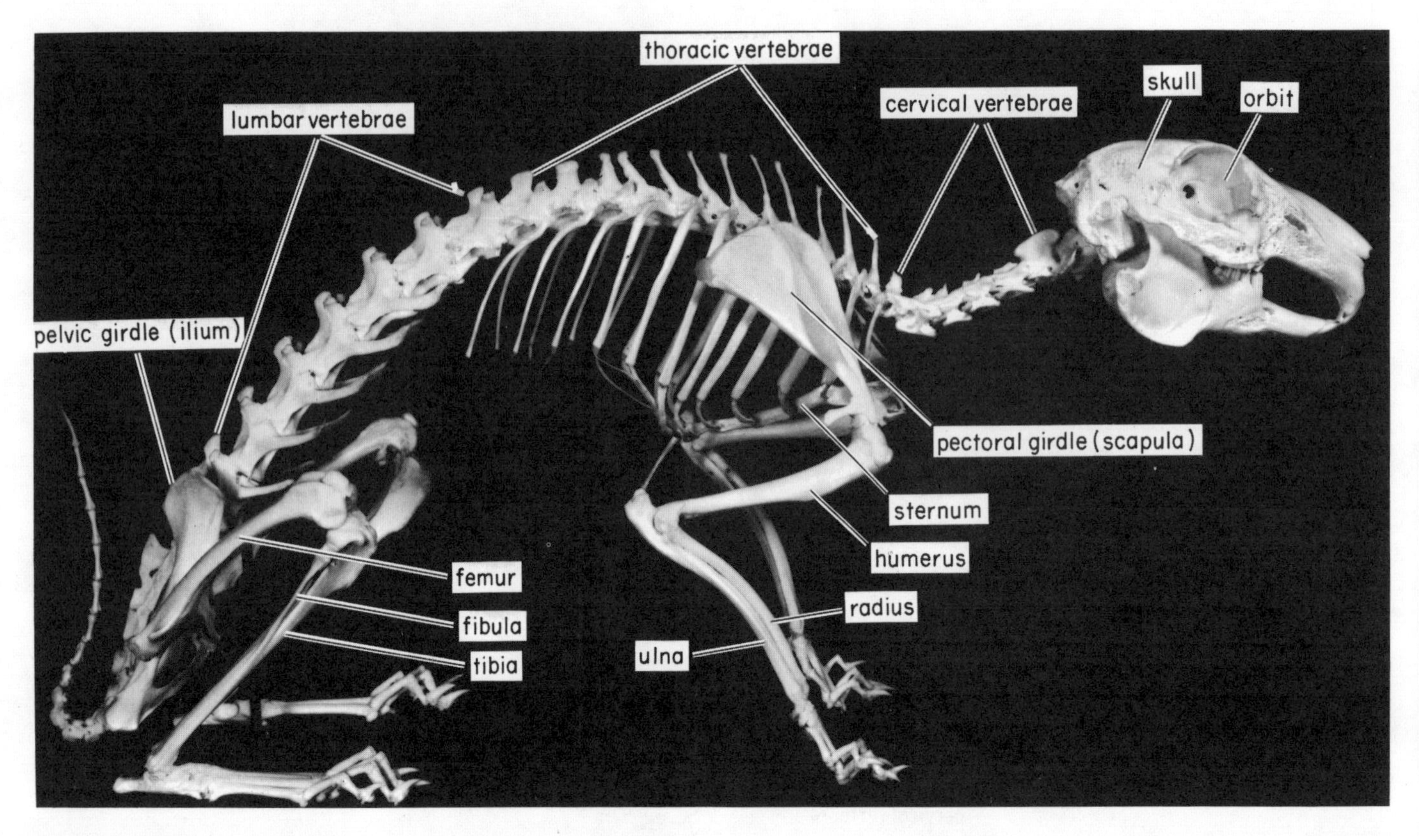

Fig. 27:6 Rabbit skeleton.

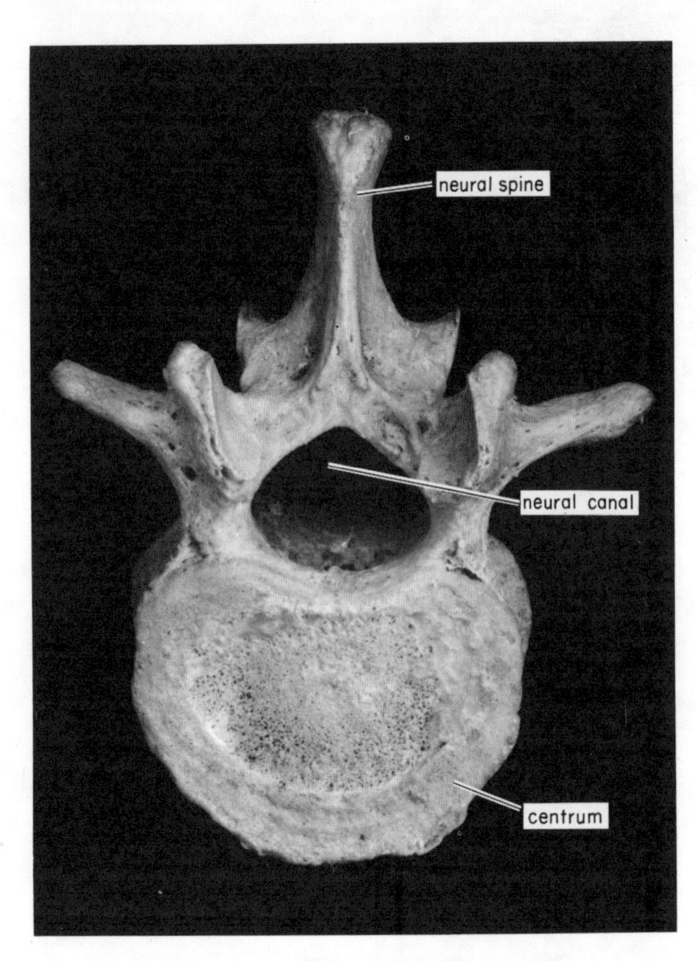

Fig. 27:7 Human lumbar vertebra, end-on.

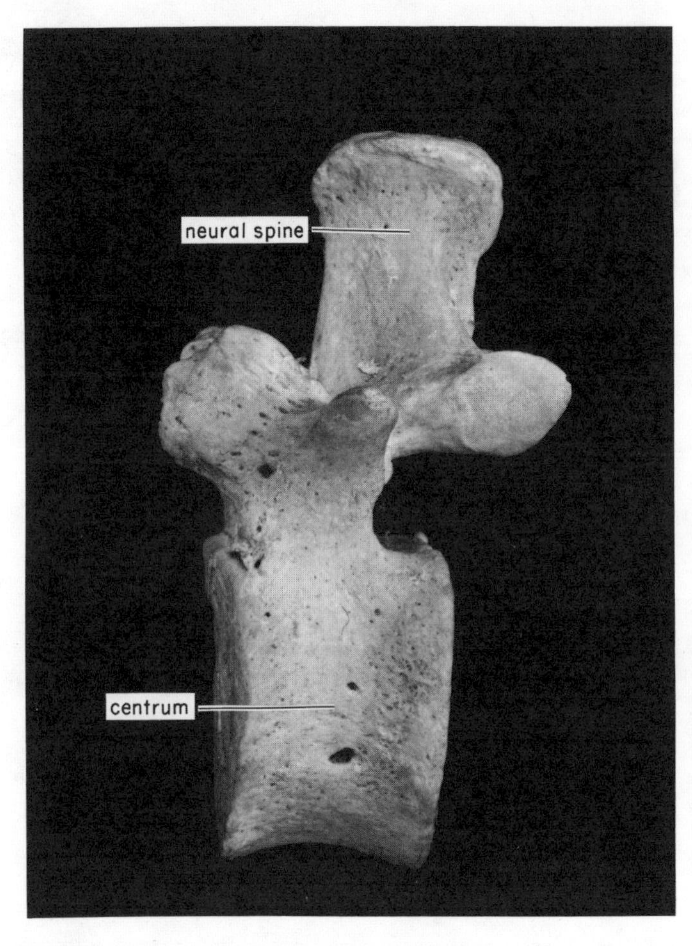

Fig. 27:8 Human lumbar vertebra, side view.

thoracic cage must be strong, but it must also allow movement for breathing (p. 94). This flexibility is aided by the ribs having a cartilaginous portion where they meet the sternum.

The pelvic girdle also supports and protects the organs in the lower part of the abdomen; this is particularly important during pregnancy.

Movement

We saw, when studying insects, that a rigid exoskeleton made locomotion and movement of the parts impossible without the presence of joints. The same applies to our own endoskeleton. We have about 200 separate bones in our body (more than half in hands and feet) and these meet each other at joints. There are two main kinds of joints:

1. **Fused joints** where the bones are rigidly fixed to each other by a pad of fibrous tissue, as in the bones of the skull (Fig. 27:9).
2. **Synovial joints** where friction between the bone surfaces is greatly reduced by the presence of a lubricating **synovial fluid**. These can be classified further according to the degree of movement the joint allows:

a) **Sliding joints** where only a very limited amount of movement is possible, as between the joints of the wrist and ankle and some vertebrae (Fig. 27:10).
b) **Hinge joints** which allow considerable movement, but in two planes only.
c) **Ball and socket joints** which allow most movement of all, in three planes.

> Test out the joints in various parts of your body to see which categories they come under. Include the shoulder, elbow, base of thumb, finger, hip and knee. In some cases it is best to hold one part still while you move the other. List your results.

A ball and socket joint is a beautiful piece of engineering (Fig. 27:11). At a butcher's you will often see the ball of a joint exposed as a glistening white knob. This covering is composed of an extremely smooth cartilage and it fits into the socket which has a similar surface. The **synovial fluid** which lubricates the junction between these surfaces is secreted by the **synovial membranes.** The articulating bones are held in position by strong ligaments which are sufficiently elastic to allow just enough adjustment when the bones move. There is also a capsular **ligament** which acts like a sleeve round the joint; this keeps the bones firmly in place and helps to contain the synovial fluid.

If you **dislocate** your shoulder or finger the bones are displaced at the joint. Dislocations stretch the ligaments abnormally and may damage them and as a result the joints are slightly looser afterwards. Unfortunately, this makes further dislocation of the same joint more likely. If this happens repeatedly it is sometimes possible for a surgeon to 'tighten' them.

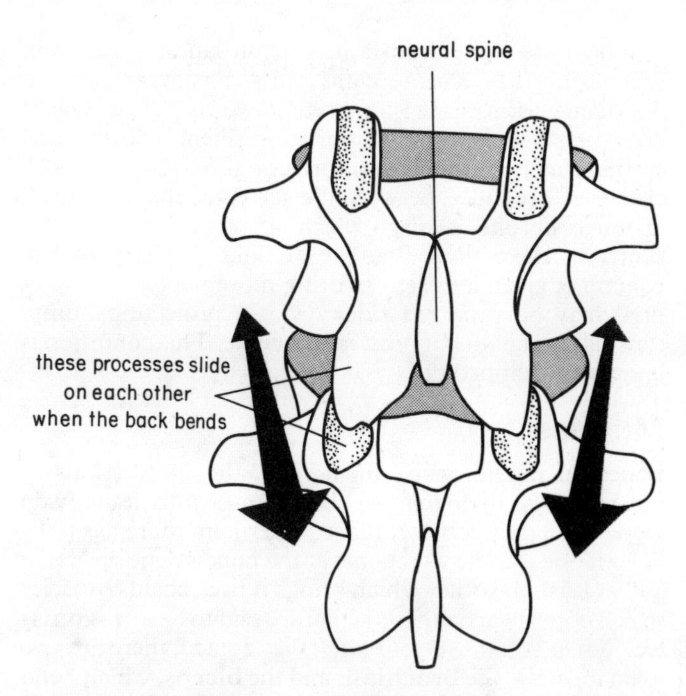

Fig. 27:10 Sliding joint between lumbar vertebrae, dorsal view.

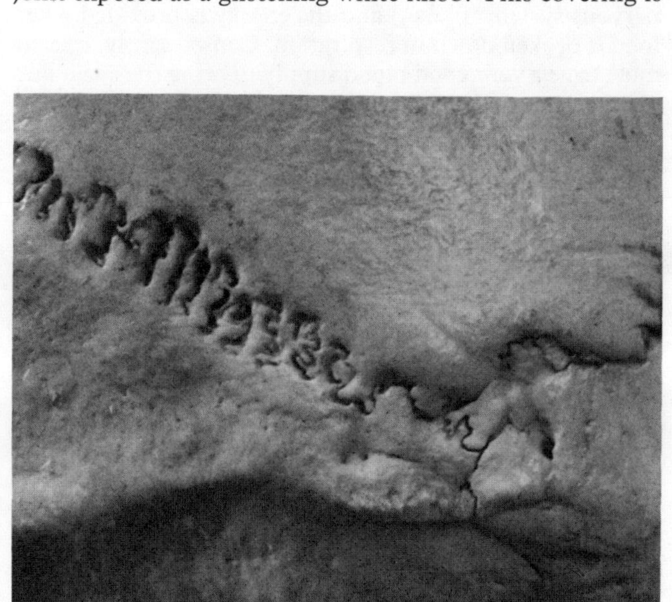

Fig. 27:9 Fused joint between bones of human skull.

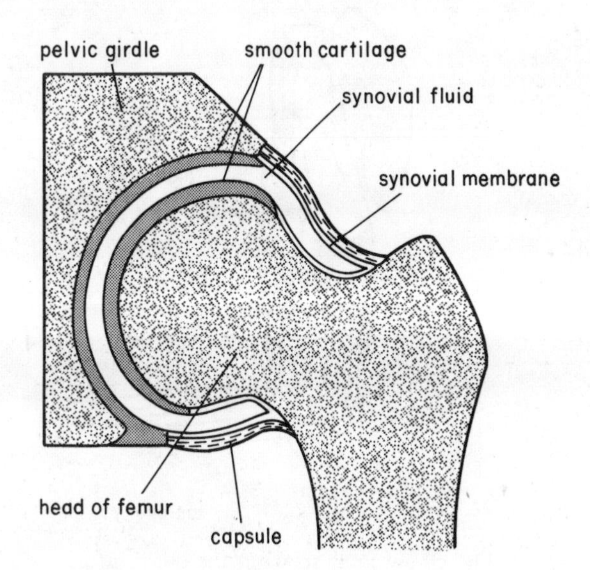

Fig. 27:11 Ball and socket joint of the hip.

When you jump from a height, on landing the main joints concerned have to withstand a considerable shock. The danger of damage is reduced in two ways. Cartilage is not so rigid as bone, having more resilience, so that the cartilage of the articulating surfaces also acts as a shock absorber. Secondly, between the vertebrae there are discs of tough fibrous cartilage which act as cushions. In the centre of these discs is a bag of fluid. If the vertebral column is given a violent twisting movement, this bag of fluid may be squeezed sideways as a projecting bump, causing great pain if it presses on a nerve. This condition is known as a slipped disc.

Muscles

Bones do not move by themselves; the contractions of muscles cause them to move. It is because muscles only do work when they contract that at every joint there has to be at least one muscle which moves the bone in one direction and at least one other which brings it back again. Muscles which oppose each other's action are said to be **antagonistic**. When you move your arm (Fig. 27:12), there are two main muscles, the **brachialis** and the **biceps**, which bend the arm (for this reason they are described as **flexor** muscles), and a **triceps** muscle at the back which straightens it again (this is called an **extensor** muscle). Most joints in the body have at least one flexor and one extensor muscle. When the flexor muscle contracts the extensor is in a relaxed state, and vice versa.

> Test this action for yourself by holding your right arm straight with the palm upwards. Grasp the upper arm with your left hand so that your fingers touch the biceps from above and your thumb touches the triceps from below. Now slowly raise your arm. What is happening to the muscles above? Now straighten your arm. What happens to the muscle below?

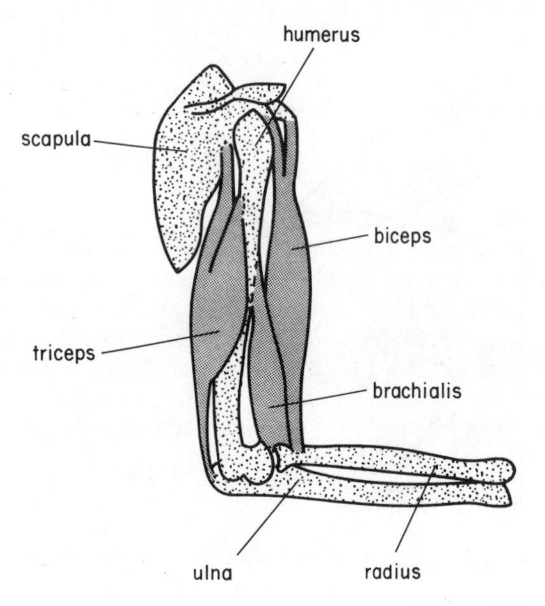

Fig. 27:12 The elbow joint showing the positions of the antagonistic muscles.

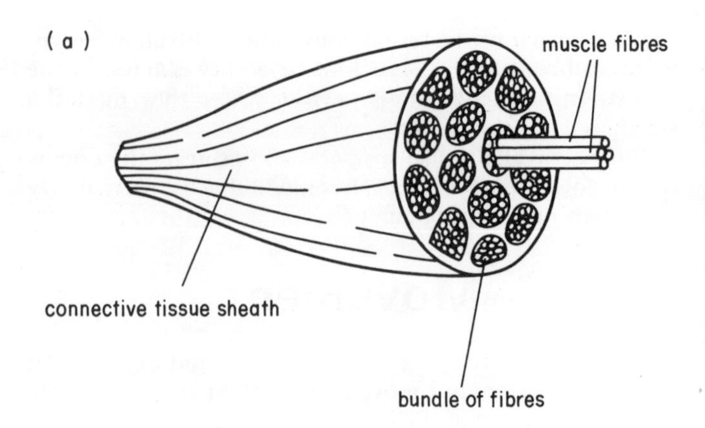

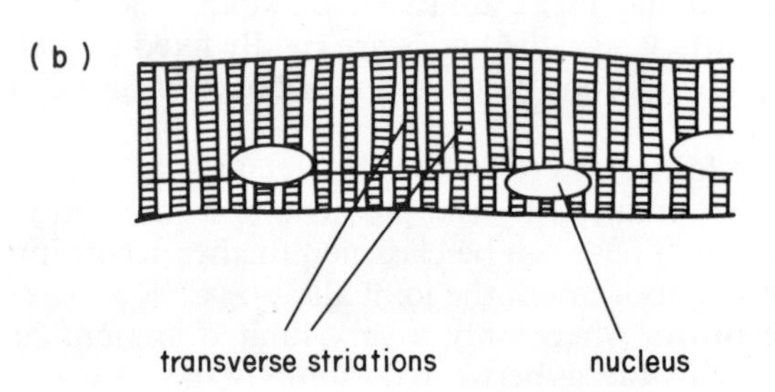

Fig. 27:13 The structure of a muscle: a) part of a muscle b) part of two muscle fibres much enlarged.

You will have noticed that when muscles contract they also become fatter. Each muscle (Fig. 27:13) is composed of many bundles of elongated fibres bound together by connective tissue which also forms a sheath round the whole muscle. The sheath is extended at each end into a **tendon** which attaches the muscle to the bone. When a muscle contracts, its individual fibres become shorter and fatter due to the special properties of two proteins which they contain called **actin** and **myosin**. The stimulus which causes them to contract comes from the central nervous system (p. 223) and the energy is provided when food is broken down in respiration. Consequently, muscle must have a very good blood supply to bring the sugar and

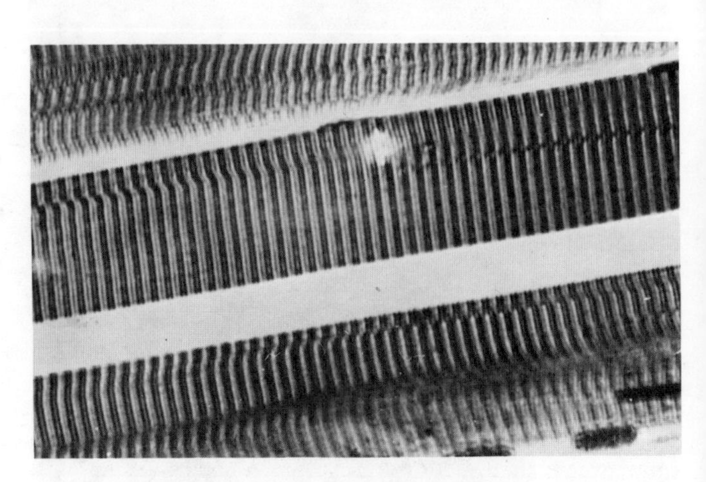

Fig. 27:14 High power photomicrograph of muscle fibres separated out.

oxygen needed for respiration and to take away the waste products. Under the microscope you can see the blood capillaries supplying the various fibres. Much can be learnt about the structure of muscle by observing meat obtained from the butcher, as most meat is animal muscle. A good carver will cut red meat so that the small bundles of muscle fibres are sliced transversely; the connective tissue between them allows thin slices to be made.

Lever systems

The movements caused by muscles at a joint can be explained in terms of lever systems. There are three kinds of levers according to the position of the pivot or fulcrum and the place where the force is applied. All three types of levers are used in the body. Fig. 27:15 gives an example of each.

For a lever system to work efficiently one end of the muscle has to be attached to something rigid so that when the muscle contracts the bone to which it is attached at the other end is the only part to move. The fixed end of the muscle is called the **origin** and the other end is known as the **insertion**. You will see that the biceps muscle has its origin in two tendons which are attached to the firm scapula and its insertion is on the radius. The brachialis has its origin on the humerus and its insertion on the ulna.

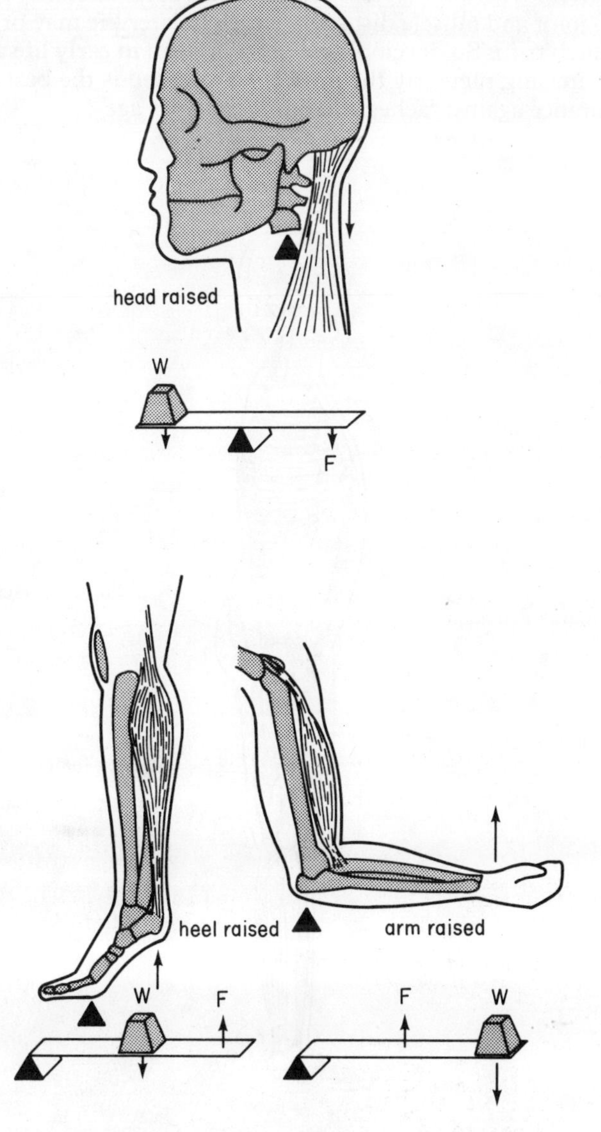

Fig. 27:15 Examples of lever systems in man. In each case the contraction of the muscle provides the force and moves the bone, and the joint acts as a fulcrum. W = weight, F = force, ▲ = fulcrum.

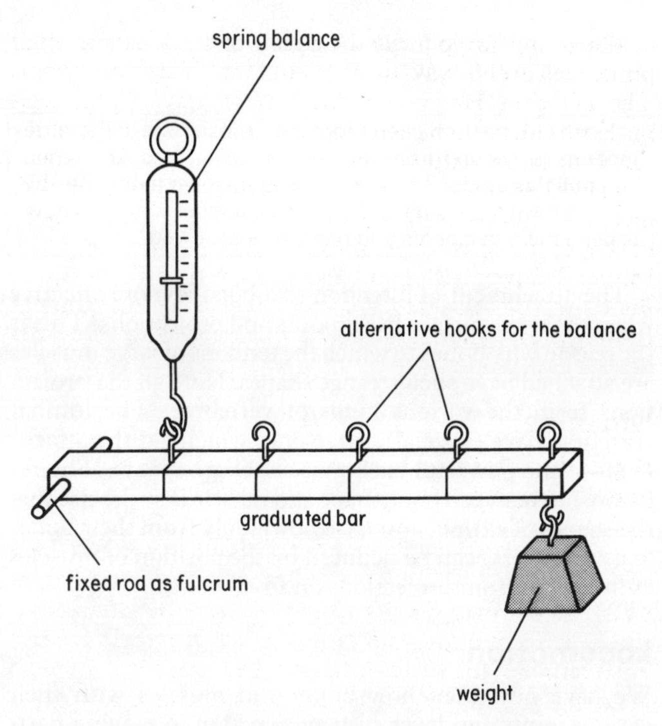

Fig. 27:16 A model which simulates the lever action of the forearm.

Some forms of lever have a greater mechanical advantage than others in that less force has to be exerted to move a particular weight. You can work this out for yourself by constructing a simple model of your elbow joint.

> Set up the apparatus as in Fig. 27:16. The ruler represents the arm which is pivoted at the elbow and carries a weight in the hand. The spring balance takes the place of the muscle and the force needed to keep the arm in a horizontal position can be read off. Record what force is needed to keep the arm horizontal when the spring balance is attached at different equally-spaced positions along the ruler. Repeat with different weights.

In the light of your results consider Fig. 27:12 again, noting the points of attachment of the muscles. Would another arrangement have given a greater mechanical advantage? If so, why do you think the attachment is where it is? One clue is to consider the distance the hand would move for different points of insertion, assuming that the brachialis and biceps muscles contract the same amount each time. Is greater mobility attained by having the point of insertion near the elbow joint?

Tendons

These connections between muscles and bones are composed of very fine fibres which are extremely strong and do not stretch. In the butcher's you see them as shiny white sheets or cords attached to the meat (muscle). You can demonstrate for yourself the action of tendons and their properties in this way:

> Examine a fresh chicken's foot. You will see one of the large tendons projecting from the cut end. What happens when you pull this tendon? Follow the tendon down to the toes by cutting away the skin carefully with a scalpel. Can you now explain the movements you previously observed?

The attachment of a tendon to a bone is more effective when the bone surface has grooves and projections. This is the reason why bones to which the tendons of large muscles are attached have such strange shapes. Note all the projections from the various kinds of vertebrae. The lumbar vertebrae have especially large ones which aid the attachment of the powerful back muscles (Fig. 27:17). The relationship between bone shape and muscle is so precise that the shapes of extinct animals known only from their bones (e.g. dinosaurs) can be deduced by the position of muscles as indicated from projections on fossil bones.

Locomotion

We have now seen how bones and muscles with their various joints and lever systems function in moving parts of the body. Let us consider further how this leads to locomotion of the whole organism.

In vertebrates, movement results from the backward thrust of the limbs against the ground, water or air according to where the animal is living. In order to move the whole body this force has to be transmitted to the spine. This is done through the two girdles. As we walk and run in an upright position it is only the pelvic girdle which plays a major part. This is fused to the sacral region of the vertebral column by very strong joints.

The pectoral girdle is *not* fused to the spine, but instead there are strong muscles which attach it to the thoracic vertebrae. It is thus less effective in transmitting the force from the limbs to the body. It is therefore not surprising that in animals which run on all four feet the hind limbs produce the greatest thrust. The pectoral girdle with its muscular attachment to the back bone provides an excellent shock absorber in these animals when they land after a leap.

There are many factors which influence the effectiveness of the thrust and hence the speed of the animal or its jumping ability. Consider some of the mammals which are noted for speed, such as the cheetah or greyhound, and others for their jumping ability such as the kangaroo, rabbit and gazelle. Which of the following factors do you consider are important in each case:

1. The weight of the animal?
2. The distribution of the weight?
3. The length of the bones?
4. The number of joints, i.e. levers?

Remember that muscle itself is heavy.

Check your conclusions by considering the build of Olympic sprinters and high jumpers. Is there any skeletal reason why coloured athletes are so often the world's greatest sprinters?

Posture

Muscles are important not only for movement but also for maintaining posture. You can test this for yourself.

> Stand upright with your hands at your sides and close your eyes. As you consciously keep yourself upright you will feel the muscles in various parts of your legs contracting to keep you in position. Usually this happens quite unconsciously.

Figure 27:17 shows how the main muscles are used for this purpose: notice how they support the main joints concerned—the hips, knees and ankles.

Balancing the head on the top of the spine (when we stand upright) is helped by the slight curvature of the spine. The head is very heavy and its position is controlled by the large neck muscles. When we work at a desk we sometimes hunch ourselves up; this puts a lot of strain on these muscles and we often compensate by propping our head up with our arm. If our posture is right this should not be necessary.

Incorrect posture puts extra and uneven strains on certain muscles and this is liable to cause distortion of the vertebral column. When this becomes habitual it may lead to joint and muscle disorders which in later life may bring much pain. So developing a good posture in early life and exercising regularly the muscles concerned is the best insurance against 'aches and pains' in older age.

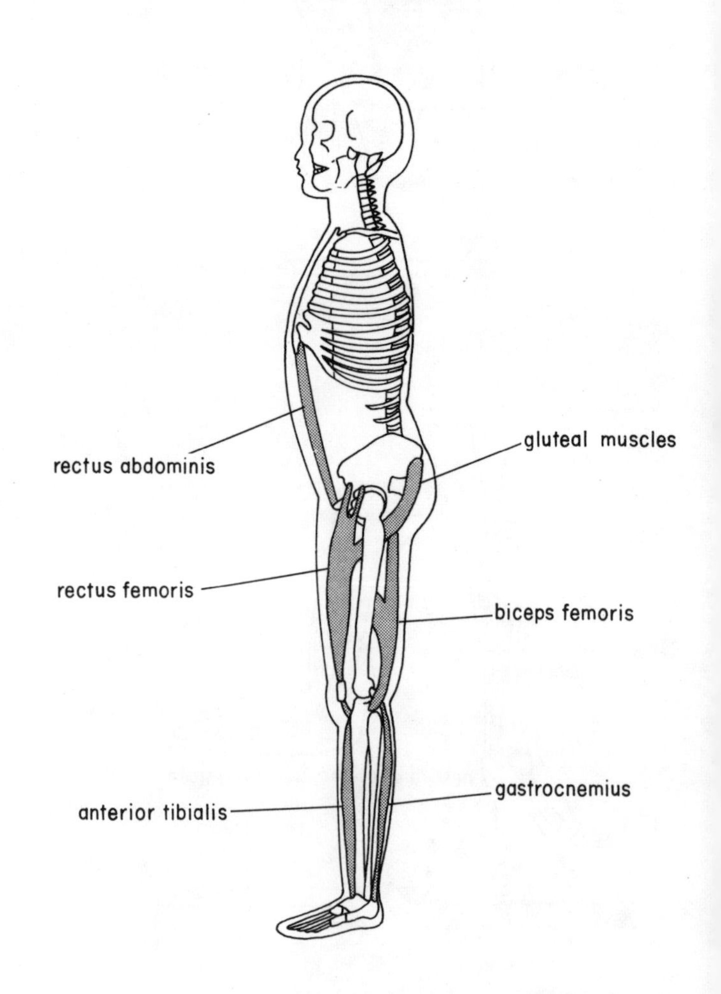

Fig. 27:17 Diagram showing the main muscles concerned in keeping an upright posture.

It is also important that furniture should be designed to give support to the body in the right places, so that relaxation of muscles can occur evenly. Beds which dip in the middle and chairs with sloping backs and no head rests are not the best ones to use.

28

Sense organs

Responding to the environment

An organism cannot exist in isolation; all the time it is subjected to many influences which come from its surroundings. These influences constitute the **environment** in which it lives. Many of these factors are physical, such as temperature, light, humidity and the nature of the soil, but equally important are the activities of all the other living things around it. As we saw earlier all organisms, wherever they live, are adapted in their structure and behaviour to the particular conditions of their environment. Most people in the Western world live in towns and cities, an environment of bricks and concrete, parks and gardens, cats and starlings, cars and buses, and above all, many other people. All these factors which influence us are changing constantly and we are responding all the time in one way or another to these changes. We have, in addition, an internal environment which is also changing—the composition and temperature of our blood, the quantities of hormones being secreted by our glands, the activity of our cells and so on. We are constantly adjusting to these changes taking place inside us, and we discussed some of the mechanisms involved in Chapter 26.

All these changes, both external and internal, are known as **stimuli**. For an organism to respond to these stimuli there are three requirements:

1. A means of detecting the changes. This is the function of the **receptors** or **sense organs**. They are the receivers of information from the environment. The main sense organs are large aggregations of cells which are sensitive to particular stimuli and include the eyes, ears, nose and tongue.
2. A means of acting on the information received. This is the function of the **effector** organs such as muscles which bring about movement and glands which secrete fluids.
3. A linking system within the body which ensures that the stimulus received is followed by an appropriate response. This is the function of both the nervous and endocrine systems.

In this and the following chapter we shall be studying these aspects in more detail, after which we will proceed to a study of **behaviour**—the response of an organism to all the information it receives from its environment, both internal and external.

What sort of stimuli do we receive?

Everything we know about the world we live in results from the information we receive from our sense organs. It is too simple to say, as **Aristotle** did, that we have just five senses—sight, hearing, smell, taste and touch—for the biologist recognises more than these. For example, a blind person when he touches an object not only discovers that it is there, but he can also detect its size and shape, its texture and its temperature, and if he touches a liquid, he knows whether it is slippery or viscous. He can distinguish between touching and being touched and between different intensities and pressures.

Consider your other sense organs—your eyes and ears, for example. What information do they provide apart from detecting light and sound?

The more information we receive from our environment the greater the possibilities of our making the most appropriate responses. We receive no information from our sense organs of touch or pressure, taste or pain until contact with the object is made. But with sight, hearing and smell it is quite different as these stimuli come to us from a distance—sight and sound more rapidly than smell. Thus some sense organs extend our knowledge of the environment far beyond the limits of our immediate surroundings.

We can distinguish at least six kinds of sense organs according to the type of stimuli they receive:

Stimulus	Receptor	Function
1. Light	Eye	Seeing
2. Sound	Ear	Hearing
3. Chemicals	Nose and tongue	Smelling and tasting
4. Temperature	Skin	Detecting temperature (both hot and cold)
5. Mechanical stimuli	Skin and muscles	Feeling and gauging pressures
6. Gravity	Ear	Balancing

The eye

Kinds of eyes

Light receptors in different animals range from the light-sensitive areas in the cytoplasm of *Euglena* and the scattered cells in the skin of the earthworm, to the complicated compound eyes of insects and the highly efficient camera-like eyes of molluscs such as squids and octopuses and those of vertebrates. The simplest structures can only detect variations in light intensity, but the highest forms produce images which provide information about the size, shape and even the colour of the object, as well as its distance from the observer. Let us now study our own eyes.

How the eyes are protected

Feel the bone all round the eye; this forms the rim of the eye socket or **orbit** in which the eye is sunk. A boxer would quickly be blinded without this protection. Behind the eye is a mass of fat which acts as a cushion between the eye and the bony orbit. People who are very ill or are starving have sunken eyes because this fat has been used up.

The eyelids also protect the eyes from foreign particles; in a dust storm one would almost close the eyelids to add to their efficiency. In addition, the eyeball is covered externally by a very thin transparent membrane, the **conjunctiva**. If dust gets into the eye this membrane becomes inflamed and pink.

The eyebrows and lashes act like the lens hood on a

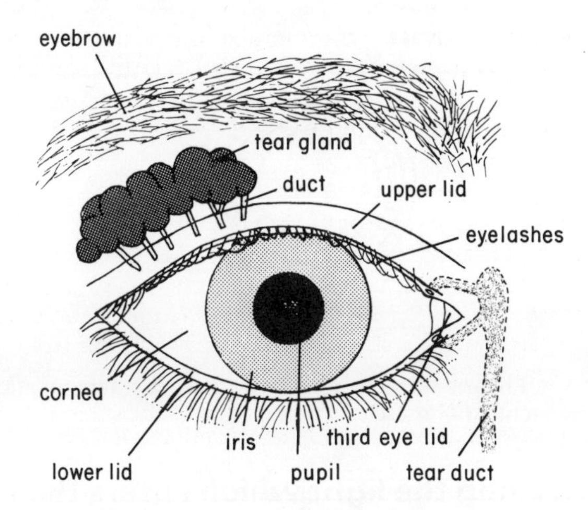

Fig. 28 : 1 The human eye showing the position of the tear gland and ducts.

camera, protecting the eye from glare. They also help to prevent rain or sweat from entering the eye.

Blinking not only protects the eye from mechanical injury, but is protective in another way. Every time you blink, fluid from the tear gland under the upper eyelid is spread over the eye surface, excess draining away through the tear duct to the back of the nose (Fig. 28 : 1). Ask your neighbour to pull down the lower lid and examine the corner of the eye nearest the nose. You will see the opening of the tear duct and also the pink mass which represents the remains of a third eyelid (it is functional in birds). Tears not only wash away dust but being antiseptic they help to prevent bacterial infection of the eye.

Eye movements

Try these movements:

> Stand in front of your partner who must keep his head quite still, and ask him to follow various movements of your hand with his eyes. Watching his eyes all the time, first move your hand vertically up and down, then horizontally and finally diagonally. Note the extent of the movement of the eyeball that is possible.

These movements are brought about by six muscles, each of which is attached at one end to the outside of the eyeball and to the orbit at the other. They act like reins guiding the head of a horse; when one contracts and pulls the eyeball in a certain direction, another which is antagonistic to it, relaxes. Four of these, called **rectus** muscles, act at right angles to each other, the other two are arranged obliquely. Look at Fig. 28 : 2 and work out which muscles are used to bring about the eye movements you have just carried out. You will also find that you can roll the eye, which involves the contraction of all the muscles in a sequence.

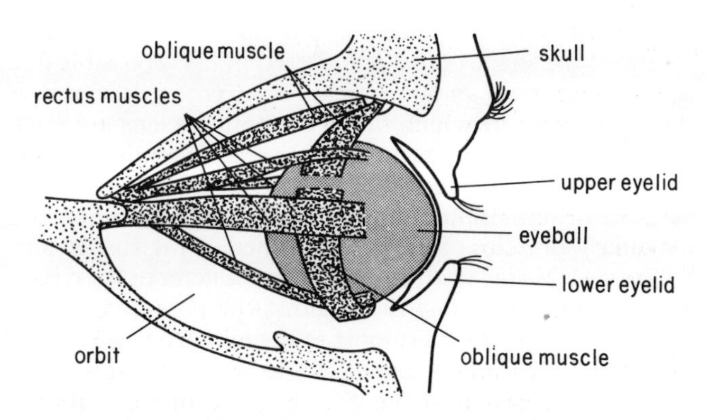

Fig. 28 : 2 Right eye exposed to show the attachment of the eye muscles.

> Now shut one eye and place your fingers gently on the lid while you roll the other. What movements can you feel in the closed eye? What do you conclude from this?

Structure of the eye

If you look your partner in the eye or look at your own eye in a mirror you will see its external features. The white part is the **sclerotic**, a tough, protective coat which surrounds the eye. In front it becomes transparent to let the light through; this part is called the **cornea**. The cornea gradually becomes opaque after death. If, for some reason, this happens in life it is possible for a surgeon to graft a fresh cornea in its place. People sometimes leave their eyes to a hospital so that when they die their eyes can be used to help somebody who needs the operation. The coloured part of the eye is the **iris**, a sheet of muscle which controls the size of the hole in the middle of which is the **pupil**.

Attached to the internal surface of the sclerotic lies the **choroid** (Fig. 28 : 3). It is black in colour due to the pigment deposited within it. The black prevents reflection of light within the eye. The choroid is well supplied with blood vessels and would look pink if it were not for the pigment. This is why albinos who lack pigment have pink eyes.

The layer of the eye which is sensitive to light is the **retina**. This is the transparent, innermost layer lying on the surface of the choroid from which it receives oxygen and nutrients brought by the blood. The **lens** is composed of layers of transparent material surrounded by an elastic membrane which moulds it into a biconvex shape. The

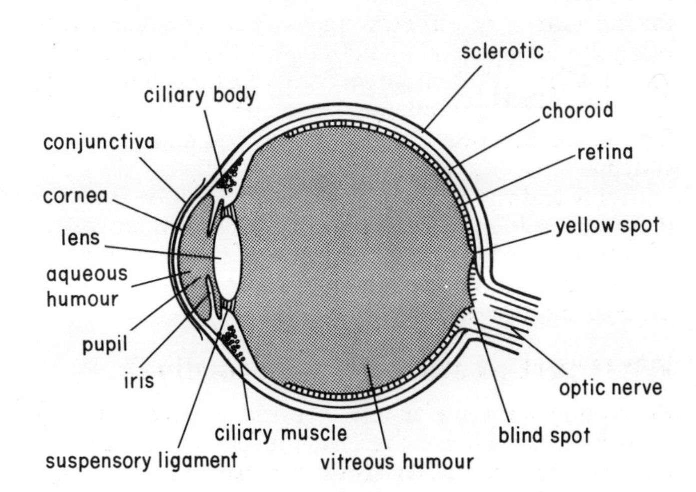

Fig. 28 : 3 Diagram of a horizontal section through the human eye.

lens is attached by a **suspensory ligament** of tough fibres to the ring of **ciliary muscles**. In old age, people sometimes suffer from **cataract**. This is a form of blindness due to the lens becoming opaque. It may be cured by the complete removal of the lens. With the help of glasses sight is then possible.

The main cavity of the eye is filled with a jelly-like **vitreous humour**; a more watery **aqueous humour** fills the cavity in front of the lens.

Examine a fresh sheep's eye.

1. Remove the fat from the back of the eye and look for the stumps of the six muscles which move the eye. Most of their length will probably have been cut away when the eye was removed but you should see where they were attached to the eyeball. Note the white **optic nerve**. It is a cord which projects from the back of the eye and takes the nerve impulses to the brain.
2. Try to demonstrate the kind of image produced by the eye by cutting a small window at the back of the eye exactly opposite the pupil (Fig. 28:4). Place a piece of tracing paper over the hole you have made and point the eye towards an electric light bulb. If the eye is fresh you should see the image of the bulb on the paper. Is it upright or inverted?
3. Remove the back third of the eye by making an incision through the sclerotic and cutting right round with scissors (Fig. 28:4). Separate the two portions, place the back part in a dish of water and examine it. Can you see in the undamaged part very fine lines radiating towards the position of the optic nerve? These are nerve fibres lying on the surface of the retina which lead to the optic nerve.
4. Examine the other portion of the eye. Squeeze it gently and note the jelly-like vitreous humour which comes out. This will be followed by the lens which should be released very slowly to prevent damage to the surrounding tissues.
5. Wash out the front part of the eye and note the ring of ciliary muscles to which, in life, the lens was attached. Hold the eye up to the light and note the position of the pigmented iris and see how transparent the cornea is.

If an eye is kept in a deep freeze it is possible to cut it lengthwise with a fretsaw in the plane of the optic nerve. This will show you the position of all the parts you have observed and you can compare them with Fig. 28:3. You will now be in a better position to understand how the eye works.

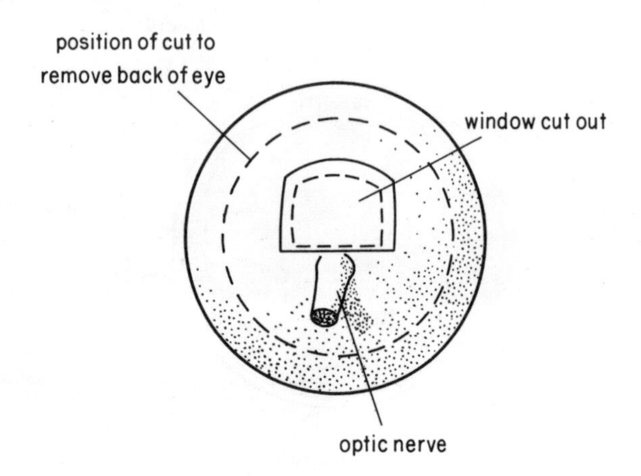

Fig. 28:4 Eyeball seen from the back to show where cuts should be made.

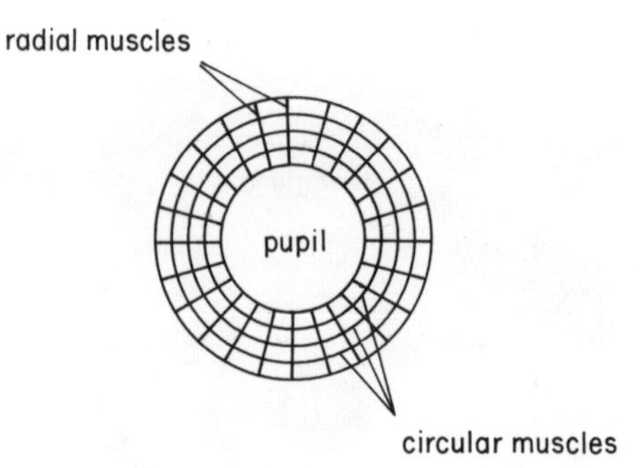

Fig. 28:5 The iris seen from in front showing diagrammatically the arrangement of muscles.

Controlling the light which enters the eye

This is done by adjusting the size of the pupil by means of the iris. The latter is composed of muscle fibres arranged both radially and in a circular manner (Fig. 28:5). When the circular muscles contract and the radial muscles relax the pupil becomes smaller; when the radial muscles contract and the circular muscles relax it becomes dilated. In bright light the pupil is reduced to a tiny aperture, so protecting the retina from damage, while at night the pupils are very dilated allowing as much light as possible to enter. The larger the eye, the more the pupil can dilate and the more light is able to enter. This is why nocturnal animals usually have large eyes. Test the reaction of your own eyes to light in this way:

This experiment works best in a dimly-lit room.
First, notice the size of your partner's pupils, then flash a torch near the eye and note any change. Now hold a book between the eyes to shade one eye while you flash the light on the other. What happens to the pupil of the shaded eye? What do you conclude from this?

Focusing

When light rays enter the eye they are refracted at the corneal surface even more than by the lens (Fig. 28:6) and are focused on to the retina. A *glass* biconvex lens has a definite focal length and an object is only in focus if it is a fixed distance away; the fatter the lens the closer this distance becomes. The focal length of the lens of the eye can be adjusted to focus on any object between about 12 cm and infinity by altering its shape. This property of adjustment is called **accommodation**.

How is the shape of the lens altered? This is possible due to the elasticity of its outer membrane. When the fibres of the suspensory ligament which support the lens are slack the lens will be thick (for near objects). When these fibres are tightened the lens will become thinner (for longer distances). The slackness occurs when the circular fibres of the ciliary muscles (which are attached to the suspensory ligaments) *contract* and so reduce the diameter of the ciliary muscle ring. They contract against the pressure exerted outwards by the fluid contents of the eye. However, when the circular muscles of the ciliary muscles are *relaxed* the ciliary muscle ring becomes larger due to the outward pressure of the fluid of the eyeball. This pulls on the suspensory ligament which in turn pulls the lens into a

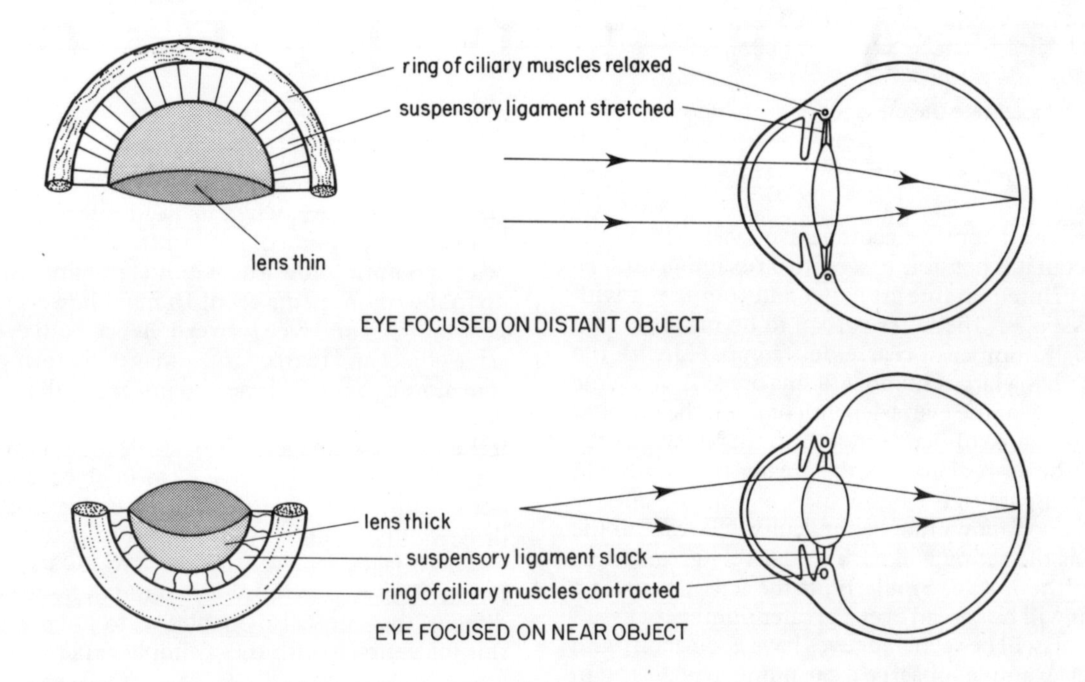

Fig. 28:6 The mechanism of accommodation: (left) relevant organs as seen from the back of the eye (right) in section.

thinner shape. Therefore relaxed ciliary muscles result in distance focusing, contracted ones result in near focusing. This explains why prolonged close work is tiring to the eyes.

The retina

The retina contains sensory cells of two kinds, **rods** and **cones**, named because of their characteristic shapes (Fig. 28:7). When these are stimulated by light, impulses pass along the nerve fibres which link them to the brain. These fibres lie on the surface of the retina and converge to form the optic nerve which leads to the brain. So when we look at an object, the brain receives a mass of impulses from

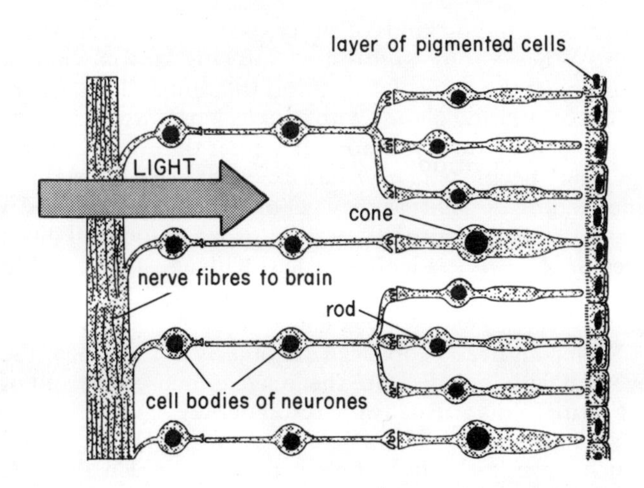

Fig. 28:7 Diagram showing the arrangement of cells of the retina. Note how the light has to pass through the nerve fibres before reaching these cells.

the retina and these are interpreted in the form of an image of the object seen.

Rods are more sensitive to light of low intensity and are more suitable for night vision, but the image interpreted by the brain lacks detail; cones, by contrast, need stronger light to activate them but they produce a much more detailed image and are more suited to day vision. Their distribution within the retina is not uniform. Rods are much more concentrated round the periphery while the cones are more numerous towards the centre. Thus at night when the pupil is dilated more rods will be exposed, but in bright light when the pupil is contracted the cones will mainly be activated. The **fovea** or **yellow spot** which is the point of principal focus is composed of cones only. When we look at any object very carefully we place it in just the right place so that it is focused on the fovea. If you look intently at an object on a very dark night you may see nothing, but if you look to one side of it, you may then just see it vaguely because the image does not fall on the cones in the fovea, but on an area nearby which contains rods. What proportion of rods do you think there would be in the retinas of nocturnal animals such as bats and owls?

Blind spot

Where the optic nerve leaves the eye there are no retinal cells present and so no image is formed. You can demonstrate the presence of this **blind spot** for yourself (Fig. 28:8):

> Hold the book at arm's length, close your right eye and look at the black spot with your left; now move your left eye to the right along the series of letters until the spot disappears. What letter were you looking at when it disappeared?
>
> Now hold the book nearer and repeat. Was it the same letter? Repeat for different distances. What do you conclude from this?

● A B C D E F G

Fig. 28:8 To demonstrate the presence of the blind spot.

Acuity

This term is used for the acuteness of vision, i.e. the amount of detail it is possible to see. If you examine under a lens a photo printed on fine art paper and compare it with one in a newspaper, the latter is seen to be made up of a relatively small number of coarse dots compared with the former which has a larger number of fine ones. In the same sort of way acuity of the eye is dependent upon the number of cones per unit area of that part of the retina on which the image falls. The more there are, the greater the number of impulses which pass to the brain and the more detailed is the image. It also follows that better acuity will result if the eye is large, as the retina will have a relatively greater area for receiving the image. Similarly, if the lens produces a large image it will be spread over a greater number of cones and more detail will be seen. Our eyes have good acuity and we might, for example, observe a car number plate 100 m away and just succeed in counting how many numbers and letters were on it, but at 50 m we might be able to read accurately what the numbers and letters actually were.

However some birds such as hawks and eagles have an acuity eight times better than ours which is equivalent to their reading a car number plate clearly at 400 m! They also possess a second fovea of concentrated cones which receives the image of objects in the forward line of flight. This enables a hawk to see stereoscopically and thus to range-find accurately as it approaches its target.

Colour vision

Colour perception in mammals is confined to man and other primates; a bull, for example, does not react to a red rag because it is red, but because it is moving. Most birds, reptiles and fish which have been tested can detect colours, and so can bees and butterflies, but there is variation in the range of colours perceived. Bees can see ultra violet as a colour, but are blind to red. We cannot imagine what ultra violet would look like as ultra violet rays are filtered off by our lenses because the lenses are slightly yellow. In old age this yellow becomes more pronounced with the result that old people have difficulty in seeing some of the violet hues visible to younger people.

It is the cones in our eyes that distinguish colour. As these need light of high intensity to stimulate them, it follows that we cannot see colours in poor light. At dusk everything becomes black or white or shades of grey. It is not known exactly how colour is detected, but put very simply, a likely theory is that there are three kinds of cone which are stimulated by different wavelengths of light corresponding to the red, blue and green parts of the spectrum. Thus the brain builds up a colour picture according to the number of impulses received from the three kinds of cone.

Stereoscopic (binocular) vision

When the eyes are at the side of the head, as in hares and horses, the field of view is very large because both eyes cover different areas and there is little or no overlap. However, in a hare, when its head is facing forwards there is a blind area immediately in front of it, so if the hare's head is pointing towards you it is possible to walk quietly up to it without being seen! In man, however, the eyes are near together and face forward, hence both eyes can see the same object and two separate images are formed. The brain somehow combines these two images so that we do not 'see double'. However, occasionally, if there is an eye defect or if the brain is damaged it is possible to see two images. This may be experienced if you look at an object and gently press one eye upwards. The effect of excessive alcohol and other drugs on the brain may also produce double vision.

The images received by the two eyes are slightly different. This makes the object stand out—we see it in three dimensions—and also enables us to judge distances. Test this for yourself with these simple experiments:

1. Hold a ballpoint pen nearly at arm's length and *quickly* try to put the cap on the pen. Now shut one eye and repeat the operation. Which method is more precise?
2. Close your eyes and ask your neighbour to put a small object a few feet away from you at an angle of about 45° from the direction in which your head is facing, keep your head in the same position and open your left eye and estimate how far away the object is; repeat for the right eye only; finally open both and estimate the distance once more.

Measure the distance and see which estimate is the most accurate.

How important do you think binocular vision would be to: a) A mammal which eats grass, b) A carnivore which feeds on small rodents, c) A mammal which leaps about in trees? Test your answer by thinking of different animals in these categories and see how their eyes are placed in their heads.

Correction of faulty vision

The most common kinds of faulty vision are long and short sight. When a **long-sighted** person (Fig. 28:9a) views a near object the image falls behind the retina. This is corrected by wearing glasses with convex lenses which help to converge the light rays before they enter the eye. In middle age many people who have had excellent eyesight begin to find reading difficult; as they get older they have to hold a book further and further away in order to focus. This is because the lens gradually hardens with age, it loses some of its elasticity and so its power of accommodation is reduced. Glasses with convex lenses correct this.

Short-sighted people can see objects in focus near the eye, but with long distance the image is focused in front of the retina. This is usually the result of the eyeball being too long (Fig. 28:9b). This is corrected with concave lenses which diverge the light rays before they reach the eye. Some forms of short sight improve with age.

Astigmatism is a condition where the cornea is uneven; it refracts the light rays in an abnormal way and so produces a distorted image. A regular astigmatism occurs

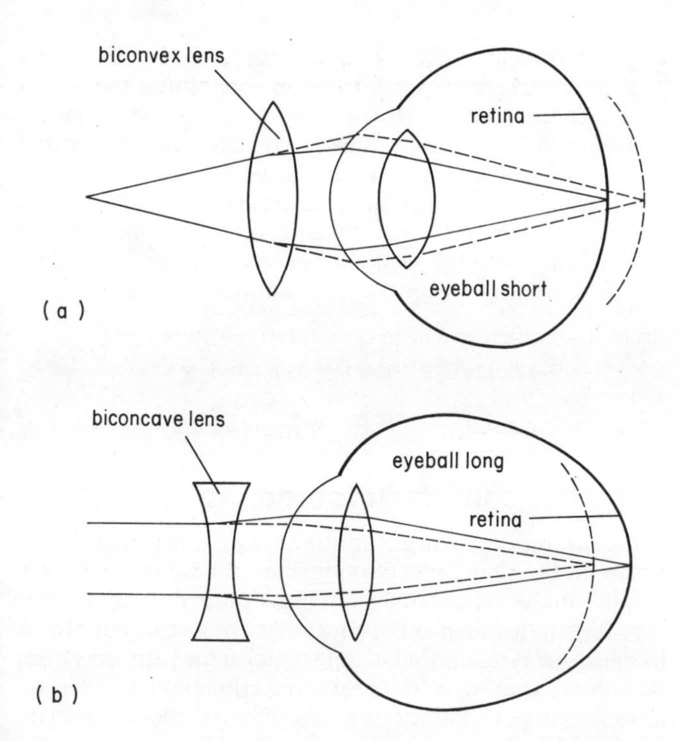

Fig. 28 : 9 a) Lens corrections for long sight. Dotted lines indicate that focus of near object is behind retina; this is remedied by a biconvex lens. b) Lens correction for short sight. Dotted lines indicate that focus of distant object falls in front of retina; this is remedied by a concave lens.

when the cornea has a different curvature in the horizontal and vertical planes; this type of defect can be corrected with the help of glasses having a lens of appropriate shape to compensate for the distortion.

The ear

The ear provides the brain with information about the sounds around us, their pitch and loudness and the direction from which they come, but the ear is also an organ of balance and helps in maintaining posture.

Its structure is shown simply in Fig. 28 : 10. It is divided into three regions—an outer, a middle and an inner ear—all three parts being concerned with hearing, but only the inner ear with balance and posture.

Outer ear

This consists of the part you can see, the **pinna**, and a tube which passes into the skull and ends at a membrane, the **tympanum** or **ear drum**. The pinna helps to concentrate sound waves and direct them towards the tympanum, but in man, because the pinna is immovable, it is not nearly so efficient as in most mammals. A deaf person may cup his hand behind his ear to improve its collecting capacity and at one time ear trumpets were used for the same purpose.

Just as with binocular vision, when two eyes are focused from different parts of the head on a single object making it possible to judge its distance, many animals with their two ears can judge the position from which a sound comes by moving them independently until each receives the maximum sound. In this way cats and foxes can leap on a mouse in the dark, judging its position by sound. In the serval cat this has become so highly developed that its huge ears act like radar screens, magnifying the slightest sounds and enabling it to leap on a hare feeding unseen in long grass from a distance of at least 3 metres.

Our ears cannot move, but we can locate the direction of a sound nevertheless; this is because the sound is heard more loudly by the ear nearest to it and also fractionally

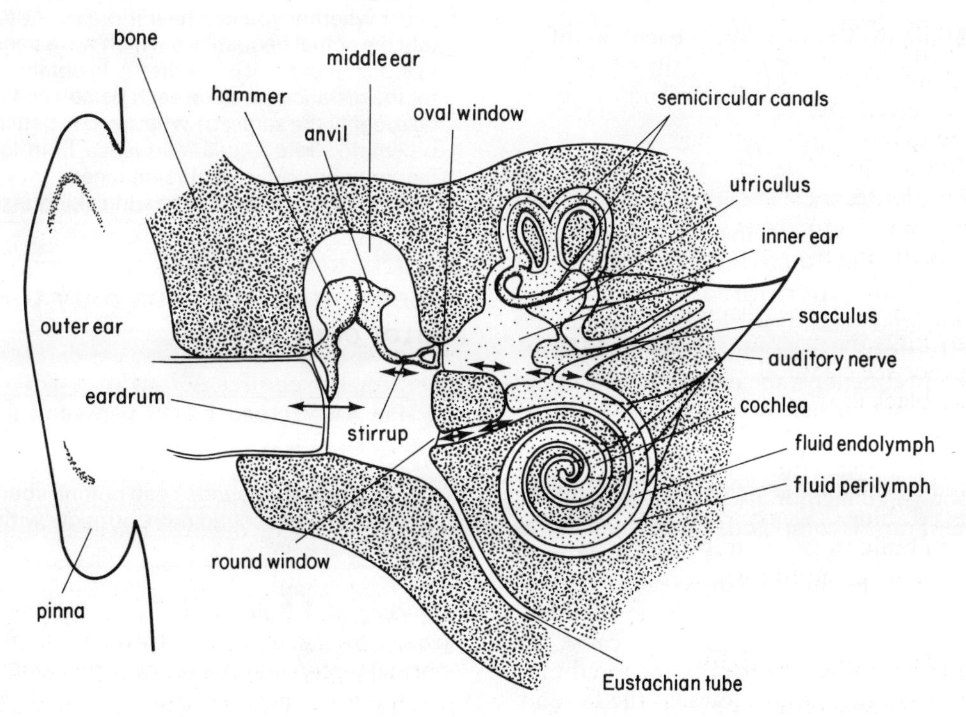

Fig. 28 : 10 The human ear.

earlier. A sound is always difficult to locate if it comes from a source equidistant from both ears. That is why we turn our heads to one side to make sure of the direction from which the sound is coming.

Test this for yourself:

> One person is blindfolded and asked to indicate the position of a ticking clock by pointing in the direction from which the sound is thought to come. He must keep his head still all the time. The clock should be held in ten positions where it is equidistant from both ears and in ten positions where it is at different distances. How often is the estimate of direction correct for each category?

The middle ear

This is an air-filled cavity surrounded by bone. Its main function is to transmit the sound vibrations which cause the tympanum to vibrate to the inner ear by means of three small bones called the **hammer**, **anvil** and **stirrup**, so named because of their characteristic shapes. The hammer touches the tympanum and the stirrup is in contact with a membrane covering the **oval window** of the inner car. Because this membrane is much smaller than the tympanum the force of the vibration it receives is much greater; it is further increased by the lever action of the ear bones with the result that the inner ear receives sounds amplified about 22 times.

The only opening of the middle ear to the outside world is via the **Eustachian tube** which opens at the side of the throat. Usually this opening is kept closed, but when we swallow or yawn it opens. It is through this opening that air can pass, thus keeping the pressure equal on both sides of the tympanum. We notice this when gaining height rapidly in an aircraft or even when we are going up a long steep hill in a car: we feel our ears pop as equal pressure is restored. Why would the chewing of sweets help to prevent this from happening?

If a person has a sore throat due to some bacterial infection the Eustachian tube is a possible route through which the infection may spread to the middle ear and cause ear ache.

The inner ear

This is the part of the ear where the sensory cells are situated and from which impulses pass to the brain via the nerves. It consists of the **labyrinth**, a delicate hollow structure filled with fluid (**endolymph**) and surrounded by more fluid (**perilymph**); it is deeply embedded in bone. Its parts have different functions; the **cochlea** is the organ of hearing and the **sacculus**, **utriculus** and **semicircular canals** are for maintaining balance and posture.

The cochlea is a much-coiled tube with a blind end. In it there is a long ribbon-like membrane which passes along its length. This membrane is composed of transverse fibres of varying lengths. Vibrations received at the oval window are transmitted through the fluids of the cochlea causing the transverse fibres of the membrane to vibrate at certain places according to the frequency. High notes cause the short fibres of the front part of the membrane to vibrate, low notes stimulate the longer fibres towards the far end. These fibres are in contact with cells bearing sensory hairs, which in turn have nerve connections to the brain. In this way the brain receives information concerning the sound received, according to which sensory cells are stimulated.

As the fluid in the cochlea is virtually incompressible there has to be another membrane which can vibrate and compensate for changes in pressure; such a membrane is present covering the **round window** which thus vibrates in sympathy with the oval window membrane.

The sequence of events can be summarised as follows:

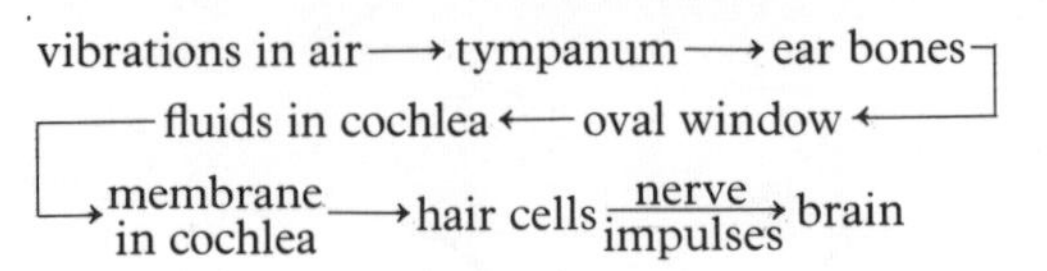

Range of sounds detected

When young we can hear sounds ranging from about 60 Hz to 20,000 Hz. One hertz is a single to-and-fro vibration per second. But as we get older we find it more difficult to hear very high notes such as the squeak of a mouse or bat; this is because our tympanum becomes thicker and our ear bones do not transmit high frequency vibrations so well. We are most sensitive to sounds of a frequency of about 3000 Hz. It is interesting that this corresponds to the piercing sound of a child or woman screaming for help. Other mammals are most sensitive to different ranges. A cat, for example, is sensitive to the very high ranges such as squeaks made by mice, but mice can hear still higher warning noises from other mice which a cat cannot hear! Bats, whales and probably many other species communicate with higher frequencies still, far beyond our powers of hearing; they also use these sounds for echo-location (p. 79).

> Find out your own range of audible sounds by using an audio-signal generator. This gives out sounds of known frequencies over a wide range through a loudspeaker. As each sound of a particular frequency is emitted record on paper whether you can hear it or not. At the end you will be told the actual frequencies which were transmitted. Consider: a) whether or not it is important, in obtaining accurate results, for the distance between each person and the loudspeaker to be roughly the same, b) whether it is better for the sounds to be emitted in a regular sequence from lower frequency to higher or in a more random manner.
>
> How much variation in hearing ability is there in your class?

Does sound reach us only through our outer ear?

> 1. Put a watch which ticks between your teeth and block your ears. Can you pick up the vibrations?
> 2. Put your first fingers into both your ears and start to hum. Can you hear the sound? Keep putting your fingers in and out. Do you hear the sound more strongly with your fingers in or out of your ears?

When you hum, your lips are closed and most sound passes to the inner ear through the skull itself. Under normal circumstances we hear the sound of our own voice via our inner ear and through our skull, but another person only hears our voice through the outer ear, that is why we hardly recognise our own voice when we play it back on a tape recorder, but other people do.

Even external sounds are transmitted to some extent through the skull. A hearing aid placed just behind the ear works on this principle; it amplifies the vibrations it receives from the air, and because it is in contact with the skin which covers the bone the hearing aid enables the vibrations to be transmitted to the inner ear.

Balance and posture

In the labyrinth there are structures which respond to gravity in such a way that when our head is tilted we are made aware of its position even if we are blindfolded and our head is quite still. In the wall of the utriculus and sacculus there are special cells with fine projecting hairs; particles of calcium carbonate called **otoliths**, embedded in jelly, make contact with these hairs (Fig. 28:11). When the head is upright the otoliths press downwards on the hairs due to gravity. If we stand on our head, the pressure on them is much reduced and if we tilt our head the otoliths bend the hairs to the side. As a result of these various stimuli the brain obtains information about the head's position and responds by causing appropriate muscles to contract which bring the head back to its normal position. Thus posture can be adjusted quite unconsciously.

In addition to these organs which are concerned with posture there are others which respond to directional movements; these are the semi-circular canals. There are three of these arranged in three planes at right angles to each other, two vertical and one horizontal. One end of each canal is swollen and contains a group of sensory cells whose projecting hairs are embedded in a cone of jelly. When we move our head from a resting position in a particular direction the fluid inside the canal which lies in the same plane as the movement exerts pressure on the cone of jelly, thus bending the hairs. As a result impulses are sent along nerves to the brain. In this way the brain receives information about **accelerating** movements in a particular plane or **rotational** movements. If speed is constant in one direction pressure on the hair cells will be regular and no further impulses will be received by the brain. You should now be able to work out what happens when you twirl your body round rapidly and stop suddenly. What kind of sensation do you feel?

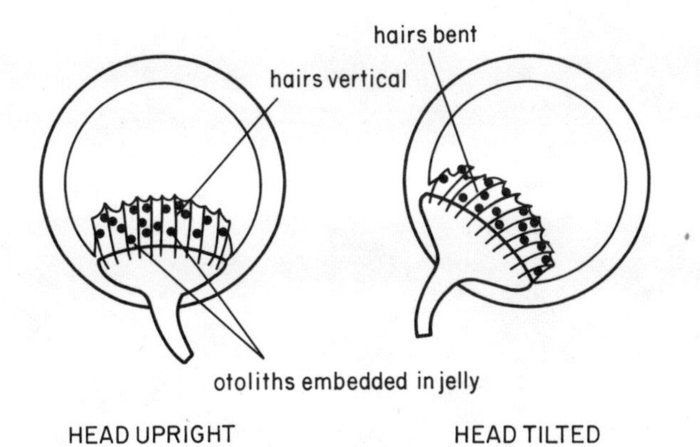

Fig. 28:11 Diagram illustrating otolith action.

Sense organs which receive chemical stimuli

The olfactory organs

These are the organs of smell and consist of sensory cells which line the roof of the nasal cavities. When we take in air through our nostrils many molecules of volatile substances pass in with it; these dissolve in the mucilage covering the sensory cells, stimulating them and causing impulses to pass to the brain. We also smell substances which we take into our mouth, as molecules from them can pass into our nasal chambers via the back of the throat (Fig. 28:12). We usually say that we can taste them, but really we are smelling them. When we have a bad cold this back entrance to the nasal cavities may become blocked with mucus and we can no longer detect the flavour of our food.

We have a very poor sense of smell compared with most mammals, but even so most of us can easily distinguish over 1000 different odours while some people can distinguish up to 4000.

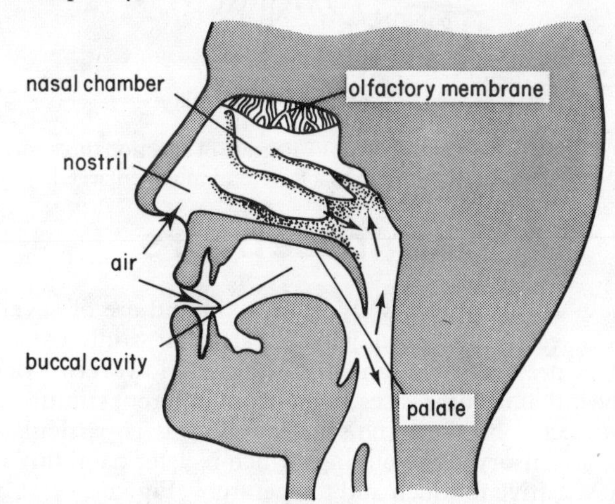

Fig. 28:12 The organs concerned with smell.

The organs of taste

These are called taste buds, which are groups of cells situated on the tongue (Fig. 28:13). As we have seen, these do not detect flavours (which are smelt), but are sensitive only to substances which are sweet, sour, salt or bitter. Their chief function appears to be to give warning regarding the suitability or unsuitability of food before it is swallowed.

The distribution of taste buds on the tongue is not even. You could map out the position of those concerned with the four sensations in this way:

Use solutions of sugar (sweet), very dilute hydrochloric acid (sour), table salt (salt) and quinine (bitter). Work in pairs and use one solution at a time. One partner should put his tongue out while the other transfers a drop of solution on the end of a glass rod to one of the four regions of the tongue to be tested—back, tip, centre and sides. Repeat for the three other regions, rinsing out the mouth between each application. The tongue should be kept still while each drop is being tested. Record which regions can detect each solution.

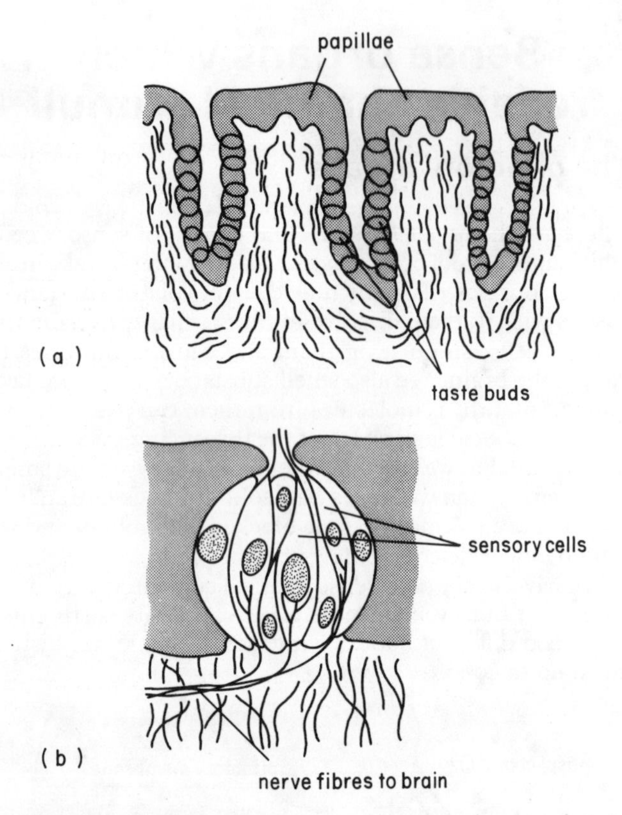

Fig. 28 : 13 The organs of taste: a) diagram of a section through the upper surface of the tongue b) a taste bud much enlarged.

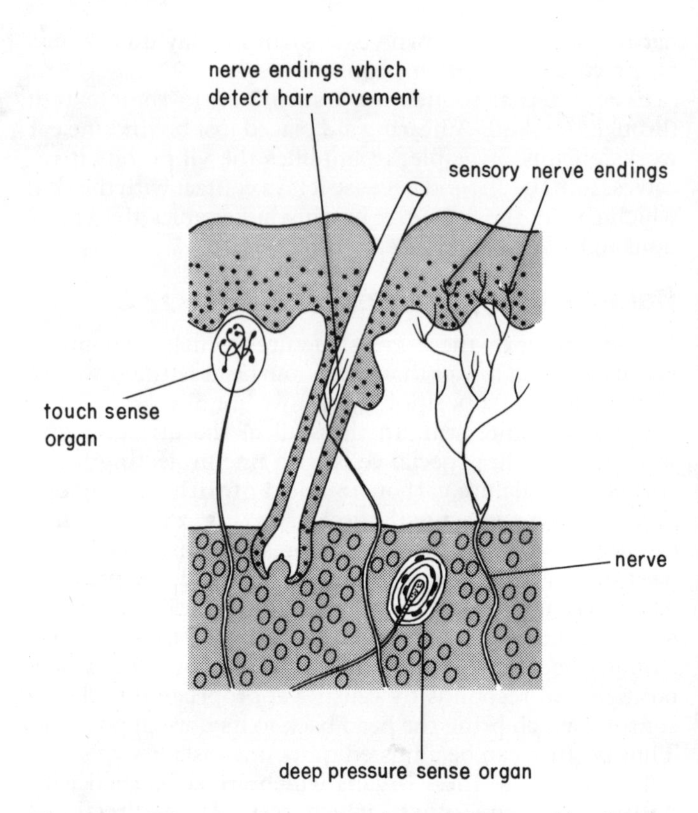

Fig. 28 : 14 Diagram of a section through human skin showing sense organs.

Skin receptors

These are situated mainly in the skin and are of several kinds, each concerned with a particular stimulus, e.g. touch, pressure, pain and temperature. However, it does appear that in some cases a number of different stimuli can be detected by the same sense organ; this is particularly true of sensory nerve endings which register pain, but are also sensitive to touch and temperature (Fig. 28 : 14). Can you think why it is advantageous for the brain to monitor pain? Although widely distributed over the skin some areas have concentrations of sensory cells of a particular kind, making such places particularly sensitive; thus the lips and the tips of the fingers are very sensitive to touch and the upper arm to temperature. This is why a mother will often judge the temperature of a baby's bath by putting her elbow rather than her hand in the water. An object feels warm because our sensory cells detect a flow of energy from the object to us; it feels cold when the flow goes in the other direction. Test this for yourself:

> Put one hand in ice-cold water and another in warm water. Now touch a stone which is at room temperature with each hand in turn. Which hand's reaction are you going to believe?

Internal receptors

The brain not only needs to know about the external environment, but it also needs information about the changes constantly occurring inside the body. This is provided by internal receptors which monitor such changes as the level of sugar or carbon dioxide in the blood, the fluctuations in body temperature and variation in the osmotic pressure of body fluids. Others called **proprioceptors** are embedded within muscles and tendons and act like strain gauges, sending stimuli to the brain according to the degree of tension. This helps co-ordination of movement and is another means of maintaining posture (p. 212). Many animals can orientate themselves in relation to the earth's magnetic field. This means that they must have sense organs which act as magnetic detectors. It has recently been discovered that bees, certain species of birds and even some bacteria have microscopic units of magnetite within their bodies which enable them to orientate. There is evidence that such a mechanism may also occur in man and assists in direction finding.

29

Internal lines of communication

We have seen in the previous chapter how information is constantly being received by the sense organs about the changes which are occurring both outside and inside our bodies. Now we will consider how appropriate action may be taken in response to the information received.

There are two linking or co-ordinating systems between the **receptors** which detect the stimuli and the **effectors** which react in consequence: these are the nervous and endocrine systems. The nervous system co-ordinates the activities of the body by means of a complex system of nerves, while the endocrine system does so by secreting hormones into the blood.

The nervous system

This consists of:

1. The **central nervous system** (CNS), composed of the brain and spinal cord. This part of the system co-ordinates the impulses received from the receptors and transmits other impulses to the effectors which then act in response.
2. The **peripheral nervous system**, composed of paired **cranial** nerves coming from the brain and paired **spinal** nerves from the spinal cord. These nerves are the living lines of communication between the receptors, the central nervous system and the effectors.
3. The **autonomic nervous system** which is concerned with the body's automatic (involuntary) activities, such as the contractions of the alimentary canal and the beating of the heart.

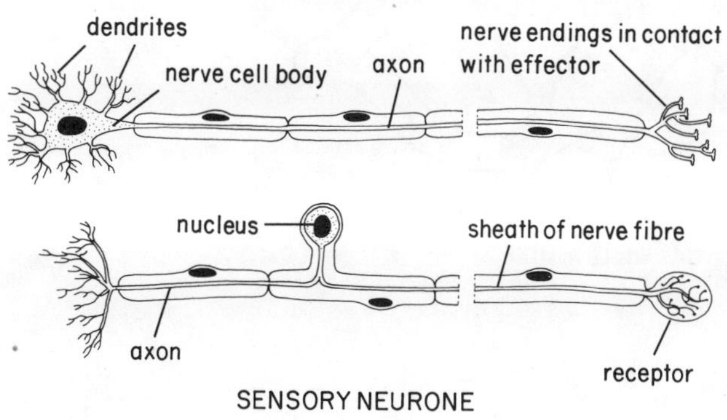

Fig. 29:1 Diagram showing the structure of a motor and a sensory neurone. They are usually much longer than shown.

The neurone

The nervous system is made up of units called nerve cells or **neurones**. There are three kinds:

1. **Afferent** or **sensory** neurones which transmit impulses from a receptor to the central nervous system.
2. **Efferent** or **motor** neurones which transmit impulses from the central nervous to the effectors.
3. **Association** neurones which link the afferent and efferent neurones; these lie within the brain or spinal cord.

Each neurone (Fig. 29:1) consists of a **cell body** composed of cytoplasm and a nucleus and a number of cytoplasmic extensions. Of the latter, one is a long thin process called an **axon**, while the others are shorter and end in many fine **dendrites**. Neurones include some of the longest cells in the body because their cell bodies are situated mainly in the brain and spinal cord, but their axons may reach the furthest extremities of the body.

The cytoplasm of one neurone is not continuous with that of another, but the dendrites of one become very closely associated with the terminal branches of another neurone or with its cell body and the impulse is able to pass across the gap. These gaps between interconnecting neurones are called **synapses**.

The nerve impulse

A neurone is able to transmit an electrical impulse very rapidly. The impulse is not the same as an electric current passing down a wire but takes the form of a wave of electrical disturbance along the neurone. The impulse normally travels along a particular neurone in one direction only and it does so at a speed of up to 100 m/s in man.

Nerves

These are aggregations of axons (collectively called nerve fibres) bound together like wires in a cable (Fig. 29:2). The fibres of cranial and spinal nerves are insulated from each other by a fatty sheath and are called **medullated** fibres.

Simple reflex actions

These are the simplest examples of nervous co-ordination. We have already come across some of the simple reflexes which are associated with keeping the head upright. You will remember that stimuli received by the eyes, ears and 'strain gauges' in the neck muscles (pp. 221 and 222) all

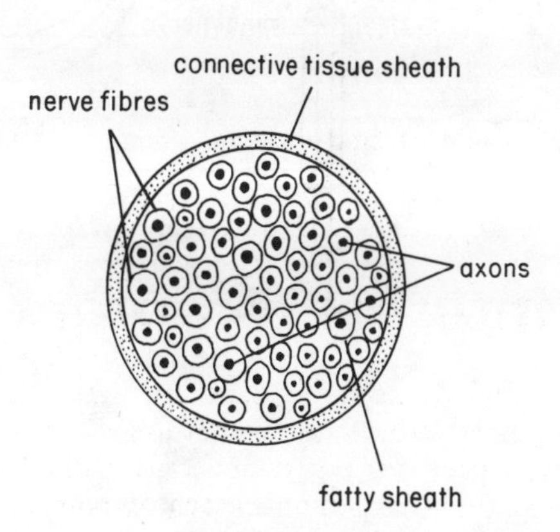

Fig. 29:2 Diagram of a transverse section through a small nerve.

bring about compensating head-righting actions. A doctor can test your reflexes in various ways. In one method he asks you to sit on a stool with your right leg crossed over your left so that it hangs freely; he will then tap the tendon just below the right knee cap. If the reflex is working correctly, the leg jerks forward. If there is no response the doctor will know that there must be something wrong with the nerves of the leg or the spinal cord itself.

You could test your neighbour's reflexes in a similar manner. Is it possible to prevent this response?

There are many other simple reflexes: a) When you touch something hot the finger is quickly withdrawn from the stimulus. b) When a light is shone in your eye, the pupil contracts. c) When pepper gets up your nose, you sneeze. d) When a crumb goes down 'the wrong way', i.e. enters the larynx, you choke. Can you think of other examples?

Simple reflexes have these important characteristics:

1. They are inherited, so they do not have to be learnt and are not forgotten.
2. They are not under the control of the will and so are quite automatic, although some reflexes can be over-ridden by willpower.
3. For a given stimulus, the response is always the same.

The reason why the last statement is true is because the impulse travels along the same nervous pathway. This pathway is called a **reflex arc**. In its simplest form, as in the knee jerk reflex, each reflex arc consists of two neurones, an afferent and an efferent, which associate at a synapse in the central nervous system. However, most reflex arcs (Fig. 29:3) have, in addition to the afferent and efferent neurones, one or more association neurones which enable a stimulus such as a prick, received at a single point, to be transmitted via many efferent neurones, causing whole systems of muscles to move the complete arm. At the same time, other neurones send impulses along the spinal cord to the brain and so make us conscious of the prick. Impulses can also be sent from the brain down the spinal cord, and these co-ordinate the action of motor neurones in other spinal nerves. This allows the brain to control any further action which is necessary. For example, sometimes, when we inadvertently touch something hot, we draw the hand away by a reflex action, but if immediately afterwards our brain tells us the object was not hot enough to cause pain we may consciously put our finger back on to it again.

Reflex arcs involving sense organs in the head pass through the brain and are called **cranial reflexes**, e.g. we automatically blink when an object approaches the eye. Those which involve sense organs from the neck downwards pass through the spinal cord and are called **spinal reflexes**. In some animals it is possible for the latter to continue to work for a time after the brain has been destroyed. Refer also to conditioned reflexes (p. 237).

From the brain arise 12 pairs of **cranial** nerves and from the spinal cord 31 pairs of **spinal** nerves. The latter emerge through holes between adjacent vertebrae in a very regular manner down the spine (Fig. 29:4).

You will see from Fig. 29:3 that the spinal nerves have a dorsal and a ventral root, and that afferent neurones pass into the dorsal root, their cell bodies being confined to a swelling, the **dorsal root ganglion**. The efferent neurones have their cell bodies in the spinal cord itself and their axons pass out along the ventral root.

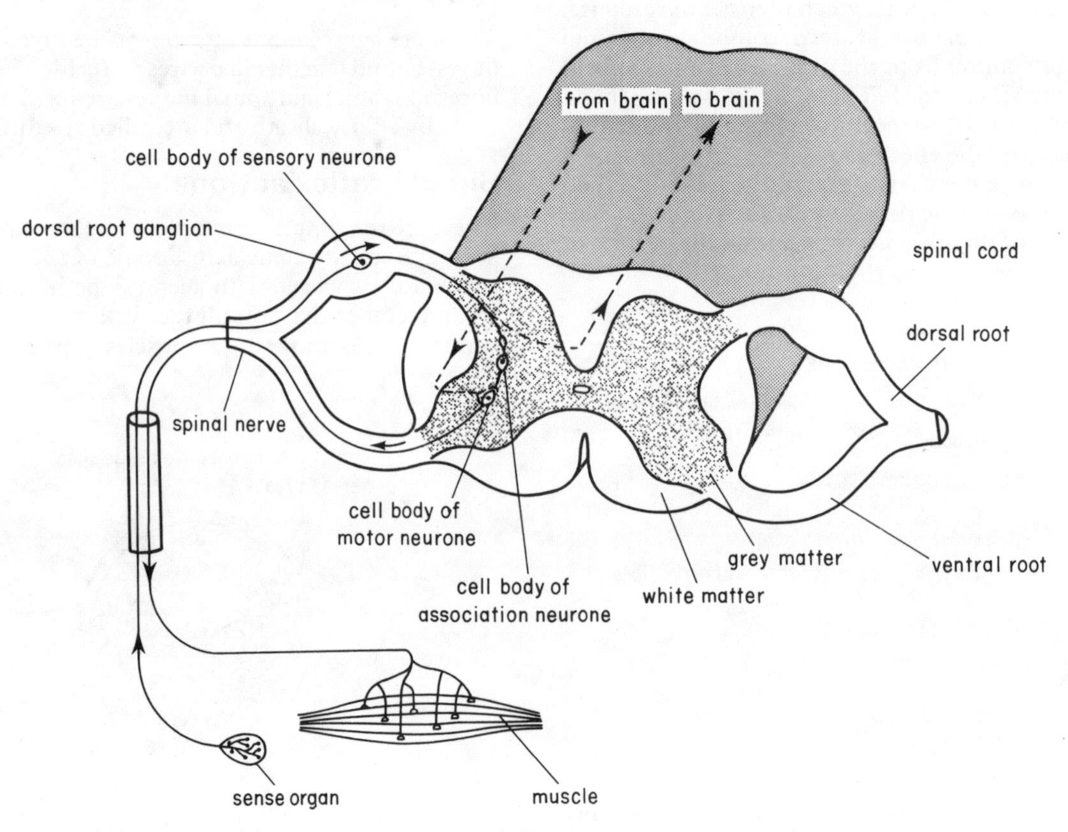

Fig. 29:3 The structures concerned in a reflex arc involving the spinal cord.

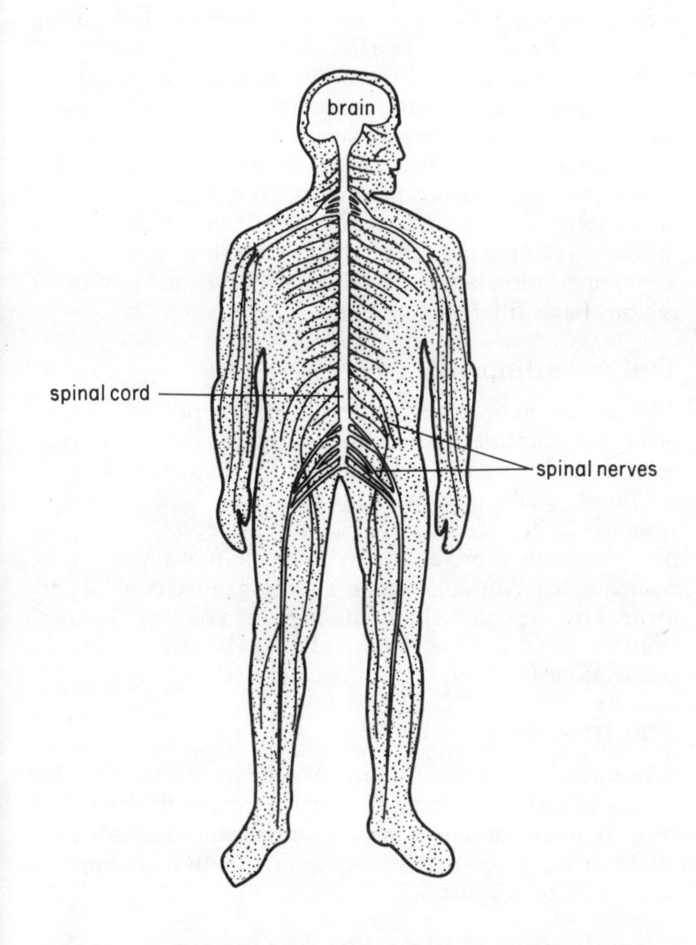

Fig. 29:4 Diagram of the central nervous system and the spinal nerves of man.

The central nervous system

The spinal cord looks fairly solid in section (Fig. 29:3) but it is in fact a hollow tube, as there is a canal in the centre which is continuous throughout its length. This canal contains cerebro-spinal fluid and is continuous with certain cavities in the brain called **ventricles**.

The brain can be looked upon as a highly specialised part of the spinal cord. In the embryo it arises as three swellings at the anterior end of the spinal cord, each part differentiating later into the main structures of the adult brain.

Both brain and spinal cord are made up of nervous tissue of two kinds, **grey matter** and **white matter**. The grey matter consists largely of nerve cell bodies, and the white matter of nerve fibres surrounded by their medullary sheaths which cause the white appearance. In the brain the grey matter tends to be situated near the outside; in the spinal cord it is confined more to the centre.

The brain

As the brain develops from the anterior part of the spinal cord, it is not surprising that it has rather similar functions; however, these are carried out on a much more complex scale. Like the spinal cord, the brain receives impulses

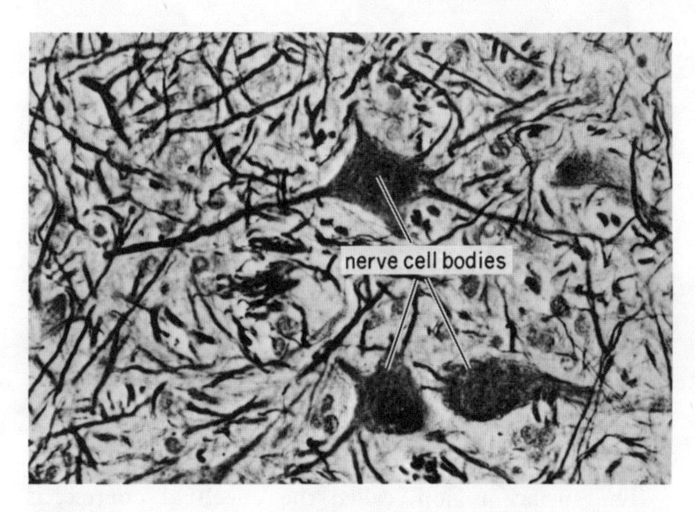

Fig. 29:5 High power photomicrograph of brain tissue showing neurones.

from sense organs, especially the nose, eyes and ears. It also sends impulses to the muscles and glands of the head, and to those in other parts of the body via the white matter of the spinal cord. The brain differs greatly from the spinal cord, however, in the astronomical number of association neurones that it contains. These allow an almost infinite number of cross-connections to occur, and consequently great variation in behaviour is possible. How are all these impulses sorted out? How are the pathways for the stimuli determined? What makes the brain act as a completely co-ordinated structure and not as a lot of isolated centres? What actually happens when we 'make up our mind' about something? What process is involved when we learn? How is information stored so that we can act according to our experience? The answers to these intriguing questions are only partially known, but brain research is providing

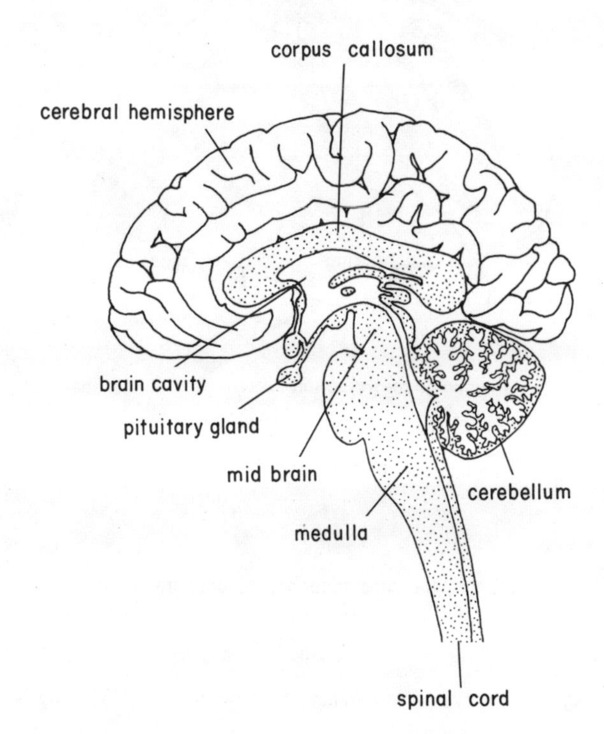

Fig. 29:6 Vertical section through a human brain.

useful clues towards the solution of some of them. A brief study of the structure of the brain should help you to understand some of the principles involved. Let us consider in particular the cerebrum, the cerebellum and the medulla (Fig. 29:6).

The cerebrum

In the course of evolution of the vertebrate brain the cerebrum has enlarged more and more and in man forms by far the largest part of the brain. It consists of two lobes or **cerebral hemispheres** partially separated by a deep cleft in the midline. The two hemispheres are connected at a deeper level by the **corpus callosum**, a broad sheet of white fibres which helps to co-ordinate the left and right sides.

The surface region, called the **cerebral cortex**, is thrown into intricate folds and grooves which give it a far greater surface area. This is important because the cortex consists of grey matter and the more there is of it, the more nerve cell bodies it can contain. In fact, about 90% of all the nerve cell bodies in the brain occur here.

The cerebral cortex is by far the most important **association centre** of the brain. All the time information from our main sense organs streams into it and is sorted out in the light of our past experiences, as a result of which motor impulses are discharged from it along pathways of white fibres and cause the appropriate action to be taken. When learning takes place it is possible that certain neurones may become associated in some way to form a particular pathway which may be followed once more if the same stimulus is received. But the exact processes of learning and storing memories are not yet understood.

When certain parts of the brain are given small electric shocks, particular sensations are felt in the body or different muscular movements occur. In this way it has been discovered that specific areas of the cortex deal with skin sensations from various parts of the body and other areas are concerned with the control of muscular movements in different regions. Fig. 29:7 shows how these sensory and motor areas form bands across the cerebrum and how other regions have different functions.

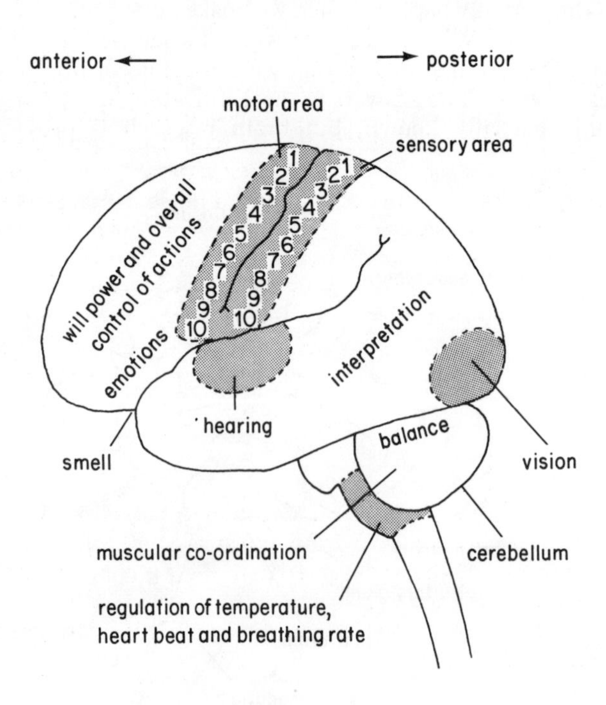

Key to corresponding motor and sensory areas:

foot   2. leg   3. trunk   4. arm.   5. hand
fingers   7. eye   8. face   9. lips   10. tongue

Fig. 29:7 Side view of human brain showing the main regions.

The cerebellum

This region is concerned with balance and posture and co-ordinates muscular movement so that all the appropriate reflexes take place at the same time. When a child is learning how to walk, its movements are at first unco-ordinated; similarly, when we learn to ride a bicycle, drive a car, or play the piano there are many intricate movements to be co-ordinated. Muscular control of these movements is performed by the cerebellum, although, as you would expect from the fact that these activities have to be learnt, there are connections between it and the cerebrum.

The medulla

This forms the brain stalk and links the rest of the brain, by means of large numbers of nerve fibres, with the spinal cord. In it are the centres which control automatically such vital functions as rate of breathing, regulation of temperature and rate of heart beat.

Circulation of blood to the brain

There is an excellent blood supply to the brain as the nerve cells require a constant supply of glucose and oxygen. If the blood supply becomes inadequate for some reason, **fainting** occurs. This sometimes happens when soldiers are on parade because they are standing very rigidly for a long time with very tense muscles. A person who has fainted should be placed so that the head is low in relation to the rest of the body in order that gravity may aid the pumping action of the heart and help the circulation to the brain. As a result of age or injury, small arteries in the brain may become damaged and so cause local haemorrhage, or a blood clot may form within an artery. In both cases, brain tissue will be damaged. This is the cause of a **stroke** which, if severe, may cause paralysis or death.

The autonomic nervous system

This system is concerned with many of the actions which go on within the body quite unconsciously. It has two divisions, **sympathetic** and **parasympathetic**.

Sympathetic fibres leave the brain as a pair of nerve chains which pass down the body on either side of the vertebral column. On these chains are ganglia which contain the cell bodies and axons of sympathetic neurones; some of these neurones transmit stimuli to various internal organs while others make links with the spinal cord. A further group of ganglia occurs in the abdominal region and comprises the **solar plexus** from which nerves pass to the gut and other abdominal organs. A blow in this region can produce a violent and painful effect.

Parasympathetic fibres come mainly from the base of the brain and pass to the organs via the tenth pair of cranial nerves.

Sympathetic and parasympathetic fibres supply the same organs, but their effects are usually antagonistic. This provides a means of regulating an action according to changing circumstances. For example, if the sympathetic nerve fibres to the heart are stimulated the heart beats faster; if the parasympathetic fibres are stimulated the heart rate is depressed. In a similar way, stimulation of sympathetic nerves causes the blood vessels of the skin and gut to constrict and the iris of the eye to dilate, while stimulation of parasympathetic nerves produces the opposite effects. These effects all concern the action of **involuntary** muscle—the kind of muscle found in the walls of the gut, uterus and blood vessels—the contractions of which are slow and rhythmical. In addition, the autonomic nervous system controls the secretion of glands, such as the salivary and sweat glands, and such endocrine organs as the adrenal glands. In general terms, it may be said that when sympathetic nerves are stimulated, the overall effect is to mobilise the body for instant emergency action, while the stimulation of the parasympathetic nerves have a calming effect on the body.

Hormonal control

This is the second linking system. We have already seen that certain hormones are concerned with the regulation of metabolic processes, growth and reproduction (Ch. 22). Some of the reproductive hormones also have a co-ordinating effect such as those concerned with the ovarian cycle, while others from the gut wall help to co-ordinate the digestive process. But it is the effects of adrenaline (p. 162) that show such a striking resemblance to those of the sympathetic nervous system. This is because the two actions are closely linked and one enhances the other. The action of adrenaline is somewhat intermediate in character between hormonal and nervous co-ordination. Most hormones are slow in their action and spread their influence over much longer periods of time; they are also more concerned with the regulation of metabolic processes. Nervous co-ordination, on the other hand, is largely concerned with momentary stimuli, action is more rapid and the response is more specific. Consequently, hormonal and nervous co-ordination both ensure that all the organs and organ systems of the body work together in an appropriate manner and at the right moment. This enables the body to behave as a whole organism rather than as a collection of independent parts.

The effect of drugs on the nervous system

The term drug is used for any chemical which alters the functioning of the body. In medicine drugs play a vital part in saving life and combating ill health. Some drugs specifically affect the nervous system and these range from the comparatively harmless caffeine in tea and coffee to dangerous killers such as heroin, cocaine and morphine.

It is the more dangerous ones that concern us here. Their effects on the nervous system vary greatly, but they may be divided into:

1. Sedatives or depressants, e.g. alcohol, barbiturates and other sleeping pills.
2. Stimulants or pep pills, e.g. amphetamines.
3. Mood changers and those which cause hallucinations (illusions), e.g. cannabis and LSD.
4. Pain killers, e.g. heroin and morphine.

While some of these drugs, when given under medical supervision, are very beneficial in treating certain conditions, their use can be greatly abused and much harm may result. Normal, healthy people do not need drugs.

Some drugs temporarily produce pleasurable sensations and many people experiment with them for that reason. This is dangerous because it may lead to **dependence** on that drug or to experimentation with much more lethal kinds. A person is incapable of knowing whether he is likely to become dependent until this state has been reached. It is then often very difficult for a cure to be effected. You may know people who have become dependent upon drugs such as nicotine and alcohol and are unable to give them up.

Dependence on a drug may be of two kinds:

1. Psychological dependence

This may have a social origin, for example when a person depends on a drug in order to fit into a particular kind of life. It is quite natural for adolescents to feel over-anxious, shy or incapable of behaving like others, and this can lead to dependence on drugs to help. When the effect wears off a marked reaction is suffered with a feeling of need for more. Dependence may also occur when a person comes to rely on a drug to provide enjoyment. This is characteristic of people who have not learnt to enjoy normal, healthy pleasure or make satisfying friendships. They crave for the extreme sensations which some drugs produce and after stimulation find life utterly boring. This may lead to vandalism and violence. Others become dependent because they cannot face up to personal problems and use drugs as a way out.

2. Physical dependence

This occurs when a person becomes dependent on a drug because, without it, painful physical symptoms develop and these become worse and worse. This is characteristic of the heroin addict.

The effects of some of the common drugs can be summarised as follows.

Alcohol

There are several fallacies commonly held about the action of alcohol. It is not a stimulant, but slows down or inhibits brain action. It does not give you strength, although it may give you the illusion of it. It does not help you to concen-

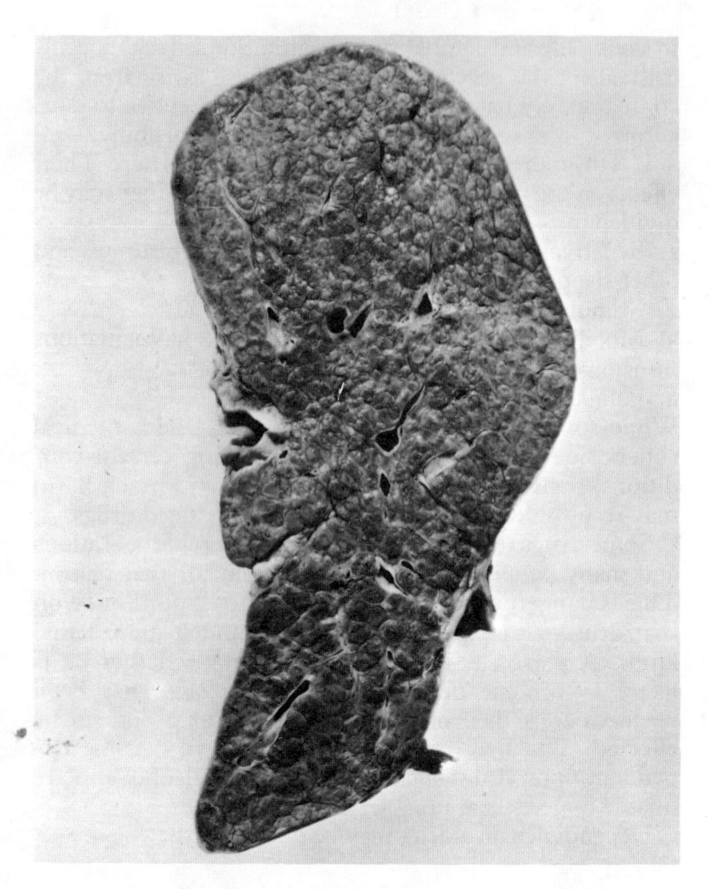

Fig. 29:8 Section of human liver showing cirrhosis—a condition resulting from excessive drinking of alcohol. The normal liver cells are destroyed and replaced by fibrous tissue.

trate, but makes you more muddled. It does not make you warmer although it makes you *feel* warmer.

Its main action is to affect the higher centres of the brain. This may have the effect of removing shyness and self-criticism, and by removing inhibitions lead on to a feeling of bravado and a consequent loss of self-control. The sense of judgement is greatly impaired and reactions become slower. Alcohol also affects the regions of the brain concerned with speech, which becomes slow and blurred in consequence. Co-ordination of movement is impaired and this accounts for the staggering gait of the drunk. Alcohol is absorbed quickly through the stomach wall and when large quantities are consumed over a short period unconsciousness and sometimes death may result.

There is a considerable variation in the effect that a given amount of alcohol has on different people. Those not used to it may be affected to a greater degree; this is also true if it is taken when the stomach is empty.

The habit of drinking alcohol can lead to addiction. People dependent on it are called **alcoholics**. There are about 500,000 known alcoholics in Britain alone and probably another half million who are in need of help over dependence on alcohol.

The social implications of too much alcohol include:

1. A breakdown of human relationships, often resulting in broken homes and mental illness.
2. An increase in the number of road accidents. Even a small amount can affect judgement and speed of reaction and, although individuals differ in their reactions to quantities of alcohol, all are seriously affected by a level of 0·08% in the blood. This is the level above which people can be convicted for driving under the influence of alcohol. The World Health Organisation (WHO) estimates that alcohol is the most important cause of road accident deaths for young people. In 1980, one third of the drivers killed in road accidents had illegal alcohol levels in their blood.
3. An increase in the number of unwanted pregnancies due to the breakdown of inhibitions and self-control, and the greater liability of seduction.
4. An increase in violence, anti-social behaviour and crime.
5. Maternal misuse of alcohol is an important cause of birth abnormalities and mental handicaps. Recent research has also shown that taking only a little alcohol daily may precipitate a miscarriage.

To illustrate the seriousness of the problem: in 1980 the Minister for Health stated that in the previous 10–15 years in Britain, deaths from alcoholism had trebled, deaths from cirrhosis had risen by a third and admissions to mental hospitals for treatment of alcoholism had doubled for men and trebled for women, with girls under 25 years old showing the fastest increase of all. There were also more than 100,000 drunkenness offences a year in England and Wales.

Nicotine

We have referred to the effect of nicotine in tobacco smoke on the circulatory system and its consequent effect on heart disease (p. 99), but as a drug it also affects the nervous system, acting both as a stimulant and a depressant according to circumstances. Dependence on nicotine is easily acquired, and most smokers experience difficulty in giving up the habit.

Barbiturates

These are sedatives and are used as sleeping pills. In larger doses their effects are very similar to alcohol, slowing down the brain's action and affecting judgement and co-ordination.

The amphetamines

These are the usual ingredients of pep pills. They cause nervous excitement and sleeplessness, enabling a person to avoid having to sleep, even if very tired. However, after the effect has worn off, it is followed by extreme fatigue, depression and irritability. Habitual use leads to lack of concentration, poor health and inability to cope with a job.

Cannabis (or **pot** or **marijuana** or **grass**)

This affects the mood of a person. It acts at first as a stimulant, producing a feeling of elation; self-confidence is increased and this often leads to irresponsible action. Different people react to the drug in various ways; usually there is a change in perception, so that tastes appear different, time appears to slow down, colours may appear more vivid and still objects may appear to be moving rhythmically. After a time its effect changes and it acts as a depressant, producing drowsiness and sleep.

Research workers differ in their conclusions as to whether

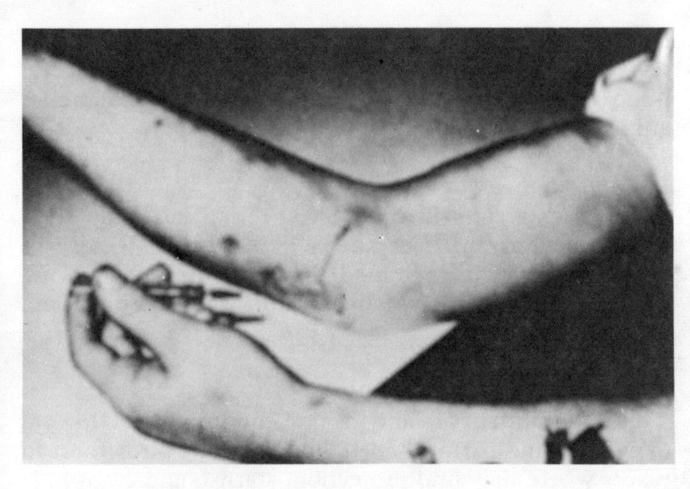

Fig. 29:9 A heroin addict injecting himself.

cannabis damages the brain cells permanently or not. One of its dangers is that it may lead to the taking of much more serious drugs such as LSD or heroin.

LSD (or acid)

This is a very dangerous stimulant which may damage the brain permanently if taken regularly. It causes hallucinations (illusions) which have resulted in murders and suicides. It may cause grave emotional reactions and lead to serious mental illness.

Heroin

This drug may be used by doctors as a painkiller in severe cases as it inhibits the pain centres of the brain. It is taken by mouth or by injection. Its misuse is extremely dangerous as it is highly addictive and as the body becomes used to it, more and more is needed to satisfy the craving. If the dose is delayed 'withdrawal symptoms' occur; these include pain in the limbs and abdomen and violent twitching accompanied by hallucinations and nightmares. These conditions are so unpleasant and alarming that the addict becomes more and more afraid of not being able to get the next 'fix' to relieve the symptoms for another brief period. This process frequently leads to criminal behaviour in an attempt to obtain more heroin.

Addiction is a tragic condition and leads to complete breakdown of normal life; typically, death occurs in the mid-thirties. Heroin addiction in the Western world has risen considerably in recent years due largely to the great increases in the quantity of heroin smuggled into Western Europe and America from countries where the poppy, from which it is extracted, is grown on a large scale.

In Britain, in 1981, there were 5,000 registered addicts but the total was considered to be nearer 20,000, as many did not register.

30

Behaviour

We can define behaviour as the observable reactions of an organism to changes in its environment. These changes may come from outside the body, such as changes in temperature, humidity or light intensity, and also from inside, such as feelings of hunger or pain. Think of your own reactions to these factors.

When we study behaviour we ask ourselves questions like 'What stimulus caused that response?' 'What purpose did the response serve?' 'How did this response come about, was it inherited or was it the result of learning?'

When we watch a thrush hammering a snail against a stone, we recognise at once that this kind of behaviour is useful to the bird because after cracking the shell it can peck out the edible part. When a fly takes off as soon as your hand approaches, it is immediately placed out of danger. When a lion approaches a herd of buffalo they all bunch together and face it. These examples show that animals tend to behave in ways which are favourable to their existence, i.e. their behaviour is an aid to their survival.

In order to survive, an animal has to carry out various vital activities such as feeding, reproducing and protecting itself; consequently, attempts have been made to classify behaviour patterns according to the functions they serve. Nine categories have been described, although the simpler animals do not carry out all of them:

1. Eating behaviour

This varies greatly in different species; dogs gulp their food; other mammals crush their food before it is swallowed. Bees suck up their food, flies lick it up. Anteaters use their tongues, sharks use their teeth; we use a knife and fork.

2. Shelter-seeking behaviour

This is the tendency to find the best conditions and to avoid dangerous and harmful ones. For example, birds roost in places where they find protection, starfish and crabs hide under boulders and shrimps bury themselves in sand. We like to take the chair nearest the fire, or put up an umbrella if it rains. In driving snow cattle will move steadily into it until they find shelter behind a hedge or wall. In contrast, most breeds of sheep move with their backs to the snow until they reach a hedge and so are more likely to be buried in snow drifts than cattle.

3. Aggressive behaviour

This includes fighting and **competing for dominance**. Stags show this at the mating season when they clash their antlers with those of rival stags and try to force them off their territories. Many birds show it when they defend their territories by displaying (e.g. robins puff out their red breasts) or flying at intruders and chasing them away. We show it when we compete in various sports, lose our tempers and fight wars.

Fig. 30:1 Gannet colony on Grassholm, South Wales.

4. Sexual behaviour
Many animals have elaborate courtship patterns of behaviour. The peacock erects its strikingly-coloured tail, bower birds decorate their nests with flowers, spiders and scorpions perform dances and newts vibrate their tails. We make ourselves more conspicuous in dress and conduct in order to attract the opposite sex.

5. Care-giving behaviour
Many animals protect their young in various ways. Birds build nests, incubate the eggs and feed the young; mammals suckle their young, protect them from predators and bring them food. We build houses for our families, protect our children, find food for them and teach them.

6. Care-soliciting behaviour
This is asking for something which is wanted or needed. Young birds show this type of behaviour when their parents return to the nest to feed them. They stretch up their necks, open their mouths wide and display their gape which is often brightly coloured and so attracts attention; at the same time they often make loud noises. Similarly, a human baby cries to attract its mother's attention and older children pester their parents for sweets or ice creams.

7. Eliminative behaviour
This relates to the elimination of waste products; such behaviour is seen when cats dig holes for their faeces and parent birds remove the faecal pellets of nestlings in their bills, thus preventing the fouling of the nest.

8. Contagious behaviour
This is sometimes described as mutual mimicking; it occurs when two or more of a species do the same thing. This includes the flocking behaviour of birds, shoaling in fish and herding in cattle. We notice it when one person yawns and others do the same. Watch a group of two or three people talking together and see how unconsciously they imitate each other by such movements as folding the arms. It becomes more obvious among spectators at football matches.

9. Exploratory behaviour
This is the tendency to explore an unfamiliar habitat. When a mouse is put in a cage it will quickly explore every part of it. A dog or cat will do the same if put in a room it has not been in before. It is very marked in us—we show great curiosity. When we go to a new place for a holiday we want to investigate the whole area. Our desire to travel, climb mountains and do research are other examples.

Social behaviour is the term used to describe any of the above categories when two or more individuals of the same species are involved. This becomes very marked when many members of a species live together in large social groups or societies such as the ants, bees and wasps—not forgetting ourselves.

Keeping this classification in mind, you would find it interesting to analyse your own behaviour during a day; try to decide into which category each main behaviour pattern comes.

We will now study one aspect in more detail—shelter-seeking behaviour.

Shelter-seeking behaviour

In a forest the animals are distributed throughout the habitat. An experienced naturalist would look in a definite place for a particular species. He would look underneath leaves of a certain kind of tree for one, under the bark for another, in the soil for something else, under an old log, in the leaf litter, amongst the mosses and lichens, and so on. One of the reasons why animals are found in certain places is because they exhibit shelter-seeking behaviour. If you find a caterpillar on the underside of the leaf and put it on the upper surface, it will crawl back again; lift up a log or a stone and many creatures which are at first visible will soon disappear from view. This is because they react to such factors as light and move until they are in darkness again. There are many other factors which influence their directional movements, such as humidity, wind, temperature and scent. Usually, an animal's behaviour is determined by more than one of these factors acting together. If you come across a fox or badger hole in a bank or wood, you may see flies constantly going in and out of the entrance; this is a good sign that an animal is living there. They fly *in* because they are attracted to the smell of the animal, but in doing so they get into the dark; this causes them to fly *out* again towards the light, hence they shuttle to and fro!

We can now investigate for ourselves why an animal such as a woodlouse is to be found in leaf litter, under stones or logs or under loose bark. These are all damp, dark places, so we might investigate both of these factors.

1. The effect of humidity
Use the apparatus shown in Fig. 30:2, or devise something similar: it is known as a choice chamber. The humidity of the two chambers can be altered by putting water under one chamber and calcium chloride or silica gel under the other (these chemicals absorb water).

Place a strip of dry (blue) cobalt chloride paper in each and put the sheet of glass on top. When the strip in the chamber above the water turns pink you will know that a distinct difference in humidity within the two chambers has been reached. Now quickly introduce an equal number of woodlice into each chamber—about 5 will do—by slipping the glass to one side just enough to put them in; in this way very little mixing with the outside air will occur. In order to eliminate the possible effect of light on the behaviour of the woodlice, place the chambers in such a position that they receive the same amount, i.e. they are at the same distance from the main light source.

Every minute record the number of woodlice in each chamber over a period of 10 minutes. Also try to estimate whether the woodlice are more active in one than in the other. Think how you could do this accurately.

What conclusions can you draw from this experiment?

2. The effect of light
Carry out a similar type of experiment to show the effect of light and darkness, by shining a bench lamp on one chamber and covering the other with a black cloth. How would you eliminate the heating effect of the lamp? Keep the humidity the same in each and quickly count the number of woodlice in each chamber at minute intervals as before. What do you conclude from your results?

In terms of activity, it has been found that animals often move faster when experiencing adverse conditions. This enables them to find favourable conditions more quickly.

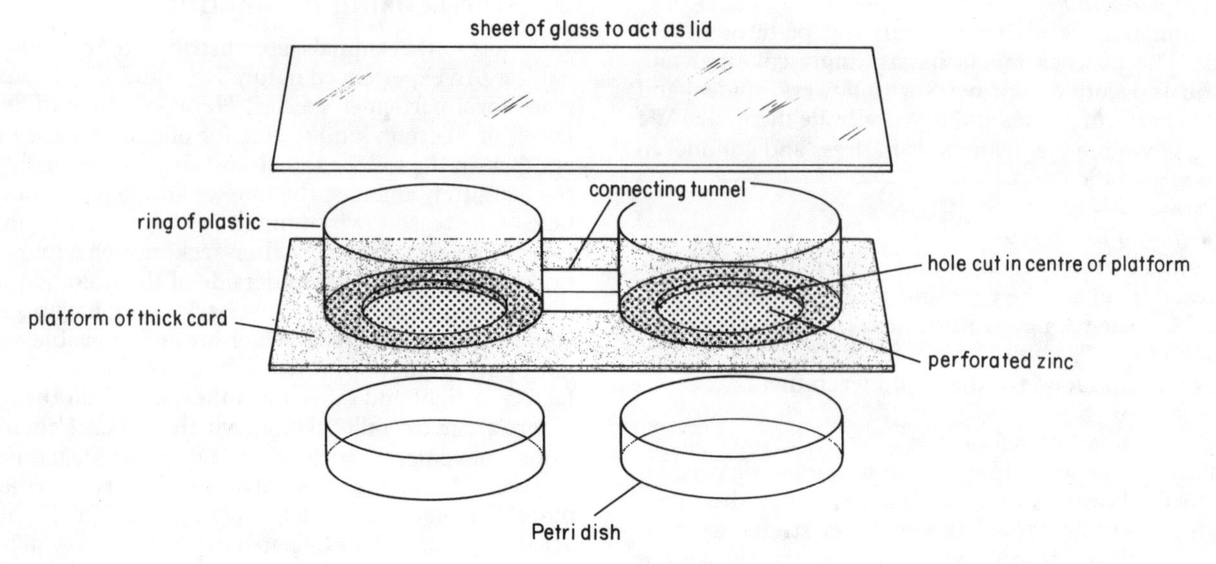

Fig. 30:2 Choice chamber. Component parts have been separated for clarity.

As well as, light and humidity, there could, of course, be other influential factors, and some might be more important than others in producing a response in a given situation. One of these other factors is contact. Earwigs, millipedes, centipedes and woodlice are often found in similar situations and react to light and humidity changes in a similar way, but once in a dark and humid place they often stop at a place where much of their body surface is in contact with something firm. This is why you often find them in cracks or tunnels. Woodlice and earwigs are often found clustered together; could this be due to contact or because of the greater humidity resulting? All sorts of questions come into your mind when you try to analyse behaviour patterns! What factors do you think cause trout to face upstream, salmon to find their way back to a particular river and fleas and mosquitoes to find a victim?

Sign stimuli

Not all rigid behaviour patterns are responses to such factors as light, temperature or humidity. Some are the result of sign stimuli. For example, a young herring-gull chick will peck at the red spot on its parent's bill, causing the mother to regurgitate food from its crop. If you present to a chick a gull's head with a red spot on it, cut out roughly in cardboard, it will peck at the artificial spot just the same (Fig. 30:3). Experiments have been done to discover the chick's reactions when the colour of the spot and the shape of the head and beak are varied, and to find out which colours and shapes act as the best sign stimuli.

David Lack, a pioneer in the field of bird behaviour, described in his book, *The Life of the Robin*, how he investigated the territorial behaviour of this species. He had watched robins displaying and chasing other robins off their territories and he wondered how a robin would react if he placed a stuffed one on its territory. So he wired a stuffed specimen to a branch near a robin's nest which contained young. When the parent returned and saw the stuffed one, it flew towards it and displayed its red breast—getting no response, it attacked it fiercely and pecked it to pieces! Was it the red breast that caused the robin to attack? To find out, Lack repeated the experiment using another stuffed robin, but this time he covered over the red feathers with brown paint. The live bird took no notice of it! Surprisingly, he found that even an isolated bunch of red feathers would cause a robin to attack. So it was the red colour that acted as the sign stimulus which initiated the attack.

Do you see any significance in the fact that *young* robins have no red breast?

Sticklebacks also use sign stimuli. At the breeding season the male develops a red underside which acts as a deterrent to other males which swim near. If the intruder persists it will be attacked and driven off.

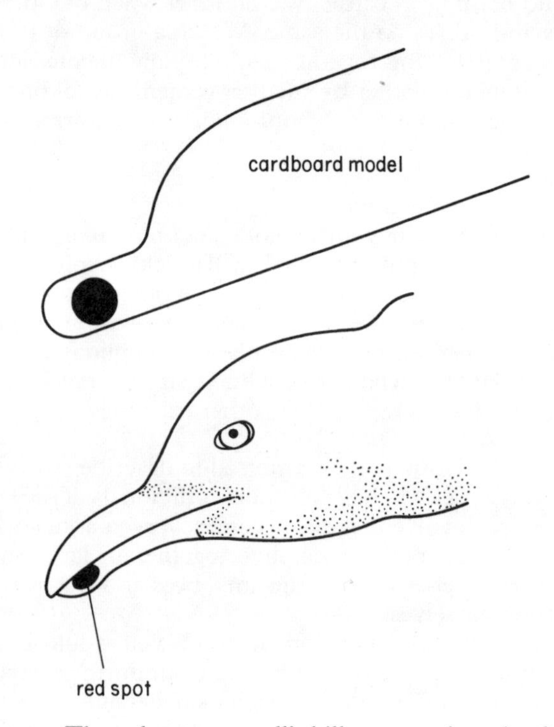

Fig. 30:3 The red spot on a gull's bill acts as a sign stimulus to a young chick which then pecks at it to obtain food. Cardboard models of various shapes have the same effect.

If you have in an aquarium a male which is in breeding condition, you can test this reaction for yourself by putting a mirror in the water. What happens when it sees its own image in the mirror? You could also make a simple model of a fish and paint its underside red and dangle it on a wire in front of the live male and note its reactions. If sticklebacks in breeding condition are unobtainable you could investigate in a similar way the reactions of such tropical fish as Siamese fighters.

Niko Tinbergen, a leading authority on animal behaviour, and other research workers were helped by experiments such as these to discover that the whole process of reproduction in the stickleback could be seen as a series of reactions to sign stimuli, each one triggering off the next. The main steps are as follows:

Once its territory is secure the male stickleback constructs a nest. It makes a shallow pit in the sand, collects small pieces of weed and glues them together by means of a sticky secretion. It then forces its way through this nest, so moulding it into the shape of a tunnel. If a female swims near, her bulging abdomen (due to the eggs inside) stimulates the male to perform a zig-zag dance round her, thus displaying his red underside. If the female is ready to lay she responds by curving her head and tail upwards. This behaviour stimulates the male to swim to the nest, causing her to follow. The male then prods the entrance with his snout, causing the female to push past him into the nest. The sight of the female's tail projecting from the nest causes the male to prod at it with his snout; this causes her to lay. When she swims off, he enters and fertilizes the eggs. The male then guards them persistently until they hatch, fanning the water with his tail; this helps to keep the eggs aerated. After they hatch, the male guards the young and if they stray sucks them into his mouth and spits them back into the nest (Fig. 30:4).

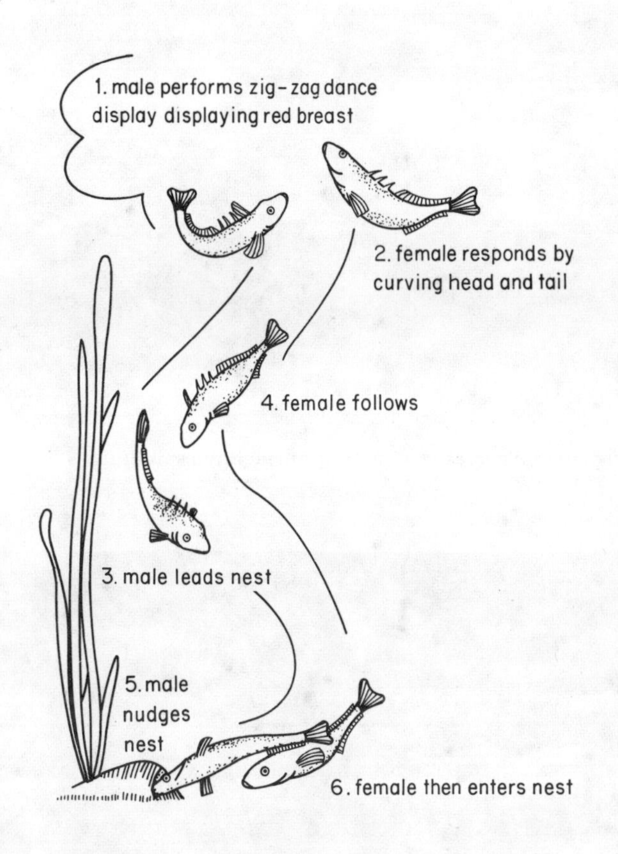

Fig. 30:4 The succession of sign stimuli used during stickleback courtship.

Instinctive behaviour

Behaviour patterns such as these are not learnt but, in some way not understood, are passed from one generation to the next; they are thus said to be **innate**. Such reactions are characteristic of what is often called **instinctive behaviour**. However, there is no animal which shows no variation in its behaviour; all appear to be able to learn from experience to some extent. Spiders are good animals to study in this respect.

Find a specimen of the common house spider *Tegenaria* sp. It is the large one which you sometimes find in baths or sinks. You can often find these spiders in sheds, cellars or among logs. Put one in a cardboard box, using a piece of glass as a lid. Put in one corner a small box with one side removed to provide a dark place for the spider to live in. After a day or so it will have made a sheet web all over the box. Now remove the glass, take a tuning fork and touch the web with the vibrating fork. The spider should rush out and attack it as if it were a fly. This is an innate response to vibrations of the web. Finding that the fork is not a victim, it will retire to its home. Repeat the procedure at two minute intervals. How long does it take the spider to learn not to react?

Fig. 30:5 Common house spider (*Tegenaria* sp.). × 3

In these examples we have shown how animals respond to a particular stimulus in a definite manner, but in many instances animals are subjected to more than one stimulus at the same time and these may be conflicting. For example,

Fig. 30:6 Aggressive behaviour: male hippos fighting.

Fig. 30:7 Feeding behaviour: lions on a kill.

Fig. 30:8 Shelter-seeking behaviour: snails found beneath a slate.

Fig. 30:9 Sexual behaviour: ostrich courtship display.

Fig. 30:10 Care-giving behaviour: monkey grooming its young.

Fig. 30:11 Care-soliciting behaviour: chicks begging from mother.

Fig. 30:12 Contagious behaviour: elephants following each other in single file when crossing a river.

Fig. 30:13 Instinctive behaviour: chick pecking at the red spot on a cardboard model of a gull's head.

if food is placed on the lawn and a cat is nearby, a hungry bird on seeing both the food and the cat is faced with conflicting stimuli—to approach and eat the food and to fly away from the cat. Clearly, it cannot respond to both stimuli at the same time. In this instance the fear reaction usually suppresses the response to food. So there must be some form of internal control which determines which

Fig. 30:14 Social behaviour: troupe of olive baboons.

pattern of behaviour is followed. However, in many other instances an animal may be stimulated strongly in conflicting ways but neither behaviour pattern becomes dominant; instead the animal does something quite irrelevant to the situation. This is called a **displacement activity**. For example, a bird when confronted with a rival may be equally stimulated to fight or fly away—in the event it does neither, but starts to preen its feathers. A cat under similar circumstances may start to lick itself. We show similar displacement activities when we have conflicting emotions—we scratch our heads, tap our fingers on the desk, or adjust our clothes.

Rigid and adaptable behaviour

The types of behaviour we have considered so far have been rigid, although each is capable of being modified to a very limited extent as a result of learning. Some animals, by contrast, rely much less on rigid, innate responses; they learn rapidly, and their behaviour is governed largely by the experiences they have built up. We can describe this kind of behaviour as **adaptable**; it is particularly characteristic of mammals, and especially man.

An animal which relies on rigid, innate patterns of behaviour has a set of ready-made answers to all the usual problems it is likely to meet, and this is true as much for the young as for the adult. This is of great survival value when conditions are normal, but when the animal is faced with unusual circumstances it may be disastrous.

An animal which relies mainly on learning is at a disadvantage when young, as it takes time to gain experience. However, adaptable behaviour is usually associated with a high degree of parental care; and this helps the young to survive until they have gained the necessary experience. Once the early stages are passed, adaptable behaviour allows an animal to cope better with unusual and difficult circumstances.

Innate or learnt?

Many animals, such as the majority of insects, never see their parents, as the latter are dead by the time they hatch and so they have no chance to learn from them; thus most of their behaviour is innate. With vertebrates there is a greater opportunity for learning. It is not always easy to know if a particular type of behaviour is innate or learnt; for

Fig. 30:15 Exploratory behaviour: badger cub emerging from its selt for the first time.

example, when a bird sings does it inherit the song or learn it? In an experiment to investigate this, various species of birds were reared in incubators and isolated completely from other members of their species. When they became mature and started to sing, the songs of some species were exactly the same, proving that these songs were innate; but in others the song was a very poor replica of the normal. However, in the latter case if the young birds were allowed to hear an adult sing just once at about hatching time, it was found that they sang the complete song when they became adult. This is an example of a type of learning called

Fig. 30:16 Defensive behaviour: the oak beauty caterpillar resembles a twig, complete with buds.

imprinting, which we will consider next.

Learning

We can distinguish several different types of learning:

1. Imprinting and early learning
Imprinting is extremely quick learning and is characteristic of some birds and mammals during a brief period when they are extremely young. **Konrad Lorenz**, an Austrian scientist famous for his studies of animal behaviour, discovered that newly-hatched ducklings accepted as their mother the first large moving object that they saw in the vicinity of the nest. Normally this would, of course, be their mother, but by raising the eggs in an incubator, Lorenz made sure that he was the first object to be seen when they hatched. Consequently, they followed him about everywhere, just as if he was the mother duck!

In the same way many mammals, if brought up artifically on the bottle, will treat their owner as if he or she were their mother. The nursery rhyme:

> 'Mary had a little lamb,
> Its fleece was white as snow,
> And everywhere that Mary went
> The lamb was sure to go'.

was biologically accurate! The period of imprinting often coincides with the time when the eyes are first becoming functional.

The learning which occurs early on in the life of a mammal is undoubtedly very important in the development of the personality of the individual concerned. Baby rhesus monkeys, for example, when brought up artificially without access to their mothers, were found to be incapable of normal relationships with monkeys of either sex when they grew up. If, when small, they were given something furry or soft to hug, they would run to this when frightened, as if it were their mother. By having this substitute mother, they partly overcame the handicap shown when they were completely deprived of their own mothers.

Does this kind of behaviour apply to humans too? Although not easy to prove because of the difficulty of making adequate controlled experiments, more and more evidence is accumulating that what happens to a child between the ages of a few months and two or three years is of very great importance. A young child needs such basic comforts as food, warmth and bodily contact from its mother or foster mother. If the child receives and can rely on these comforts when required, and is not left to cry for long periods on its own, it feels more secure and is more likely to grow up to be more responsive, more eager to learn and to have easier relationships with others. If these normal comforts are denied to a young child he is more likely to grow up into a 'difficult' personality. In other words, what a child needs most of all is to be loved and to have the security of a home and happy family background. But being loved is not the same thing as being given everything that is wanted. Wise and consistent discipline is an essential ingredient of love, as only this will lead to self-discipline later on.

2. Habituation
This is brought about when a natural response is lost.

Young animals are frightened by loud noises, but later they learn to discriminate and only react to those they have learnt are significant. If we live by a railway or busy road we soon become habituated to the noise and become unconscious of it, but a visitor from the country might find the noise most disturbing. Nocturnal animals have been filmed successfully, using artificial light which normally would have frightened them, by gradually allowing them to become habituated to increasingly intense light over a long period beforehand.

3. Conditioning

The Russian scientist **Pavlov** (1849–1936) performed many famous experiments on the learning behaviour of dogs; he used the simple reflex, whereby saliva is secreted when food is taken into the mouth, as a basis for his work. This reflex is innate, but he demonstrated that if he rang a bell every time he fed the dog there would come a time when, by ringing the bell without giving any food, saliva would be secreted just the same. In this way the original stimulus is replaced by a different one, but the response remains the same. This is the essence of a **conditioned** reflex.

4. Trial-and-error behaviour

This type of learning may be looked upon as an extension of the last. It occurs when an association is built up between a certain action and a reward or punishment. For example, young birds rapidly learn to avoid eating insects which are yellow and black once they have taken such a specimen which is either bad tasting or can sting.

Pets brought up in the home, or animals trained in a circus, associate reward or punishment with certain actions and so learn to behave in a particular way. In this manner they are taught to do most complicated actions. You will have noticed how the circus trainer will give a reward whenever the animal carries out the action successfully, but withholds the reward if it does not. The same applies to us; small children are helped to behave in a particular way by praising them or rewarding them when they do well. If we ride a bicycle recklessly and consequently have an accident, we learn to ride more carefully as a result of the experience. The expression 'learning the hard way' embraces many examples of trial-and-error behaviour. Can you think of other examples?

5. Insight learning

This is the highest type of learning and is best shown in apes and man. Insight learning involves reasoning—the situation is considered in the light of past experience and the problem is dealt with in a particular way. For example, a chimpanzee was put in a cage with a bunch of bananas hung out of reach, and some boxes were left on the floor. The ape had never been in this situation before. The chimpanzee looked at the bananas and considered the problem, then it piled the boxes on top of each other, climbed up and reached the bananas. This is an example of **intelligence**, which may be defined as the ability to organise behaviour in the light of experience.

Insight learning is of great survival value because it enables an animal to adapt to changing circumstances; it is developed to a very high degree in man and is the basis of many of his great achievements.

Behaviour and social organisation

Some animals have an increased chance of survival if they live together in societies. Consequently any aspect of behaviour which helps to maintain the organisation of the group is important. We saw on p. 56 that the social behaviour of bees is remarkably complex, the behaviour of each individual being related to the well-being of the whole colony. In this society the complex role played by each bee is almost entirely innate, for its life span is too short for it to learn all that is needed and its brain is not complex enough.

As we have seen, in many of the higher vertebrates learning about the environment through processes of habituation and trial-and-error behaviour constitutes an important part of an individual's development. In the case of social vertebrates such as jackdaws, wolves, monkeys and humans, the environment includes other members of the same species. Because individuals in such societies may vary in age, size and fitness, there is often a hierarchy, or '**peck order**'. In birds such as hens or pigeons it is linear; this means that there is a dominant bird A which can peck any other without being pecked, B which can peck all but A, C all but A and B, and so on. The unfortunate one at the bottom of the order may be persistently pecked and fail to obtain enough food in consequence. Such a system ensures that in times of hardship the strongest members of the group are the first to have access to food. An individual soon learns his position in such a hierarchy through fighting or trials of strength, but once determined the situation is usually accepted and so there is less aggression.

In mammals the relationships are seldom linear, but there are often dominant males as in cattle and deer herds, and dominant females in sheep. The dominant lion in a pride has the privilege of eating first and obtaining the best pieces; the juveniles have to wait their turn. We notice similar signs of a hierarchy in human society, for example in a family, class, school or team.

In highly developed animal societies not only do such mechanisms exist to restrict aggression, but care-giving behaviour between individuals makes a positive contribution to the harmony of the group. This is developed to its highest levels in humans with the concepts of unselfishness and love. It would seem that the greatest problem of human society today is learning how to live together; without the application of these higher aspects of care-giving behaviour civilization will have no future, as aggressive and anti-social behaviour will ultimately lead to the break-up of society.

31

Living together

Ecology

In the next two chapters we shall be studying the principles of ecology. This is a branch of biology which deals with the relationships between living organisms and their environment.

When studying the vertebrates, we saw how fish were adapted to an aquatic environment, how birds were adapted for life in the air and how mammals fitted into a variety of contrasting habitats. But fish, birds and mammals, and any other group for that matter, cannot live in isolation; they are dependent on many other kinds of organisms for their vital needs. So in every habitat there is a great variety of plants and animals all living together and, to a varying extent, dependent upon each other. Groups of animals and plants which live together in a particular habitat are called **communities**.

The size of the community varies with the size of the habitat. We talk of terrestrial communities of organisms, such as those living in deserts, forests and in the soil; aquatic communities, including those in a puddle, a lake or an ocean; there are communities to be found in a pile of dung, in the intestine of a living animal or in the rotting body of a mouse. You will be able to think of many others.

Each community is characteristic of the particular habitat in which it is living; thus the flora and fauna of a rain puddle are very different from that of a rock pool, and the life of a pine forest is very different from that of an oak forest. This is because the organisms comprising a community are specially adapted to live under these particular conditions, but in other habitats, like a 'fish out of water', they would probably not survive.

Because of this close relationship between community and habitat they are always studied together, and we refer to a community and its associated habitat as an **ecosystem**. Ecology is the study of ecosystems.

Ecology is a subject of very great importance to all of us, because the principles involved can lead us to an understanding of how man himself can live in balance with all other living things, utilising in the best way the limited resources at his disposal without destroying the environment. Man's survival depends basically upon whether he can apply the principles of ecology to himself.

Ecosystems

All life on our planet is confined to a thin envelope consisting of the atmosphere, oceans and earth's crust. This region—the world of living things—is termed the **biosphere**.

Within the biosphere there are a number of major eco-

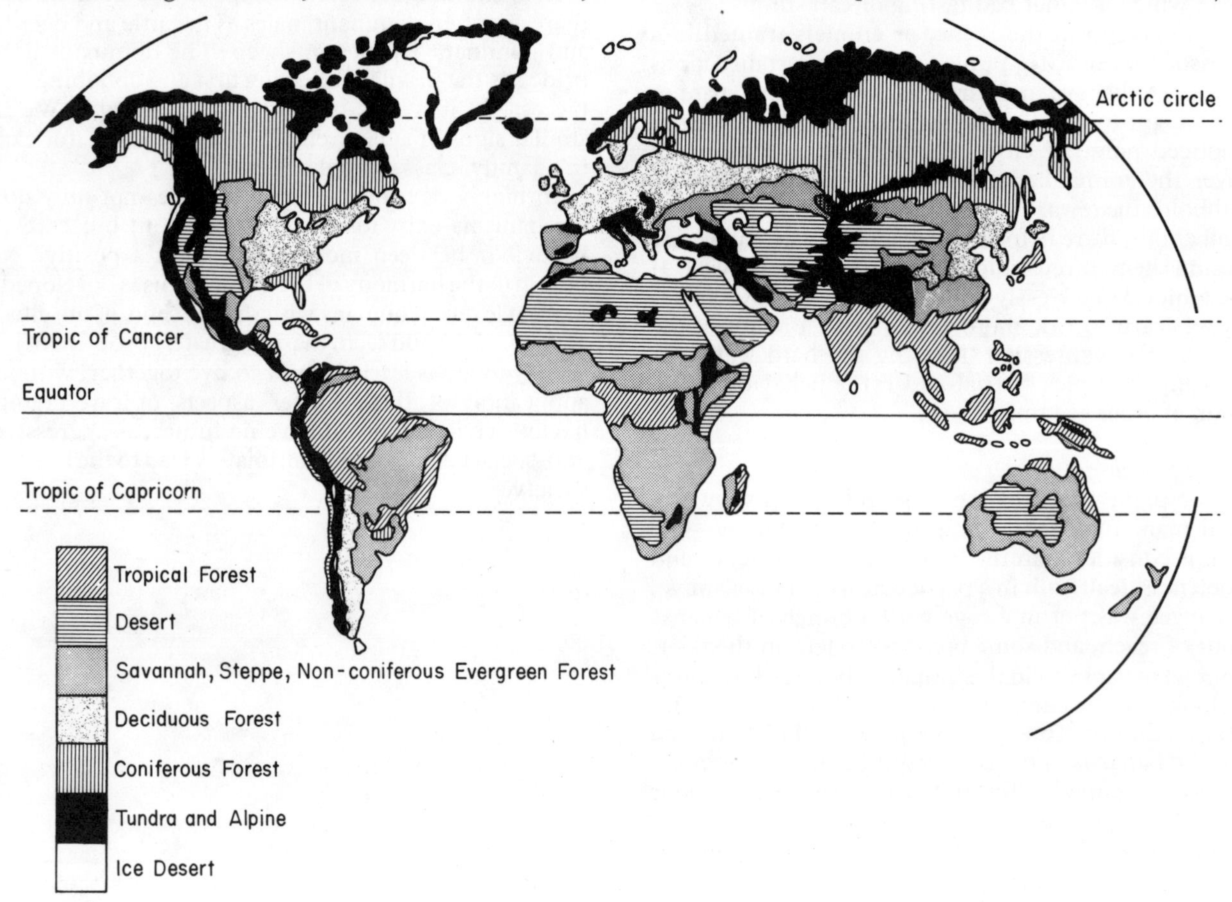

Fig. 31 : 1 Map showing the major terrestrial ecosystems.

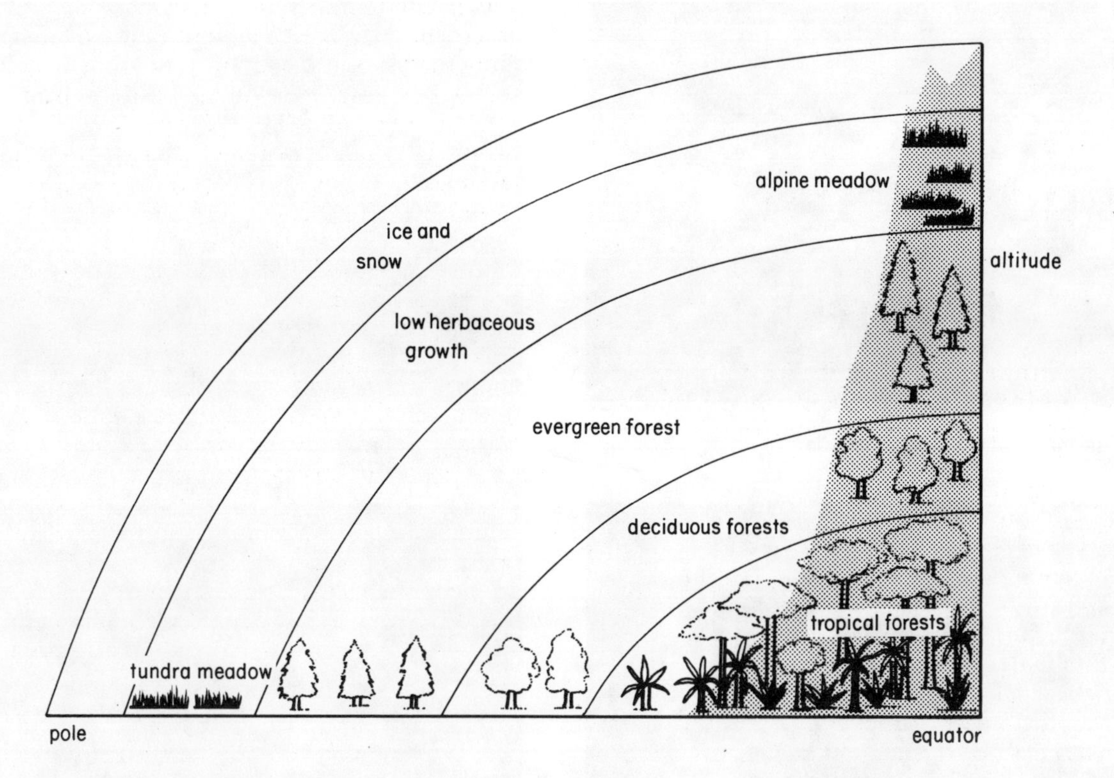

Fig. 31:2 Diagram showing how mountain climatic regions can be compared with the horizontal climatic regions of the earth.

systems, the terrestrial ones being determined largely by the variations in climatic conditions between the Poles and Equator (Fig. 31:1). In a similar way, if you climb a mountain such as Kilimanjaro (5811 m) in Equatorial Africa, you quickly go through a comparable system of ecosystems, starting with tropical rain forest at the base and ending with perpetual snow and ice at the summit (Fig. 31:2).

The main climatic influences which determine these ecosystems are rainfall, temperature and the availability of light from the sun. For instance, forests are usually associated with high rainfall, but the type is influenced by temperature and light; the same applies to deserts which occur in regions where rainfall is extremely low.

Let us consider the characteristics of some of the major ecosystems.

1. Tundra

This region, which adjoins the permanent ice of the polar region, is devoid of all trees, but stunted shrubs such as birch and sallow occur in its more southern parts. The ground flora include many lichens, mosses and sedges. The soil is frozen for most of the year, but the top layer melts during the summer, allowing a short growing season of about two months. The fauna include polar bear, Arctic fox, reindeer, Arctic hare, lemming, snowy owl and ptarmigan.

2. Coniferous forest

This region occurs south of the tundra. Here the winters are not so long and the greater summer warmth allows trees to develop. In the north they occur only in sheltered places, but further south are extensive forests dominated by spruce, pine and other conifers. The fauna include lynx, wolverine, wolf, elk, red squirrel and grouse.

3. Deciduous forest

This region lies to the south of the coniferous belt. Most of Britain lies within this region. It has been greatly reduced in size and modified by man's activities, but what remains is dominated by such trees as oak, beech, birch, ash and alder with many shrubs and herbaceous plants. The fauna include fox, badger, red deer, roe deer, mole, mice and voles.

4. Savannah

This is a tropical region dominated by grasses with scattered trees and fire-resisting thorny shrubs. The fauna include a great diversity of grazers and browsers such as antelopes, buffalo, zebra, elephant and rhinoceros; the carnivores include lion, cheetah, hyaena, mongooses and many rodents such as mice and ground squirrels.

5. Tropical forest

This occurs in the equatorial region where rainfall is heavy. It consists of lush forest vegetation with tall trees and woody vines with stems that climb up and hang down from trees, called **lianas**. The fauna include chimpanzee, monkeys, okapi, forest elephant, small antelopes, hornbills, woodpeckers and many other species of birds.

6. Deserts

Deserts are regions of very low and sporadic rainfall. With low humidity the sun's rays easily penetrate the atmosphere making ground temperatures very high, but nights are often cold by contrast as the earth loses heat rapidly. Many plants such as cacti and thorny shrubs are adapted to such conditions by having deep roots and water storage tissue. The fauna is remarkably varied. Many animals are nocturnal, avoiding the heat of the day by burrowing: they

Fig. 31:3 Tundra: Hudson's Bay, Canada.

Fig. 31:4 Coniferous forest: England.

include desert foxes, snakes and lizards and the many herbivores which form their prey, such as many kinds of insects, jerboas and kangaroo rats.

Natural variation within a major ecosystem

Although each major ecosystem has its characteristic climatic conditions and typical flora and fauna, it becomes divided into smaller ecosystems which differ markedly from the basic type because of variation in local conditions and especially because of the influence of man. We have already seen how change in altitude influences vegetation and therefore the fauna too, but there are many other

Fig. 31:5 Deciduous forest: England.

factors which cause variation in the ecosystem, such as the influence of the sea and inland waters, the wetness of the soil, the degree of exposure to wind and the geology of the area, to name just a few. You will understand the importance of some of these factors if you consider in general terms the kind of changes occurring in the vegetation when:

1. A river, such as the Nile, runs through a desert region.
2. Low-lying land is subjected to periodic flooding: a) by salt water, b) by fresh water.
3. A region is subjected to very high winds, such as on an island.
4. The land slopes steeply as in a mountainous region.
5. There is a change from a sandy to a calcareous soil.
6. There is a change in the amount of exposure to the air due to tidal effects on a rocky shore.

The effect of man on ecosystems

In many areas man has changed the natural ecosystems very greatly by damming rivers, draining marshes, reclaiming land from the sea, cutting down forests, ploughing up land and growing crops, and by building towns, cities, canals and motorways. These changes have greatly altered the communities of plants and animals living there.

Take the development of a large town, for example. There will be three kinds of change: a) Some plants and animal species will die out. b) Some will adapt to the new conditions sufficiently to survive in reduced numbers. c) Some will benefit by the new conditions and will increase in numbers. Many of these changes will vary accord-

Fig. 31:6 Savannah (with Thomson's gazelles): Serengeti.

Fig. 31:7 Tropical montane forest (with lichens): Muhavura.

Fig. 31:8 Desert with yucca cacti: Arizona.

ing to where the town is but, as you would expect there will be some organisms in the last category which benefit from being in close association with man and his buildings, such as sparrows, starlings, pigeons, rats, mice and many plants we describe as weeds. Can you add to this list and think out which plants and animals in your area would come into the first two categories?

Interactions within the ecosystem

These interactions are complex, but the general principles are illustrated in Fig. 31:9.

1. The physical environment is determined mainly by the

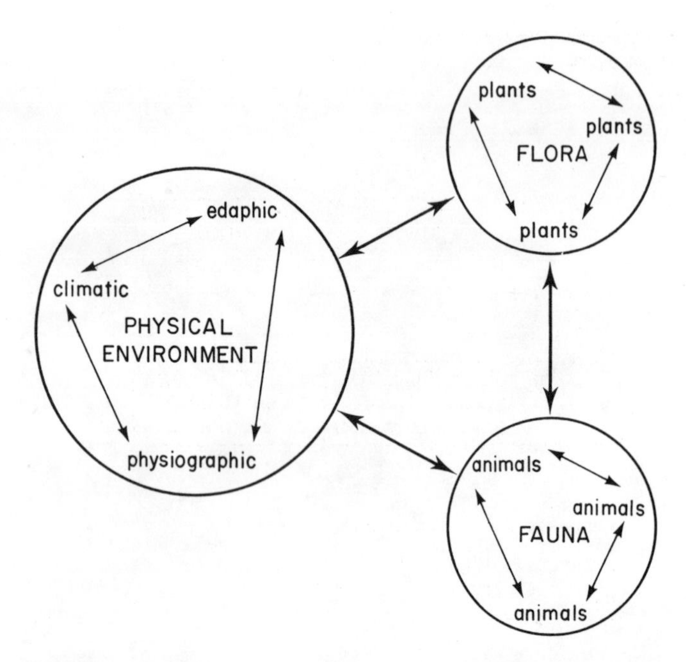

Fig. 31 : 9 Diagram illustrating the various interactions within the ecosystem.

interaction of three kinds of factors: a) **climatic**, e.g. rainfall, temperature, light; b) **edaphic** (soil), e.g. type of soil, water content of soil, c) **physiographic**, e.g. altitude, slope.

2. The physical environment interacts with the flora and largely determines its composition. We have already mentioned that rainfall and temperature are particularly important in determining the presence of forest or desert, but this is a two-way relationship as the flora itself can affect the physical environment; for example, the presence of forest may increase the rainfall because of the high rate of transpiration from the leaves.

3. The physical environment also affects the fauna. In the Arctic, for instance, the intense cold restricts the number of terrestrial species to warm-blooded birds and mammals. On the sea-shore the distribution of a number of species is affected by the strength of the waves. The fauna can also affect the physical environment; for instance, when hippopotamuses are too numerous in parts of Africa, their grazing is so severe that much soil erosion takes place during tropical rainstorms. You will also remember how the activities of earthworms can alter the nature of the soil (p. 28).

4. The flora interacts with the fauna in a great many ways. All animals are dependent upon green plants for food and oxygen, either directly or indirectly, and the species of plants which are present will determine the kinds of animals which can live there. As well as using it for food the fauna will also affect the flora through factors such as trampling, manuring and pollination. These relationships will be treated more fully later.

5. There are also interactions between the plants themselves. There will be competition between species for light and mineral salts; some species will act as parasites on other plants, while species such as those which form lichens will live together in symbiosis (p. 245).

6. Finally, there are many interactions between the animals present: between predator and prey, parasite and host and through many aspects of competition both between different species and between members of the same species.

The last three categories listed above are described as **biotic** factors as they concern the interactions of living organisms within the ecosystem.

Food relationships

All living organisms need energy. The ultimate source of all energy is sunlight and only one type of organism can harness this energy for its own use—the green plant. This can be done because chlorophyll absorbs the light energy which is used to build up sugars in photosynthesis. In addition, the green plant, with the help of nutrients from the soil, can build up more complex substances such as proteins. All these materials synthesised by green plants become the *only* source of food for non-green plants and animals, directly or indirectly. It is this food that becomes the source of energy for all these organisms and which provides material for their growth. Thus green plants in any ecosystem are called the **producers** and the remaining organisms which are all dependent upon them are the **consumers**.

Animals which feed directly on plants are called **first-order consumers** (herbivores); elephants, antelopes, rabbits, snails, caterpillars and aphids all come into this category. Within their bodies the food is digested, absorbed and built up into their own tissues, which then become the food of the animals which feed on them; these are the **second-order consumers** (carnivores). They range in size from lions which feed on antelopes, foxes which prey on rabbits, warblers which eat caterpillars to ladybirds which feed on aphids. **Third-order consumers** may be present in some ecosystems; they feed on the smaller of these carnivores, e.g. the hawks which prey on warblers. Thus organisms can be classified ecologically according to what **trophic** (feeding) level they belong.

These relationships based on feeding habits are called **food chains**, e.g.

Oaktree→aphid→ladybird→warbler→hawk.

Grass→antelope→lion.

But these links are never as simple or rigid as the word 'chain' suggests. For example, aphids are eaten by many insectivorous birds in addition to warblers, and also by ladybirds and other insects; hawks, on the other hand, prey upon a considerable variety of birds and small mammals. So the term **food web** is often a better one to use when being precise, as it suggests a far greater number of possible links and reflects the fact that the whole community is a complex inter-connected unit. Thus the original energy from the sun flows through the whole ecosystem from one trophic level to another.

Figure 31 : 10 is a much simplified diagram to show some of the feeding relationships amongst organisms living in deciduous woodland. You will see from the diagram that animals fit into special positions within the food web; each is described as its **niche**. For instance, there is a niche for insects such as aphids which suck up the juices of leaves by means of a proboscis; another niche for insects such as

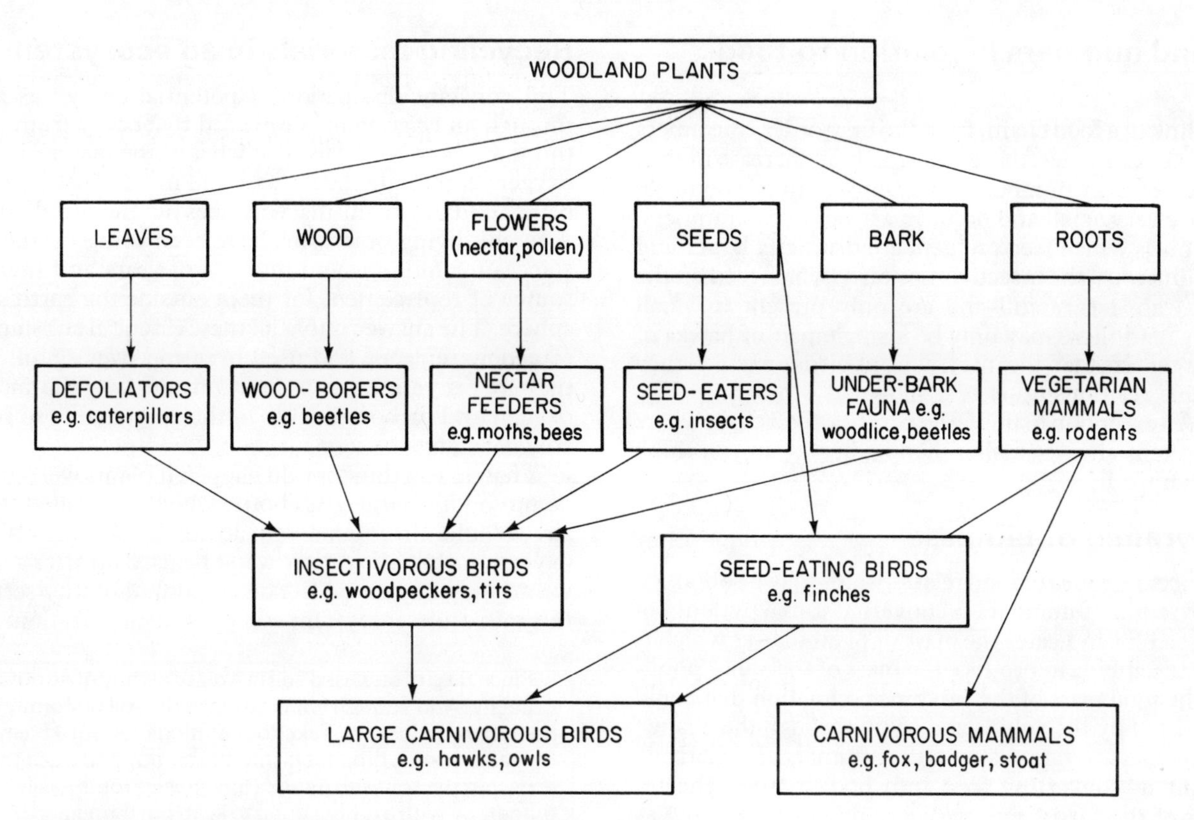

Fig. 31 : 10 A food web characteristic of a woodland ecosystem (simplified).

caterpillars which have strong jaws for biting off pieces of leaf; and a niche for relatively large animals such as deer which browse on the vegetation. All these animals feed on leaves but they differ both in size and in the manner in which they feed. So the term 'niche' denotes not only the animal's position in the food web and what it eats, but also its mode of life. Just as a habitat is the place where an animal lives, so a niche describes its occupation—the way it 'goes about its business and earns its living'.

Food chains involving scavengers

When plants and animals die or produce waste substances, their products form the food of other species which are known as **scavengers**. Thus dead plant material, **humus**, may be eaten by earthworms or woodlice; dead animal matter, **carrion**, by maggots, beetle larvae or in some ecosystems, vultures; and the dung of animals by flies or dung beetles. These scavengers in turn become the prey of carnivorous birds and mammals, which again may be eaten by still larger carnivores.

Much dead material is not eaten but decays, due to the activity of fungi and bacteria—these are described as **decomposers**; they play an important part in returning valuable material to the soil in a form which can be utilised by green plants once more (p. 245).

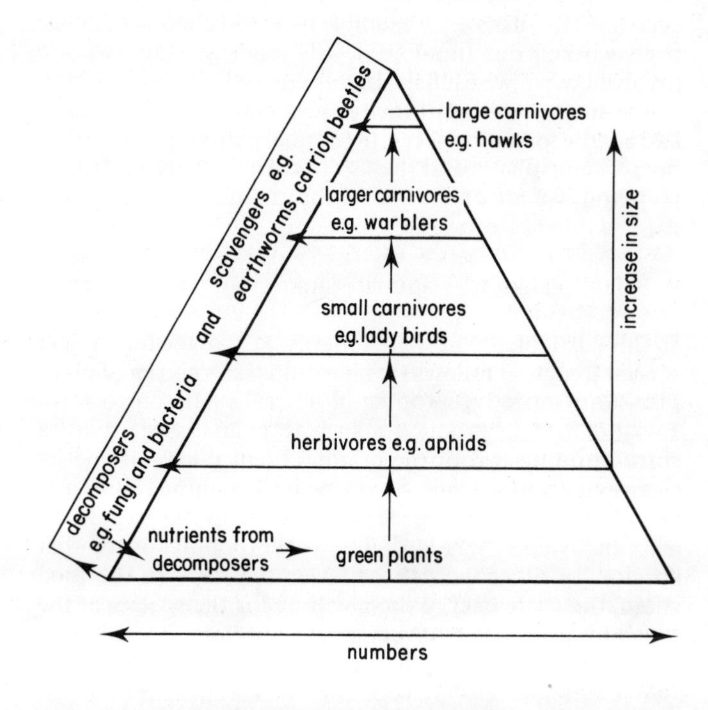

Fig. 31 : 11 Diagram to illustrate the principle of the pyramid of numbers and the passage of energy from one trophic layer to another. N.B. The **number** of green plants is not necessarily greater than that of herbivores but their biomass is.

Food chains involving parasites

When studying the large white butterfly (p. 48) we saw how their caterpillars were parasitised by the ichneumon *Apanteles*. The latter is the start of a parasite chain as *Apanteles* may be parasitised by another ichneumon, *Hemiteles*, and these in turn by a chalcid wasp.

Many food chains involving parasites are more simple. For example, the ticks which feed on the blood of buffaloes are themselves eaten by oxpecker birds which may in turn be eaten by a predatory bird such as a hawk.

Size and numbers in relation to food chains

At each link in a food chain, from the first-order consumers to the large carnivores, there is normally an increase in size, but a decrease in number. For example, in a wood, the aphids are very small and occur in astronomical numbers, the ladybirds which feed on them are distinctly larger and not so numerous, the insectivorous birds which feed on the ladybirds are larger still and are only present in small numbers, and there may only be a single pair of hawks of much larger size than the insectivorous birds on which they prey. This relationship is best shown as a pyramid (Fig. 31:11). An exception to this is the parasitic food chain where the parasite is always smaller than its host. Can you think why this must be so?

The pyramid of biomass

A more accurate idea of food relationships may be obtained if this pyramid of numbers is converted into a pyramid of **biomass**. This indicates the mass of plant matter which is used by the aphids to produce the mass of the aphid population, the total mass of the ladybird population that could be supported by the aphids and so on through the chain.

The biomass pyramid shows that animals are relatively inefficient in converting food into body tissues, the remainder of the food being undigested and passing out as waste, or broken down in respiration to supply energy for such activities as feeding. Many animals convert no more than 10% of their food into body tissues, some herbivores even less. Let us take an example of a food chain which has been worked out in some detail—one in which we are involved when we eat fish. In this chain the plant plankton in the surface waters of the sea trap energy from sunlight and are the food producers; the animal plankton feed on the microscopic plants and the fish in turn feed on the animal plankton; we are at the end of the chain when we eat the fish:

plant plankton → animal plankton → fish → man.

In this particular food chain roughly 90% of the food is lost at each step, so it follows that it would take 1000 kg of plant plankton to produce 100 kg of animal plankton to form 10 kg of fish to produce 1 kg of human tissues, with a corresponding loss of the original plant potential energy that came from the sun. So the nearer an animal species is to the original plant source in a food chain the greater the amount of energy is available to the population of that species. In other words, the fewer the steps in the food chain, the more energy there will be for the species at the top. This principle is of very great importance to man with his problem of feeding the world's expanding population.

You can now see why vegetable food is cheaper than fish or meat, because it is at the bottom of the food chain, and why it is so wasteful to turn fish into fishmeal for cattle so that people can eat beef rather than eating the fish direct. The principle also explains why it is more economic to breed vegetarian fish for human consumption rather than carnivorous ones such as trout.

Recycling materials in an ecosystem

This constant dissipation of potential energy as it flows through an ecosystem is replaced by energy from the sun through photosynthesis. But what of the materials such as oxygen, carbon dioxide, water and mineral salts which are being used by organisms; why does not the supply run out? After all, living organisms have been using up these substances for millions and millions of years and there is no source of replacement for them outside the earth's atmosphere. The answer is that all these essential substances are somehow replaced and used over and over again; that is, they are **recycled**. Let us first consider why the amount of oxygen and carbon dioxide in the air remains so remarkably constant.

What do you think would happen if plants were grown in damp soil in a large glass bottle which was sealed from the air? Would the plants remain alive? Would either the oxygen, carbon dioxide or water be used up after a time, or would they be recycled? You could find out by setting up this miniature ecosystem as a class demonstration:

> Place 5 kg of sterilised soil in a glass carboy of about 25 litre capacity. Add enough water to make the soil uniformly damp, but not waterlogged. Take four cuttings (about 15 cm long) of *Tradescantia*, drop them into the carboy and push their cut ends into the soil with a stick (this species roots easily). Close the carboy with a rubber bung to keep out the air and place the apparatus in good light, but not direct sunlight. Write the date on the carboy at the start of the experiment. Note whether the plants grow and, if possible, how long they live.

'Bottled gardens' of this kind sometimes flourish for many months and sometimes years. Attempt to explain how recycling is going on in a 'bottled garden' by considering the processes of photosythesis, respiration, water absorption and transpiration. Why is it important not to put the apparatus in full sunlight?

We will now study in more detail some of these recycling processes, considering the whole world as an ecosystem.

The recycling of water

Over a long period of time there is a rough balance between the water which is being precipitated as rain, dew, hail or snow from the atmosphere and that which is evaporated from the surface of the land, oceans and other bodies of

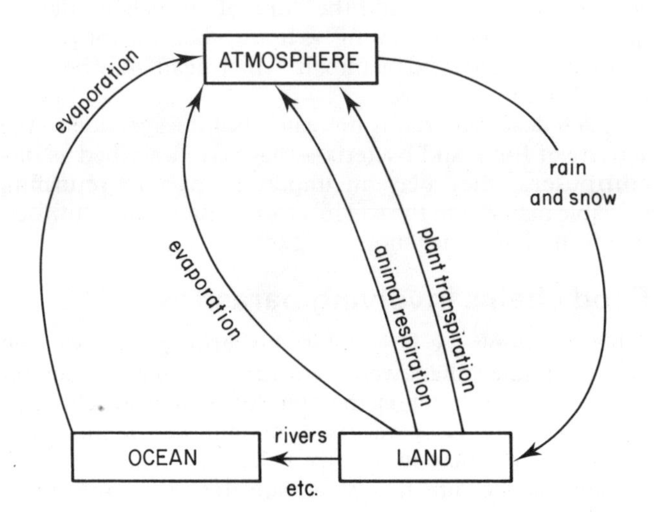

Fig. 31:12 Diagram showing the recycling of water.

standing water. A variation of this cycle occurs when water is taken in by plants and animals and passed on to others through food webs, or given out once more in transpiration, excretion, respiration and other processes. But again there is a balance between what is taken in and given out over a long period of time (Fig. 31 : 12).

The recycling of oxygen and carbon

The two major processes involved in recycling these substances are photosynthesis and respiration. In photosynthesis oxygen is evolved as a by-product and this is the only source of oxygen which can be used for the respiratory break-down of food by both plants and animals. Over millions of years a balance has been achieved, between the amount of oxygen released by green plants and the amount used up in respiration and combustion. The same happens on a small scale in a 'bottled garden'.

The carbon dioxide in the atmosphere is the source of carbon for all plants and animals. The only method of extracting it is through photosynthesis, when it is built up into carbohydrates and later into other organic compounds such as fats and proteins. These substances are passed from the producers to the consumers through the food webs. At each stage they are digested and the products are built up into the organic substances of the organism's own body. But how is carbon dioxide passed back into the atmosphere to keep up the supply? There are three main processes which bring this about:

1. During the respiration of plants and animals some organic substances are broken down for energy release and carbon dioxide is given out.
2. During decay, when these organic substances are broken down by bacteria and fungi, there is a release of carbon dioxide.
3. During the combustion of carbon compounds such as wood, peat, coal and petroleum. We can think of these substances as reservoirs of carbon stored up over varying lengths of time, in the case of coal and petroleum, for many millions of years.

One of the problems facing man today is whether this cycle is being upset by the enormous amount of fuel combustion that is taking place to provide energy for industry and transport—particularly the motor car. These processes are putting back immense quantities of carbon dioxide into the atmosphere, the long term consequences of which are uncertain (p. 261).

The recycling of carbon is summarised in Fig. 31 : 13.

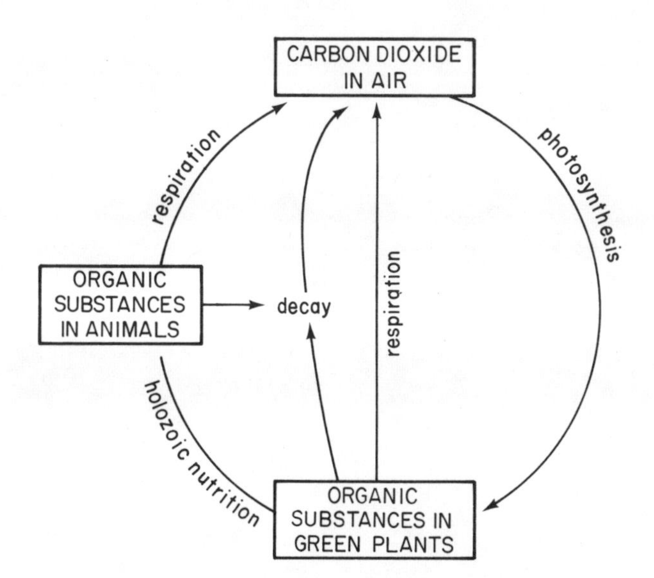

Fig. 31 : 13 Diagram showing the recycling of carbon.

The recycling of nitrogen

The nitrates in the soil are the key substances in this cycle because from them plants build up the proteins which are essential for life and from which all animals derive their proteins. The supply of nitrates in the soil would quickly be exhausted if it were not for several important processes going on continuously.

1. During decay the proteins of dead plants or animals are broken down in stages. First they are turned into ammonium compounds by certain kinds of **saprophytic** bacteria. Then the ammonium compounds are acted upon by **nitrifying** bacteria of two kinds, one kind oxidising them into nitrites and the other oxidising the nitrites further into nitrates.
2. During excretion nitrogenous waste in the form of various ammonium compounds is returned to the soil or water where it is then acted on by the nitrifying bacteria as above.
3. Certain bacteria can extract atmospheric nitrogen and build up nitrates from it. They are called **nitrogen-fixing** bacteria. Some species live freely in the soil while others live in nodules which grow as swellings on the roots of plants belonging to the *Fabaceae*—a family which includes clover, beans, peas, vetches and sainfoin (alfalfa) (Fig. 31 : 14). Each nodule contains millions of these bacteria living in **symbiosis** with the host plant. Symbiosis is a relationship between two organisms when they live together in harmony, each benefiting the other. In this case the

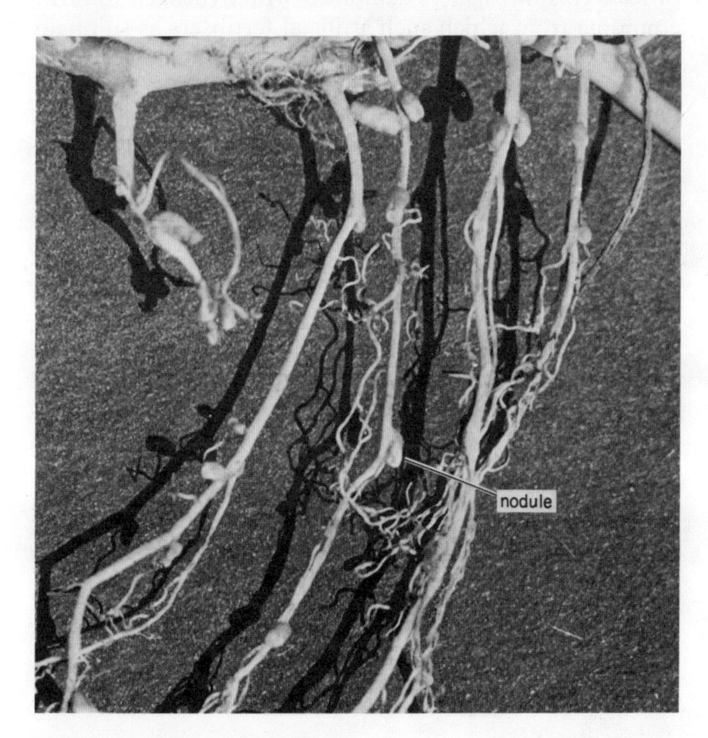

Fig. 31 : 14 Root nodules on clover.

host plant gains nitrogen compounds which the bacteria synthesise and the bacteria receive sugars made by the host plant.

Farmers make use of these plants with their nitrogen-fixing bacteria by growing them first as a hay crop and then ploughing in the roots; this increases the nitrates in the soil.

Unfortunately for the farmer some soils, particularly those that have become sour (acid), contain **denitrifying**

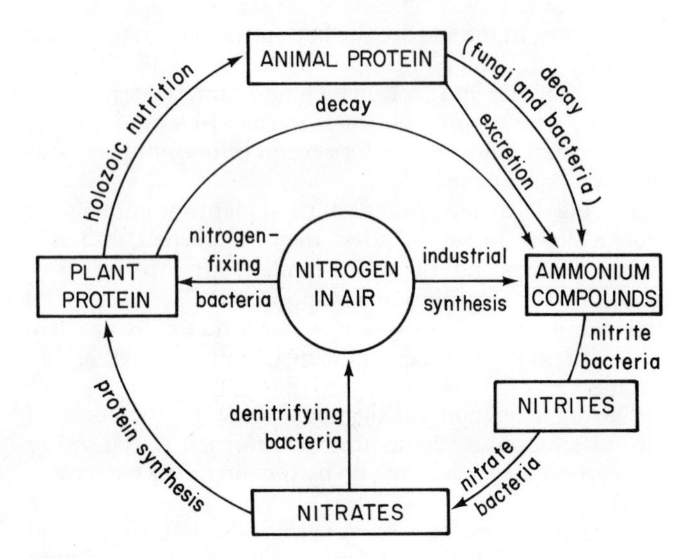

Fig. 31:15 Diagram showing the recycling of nitrogen.

bacteria which have the opposite effect, acting on the nitrates and releasing the nitrogen from them into the atmosphere.

Although air contains about 79% nitrogen, only the nitrogen-fixing bacteria can use this potentially large source. However, by an extremely important industrial process this nitrogen is combined with hydrogen to form ammonia from which such artificial fertilisers as sulphate of ammonia are formed. These fertilisers are used to replace the nitrates lost to the soil when crops are grown. In a natural ecosystem, nitrates used by plants are replaced when the plant dies or decays, but with food crops the product is removed and so nothing is returned. Because of this, manure or fertilisers have to be added.

In addition, during thunderstorms, some of the oxygen and nitrogen in the air is converted by the high temperature of the lightning into oxides of nitrogen which dissolve in rainwater to form nitrous and nitric acids. These combine with soil chemicals to form nitrites and nitrates.

The details of recycling of nitrogen are summarised in Fig. 31:15.

32

Studying ecosystems

How to study an ecosystem

So far we have studied some of the theoretical aspects of ecosystems; we can now use these principles as a basis for studying an ecosystem for ourselves. To do so we might follow four lines of investigation which could be expressed by these questions:

1. What is the physical environment like as a place for organisms to live in?
2. What kinds of organisms live there? How are they distributed and in what numbers?
3. How are these organisms adapted in structure and behaviour for living there?
4. What interactions are there between the organisms in terms of food relationships, competition, etc.?

Let us study soil as an example of an ecosystem beginning with an investigation of the physical and chemical properties.

Soil

How can soil be analysed?

What is soil like as a place to live in? For organisms to survive, the soil must contain adequate food, water and air and it must be in a form which allows it to be penetrated by animals and by the roots of plants. Soils differ greatly in composition and these differences determine the organisms which can live there. Therefore we will take two contrasting types—sandy and clay soils—and analyse them to determine to what extent they provide the necessities for life. Working in pairs, one of you should analyse the sandy soil, and the other the clay. Each sample should be obtained from the top 15 cm and placed in a polythene bag to prevent evaporation of water.

a) What solid matter is it made up of?

> Quarter fill for each type of soil a tall narrow jar with a screw top, or a gas jar with a tightly fitting cork; add water until the jar is about three-quarters full, put on the top, shake vigorously and leave on the bench to settle. If the soil contains particles of various sizes, the largest and heaviest will settle first and you should see a gradation (Fig. 32:1). Soil particles are classified according to size: those above 2 mm in diameter, gravel; 0·1–2 mm, sand; less than 0·1 mm silt and clay. The latter is so fine that it may take a long time to settle. Obtain a rough measurement of the depth of the various layers corresponding to the three categories given above. How different are the two samples? Are there any particles which float? These will probably comprise the dead remains of plants, collectively called humus.

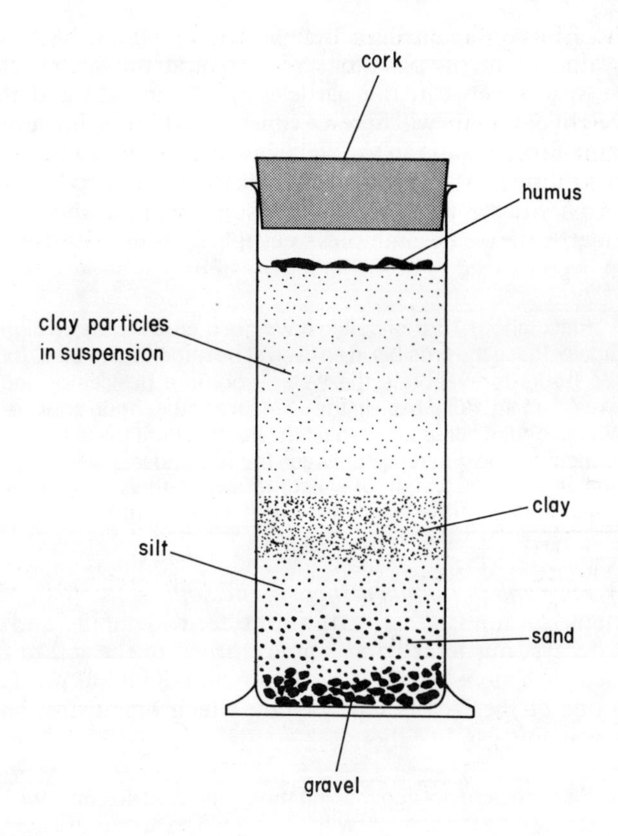

Fig. 32:1 A quick method of determining the composition of soil.

b) How much air is present in the two samples?

To obtain an accurate figure, the analysis should be done on undisturbed soil, but a rough estimate, using your sample, may be made as follows:

> Place about 50 cm³ of soil in a measuring cylinder, tap it gently on the bench so that the soil consolidates and top up with more to reach the 50 cm³ mark again. Now add 50 cm³ of water and stir gently with a stiff piece of wire until all the air bubbles have been released from between the soil particles. Measure the final level of the water. The difference between this level and 100 cm³ is the amount of air in 50 cm³ of soil. Which sample would provide most oxygen for soil animals and plant roots?

c) Are there soluble substances dissolved in the soil water?

You could find out in this way:

> Add some distilled water to the soil sample, shake, filter and evaporate the filtrate to dryness in an evaporating dish. Is there a residue left behind? If so, this could contain the mineral salts which may be taken up by plant roots through the root hairs. Sodium, for example, could be tested by putting some of the residue on to the end of a platinum or nichrome wire and placing it in the flame of a bunsen burner. A yellow flame would indicate that sodium was present.

d) How much water is there in the two samples?

The water content of any soil sample varies according to weather conditions; just after rain, it obviously contains much more than during a drought. If a soil sample is spread out in a warm, dry place for several days, all the water in the air spaces between the particles will evaporate and the weight of the soil will become constant. There is, however, some further water in the soil which can only be removed by heating the sample at a temperature of about 100°C. To simplify our analysis we will measure the total amount of water in the two soil samples by evaporating *all* the water in an oven instead of doing so in two steps.

Place about 15 g of soil in a weighed crucible and weigh again. Place the crucible in an oven at not more than 100°C for 24 hours to evaporate the water, cool in a desiccator and weigh again. Put it back in the oven for another hour, cool, reweigh and, if necessary, repeat the heating until the weight is constant. The difference between the first and last weighings will be the total weight of water that was in the soil samples. Keep the oven-dried soil for the next experiment.

e) How much humus is there in the soil?

Many soil animals, e.g. earthworms, feed on humus, and as it decays, nutrients from it are returned to the soil to be used once more by plants, so the amount of humus present is one of the factors determining which organisms, and how many, can live there.

Place the crucible containing the oven-dried soil on a wire gauze and heat it strongly with a bunsen. The humus will burn up completely into carbon dioxide and water vapour. After 15 minutes, cool in a desiccator and weigh. Heat it again, cool and weigh once more; repeat if necessary until the weight is constant. Any loss in weight represents the weight of humus. Which of the samples contains the most humus? The matter that remains is the mineral content of the soil—the actual rock particles. Compare the weight and size of particles in the two samples.

f) How quickly does water drain through soil?

If drainage is fast, this is important for two reasons: a) Soluble salts in the soil, including artificial fertilisers, will quickly be removed from the soil—for example, after heavy rain. The process is known as **leaching**. b) There will have to be a greater supply of water, e.g. a higher rainfall, to provide the organisms with all they need as the top soil will dry out quickly. You can compare the drainage rates for the clay and sandy soil in this way:

Fit up two funnels of equal size (Fig. 32:2) and plug each with glass wool using forceps (do not handle). Add some clay soil to one funnel and an equal volume of sandy soil to the other. Pour on enough water to soak both lots of soil and keep adding it until it is dripping regularly from the base. To find out how much water soaks through in a given time, place a beaker under each simultaneously and keep topping up the funnels so that the levels of water above the soil remain constant (why?). When enough water has collected, remove both beakers at the same time and measure the amounts of water in each in a measuring cylinder. The difference between the two indicates the relative drainage rates of the two soil samples.

g) Soil capillarity

Water can be drawn towards the surface of the soil as well

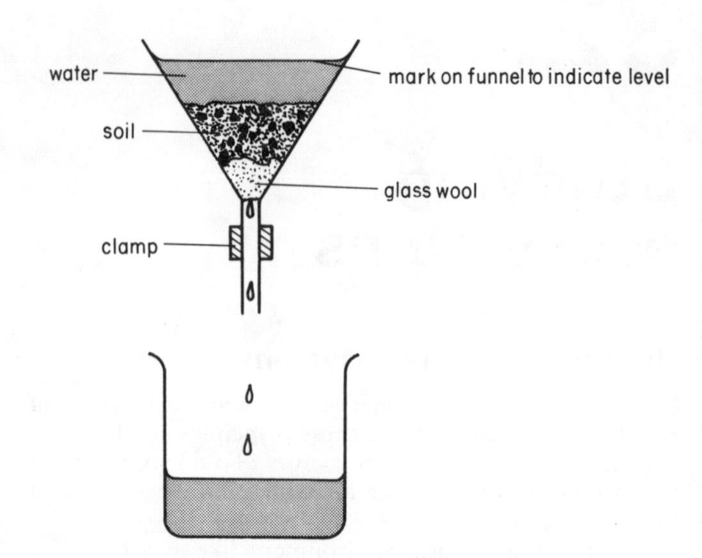

Fig. 32:2 Apparatus for comparing the drainage rates of water through different soils.

as drain down through it. After rain some of the water is retained between the soil particles, but the excess drains away through any rocks which are porous until it finally comes to an impervious layer of rock. Unable to sink any further, it accumulates to form the underground water table (Fig. 32:3). Plants can tap this reservoir even if their roots cannot reach the water table itself, because it is drawn upwards by capillarity through the spaces between the soil particles. Which of the two soil samples will be most efficient in this respect? You can find out in the following way:

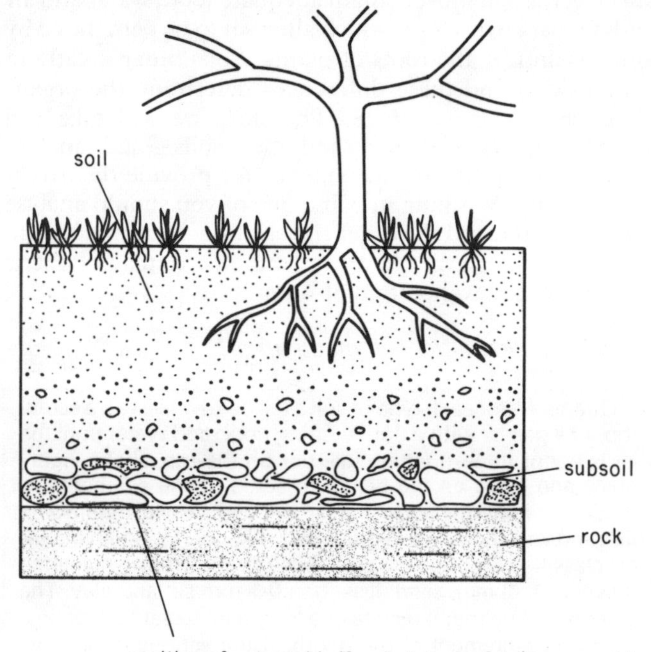

Fig. 32:3 Profile of soil and underlying rock.

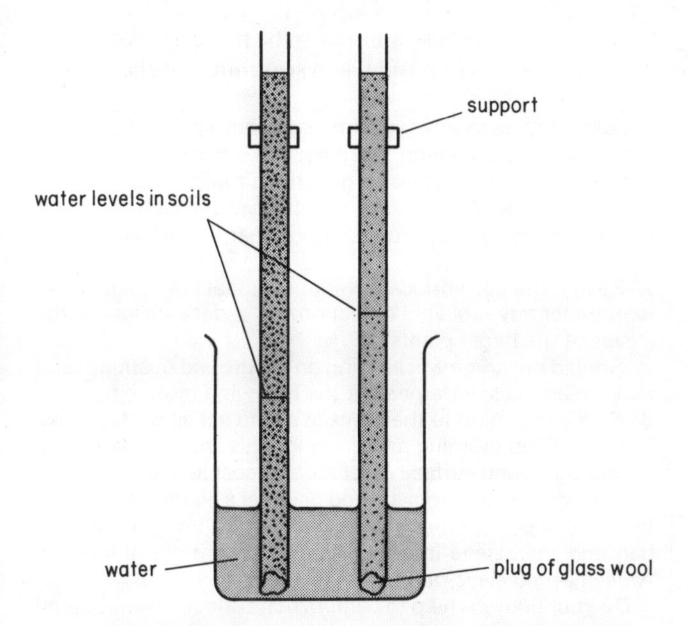

Fig. 32:4 Apparatus for comparing the capillarity of different soils.

> Plug two glass tubes about 1·5 cm in diameter with glass wool and then fill them with oven-dried powdered clay and sand samples respectively (Fig. 32:4). Ram the soil down as hard as possible. Now stand them in a beaker of water. You should be able to tell how high the water travels by the change in colour of the soil. Note the level reached in each tube every 10 minutes for the first hour and again the next day. In which does the water rise fastest to begin with? In which tube does the water reach the greatest height eventually?

We are now in a position to summarise the main differences between a sandy and a clay soil:

	Sand	*Clay*
Size of particles	Large	Small
Air content	Large air spaces	Small air spaces
Drainage	Good	Poor, i.e. easily waterlogged
Retention of water	Poor	Good
Nutrients	Easily leached out	Retained longer

If you were a farmer you would also add that a sandy soil is easy to plough and warms up quickly in the spring, so producing earlier crops; in contrast, a clay soil is heavy and difficult to work and takes longer to warm up.

If you consider these properties from the point of view of the animals and plants living in these soils, with their need for oxygen, food and water, you will see that both soils have good and bad characteristics. It is therefore not surprising that some of the best soils consist of a mixture of both clay and sand, together with a good humus content; these soils are called **loams**.

Soil texture

Apart from its component parts, the suitability of a loam as a place in which organisms can live depends upon its texture. This results from the manner in which it is built up. A good soil texture is one which is crumbly; when pressed together between the fingers it does not go into a paste but separates into small **crumbs**. These are aggregates of rock particles of various sizes incorporated with humus; they are held together by the surface tension of the water film which lines the air spaces. The larger air spaces between these crumbs of soil allow rootlets to penetrate and animals to move.

Another factor which affects the organisms which live in soil is the chemical nature of its constituents. We will take one example:

The presence or absence of lime in the soil

A soil containing a significant amount of calcium carbonate or lime is said to be **calcareous**, one lacking in lime, **non-calcareous**. Each kind has its characteristic flora, because some plants, e.g. wild clematis, dogwood and spindle only grow well when it is present and others, such as heather, cannot tolerate much lime.

Calcareous soils are usually derived from underlying rocks, such as chalk and limestone by a process of weathering. Soils formed in this way are said to be **sedentary**. However, **sedimentary** or **alluvial** soils (formed from particles brought down by rivers or glaciers) may also contain calcareous particles if the river or glacier passed over calcareous rocks during its course. You can find out if your soil sample is calcareous in the following way:

> Add dilute hydrochloric acid to the soil in a test tube. If calcium carbonate is present, carbon dioxide will be given off as bubbles. How would you test that it was carbon dioxide?

Calcareous soils are neutral or slightly alkaline (you can test this using universal indicator), but non-calcareous soils tend to be acid. Acid soils usually develop because of insufficient oxygen. In moorland, for example, the soil is poorly aerated because of its waterlogged condition. This means that aerobic bacteria are unable to flourish, so the humus is not broken down completely and collects in the form of peat. Partial breakdown of humus results in the formation of humic acids. This accounts both for the acid nature of the soil and for its lack of fertility, as potential nutrients in the humus are not returned to the soil. Heavy clay soils also tend to be acid for lack of adequate aeration.

Acid soils can be greatly improved by adding lime to the soil (usually as slaked lime). This has the effect of neutralising the humic acids. But lime also improves the texture of the soil. When worked into a clay soil it causes the clay particles to clump together into larger units—a process called **flocculation**. This makes the soil lighter and more easily worked and allows air to penetrate better. This in turn improves conditions for the aerobic bacteria which act on the humus. The process of flocculation may be demonstrated as follows:

> Take two beakers of equal size and add 10 cm^3 of powdered clay and 150 cm^3 of water to each. Add 2 cm^3 of lime to one beaker only. Stir both beakers vigorously with a glass rod and then allow the contents to settle. The speed of settling will vary according to the size of the particles. In which beaker does the clay settle first?

Soil factors

We can summarise our findings so far by saying that all soil is made up of five non-living constituents—mineral particles, water, soluble mineral salts, humus and air—but it varies considerably according to the chemical nature of its constituents, the size of the particles and the relative proportions of its component parts. Soil organisms have to be adapted to all these factors, but there are also other physical factors which are characteristic of soils to which organisms also need to be adapted. These are: a) A relatively low temperature (even if the surface of the soil becomes hot in strong sun, the warmth does not penetrate far). b) A high humidity. c) Absence of light except at the surface. We must take these factors into account when we study the organisms themselves.

Plants living in the soil

Owing to the absence of light below the soil surface no green plants can exist there, so the flora consists of teeming masses of bacteria and fungi which feed saprophytically on the humus. You have seen some of these for yourself (p. 29). However, most soils if left undisturbed for long enough become covered in green plants whose roots penetrate into it. In places such as woods and grassland, for example, this layer of vegetation is so thick and composed of so many species that it is difficult to understand how they can all find enough space and nutrients.

Examine the root systems of four of the commoner plants found living together in any one well-established habitat, such as a wood or grassland. Dig them up carefully with plenty of surrounding soil and soak them in water before washing off the soil that still clings to the roots. Compare the lengths of the roots and the general lay-out of the root systems: this should help you to explain why certain species can live close together and yet obtain enough nutrients. There are, of course, other reasons which enable them to do this which do not concern their roots; can you think of any?

Finding soil animals

Soil animals vary greatly in size, but the majority are very small. Could small size be an adaptation to living in soil? Soil animals also vary greatly in their distribution; some are mainly surface-dwellers, others occur at various levels within the soil. Let us consider the surface-dwellers first.

Surface-dwellers

If a soil organism is adapted for living in a dark, damp habitat where temperatures are low, life on the surface of the soil could be difficult, as this is a region which is exposed to light, tends to be drier and is subject to greater changes of temperature; it is also exposed to the larger predators, e.g. birds. But the soil surface is seldom uniform; in some places it is covered with leaves, in others with large stones or logs and in most places there is vegetation to provide shelter. You will remember from your choice-chamber experiments on woodlice that this species tended to move into dark and humid places (p. 231). It is therefore probable that most soil animals will be found by searching in any place where these conditions are prevalent. Also, one would expect these animals to be more active at night. Keep these points in mind when searching for the animals.

Collect representative specimens of each species for identification and examination. Keep each in a specimen tube and put in a few damp leaves to give them moisture and something to cling to. Add a label indicating where each specimen was found. Here are some methods that you could use for finding them:

1. Search the soil surface in various habitats in the daytime; look under any movable object on the surface including the leaves of the larger plants.
2. Spread out some wet sacking on the ground overnight and look for animals underneath it the following morning.
3. Sink jam jars up to their rims in soil to act as pit-fall traps. Examine them morning and evening. This should enable you to find out when surface dwellers are most active.
4. Collect leaf litter from a wood or under a hedge. Use a hand fork to scrape together a large heap; put it into a polythene bag and then sieve it, a few handfuls at a time, through a wide-meshed sieve on to a white sheet.

Do your findings help to confirm that surface-dwellers tend to be more active at night, while in the day they seek moist, dark, cool and sheltered places?

5. Identify your specimens according to their main types; the illustration in Fig. 32:5 may help you, but also consult appropriate books. Make a table, stating where each was found, together with any structural adaptations that you observe which would help them to live as surface-dwellers. Consider such aspects as shape, size and organs of locomotion. Much time would be needed to discover their feeding relationships—whether they were vegetarians, scavengers or predators—but you could devise simple feeding experiments to test their reactions to various foods. Each member of the class could study one species, and then the results should be pooled.

Animals living in the soil itself

The larger species can be found by digging, and animals such as earthworms can be drawn to the surface by soaking the soil with 2% formalin (p. 27), but the majority of soil animals are too small to find in these ways. One technique that can be used relies on the principle that if soil organisms are adapted to moist, dark, cool conditions, they can be driven out of soil by illuminating it and making it dry and warm. However, the treatment must not be too severe, otherwise they will die before they come out of the soil. To do this the **Tullgren apparatus** can be used. This consists of a large funnel (Fig. 32:6) made of metal, glass or tin foil, and a sieve (on which the soil sample is placed) with a mesh fine enough to prevent the soil from slipping through, but large enough for the animals to penetrate. An electric light bulb fitted above the funnel supplies both a source of light and heat, and so will dry out the sample as well. The animals, in moving away from these adverse conditions should pass downwards out of the soil and fall into the beaker of dilute formalin below, which will kill and preserve them.

At least three of these pieces of apparatus should be set up using soil samples from different places—the top soils of woodland or grassland contain a large variety of organisms. Break up the soil gently before placing it on the sieve so that the animals may escape more easily. Switch on the lamp and leave for several hours.

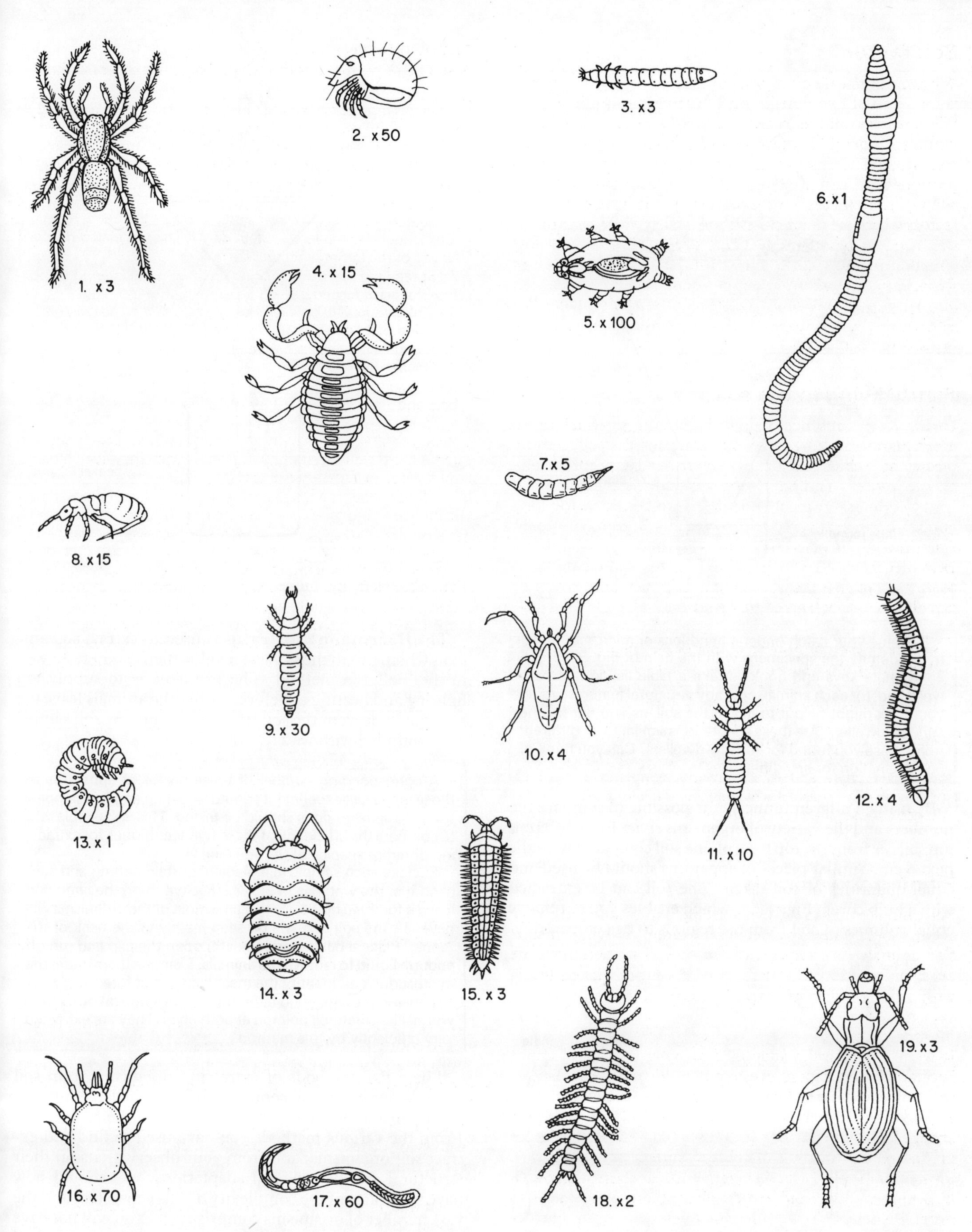

Fig. 32:5 A selection of soil organisms, mainly surface dwellers: 1. Wolf spider. 2. Armadillo mite. 3. Wire worm (click beetle larva). 4. False scorpion. 5. Tardigrade. 6. Earthworm. 7. Fly maggot. 8. Springtail. 9. Proturan. 10. Harvestman. 11. Dipluran. 12. Snake millipede. 13. Larva of cockchafer beetle. 14. Pill woodlouse. 15. Flat millipede. 16. Mite. 17. Nematode. 18. Centipede. 19. Ground beetle.

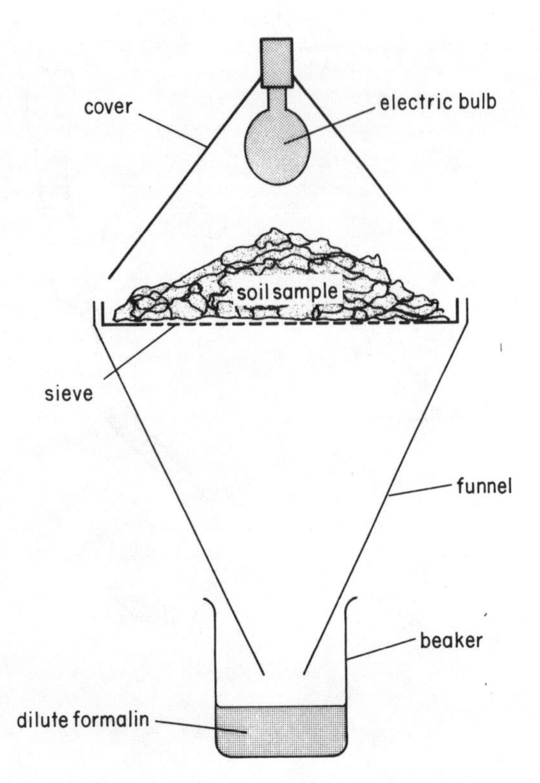

Fig. 32:6 Tullgren funnel.

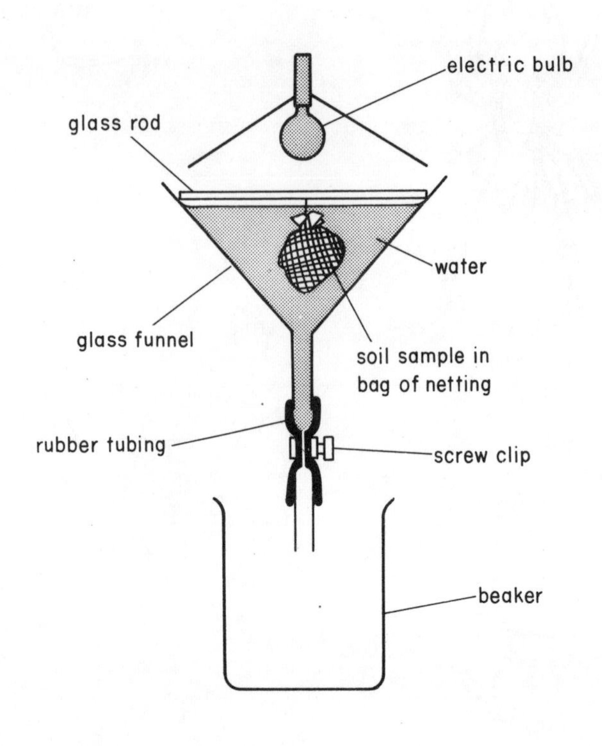

Fig. 32:8 Baermann funnel.

Examine your catch under a hand lens or microscope and try to identify the specimens with the help of Fig. 32:5 and appropriate keys and books. Make a table listing the main types and, for each animal, note any structural features which you think might help it to live in the soil, as you did for the surface-dwellers. Are these adaptations similar to, or different from, those you found for surface-dwellers? Can you think of an explanation for any differences?

By using Tullgren funnels, it is possible to compare the numbers and the variety of organisms *either* from different top soils *or* from the top 6 cm of one soil compared with the next 6 cm. Similar pieces of apparatus should be used and equal quantities of soil taken. The soil can be extracted with a bulb corer (Fig. 32:7) which enables you to remove equal volumes of soil from the regions to be compared.

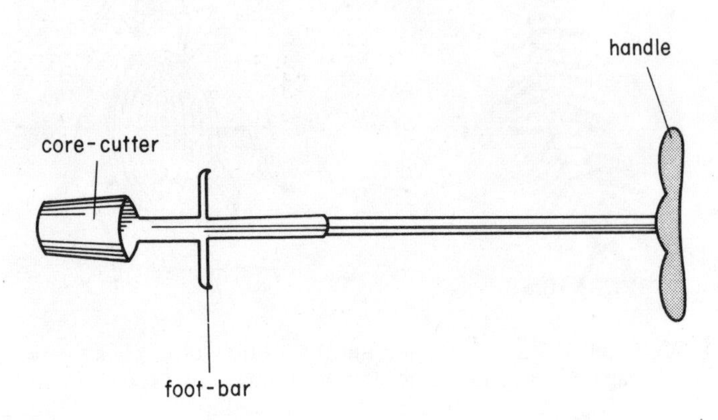

Fig. 32:7 Bulb-corer.

The **Baermann apparatus** is used to extract soil animals which live in the soil water and which do not react well to the Tullgren method. The principle is to supply top lighting and heating as before, so that the animals leave the soil, but to prevent the soil from drying up, the soil sample is surrounded with water.

A corresponding number of these pieces of apparatus to those in the last experiment should be set up and soil samples from the same regions should be tested. This will enable you to compare the efficiency of these two techniques for extracting different species of the soil fauna.

Enclose each sample in a piece of curtain netting and suspend it in the water as in Fig. 32:8. Switch on the lamp and leave it for three hours. By this time most of the organisms will have left the soil and will be seen mainly in the neck of the funnel. Place a beaker underneath, open the clip and run off enough liquid to remove the animals. Examine them under the microscope and identify the main kinds as before.

You will probably find large numbers of **nematodes**, tiny worm-like creatures pointed at both ends. They are extracted very efficiently by this method.

These two methods of extracting organisms from soil should have given you some idea both of the large number of organisms which exist in soil and of their diversity. From the various methods you have used to find and extract soil organisms, and from your observations on their structural and behavioural adaptations, you should now have some idea of the complexity of this ecosystem and the vast number of organisms comprising it. You will not have been able to work out many of the details of the food relationships of the various species, but the following points should help you to fill in some of the gaps and see how the soil organisms are all inter-related.

1. The only producers are the green plants.
2. The available plant food comes from living roots or the dead remains of plants (humus).
3. The scavengers include earthworms, millipedes, woodlice and the minute springtails.
4. The main decomposers—bacteria and fungi—and the nitrogen-fixing and nitrifying bacteria occur in soil in vast numbers.
5. The smaller carnivores include ground beetles, centipedes, spiders and certain mites.
6. The larger carnivores include moles below the surface and shrews and blackbirds above. These, with many others, link the soil with other ecosystems.
7. When animals other than those in the soil die, their bodies are eaten or decomposed and may form food material for soil organisms. The same also applies to their waste products—dung and urine.

From these facts and other data the class has collected, try to construct a diagram to show some of the interrelationships of soil organisms.

Changes within ecosystems

If it were possible to count the numbers of the various organisms present in an ecosystem such as soil over a period of years, we should probably find that although numbers varied from season to season and from year to year, they would tend to fluctuate around a mean, i.e. average numbers would be maintained over a long period. This is characteristic of a **stable** ecosystem.

Temperate woodland and tropical forest are other examples of stable ecosystems, because if left undisturbed by man conditions remain fairly constant and their communities remain in balance.

But drastic things can happen to what appears to be a stable ecosystem. Volcanic eruptions, floods, droughts, hurricanes, etc., can greatly alter an ecosystem or destroy it completely. What happens then?

Colonisation and succession

An ecosystem which has been drastically changed undergoes **colonisation**. Take, for example, an area of forest destroyed by fire. The first plants to re-appear will probably have originated from spores or seeds brought by various means, especially wind; mosses and grasses may be among them. As a result, the ground will soon be covered by herbaceous plants which compete with each other for light and space. Later, the slower-growing shrubs and trees will make their appearance and as they grow taller will shade the ground flora, much of which will die out for lack of adequate light. Later still, the trees will outstrip the shrubs and in turn shade them; in consequence some of the shrubs will die if the trees are close together, and so the

Fig. 32:9 Pond ecosystem showing gradual colonisation of open water by marsh plants.

original forest structure is restored. In this way recolonisation of a devastated area tends to produce in the end what is called a **climax** vegetation; this is a relatively stable ecosystem.

This series of changes is known as **succession** and, of course, it applies equally to the fauna which is dependent upon the plant life. Thus during this process of succession every organism which succeeds in establishing itself changes the conditions and makes the habitat either more or less suitable for other organisms; their success or failure in turn will change the conditions again until eventually, in the climax condition, a balance will be established which is relatively stable.

If man interferes with a habitat by tree-felling, ploughing, etc., and the region is then left alone, it gradually changes through a succession of communities until the original climax condition is produced. So if the human population of southern England was completely evacuated, most of the region over the following 50 years would revert to deciduous temperate forest. The same would apply to all the major world ecosystems such as tundra, savannah and tropical forest. You will see later (p. 257) how man has to battle constantly with nature to prevent this from happening, if it is in his interests to do so.

You can observe succession taking place in this way:

Dig a square metre of soil very thoroughly, carefully removing all roots and whole plants. Leave it completely alone for several months. At regular intervals count the numbers of individual plants of each species present and note the changes that occur. To reach a climax you would have to keep this up for 50 years or more, but many changes occur, even over a few months, as new arrivals compete for light and space and vegetarians such as slugs and snails take their toll.

Changes in population

The term **population density** is used in the same sense for animal and plant numbers as for humans. It refers to the number of a particular species in a given area.

In any ecosystem the population of a particular species depends upon four factors: a) Natality—those being added as a result of reproduction. b) Mortality—those being removed from it by dying. c) Immigration—those reaching it from other ecosystems. d) Emigration—those leaving it for other ecosystems. The population density will be stable if:

natality + immigration = mortality + emigration.

In practice, all populations fluctuate because of the ever-changing conditions within an ecosystem which affect these four components of the equation. Sometimes the fluctuations are cyclic, i.e. the rises and falls are fairly regular and the average population density over a long period of time is maintained. Thus the snowshoe hare population in Canada shows high or low peaks about every ten years (Fig. 32:10) and the Scandinavian lemming every three or four. However, most fluctuations are much more irregular.

In all species there is the potential to increase in numbers. A pair of rabbits under perfect conditions could theoretically produce a population of 14 million in 5 years, but in practice this does not happen because of the effects of a number of factors. Let us consider some of these:

a) Climatic factors

Extremes of weather in terms of temperature, rainfall, wind, humidity, etc., combine in various ways to affect the density of the population. Some of these are cyclic, such as the seasonal effects of summer and winter in temperate regions and the wet and dry seasons in tropical countries. But when these factors are extreme a high proportion of a population may be destroyed and it is many years before more normal numbers are restored. For example, in England, a period of unusually prolonged frost in 1962 caused the ground to become so hard that bird species which were largely dependent upon soil fauna for their food, e.g. green woodpeckers, were greatly reduced in numbers. Similarly, periods of severe drought have caused the deaths of vast numbers of antelopes on the African savannah. You will think of many other examples of the effect of climatic conditions on population densities.

b) Food supply

When food is abundant relative to the number of individuals in a population, it is of no importance in regulating population density, but when scarce it will cause numbers to drop, either because competition for what remains will increase and this may lead to starvation or death from other causes, or because there will be emigration to areas of more plentiful food supply. Insectivorous birds such as swallows, swifts and warblers migrate from temperate regions as winter approaches to tropical countries where insects are more abundant, and return in the spring when insects are available once more in sufficient numbers.

c) Space

When the population density is low the individuals are spread out and there is enough space for each to acquire

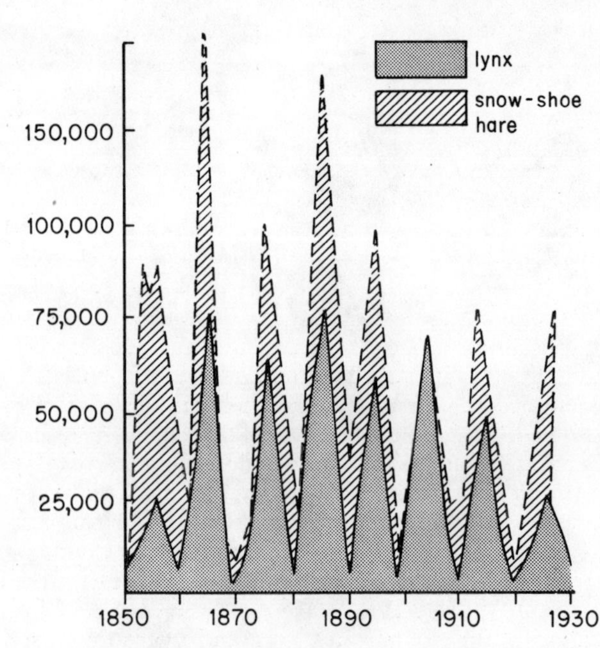

Fig. 32:10 Graph of snow-shoe hare and lynx population over 80 years based on skins obtained by the Hudson's Bay Company of Canada.

sufficient food and shelter to live successfully, but under crowded conditions, competition for food and shelter becomes greater and, in many species, behaviour becomes more aggressive. This happens in small rodents such as voles which, like snow-shoe hares, have periodic increases in population followed by dramatic crashes in numbers. In this example one of the important factors in causing this sudden drop, is stress (p. 162). When overcrowding occurs the animals are constantly being stimulated by their neighbours. There is more fighting, mothers with litters cannot care for their young adequately and there is insufficient time for resting and feeding. This constant unrest upsets the normal working of the endocrine system, the pituitary and adrenal glands become gradually exhausted and the animals die of 'shock'. After the crash in numbers, the few survivors have plenty of space, so the population builds up again and the cycle is repeated.

You might consider the effects of overcrowding on man. Does it cause more aggressive behaviour and increasing stress symptoms? Compare life in overcrowded cities with life in the country; in overcrowded schools with those possessing much greater space.

Many species avoid overcrowding by showing territorial behaviour. A **territory** (p. 75) is a part of the habitat over which an individual, pair or social group asserts a dominating influence over other members of the same species. It varies greatly in size according to the food and breeding requirements of the species. In birds, territories are defended against intruders by various forms of aggressive behaviour such as singing and displaying; in mammals, ritual fighting may occur.

Territories provide some kind of home for breeding and a refuge from enemies. The number of adequate shelters in an area often influences the density of the population. For example, it has been found that by increasing the number of nesting sites in a wood by putting up nesting boxes, the population of tits and flycatchers has increased. Conversely, when old trees have been cut down, the number of natural nesting holes has been reduced and numbers have fallen.

So the number of potential homes, as well as the available food, will influence the size of the territory and thus the population density.

d) Diseases and parasites

Epidemics may sweep through both animal and human populations and cause numbers to fall. For example, when myxomatosis was introduced into Britain over 90% of the rabbit population was destroyed. But even when some degree of immunity has been established, disease is an important factor in limiting numbers. Bovine tuberculosis, for example, is common in buffalo herds in East Africa and even if it does not cause death it will weaken the animals and so make them easier prey for lions. This is an example of how different factors often interact to affect numbers.

Parasites also play an important role in regulating numbers in some species. We have already seen an example of this with the large white butterfly and its ichneumon parasite (p. 48). When a population reaches a high level, it is much easier for the parasite to spread from one host to another, and the same applies to the organisms causing infectious diseases. This leads to the important principle that factors such as available food, space, shelter, parasites and disease exert an increasing influence as population density becomes greater. Is this also true for the human population?

e) Man

Man is one of the most important factors influencing animal and plant population densities because he has drastically manipulated the environment for his own ends. The result is that many of the world's ecosystems are no longer in a relatively stable condition and many artificial ones have been formed. We shall consider the implications of this in the next chapter.

33

Man and his environment

Man's brain has developed to an incredible extent over the past 100,000 years. This has allowed him to become more independent of animal instincts and more capable of exploratory and creative actions, but at the same time it has increased his powers of destruction. Through the development of these powers man has increasingly destroyed or modified much of the natural environment and created an artificial one of his own making.

Past history

In order to understand man's special relationship with his environment it is necessary to review his more recent evolution briefly. Some 15,000 years ago man was an unobtrusive species within the living world; his effect on the environment was relatively small and he was well adapted, like other members of the fauna, to the ecosystem of which he was part. But his brain was developing fast and he was becoming much more effective in making stone implements, so he began to make a greater impact on his environment. He started to cut down the forests and grow crops, build more durable huts and domesticate animals. This meant that he gradually changed from a nomad seeking food wherever it could be found to a village dweller surrounded by the fields he could till, the crops he could harvest and where his domesticated animals could be cared for and protected. He learnt how to store the grain that he grew and preserve the fish and meat he procured to tide him over any periods of scarcity. He improved his method of communication with other individuals by inventing language; this made co-operative action much more effective and started him along the road of **cultural inheritance**—the passing on of experience from one generation to the next. He discovered the value of fire for cooking and preserving, for clearing land before planting, for keeping warm in colder regions and, later on, for moulding metals to form more useful tools for hunting, for war and for agriculture. So man moved from the Stone Age through the Bronze Age to the Iron Age. Life became increasingly complex—there were more things to be done, more trees to be felled, huts to be built, animals to be cared for, fields to be tilled and tools to be made; also, there was pottery to be moulded and baskets to be woven. Thus man became more specialised; there was greater division of labour, bartering of products took place, roads were built to link up the villages and as man began to prosper, larger towns and cities were constructed. Our modern world is the logical development of these trends.

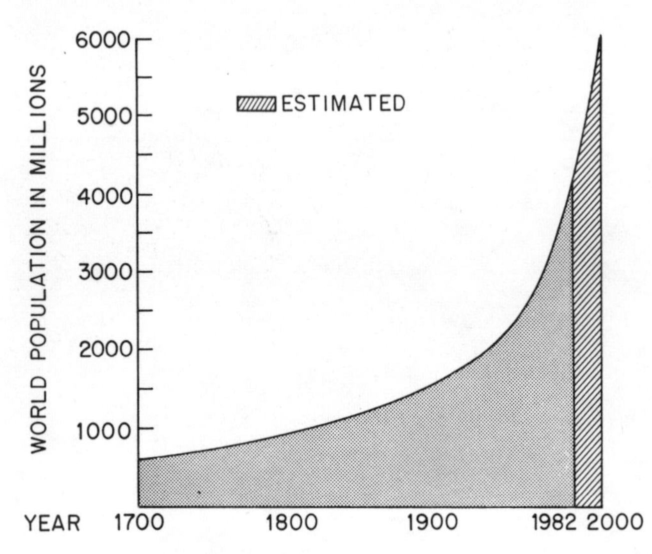

Fig. 33 : 1 Human population growth.

Population growth

With greater powers for subduing nature and harnessing natural resources for his own ends, man has increased in numbers. At first the growth rate was slow. It is impossible to estimate numbers in the distant past with any accuracy, but it is thought that the 100 million mark was not reached until about 3000 B.C. The population probably reached 500 million by A.D. 1650 and 1000 million around the middle of the 19th century. But since then the rate has accelerated to an alarming extent, due largely to man's success in tackling diseases such as malaria, cholera, plague and typhoid (Ch. 7) which in the past had been major killers. So by 1930 numbers had risen to about 2000 million, by 1960 to 3000 and by 1983 to 4400 million. It is estimated that by the end of the century it may be in the region of 6250 million! (Fig. 33 : 1). This means that at the present time babies are being born all over the world at a rate approaching 200 a minute while people are dying at the much slower rate of about 110 a minute. Of course, this rate of increase varies greatly from one country to another. In Brazil, for example, the population will have doubled in about 20 years if the present rate of increase continues. Think what this means in terms of food, water, housing, power and other amenities. All these will have to double in 20 years just to keep pace with the present standard of living. Even in countries like Britain where the increase is low, the problem is great as this country is already very overcrowded.

Another factor of great importance is that, all over the world, vast numbers of people have moved from the land to urban communities, in the hope of obtaining a better living. This has led to great overcrowding and poverty in many great cities, with hopelessly inadequate housing, food, water and sanitation, and all the social problems that go with them. Today, about one-third of the world's population lives in towns of over 20,000 people and by the end of the century it will be more than half. In highly developed countries the proportion is already much greater. This staggering increase in numbers is already having an immense effect on people everywhere. When and how will the increase level off? Clearly some means of regulating numbers is essential (p. 172).

Fig. 33:2 Overcrowding—a scene on Swanage beach.

The effects of the population explosion, together with man's technical achievements, have already led to the exploitation of the natural environment on an unprecedented scale. But fortunately there is now a greater awareness of the environmental problems facing mankind and much is already known about how these may be tackled. What is needed most is the determination for all people and nations to work together so that our knowledge is applied sensibly. Let us now examine some of the effects of man on his environment from an ecological point of view.

The effect of agriculture

When a forest is cut down and a food crop is grown in its place, a natural climax ecosystem with its vast number of species in a state of dynamic equilibrium is replaced by a **monoculture**, i.e. an unnatural concentration of a single species in one area. Look around a farm and you will see crops of various kinds grown in different fields—some to provide cereals or roots, others grass for domestic animals. Whatever the crop, it is an unstable community and if left on its own would revert to the natural climax ecosystem once more.

You know what happens when a garden is left untouched for a long time; it soon becomes a wilderness. Only by constant attention can it be kept as it is wanted. The same applies to any land which is farmed; pasture is only kept as such by routine grazing or cutting for hay; fields can only be maintained in a fit state for growing crops by regular cultural methods. The battle to prevent the natural succession from taking place brings with it many problems.

Pests and diseases

By growing crops in large concentrations man provides ideal conditions for the pests which feed on them, and the parasites, particularly fungi, which cause disease. Here is food in abundance and excellent conditions for diseases to spread from one plant to another. Consequently, multiplication is rapid and the resulting damage very great. To avoid this happening man has tried to eliminate these competitors for his crops by using toxic chemicals. Many of them have been very effective, but their use has also created new problems.

Pesticides

The perfect pesticide is one which destroys a particular pest and is completely harmless to every other form of life; no such pesticide exists or is likely to. Do you see why?

In judging the usefulness of a pesticide it is necessary to take into account both its effectiveness and the possible harm it does to other forms of life, including ourselves.

Beneficial effects

The value of such substances as DDT has been very great in controlling disease and increasing food output. Their use has greatly reduced the incidence of such killers as malaria, yellow fever, plague and typhus by destroying their insect carriers, and when sprayed on crops they have greatly increased the yield. In some countries, due to the elimination of such pests as wireworms, the output of cereal crops has increased by 60% and potatoes by 70%. Also, the spraying of orchards has led to a marked improvement in both the quantity and the quality of fruit.

Harmful effects

Pesticides are often indiscriminate in their action and vast numbers of other animals may be destroyed. Some of these may be the predators which naturally feed on the pests, others may be the food of other animals, thus causing unpredictable changes in food chains and upsetting the balance within the ecosystem. When pesticides are sprayed from the air over large areas of land these effects may be dramatic.

A further danger is that some have a cumulative effect. Pesticides vary in their length of 'life' as toxic substances. Some, such as the organophosphates (and also the herbi-

Fig. 33:3 Large-scale monoculture of wheat, Saskatchewan.

cides) are said to be **degradable** because they are broken down into harmless substances in a comparatively short time, usually under a year. Others are **non-degradable**, and include those which contain mercury, arsenic or lead. Others, such as the chlorinated hydrocarbons, DDT, aldrin and dieldrin, are extremely difficult to break down and may remain in the soil for over ten years. This has given time for many insects to build up resistance to their action, so research goes on all the time to produce different ones. These non-degradable pesticides are potentially dangerous as they accumulate in the bodies of animals and pass right through the food web, being further concentrated at each step until animals at the top of the pyramid may receive enough to do considerable harm. This has been very evident in birds such as herons, which have gradually accumulated in their bodies the DDT they acquired from fish. Similarly, when dieldrin was used as a sheep dip to eliminate parasites such as mites, ticks and blow-flies, the eagles were greatly affected as they fed on any sheep that had died. In both these instances reproduction either ceased, or the eggs had such thin shells that the developing chicks did not survive. The persistence of these chlorinated hydrocarbons may be judged by the fact that so much has been used over the past 25 years that it has found its way from the land and rivers to the oceans and has been dispersed by currents to all regions; even the penguins and polar bears which live literally poles apart have traces in their tissues. Man, being at the end of so many food chains, also accumulates these substances and it is essential to question their possible effect on him. There is no evidence at present that amounts absorbed so far are doing recognisable harm, but the danger of further accumulation must be taken very seriously indeed.

What should be done?

It is easy to say 'Ban all pesticides!' but the pests still have to be kept in check. At the present time pests still claim 10% of the world's food supply even after £1000 million has been spent on chemicals to control them every year. If pesticides were totally banned what would happen to the diseases they are controlling and the crops we so vitally need for our growing population?

The problem is not insoluble. Already some of the most harmful substances have been banned in a number of countries and the search continues for alternative methods of pest control. For the present it is obviously a sane policy never to use pesticides unnecessarily, and when they are essential, to use a degradable kind as sparingly as possible. Those of us who have gardens can all play our part in pursuing this policy.

But the long-term answer is to find other effective methods of controlling pests which have far less harmful effects and are based on sound biological principles. This more positive approach is being pursued with vigour. Here are some of the more important methods used:

1. Growing different crops on a particular piece of land in successive years

This rotation of crops reduces the build-up of pests from year to year in that area and in this way less damage is done.

2. Studying the life histories of the pests

When this is done it is sometimes possible to sow the crop at a time when least damage will be caused. For example, the larvae of the frit-fly feed on the delicate seedling stages of oats and rye, but do little damage to the older plants; thus by sowing the cereal well before the frit-fly larvae hatch, the damage is much reduced.

3. Introducing a natural predator or parasite of the pest

This is an example of **biological control**. In an ideal situation (which seldom occurs) the organism selected should be one which attacks only the pest species, as otherwise it might turn to another, perhaps useful, species and then become a pest itself! This method has been particularly useful when a plant or animal has been introduced into another country and multiplied excessively to become a pest. In this way the cabbage white butterfly, when it became a pest in New Zealand, was successfully controlled through the breeding and releasing of large numbers of its ichneumon parasite (p. 48). In a similar way, buffalo flies are being controlled in northern Australia by dung beetles. These flies breed in cattle dung and multiply very fast. They are a menace because they irritate the cattle so much that meat production falls. Domestic cattle were introduced into Australia but no indigenous dung beetles are present to feed on their dung. In consequence: a) The dung pats do not decay for months, and as about 200 million pats are produced every day, many acres of pasture are lost each year. b) The buffalo flies breed in vast numbers on the pats. c) The nitrogen in the dung is not returned to the soil to enrich it. To deal with this problem dung beetles were imported from Africa and Asia in 1967, and by 1970, 275,000 beetles had been released. One species colonised an area of more than 3000 km^2 within two years. The beetles live on dung which they roll into balls and bury; eggs are laid inside and when the larvae hatch they feed on the dung, pupate and later emerge as beetles; the cycle is then repeated. The beetles can bury a pat within 48 hours, thus preventing buffalo fly eggs from developing. So in one operation they bury the dung, remove the pest and put the

Fig. 33:4 Female dung beetle rolling dung ball for subsequent burial and egg-laying.

nitrogen back into the soil! The beetles can only feed on dung, so there is no likelihood of their becoming pests themselves.

A plant example of biological control is the prickly pear cactus which, when introduced into Australia, thrived so well that it invaded and destroyed vast areas of pasture. By introducing the larvae of a moth which feeds on it, it was brought under control within ten years.

A more controversial method of biological control was used when the myxomatosis virus was introduced to reduce the rabbit population in Australia and later in Britain. Statistically it was very successful, as it caused the deaths of over 90% of the rabbit population which in Australia had reached plague proportions in the absence of natural predators. But on humane grounds the operation was severely criticised as the animals died with distressing symptoms.

4. Rendering the males of a pest species sterile

This can be done by subjecting them to radiation and then releasing them in the wild population. This has been done successfully with the screw-worm fly which infects cattle. The insects mate normally, but the eggs are infertile.

5. The development of genetic strains which are resistant to the particular pests

Through extensive breeding experiments strains of cereals have been developed which are resistant to fungus diseases caused by rusts and mildews. Although this method holds out great promise for many species, it has been found that pests sometimes adapt successfully to the new strains.

Weeds

By ploughing land, man not only provides a good soil for his crops, but he produces conditions which are excellent for weeds. Weeds are often the first colonisers of disturbed soil. They grow fast, form flowers very early, and produce large numbers of seeds; many are annuals, and some even complete their whole life history in a few weeks. Thus they multiply very rapidly and compete strongly with the crop for space and nutrients. There are several ways of lessening this problem:

1. Crops may be sown very thickly. This works well with cereals as it leaves very little space for the weeds to develop.
2. Hand-weeding, hoeing or machine weeding may be done during the early stages of growth. This applies especially to root crops which have to be given more space. The method of weeding varies according to the labour available and the type of crop.
3. Crops may be sprayed with selective weed-killers. This method is particularly effective for cereals, as some of these substances destroy broad-leaved plants (most weeds come into this category) but do not harm grass-like species (cereals).

The removal of crops from the land

In a natural ecosystem plant products are returned to the soil and recycled, but when crops are grown, they are

Fig. 33:5 The effect of a selective weed-killer in the control of charlock in oats. No treatment was given to the strip now full of charlock.

Fig. 33:6 Soil blown off the fields into a drainage ditch: East Anglia, 1972.

usually harvested and removed, so the soil becomes impoverished. To replace the mineral nutrients removed by the crop the farmer uses artificial fertilisers, but this does not restore the humus content which is so valuable for water retention and soil texture. Previously on mixed farms where animals and crops were produced together, the manure from the animals could be used to enrich the soil for the crops, but with more specialised farming this is no longer possible. Consider the situation where the land is particularly suitable for cereal growing as in East Anglia. In order to grow cereals intensively the traditional rotation of crops has been abandoned and hedgerows and windbreaks have been torn down to increase the size of the fields and to make it easier for larger machines to operate. (In the country as a whole farmers have taken out 100,000 kilometres of hedgerow over 20 years with consequent loss of wildlife.) As a result of these practices humus has been lost, the soil has become lighter and in times of drought some of the top soil has literally been blown away. The use of heavy machinery may also make the soil more compacted and so more liable to flooding. A compacted soil also reduces the amount of fertiliser that reaches the root systems of the crops, and in wet weather the fertiliser may be washed out of the soil and reach rivers and lakes. Here the extra nitrogen and phosphorus cause rapid growth of algae and when these die the bacteria which bring about their decay use up so much oxygen from the water that fish and other aquatic animals may suffocate. This form of pollution is known as **eutrophication**; it is already a very serious problem in lakes and rivers wherever these intensive methods of farming are practised.

Fig. 33:7 Combating soil erosion in hilly country by terracing: Maharashtra, India.

Factory farming

The need for more cheaply produced animal food with less land to produce it has led to this specialised form of farming, where animals are reared indoors. The ethics of these methods are not considered here, but the results are certainly impressive. In Britain in 1970, 12 million cattle, 7 million pigs and 127 million poultry spent all or some of their lives indoors. With the production of new strains and intensive feeding methods beef cattle now mature in 10–18 months, compared with 3 years in 1946; broiler chickens are ready in under 9 weeks compared with 16–18 weeks; on average a hen lays 211 eggs per year compared with 108, and cows yield 3700 litres of milk compared with 2475. But consider the problems which result:

1. All the animals produce a colossal amount of manure which is concentrated in one place, and in most cases there is no agricultural land nearby to put it on. It is not economical to move it to arable farms which need it so badly and much of it goes into the drains and streams and pollutes them. However, there are schemes which could make farms self sufficient in energy by using the manure to provide methane gas by bacterial action.
2. When large numbers of animals are reared together they are more prone to disease. An epidemic would be disastrous to the farmer, so to prevent it, small quantities of antibiotics are used repeatedly. This action has led to the development of bacterial strains which are resistant to the antibiotics, and this may pose a real threat to the health of both animals and humans.

Fig. 33:8 Factory farming: beef cattle reared indoors.

3. To obtain these impressive yields the animals have to be fed on high protein animal feed, much of which comes from developing countries where the need for extra protein for the human population is far greater. This raises both humanitarian and political issues of great importance.

In order to take a balanced view of all these techniques which man has developed, the beneficial results have to be weighed against the harmful, but in coming to a conclusion it is wise to apply the ecological principle that there should be a true balance between what is taken out of the land and what goes back into it.

This raises big questions—moral, social and political. For example:

1. Should man try to extract from the land the *maximum* possible, irrespective of the harm to the environment the method may cause, or should he be content with less—the *optimum*—which would allow him to keep his land in balance?
2. Do people in the affluent countries actually need all the food that is being produced? Consider the E.E.C. problem of surplus in the form of 'butter mountains,' 'beef mountains' and 'wine lakes'.
3. How can the developing countries increase their food production without inheriting the same problems as the more highly developed nations?

The effect of industry and technology

The environment has not only been greatly affected by man's agricultural activities, but also by industrial development. The tremendous technological achievements of the past century have brought many benefits, including a rising standard of living, but how far has our environment suffered? How far has industry polluted the air, water and land? How much is being done to keep these essential parts of the ecosystem unharmed?

Air pollution

Industry has been greatly dependent upon combustion to provide energy and wherever fuels are burnt vast quantities of waste pass into the atmosphere. In Britain, before the Clean Air Act of 1956, it was estimated that 2·3 million tonnes of smoke, 5·2 million tonnes of sulphur dioxide, 0·8 million tonnes of grit and 0·3 million tonnes of acid were emitted into the atmosphere each year. These substances had serious effects on plants, causing an annual loss to agriculture of about £10 million, and producing a much more serious effect on the health of animals and in particular, man. Since 1956 the smoke problem has been greatly reduced, and buildings and vegetation in industrial towns and cities are no longer covered in grime. Further improvements will certainly take place with the greater use of smokeless fuel, natural gas and nuclear power, but sulphur dioxide is still a major problem as this harmful gas is produced mainly by the burning of fossil fuels, especially in power stations. Winds can cause sulphur dioxide to drift for hundreds of miles and fall as 'acid rain' damaging forests, rivers and lakes. Many of the lakes in Sweden are now devoid of wildlife as a consequence of this pollution—much of it coming from industrial plant in Britain. However, sulphur can be removed from fuel before burning, or retained after burning, although the process is costly.

Whenever combustion takes place oxygen is used up and carbon dioxide is produced. The normal recycling of carbon dioxide (p. 245) in the world ecosystem has kept the level of carbon dioxide in the atmosphere remarkably constant, but in recent years the vast increase in combustion has led to a 10% rise and it is still going up. Carbon dioxide in the atmosphere traps the sun's heat, and some scientists believe that by the year 2000 the world's average temperature may go up by 0·5°C. The danger of such a rise is that it could cause considerable melting of the world's ice, with a corresponding rise of ocean levels and consequent flooding. However, no actual rise in temperature has so far been discernible, due possibly to other counteracting factors such as the presence of atmospheric dust preventing some heat from reaching the earth. Not nearly enough is known about the consequences of these factors. Keeping the carbon dioxide cycle in mind, you will see that there is only one basic way of counteracting a rise in carbon dioxide levels—by extra photosynthesis. This is a factor to remember when considering the effects of increasing 'concrete jungle' at the expense of forests and green fields (p. 263).

Motor vehicles are also major pollutants of the atmosphere as the combustion of motor fuel produces not only carbon dioxide, but also carbon monoxide, oxides of nitrogen and lead. This is so bad in some cities that in Tokyo, for example, traffic policemen are made to breathe pure oxygen after 20 minutes of duty. The carbon monoxide in exhaust fumes is very poisonous as it combines with the haemoglobin of the blood and so prevents it from carrying oxygen. The lead is also a poison as it accumulates in the body causing very serious illness. Research is going on to discover the most effective way of removing it from petrol. This is quite possible to do, but once more, it adds to the cost and somebody has to pay.

Nuclear fall-out is another form of pollution. After an atomic test, radioactive substances may reach the ground in rain, get taken up by plants, and may then pass up through the food web, becoming concentrated at each step. For example, the radio-isotope, strontium 90, can be taken up by cattle and passed on to man in milk and cheese, and end up in his bones. Here it may affect the bone marrow where blood corpuscles are made, causing **leukaemia** or blood cancer. Similarly, caesium 137 may be picked up directly from vegetables and concentrated in various organs of the body, including the gonads. Here it may damage the carriers of hereditary characteristics, the **genes** (p. 270), and so have a harmful effect on future children. Although the amounts of these radio-active isotopes found in children are very small, and not thought to be harmful, the danger lies in their cumulative effects. Strontium 90 is radio-active for a long time—at least 28 years. This recognised danger from nuclear fall-out led to the Test-ban Treaty of 1963 which prevented further air-testing of atomic weapons by Russia and America.

Water pollution

Rivers have always been the dumping ground for man's unwanted material. It is astonishing how much waste can actually be broken down into harmless products by bacteria, but there are limits beyond which rivers become stinking sewers. Nearly all aquatic life depends on the oxygen dissolved in the water and if too much organic matter is present the bacteria causing its decay will use up all the oxygen and other organisms will suffocate. It is then that the anaerobic bacteria take over and release evil smelling gases such as hydrogen sulphide.

We have already mentioned the problems of raw human sewage and how it can be processed before being passed into the water (p. 137), the dangers of high concentrations of animal manure resulting from factory farming practice, and the run-off of artificial fertilisers which cause 'blooms' of algae to develop which may later die and decay. All these factors may help to make a river 'dead' through lack of oxygen. However, industry adds to the load of these **biodegradable** substances—those which may be broken down by bacteria. In Japan, for example, the organic wastes from pulp and paper-making have rendered the rivers round the city of Fuiji completely dead. When large rivers empty their contents into lakes and inland seas, these in turn silt up rapidly, their oxygen content falls and they too become dead. Lake Erie in North America is a notorious example, but many lakes in Europe, and even seas such as the Baltic, approach this condition and support little life. To add to the problem there are the varied chemical effluents from factories and the accidental seepage of toxic agricultural chemicals into the river system. In three years it was estimated that 15 million fish were killed by the pesticide endrin which leaked into the Mississippi, and in 1969 when 90 kg of endosulfan fell off a Rhine barge in Germany, millions of fish died further down-stream, mainly in Holland. These are extreme examples of an event which, on a smaller scale, occurs commonly.

But what is being done about it? First, there has been a tightening of the laws in many countries regarding the dumping of raw sewage into rivers and the control of effluents from factories. Second, vast sums of money are now being spent on cleaning up rivers and lakes—especially in America and Europe. Some rivers, such as the Thames, heavily polluted 15 years ago, now have many species of fish in them once more. The situation is certainly improving in many countries.

What of the oceans? They are so vast that most people looked upon them, until recently, as an inexhaustible depository, but this is far from the truth. The oceans are the sinks into which effluents from the rivers pass. Pollutants come in, but they do not pass out! If they are biodegradable, bacteria will act upon them, but if not they will accumulate.

Fig. 33:9 Oil pollution: a guillemot with oiled feathers.

Oil pollution is a particularly serious threat, not only because it fouls our beaches, but because of its effect on living things. A United Nations Committee estimated that over 2 million tonnes of oil finds its way into the oceans every year, about half of this by deliberate action. Collisions involving giant tankers and massive oil leaks from underwater drilling have already occurred, and in spite of the recognised danger, tankers become larger every year and underwater drillings are more common.

One obvious effect of oil on wildlife is the fouling of the feathers of seabirds, with their consequent death. It has been estimated that 10% of the eiderduck population in British waters died in this way when oil spilled out from an oil tanker after a collision in the Tay Estuary in 1968. When the giant oil tanker, the *Torrey Canyon*, ran aground off the Scilly Isles in 1967, the huge oil slicks were first tackled with the help of detergents, but these were found to be far more toxic to the wild life than the oil. Today, more effective and less lethal methods are used, but prevention is always better than cure.

A fact of importance when considering the effect of pollution on the oceans is that so much depends upon the **phytoplankton**—those microscopic green plants in the surface waters of the sea which are the basis of nearly all oceanic food webs. Not only are they the main producers of the oceans, but they provide through photosynthesis a quarter of all the oxygen in the atmosphere. Anything that affects these vital plants adversely on a large scale is clearly a grave hazard. It is also significant that estuaries and shorelines are the main spawning grounds of fish and these areas are liable to be the most polluted.

The oceans have also been the dumping ground for the most dangerous poisons including stock-piled nerve gases and radio-active wastes. Although sealed in containers, leakages of some poisons have been known to occur causing large-scale deaths of fish and seabirds. A great and growing hazard arises in the disposal of nuclear waste which is increasing year by year as more atomic power stations are built. Most of this waste is either sealed in stainless steel tanks if it is liquid, or in other 'safe' containers if solid, and dumped in deep water. The operation is subject to the strictest control and there is regular inspection of the water and fish catches for signs of radio-activity, but the potential danger is very great.

Nuclear power stations need vast amounts of water to cool the reactors; this warmed-up water passes back into the rivers or sea causing **thermal pollution**. The rise in temperature greatly affects the organisms living there and may destroy many of them. However, in some places an attempt has been made to use this rise in temperature to good effect, as a few species such as eels and carp grow better in warmer water.

Land pollution

Man is producing more and more waste as industry and technology advances. Consider the effect of mining and quarrying. In the United States alone, 3 million tonnes of slag are being produced annually; this indicates the size of the problem of disposal. But a lot is being done; some waste is being left underground in disused workings, dangerous tips in the vicinity of houses are being removed and others reclaimed by growing suitable plants which can withstand the toxic effects of many of these wastes. Much research has gone into finding the best plants to grow on these heaps according to their composition, in order to provide a gradual succession of vegetation and to build up a fertile soil once more.

Household and industrial waste is also increasing at an alarming rate. In addition to the garbage, food wrappings, plastic, metal and glass containers and waste paper that fill household bins every day, there is the increasing bulk of worn-out machinery, household furniture and unwanted cars to be disposed of. In the United States, to mention just a few items, 7 million cars, 48 billion metal cans and 26 billion bottles are discarded annually. All contain valuable material, but much of it is merely used for a time and then abandoned. Thus man is using more and more of the earth's resources and is wastefully throwing much of it away after use. We now realise that supplies of raw materials are not inexhaustible and that this kind of wastage cannot go on. So we are faced with two big problems: how to collect the more valuable material so that it can be recycled and how to dispose of the unusable residues without harming the environment.

Collection is the most difficult and expensive, because the waste is dispersed so widely; every house and factory has its quota, and even the roadside parking and picnic places are littered with cans, bottles and wrappings thoughtlessly thrown away by polluters of the countryside. In many countries the problem of litter is being tackled by the imposition of heavy fines, but this would not be necessary if each person took responsibility for his own litter.

Any method which separates bio-degradable waste such as garbage from other waste products is valuable, as the former can be treated and put back on the land as a useful fertiliser. In Holland over 30% of the waste from cities has been returned to the land as compost.

Disposal of waste is being tackled in many ways. These include:

1. Tipping

This is often the cheapest method and can be effective, especially in smaller communities, if suitable sites are available. The best way is for the refuse to be spread in layers about 2 m thick and then covered with 30 cm of soil before another layer is added. In this way it gradually builds up, like a layered cake, consolidates, and may eventually be reclaimed as useful land. Such tips, however, are apt to attract pests such as rats and flies and the use of pesticides to control them is usually advocated. In addition, their possible adverse effects have to be considered. Suitable sites for tipping are often hard to find and the habit of using wetlands can be strongly criticised as in many places these are valuable ecosystems in their own right for wildlife and may be useful fishing areas.

2. Burning

Modern incinerators have great advantages over tipping, which is at best a short-term policy. Cities such as Osaka, Paris and Düsseldorf have installed great plants which not only dispose of the rubbish, but may generate enough heat in the process to warm many of the buildings and also produce electricity. Metals are also recovered from the waste and the remaining ash is used for making concrete blocks for building.

3. Special processes

These include such machines as car-shredders, which crush and break up old cars so that the materials can be processed and used again.

The problem of land space

Man's history of land use has been one of haphazard development. Only in recent times has there been any real attempt to plan the best use for the available land. In many parts of the world much land has been spoiled, disfigured or wasted through lack of foresight, greed and expediency. But because of the pressure of expanding populations, massive technological advances and the necessity to grow more food, we are beginning to realise what a precious commodity land is.

Apart from the need to grow more food, there are many other uses to which land may be put. In the more densely populated countries, valuable agricultural land is being whittled away each year at an alarming rate as more houses are built and industries set up, more roads and motorways are constructed and more valleys flooded to meet the growing need for water. In the developing countries, with their fast expanding populations, some of the last of the major natural ecosystems are being destroyed as forests are cut down and scrub and hill-sides cleared to make way for agriculture. For example, an area of tropical rain forest the size of Wales is being felled every month affecting both the local and global climate. Also the vast areas of the African savannahs, with their teeming wildlife are being invaded on all sides and the larger animals are being forced into smaller and smaller areas.

But 'man does not live by bread alone'. He needs land for other uses; more and more he needs relaxation from the pressures of modern life. He needs recreation; he needs to satisfy his deep-seated craving for natural beauty as an antidote for the artificiality of the urban environment. The wildlife with which he shares this planet also have their needs. Ecology makes it abundantly clear that we are part of nature, not apart from it. The world is a single ecosystem and there is a precarious balance between all the organisms which compose it. This balance has already been considerably disturbed by man. It is essential that natural ecosystems should remain and be studied so that he may understand the complexities of this balance and that his actions may not lead to further disasters. It is unthinkable that the magnificent heritage of wildlife that has taken hundreds of millions of years to evolve should, in the course of a few decades, be destroyed for ever. The need for national parks and smaller local nature reserves becomes more and more urgent as human activities expand. At present they take up less than 1% of the earth's surface and many of them, especially in America and Africa, have become so popular as tourist attractions that human pressures on these parks are becoming a major problem. A compromise has to be reached between tourism and enjoyment, and the destruction by human pressure of the very ecosystem we wish to enjoy. This calls for careful management so that large areas are left undisturbed by the public. How then can so many conflicting claims on land-use be satisfied when land is already in such limited supply? It can only be done by the most careful planning at local, national and international levels, and by the scientific conservation of our whole environment.

The principle of **multiple land use** is an important one, particularly in regions of high population density. It can be illustrated by a scheme for the reclamation of a large area of peat moor in south-west England after the peat has been extracted for commercial purposes. The area is particularly valuable for its distinctive flora and fauna and in consequence some of it has been bought by the County Trust for Nature Conservation to prevent further peat extraction and to conserve the present ecosystem. But by far the greatest part will have its peat removed right down to the underlying clay. This will leave a vast 'hole in the ground'. The plan is to fill it with water and produce a valuable reservoir. Part of the lake thus formed will be kept as an amenity area for sailing, part will be colonised by reeds to become a wildfowl refuge and fish will be introduced for the anglers. In this way the one area can serve a variety of interests.

Conservation

In the widest sense this means the sensible utilization of natural resources; the keeping of the environment in a state of balance; the restoration and maintenance of unpolluted air, land and water and the preservation by careful management of a great variety of ecosystems with their diverse communities of plants and animals. We have already referred to many examples of conservation. But we can only conserve our environment if we face the situation squarely, realistically and optimistically.

The world problems are serious. Man is in the grip of a population explosion without parallel in human history; unless this is stabilised the pressures will continue to increase. With his great technological advances he has replaced stable natural ecosystems with unstable artificial or semi-artificial ones. He has subdued nature, but in doing so has broken most of the ecological laws and brought an

Fig. 33 : 11 Members of the British Trust for Conservation Volunteers re-claiming a pond in Britain.

infinity of problems on himself in consequence. With the increasing demands of a consumer society he has dug deeply into the world's resources; he has taken out far more than he has put back for further use, and has very considerably polluted the environment on which he depends for life—how much, nobody yet knows.

But all is certainly not gloom! Man is capable of great wisdom as well as great destruction; he can analyse the difficulties he has created and reverse the dangerous policies he has set in motion. Already there is the knowledge and the technology to do this, and in this chapter we have seen how major problems are being solved. Many countries have launched effective programmes for pollution control and the recycling of the more important basic resources; rivers are being cleaned up, dangerous pesticides have been banned and less harmful alternatives used; land is being restored, erosion is being prevented, reafforestation is taking place, national parks and nature reserves are being created and managed. A start has been made, but there is a tremendous task ahead.

Conservation in its widest sense is of paramount importance; it is a matter of the greatest urgency involving governments and industry, science and technology, but above all it concerns each one of us. When we begin to understand the problems, are really concerned about them, and take responsibility for them in some positive way, if only on the smallest scale, we are on the way towards solving them. We all have to think and act ecologically—we have to live and work in harmony with each other and with the rest of the natural world, because the world is a single ecosystem which is the precious heritage and responsibility of us all.

In 1980 the World Conservation Strategy was launched by the International Union for the Conservation of Nature and Natural Resources (IUCN). This is a comprehensive plan covering all aspects of conservation. If all governments co-operate by putting into practice the principles embodied in this plan the whole of mankind will reap the benefit. But will they? To quote from *Only One Earth* by Barbara Ward and René Dubos: 'Today in human society, we can perhaps hope to survive in all our prized diversity providing we can achieve an ultimate loyalty to our single, beautiful and vulnerable planet earth. Alone in space, alone in its life-supporting systems, powered by inconceivable energies, mediating them to use through the most delicate adjustments, wayward, unlikely, unpredictable, but nourishing, enlivening and enriching in the largest degree—is this not a precious home for all of us earthlings? Is it not worth our love? Does it not deserve all the inventiveness and courage and generosity of which we are capable to preserve it from degradation and destruction and, by so doing, to secure our own survival?'

For those who wish to become involved in the work of conservation here are some ideas:

1. Take part in such group projects as clearing footpaths and bridle ways, and restoring canals, ponds and rivers so that they may support thriving communities once more. If there is an ugly or derelict area in your neighbourhood, help to turn it into something more pleasant, such as a garden, a copse or a nature reserve.

For these projects you will need to obtain permission from owners and councils, and expert advice will be necessary. Consult such bodies as your District or County Council, the Nature Conservancy Council, the Forestry Commission and your County Trust for Nature Conservation.

2. Make your garden more attractive to wild life. Provide food and water for birds in winter and in times of drought. Erect nest boxes. If possible leave part of the garden 'wild' with plenty of cover. Grow plants which have a special attraction for bees and butterflies, and those on which the caterpillars of the butterflies feed.

3. Join a local or national organisation devoted to conservation such as the County Trusts for Nature Conservation (most have conservation corps), the Wildlife Youth Service of the World Wildlife Fund or the British Trust for Conservation Volunteers.

34

Genetics

Variations

It is obvious that all members of a species have certain characteristics in common which distinguish them from other species, but within the species there is also much variation.

Looking round the class you will readily agree that you are all different; you may also agree, probably for the wrong reasons, that this is a very good thing! We vary in many ways; in the colour of our eyes, skin and hair, in the shape of our noses, whether or not our ears are lobed, in our height, weight and the proportions of our bodies, in our blood groups, our finger prints, in our ability to recognise colours and whether or not we can roll our tongues. The same applies to other animals; dogs and cat differ greatly from one another, so do the pigeons in our cities. It is more difficult to see variation, for example, in the members of a flock of gulls, but on close inspection differences can be detected. Certainly the birds themselves have no difficulty in recognising each other. Plants also show variation. Individuals of the same species may vary in height, in the number of petals, in the colour or shape of the seeds, in their reproductive capacity and so on.

Kinds of variation

You will see from some of the examples of variation given so far that they may be grouped into two categories. In the first, variations such as a person's blood group, are quite definite and clear-cut. These are known as **discontinuous variations**. In the second, of which height is an example, variation is **continuous** as there is a complete range of intermediates between the two extremes. Consider the examples of human variation mentioned above; into which of these two categories would you place each?

Variations can also be classified according to whether they are inherited or not. It is easy to recognise that some clear-cut variations are inherited, but with those which show a continuous range it is more difficult, as environmental factors may also have an effect. For example, our general body shape and size is determined by heredity, but it is modified by the amount of food that we eat and the exercise we take. The same applies to intelligence; heredity certainly plays a most important part, but the type of upbringing, especially during the early years, is a very significant factor too. The scientific study of heredity and variation is known as **genetics**.

Mendel's experiments

Man has experimented with the breeding of plants and animals for many hundreds of years, in order to produce more useful varieties. He did so by selecting individuals for breeding which had the most useful characteristics. Sometimes the resulting progeny had the desired characters, but more often than not the opposite was the case. It was not until 1866 that **Gregor Mendel**, an Austrian monk who was both a biologist and a mathematician, published the results of his experiments which put the whole subject of inheritance on a sound scientific basis. The great significance of his work was not realised until 1900 when his paper, which had been published in a local journal, was 'rediscovered' by three eminent scientists working independently in different parts of Europe.

Mendel worked mainly on garden peas because they were easy to grow, had many distinct varieties and produced numerous seeds in a relatively short time. Garden peas have flowers which are usually self-pollinated. Mendel noticed that some pea plants had stems which were consistently tall over several generations, i.e. they bred true, whereas others were short; some had yellow seeds, others green; some had red flowers, others white. He devised simple experiments involving the **cross-breeding** of varieties having these contrasting characters. By removing the stamens early and dusting the stigmas with pollen from another variety, Mendel found that cross-pollination could easily be achieved. To make certain that no pollen was deposited from any other source, he covered the flowers with small paper bags.

Choosing plants which had bred true for a particular character, Mendel devised seven cross-breeding experiments, each involving one pair of contrasting characters. When the seeds from these crosses were ripe he then grew them all in separate plots and noted which characteristic appeared in this **first filial** or $\mathbf{F_1}$ generation. (Note: in the case of crosses involving seed or pod characters the type of F_1 could, of course, be determined as soon as pods had formed on the parent plants).

In all seven experiments *all* the progeny resembled *one* of the parents only. Thus when tall plants were crossed with short, *all* the progeny were tall. The result was the same irrespective of whether pollen was taken from the tall or short parent. He then allowed all these F_1 plants to self-pollinate and again he carefully grew all the seeds that were produced. This time the results were very different. For example, when the tall F_1 plants were self-pollinated the resulting **second filial** or $\mathbf{F_2}$ generation consisted mainly of tall plants, but there were some short ones as well. He counted them carefully and worked out the ratio. The results he obtained are shown in Table 34:1.

The character which always showed in the F_1 Mendel called the **dominant**, the one that remained hidden, but appeared once more in the F_2 he called the **recessive**. But perhaps the most significant thing about his results was that the ratios all approximated to 3:1 in the F_2.

Mendel carried the experiments further by allowing the F_2 plants to self-pollinate and seeing what happened in the F_3. For example, he found that all the short plants bred true, but of the tall ones about one-third bred true and two-thirds produced both tall and short plants in approximately the same proportion of 3:1 as in the F_2. Similar results were obtained for the other characteristics investigated. These results, using proportions only, are summarised in Fig. 34:1.

TABLE 34:1. MENDEL'S RESULTS IN CROSSES INVOLVING SEVEN PAIRS OF CONTRASTING CHARACTERS

Experiment	Type of Cross	F_1 Generation	F_2 Generation	Ratio
1.	Stem: tall (180–220 cm) × short (22–45 cm)	All tall	787 tall 277 short	2·84 : 1
2.	Seeds: yellow × green	All yellow	6022 yellow 2001 green	3·01 : 1
3.	Seeds: round × wrinkled	All round	5474 round 1850 wrinkled	2·96 : 1
4.	Pods: green × yellow	All green	428 green 152 yellow	2·82 : 1
5.	Pods: inflated × constricted	All inflated	882 inflated 299 constricted	2·95 : 1
6.	Flowers: red × white	All coloured	705 red 224 white	3·15 : 1
7.	Flowers: axial* × terminal**	All axial	651 axial 207 terminal	3·14 : 1

* axial: flowers spread all along the stem; ** terminal: flowers only at the end of the stem.

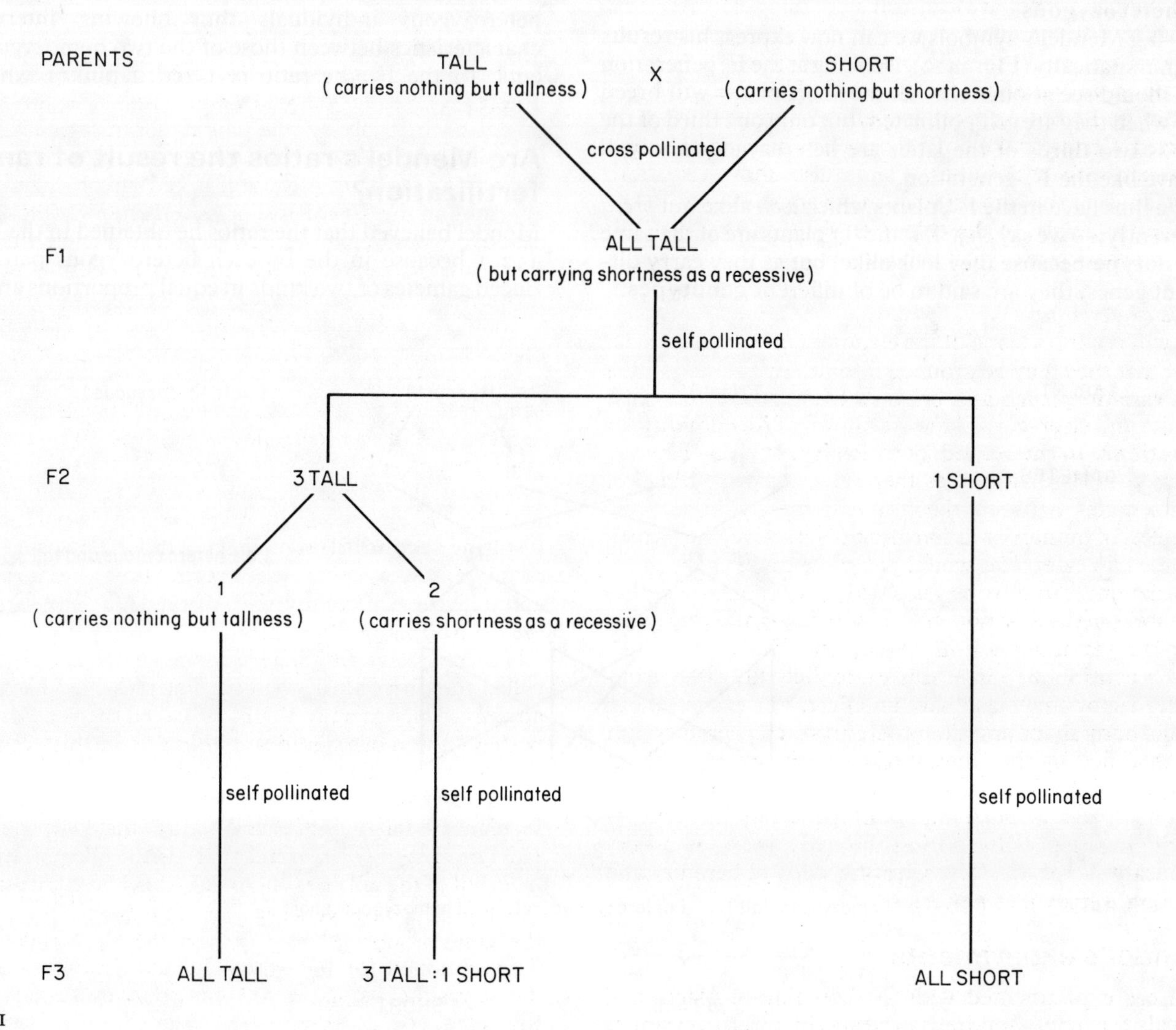

Fig. 34 : 1

How could these results be explained?

Mendel realised that there must be factors in pea plants which caused the characteristics such as tallness or shortness to develop. He had no idea what these factors were, so he used symbols. He called the factor for tallness T because it was dominant and the factor for shortness t because it was recessive.

Mendel also reasoned that these factors must occur in the parents as *pairs*, because in the F_1, although all the plants were tall they passed on the recessive character to the F_2, so there had to be a factor for tallness and another for shortness present. Consequently he called the parents TT and tt and the F_1 hybrids Tt.

Mendel also realised that if parent plants had pairs of factors, they must separate into single components when gametes were formed and pair up once more when fertilization took place, otherwise the numbers of such factors would double at each generation.

We now call these paired hereditary factors **genes**. The alternative forms of the same gene, e.g. T and t, are called **alleles**. When parents carry pairs of identical genes and therefore breed true, such as TT and tt, they are said to be **homozygous**, while those which are hybrids such as Tt, are **heterozygous**.

Using Mendel's symbols we can now express his results diagrammatically (Fig. 34:2). Looking at the F_2 generation you should see at once why all the short plants will breed true when they are self pollinated, but only one third of the tall, as two thirds of the latter are heterozygous and will behave like the F_1 generation.

We thus have in the F_2, plants which *look* alike but *breed* differently; so we say that TT and Tt plants are of the same **phenotype** because they look alike, but as they carry different genes, they are said to be of different **genotypes**.

We can now summarise Mendel's conclusions using modern terms, as follows:

1. The genes which determine a particular characteristic occur in pairs. The genes may both be the same (homozygous condition) or two alternative forms of the gene (alleles) may be present (heterozygous condition). Mendel discovered that in heterozygotes only one allele (the dominant) was expressed; the other (the recessive) remained hidden.
2. In the formation of gametes, each member of a pair of genes becomes separated, so that each gamete only carries one gene from that pair. It follows therefore that for a plant with genotype Tt, 50% of the gametes carry the T allele and 50% the t allele.
3. At fertilization the gametes fuse in a random manner and in the case of those from heterozygous parents the ratio of the progeny approximates to 3:1 in favour of the dominant character. However, this is in fact a 1:2:1 ratio as one third of the dominant phenotype will be homozygous and two thirds heterozygous.

Referring to 1. above, there are now known to be many instances where the dominance of one allele in a heterozygote is only partial. For example, some plants such as antirrhinums have both red- and white-flowered varieties; when crossed these produce pink-flowered F_1 plants, the heterozygous individuals thus showing intermediate characteristics between those of the two homozygous parents. In the F_2 the ratio is 1 red:2 pink:1 white (see Fig. 34:3).

Are Mendel's ratios the result of random fertilization?

Mendel believed that the ratios he obtained in the F_2 were 1:2:1 because in the F_1 each heterozygous parent produced gametes of two kinds in equal proportions and it was

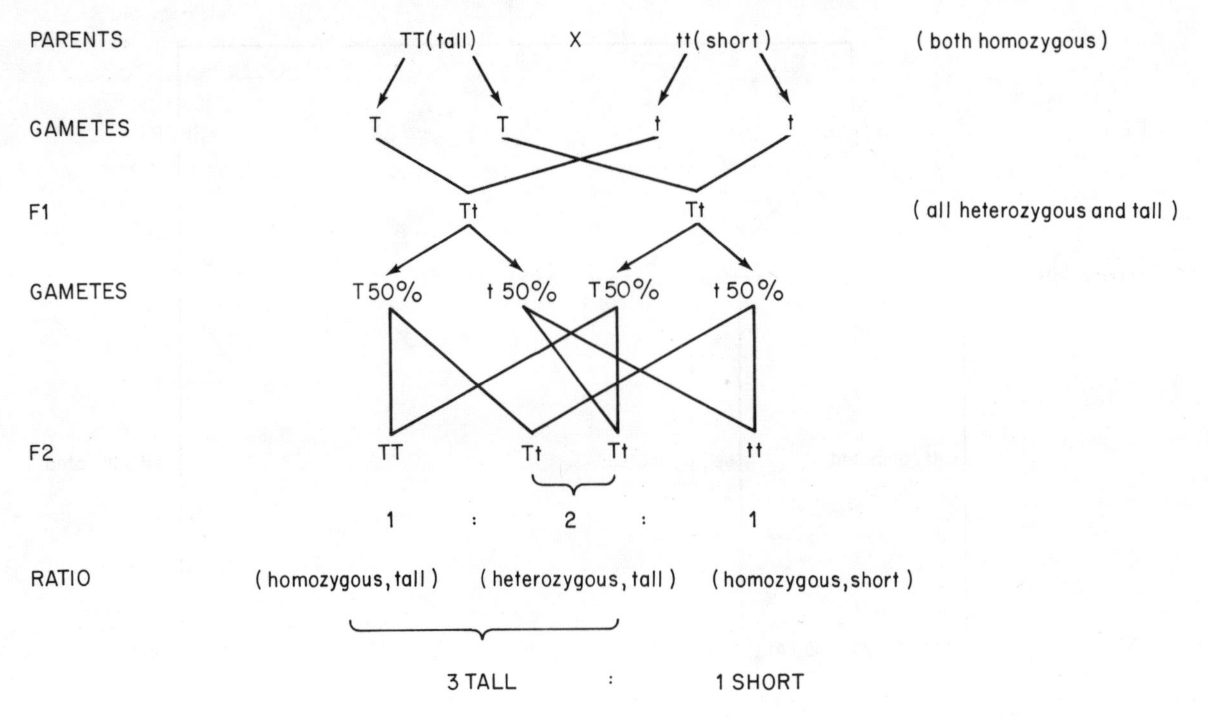

Fig. 34:2

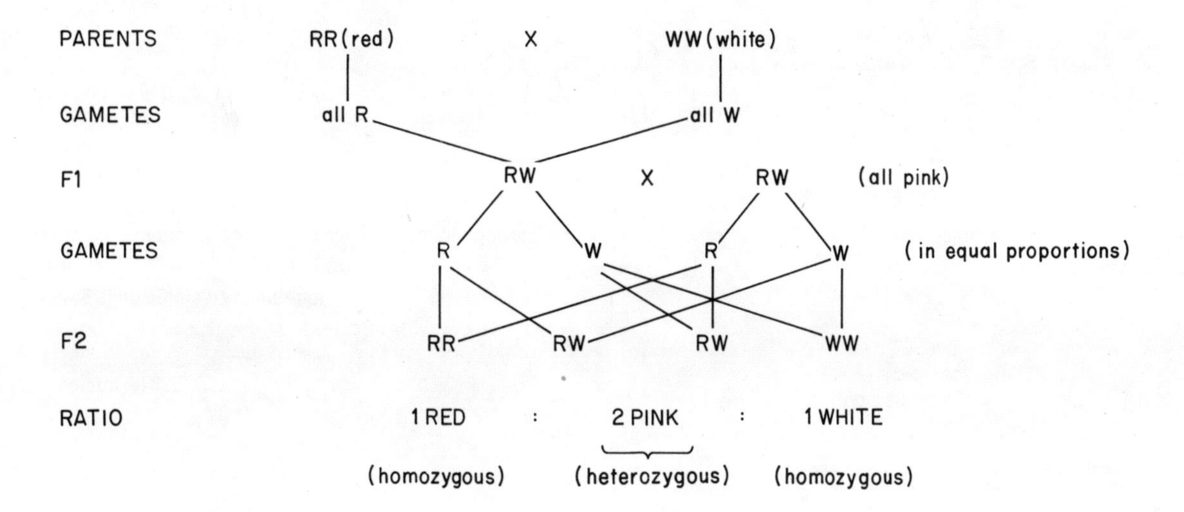

Fig. 34:3

entirely a matter of chance which fused with which. You can demonstrate this law of probability for yourself by doing some 'breeding' experiments using plastic beads.

Take two large beakers and place 100 red and 100 white beads in each. Each bead represents a gamete and each beaker will thus contain the two kinds of gametes produced by the F_1 in equal proportions. Let the beads in one beaker represent male gametes and the beads in the other the female gametes. Mix the beads thoroughly in each beaker and then extract one bead from each simultaneously without looking at them, i.e. at random, and place them together in pairs to represent the zygotes formed. Some pairs will be reds only, some white and some red and white. After you have taken out 12 pairs make a note of the ratio, repeat for 24 pairs and finally for 200 pairs. Compare your results with other members of the class. How nearly do your results approximate to 1:2:1? By using large numbers of pairs, were the results more, or less accurate?

You should now understand that, in practice, these theoretical Mendelian ratios are far more likely to be attained if the number of progeny is very large. Is this borne out by Mendel's seven experiments on peas? Are the most accurate ratios the ones where larger numbers were used? (Study Table 34:1.)

The mechanism of gene transfer

Mendel had brilliantly shown by his experiments that characteristics were inherited according to laws of probability. But during his lifetime there was no explanation of what genes were and what caused them to behave in the way they did. However, after his death it was discovered that the clue to the problem lay in the way chromosomes behaved during cell division.

We have already seen (p. 189) that each cell in the body of a plant or animal has a definite number of pairs of chromosomes in the nucleus, the number being characteristic of the species. For example, man has 23 pairs, a fruit fly 4 pairs and maize has 20 pairs. All through life when growth and repair is taking place the cells divide by **mitosis**. During this process the number of chromosomes *remains constant*. This happens because each one duplicates itself to form two identical chromosomes, one going into one daughter nucleus and the second going into the other.

However, when sexual reproduction takes place and gametes fuse, chromosome numbers would theoretically double, but in fact this does not occur because at some point in the life cycle the number is halved. This halving takes place in animals and some of the lower plants when gametes are formed. In flowering plants the chromosome number is halved during the formation of pollen grains and during the nuclear divisions in the ovule which result in the formation of the egg cell (p. 183). The type of cell division which brings about this reduction in chromosome number is called **meiosis**.

Meiosis

During meiosis the chromosome pairs become separated so that one *whole* chromosome from each pair passes into each daughter cell. Thus the number of chromosomes is reduced from 2n to n, where n is a number characteristic for the species (Fig. 34:4). In man there are 46 chromosomes (23 pairs) in the body cells, but in the eggs and sperms there are 23 single chromosomes. The term **diploid** is given to

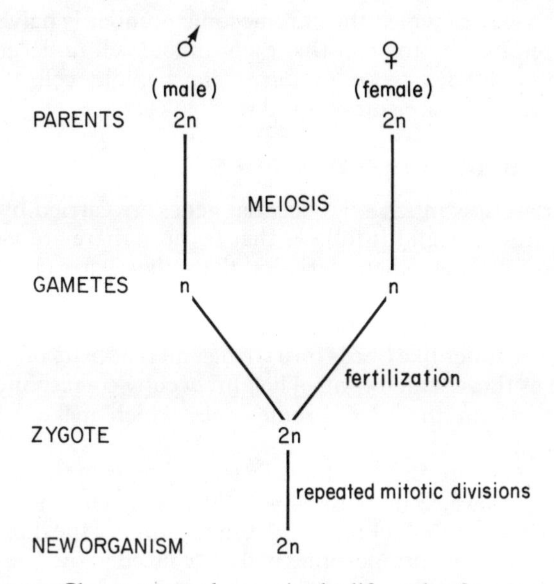

Fig. 34:4 Chromosome changes in the life cycle of an organism.

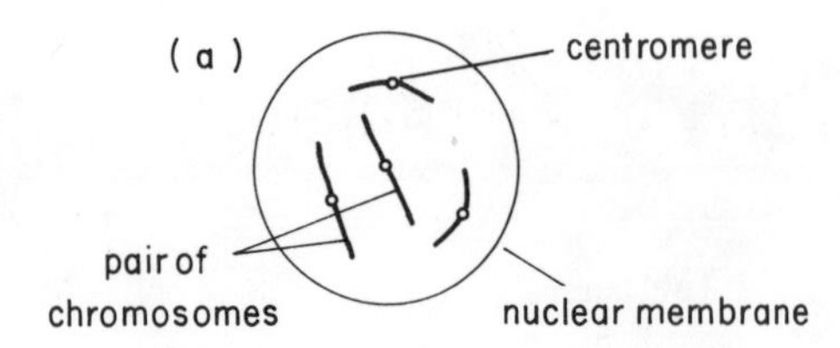

Cell showing two pairs of chromosomes. Each chromosome appears as a single thread.

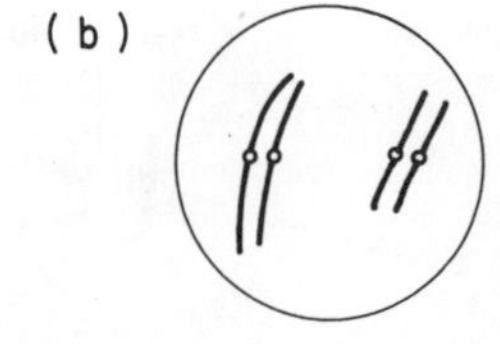

The chromosomes become associated in their pairs.

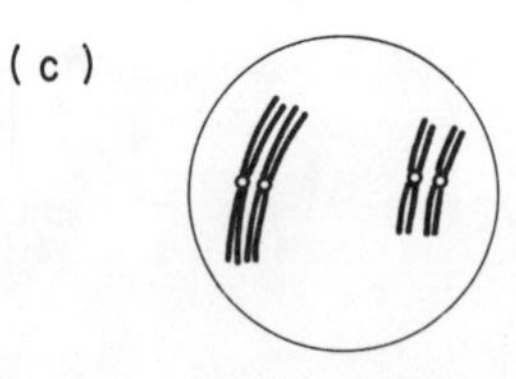

Each chromosome forms an identical copy of itself, but the two strands are held together by a single centromere.

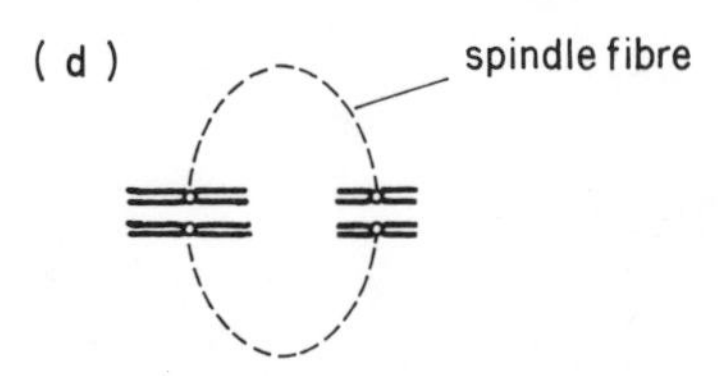

The nuclear membrane breaks down and the centromeres become attached to a spindle fibre.

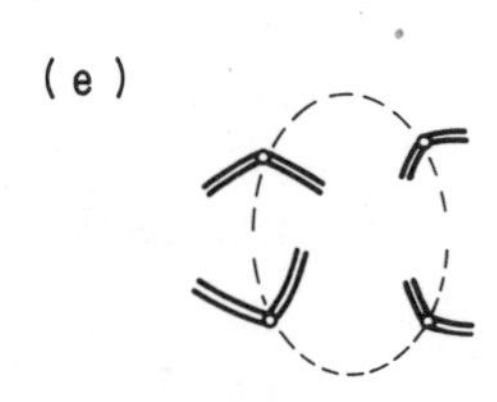

The chromosomes are drawn to opposite ends of the cell.

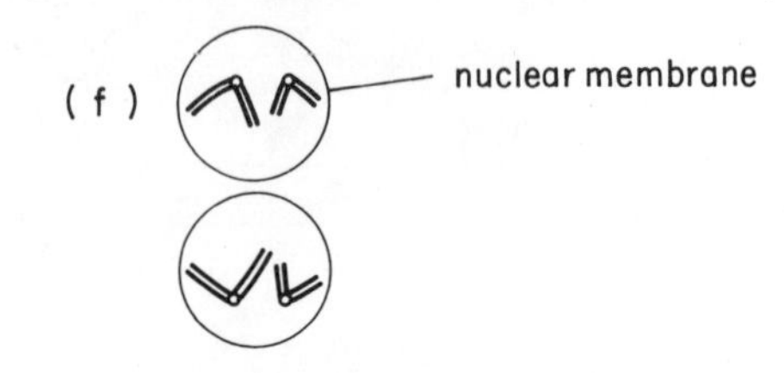

The nuclear membrane reforms and the cell divides into two. (not shown in diagram).

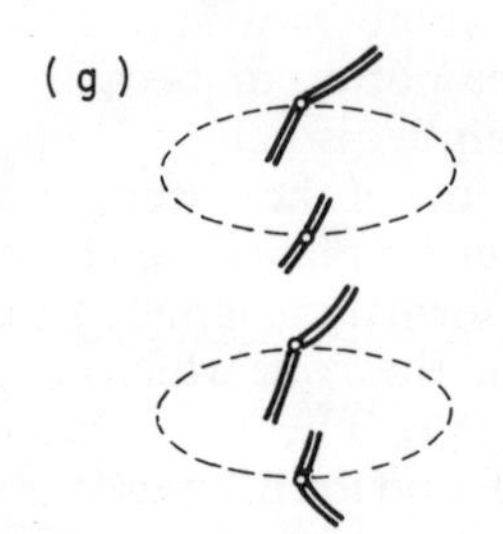

Nuclear membranes break down and a spindle forms. The chromosomes become attached to spindle fibres.

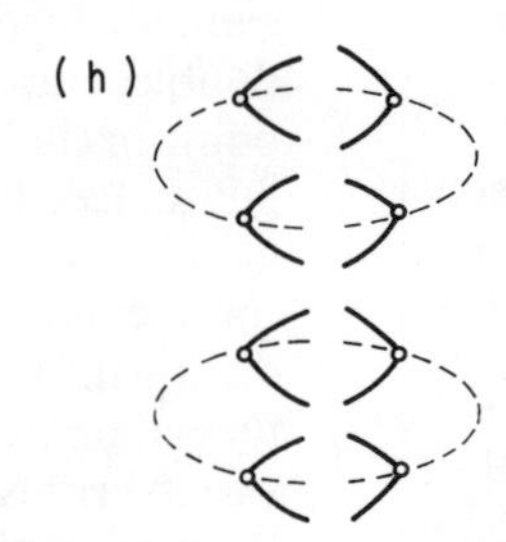

The centromeres divide and the chromosomes are drawn to opposite poles of the cell.

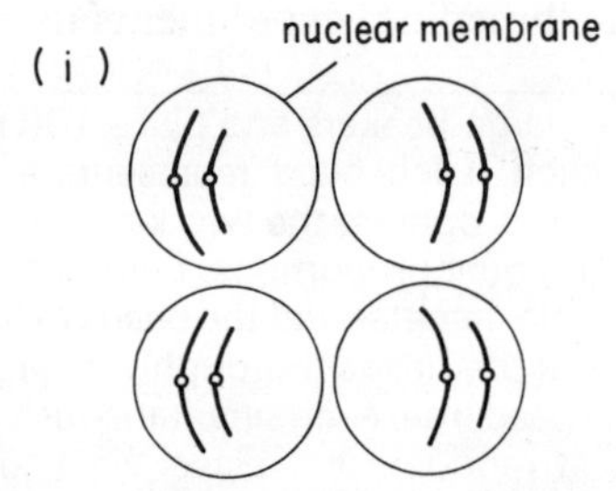

Nuclear membranes are formed around the chromosomes. Each cell divides, separating the two nuclei : thus four cells are formed , all of which contain only one member of each original pair of chromosomes.

Fig. 34:5 Stages in meiosis (nuclear division only); cell membranes not shown.

the condition where all the chromosomes are in pairs (2n), and **haploid** where only one member of each pair is present (n).

The way in which the chromosome number is halved is complicated by the fact that each diploid cell undergoing meiosis divides twice to form *four* haploid cells. The various stages are summarised in Fig. 34:5.

Genes and chromosomes

There is now much evidence that genes are carried by the chromosomes and it follows that because there are vastly more inherited characteristics than there are chromosomes, there must be a great many genes which are carried by each. It has been established that these are arranged in a linear manner like beads on a string and that pairs of genes such as those which control height, occupy corresponding positions on the two chromosomes which make a pair (Fig. 34:6).

If we now return to Mendel's experiments with tall and short peas we could depict his results using chromosomes instead of symbols (Fig. 34:7), but to simplify the diagram the number of chromosomes will be reduced to the one pair which carries this pair of genes. Let ● represent the gene for tallness and ○ the gene for shortness.

We can see therefore that the behaviour of chromosomes at meiosis and fertilization exactly confirms Mendel's hypothesis and explains how the genes become separated when gametes are formed and pair up once more in the zygote. It also demonstrates that in every chromosome pair one component has originally come from the father and one from the mother, hence there is an equal probability of father and mother contributing towards the characteristics of the resulting progeny.

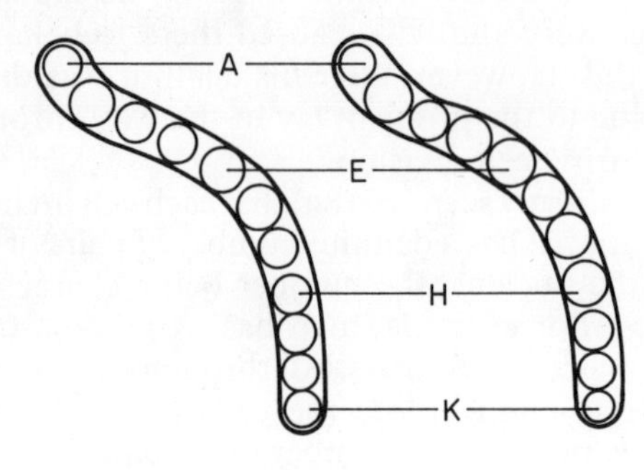

Fig. 34:6 Diagram illustrating the principle that pairs of genes occur in corresponding positions along the chromosomes.

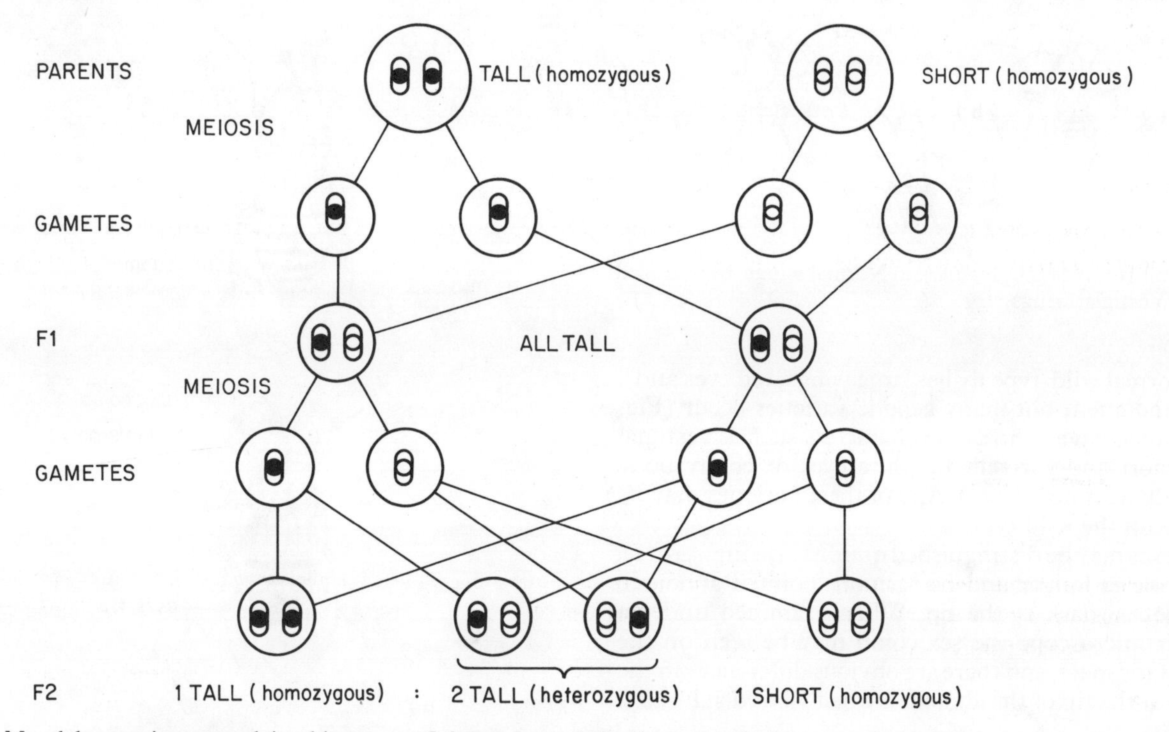

Fig. 34:7 Mendel: experiment explained in terms of chromosome behaviour at meiosis.

Sex determination

In man and the majority of animals the sex ratio of the progeny is approximately equal. This is because one pair of chromosomes is concerned with the determination of sex. In the majority of animals, in the female, this pair consists of chromosomes which are similar in size and shape and are called X chromosomes, but in the male one is a normal X chromosome and the other is much smaller and is called the Y. When the female forms gametes and the chromosomes separate in meiosis each egg will contain an X chromosome, but in the male, when sperms are formed, 50% will contain X and 50% Y. So it follows that with random fertilization there is an equal probability of a male or female zygote being formed.

In man this proportion is not achieved exactly as the probability of a Y-carrying sperm fertilizing an egg is slightly greater because it is fractionally lighter and more active than the X-carrying sperm.

It is not always the male that carries the Y chromosome; in some animals, e.g. birds and insects, it is the female, hence in these it is the egg which either contains an X or a Y chromosome, and the sperms which all contain an X.

It follows from this that the sex of the progeny is determined at fertilization, although the later development of secondary sexual characters is influenced by the sex hormones (p. 170).

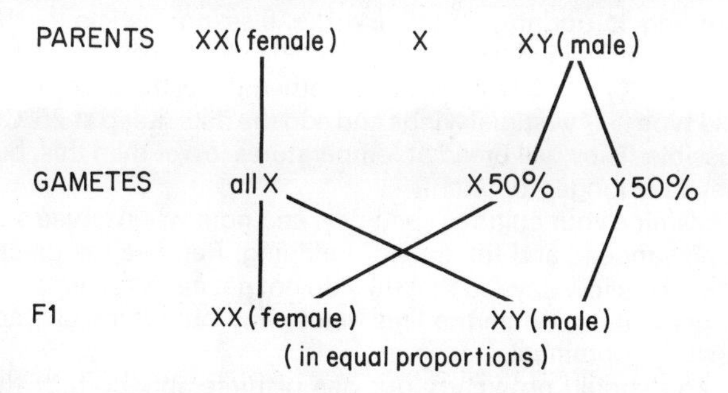

Fig. 34:8 The mechanism whereby the sex ratio is kept constant.

Breeding experiments

It is interesting to carry out breeding experiments yourself, but to be successful over a limited period of time it is essential to choose a species which breeds rapidly, is easy to keep and produces a large number of progeny. For these reasons the fruit fly *Drosophila melanogaster* is a suitable choice. Much genetic research has been done on this species.

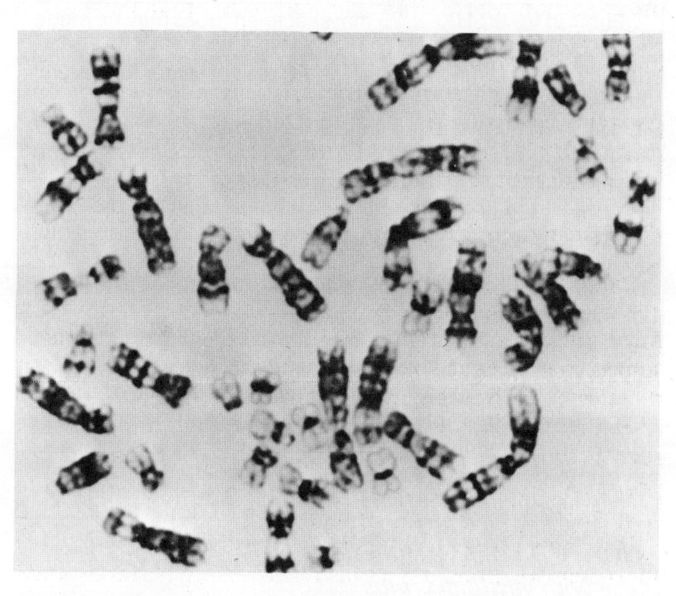

Fig. 34:9 Photograph of human chromosomes (male) from a body cell. There are 46 altogether composed of 23 pairs. Each pair is similar in size and shape except for the X and Y.

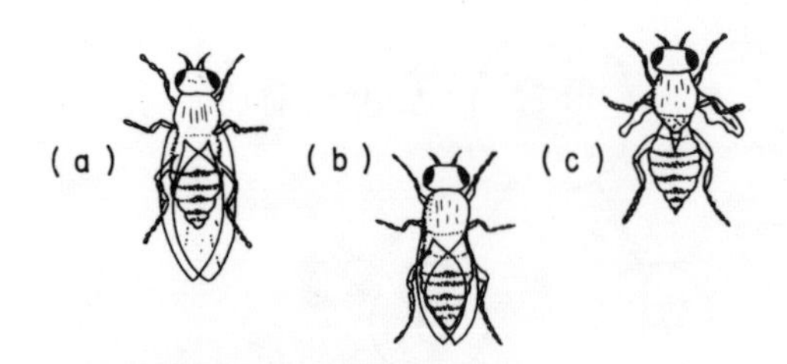

Fig. 34:10 Fruit flies (*Drosophila*): a) Normal wings. b) Reduced wings. c) Vestigial wings.

The normal wild-type fly has large wings, red eyes and a striped abdomen, but many genetic varieties occur (Fig. 34:10) which have contrasting characters such as vestigial wings (short and shrivelled), white eyes or ebony body (dark body with no stripes). Any of these varieties may be crossed with the wild type.

The sexes may be distinguished quite easily (Fig. 34:11). The female is longer and has a more pointed abdomen and is not so dark at the tip. When examined under a binocular microscope the sex comb may be seen on the foreleg of the male, and there are obvious differences in the genitalia at the tip of the abdomen when viewed ventrally.

Breeding the flies

They may be bred in wide-mouthed bottles or large specimen tubes using a specially-prepared culture medium as a food supply. A folded piece of paper towelling is placed inside (Fig. 34:12). When flies are introduced, they lay eggs on the medium and the semi-transparent larvae which soon hatch out tunnel into the food. When they are fully grown they climb up the towelling or sides of the glass to pupate. The flies hatch out in about 12 days if kept at 25°C.

Successful breeding depends on keeping all food and containers sterile. The culture medium may be prepared in the following way:

Preparing the food medium

Enough for 60 100 × 25 mm specimen tubes may be made as follows:
1. Soak 70 g of fine oatmeal in 120 cm^3 of water for several hours.
2. Dissolve 30 g of black treacle in 40 cm^3 of water.
3. Add 6 g of powdered agar to 400 cm^3 of water, stir and then heat to boiling to dissolve it.
4. Boil all the above together for 15 min adding 6 cm^3 of a 10% solution of nipagin in 95% alcohol—this is an anti-mould substance.
5. Pour the medium into the clean specimen tubes to a depth of 25 mm while still hot.
6. Sterilise the tubes and their plugs by keeping them under pressure in an autoclave for 15 minutes. Cool.
7. Push into the medium, while still warm, a piece of folded paper towelling or filter paper which has been sterilised.
8. When the medium has set add a few drops of yeast suspension and plug the tubes with the stoppers. They are now ready for culturing the flies.

Making experimental crosses involving wild type and those with vestigial wings

1. *To find out which is the dominant, normal wings (wild type) or vestigial wings.* Given pure breeding stocks of these two varieties, you will need to separate 5 virgin females of the wild type and 5 males with vestigial wings and put them together in a culture tube. Others in the class could do the reciprocal cross with 5 virgin females with vestigial wings and 5 males of the wild type.

Female flies do not mate until at least 8 hours after hatching so in order to obtain virgin females it is necessary to remove all adult flies from a culture which is hatching and then use the females which subsequently hatch during the next 8 hours. As the flies are active it is necessary to anaesthetise them lightly with ether using the technique shown in Fig. 34:12. Examine the flies on the tile under a lens, select the ones you need and place these in a culture tube. Keep the tube horizontal when the flies are put in and wait until they recover before standing it upright, otherwise the flies may stick to the medium and die. Repeat for the second stock. You now have 5 males (♂) and 5 females (♀) together. Label the tube, e.g. ♀ wild type + ♂ vestigial wings and add the date. Keep at 25°C if possible. They will breed at temperatures lower than this, but they take longer to hatch.

Examine your cultures each day, and note when larvae and pupae appear, and the date of hatching. Remove the parent flies after a few days so that they do not get mixed up with the F_1 generation. When the flies hatch, find out which characteristic is dominant.

2. You should now carry out one or preferably both of the following different crosses:

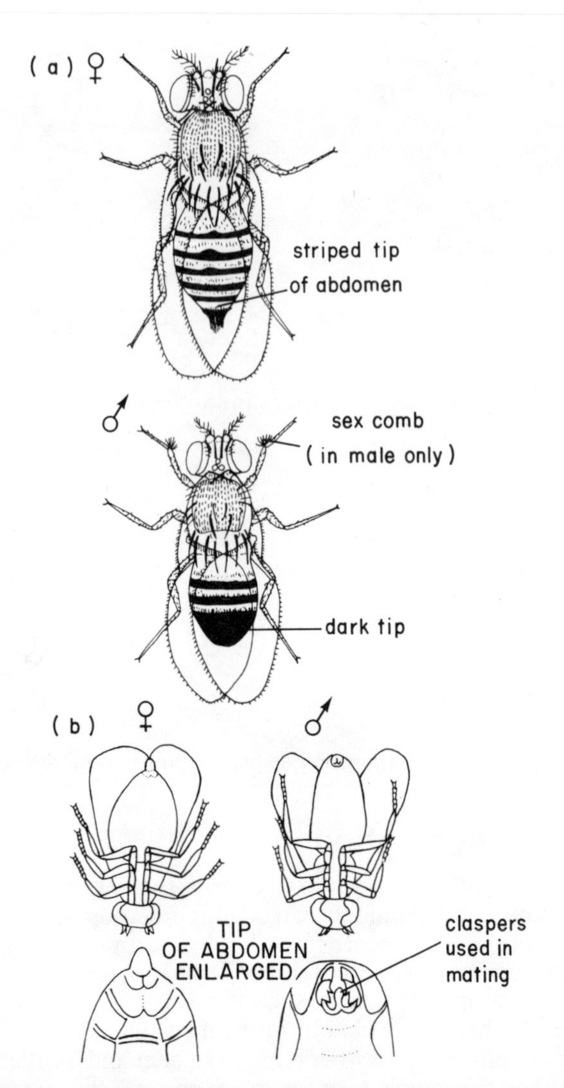

Fig. 34:11 Sex differences in *Drosophila*: a) dorsal b) ventral.

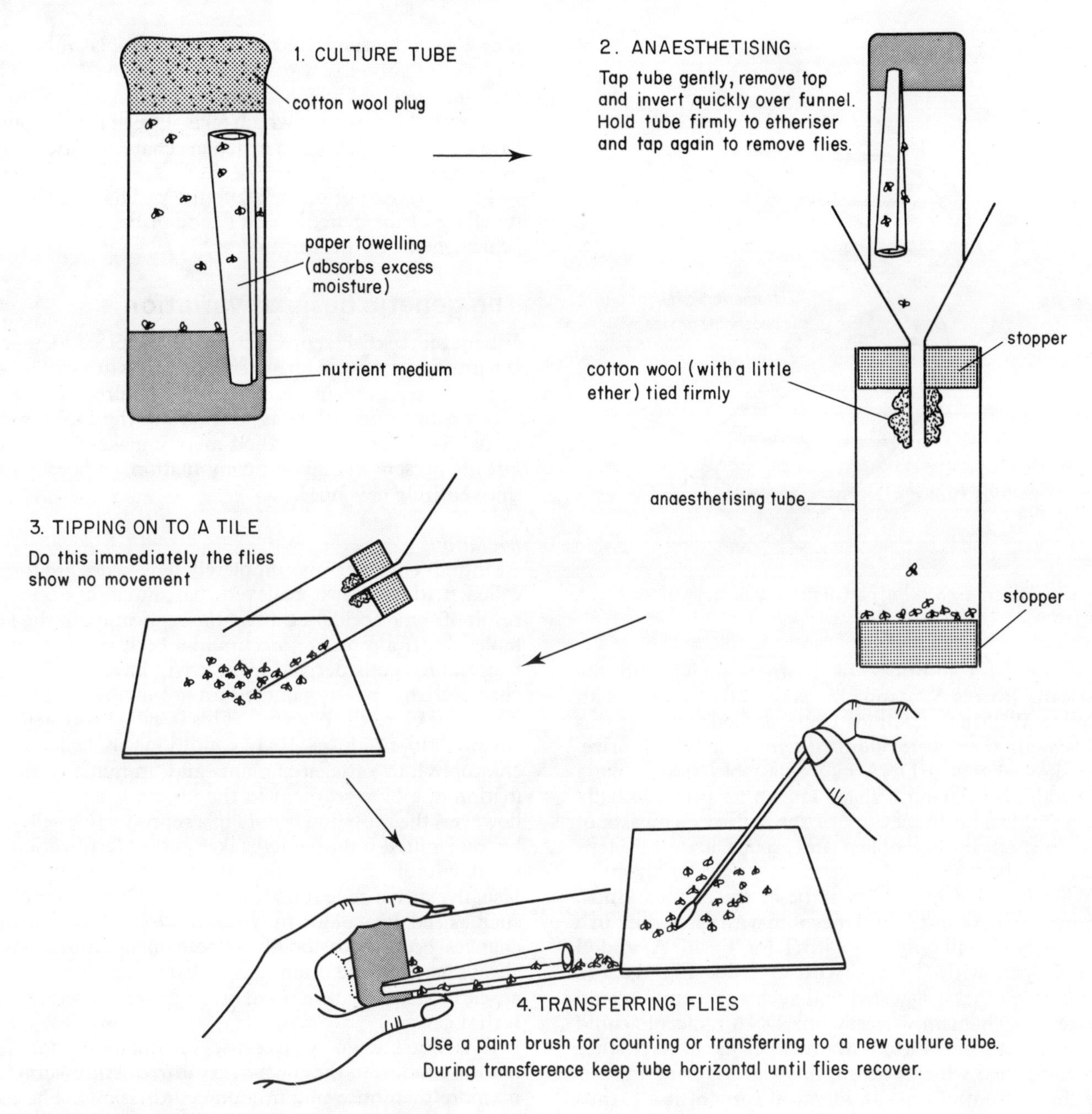

Fig. 34:12 The technique for handling *Drosophila*.

a) Breed the F_1 among themselves. Select 5 of each sex, transfer to a new culture tube and label. It is not necessary to select virgin females because the F_1 flies are all of the same genotype.
b) Make a back cross with the recessive parent. Select 5 virgin females of the F_1 and 5 males from the original vestigial-winged parent stock, put them together in a second culture tube and label.

The flies will hatch out over a number of days so it is best to sort them out in batches. Etherise each colony in turn, tip the flies on to a tile and, using a paint brush, separate them into normal and vestigial-winged types and count the number in each category. Unwanted flies can then be disposed of by putting them in 70% ethanol. Calculate the ratios, both for your own experiment and for those of the class as a whole.

You should find a considerable difference in the ratios obtained in a) and b) above. Pure-breeding wild-type parents with normal wings are usually given the symbol + +, whereas the symbol for the pure-breeding vestigial-winged parents is vg vg. The two crosses you performed were therefore: a) + vg × + vg and b) + vg × vg vg. You should now be able to work out what progeny would result in each case and in what ratio. How do the theoretical ratios compare with your own results?

Genes and the genetic code

It has been estimated that the number of genes carried on the four pairs of chromosomes in *Drosophila* may be as many as fifteen thousand. These genes contain all the information required for a fertilized egg to develop into a mature adult fly. This information is now known to be stored in chemical form in a complex molecule found in chromosomes called DNA (deoxyribose nucleic acid).

In 1953, the detailed structure of DNA was finally

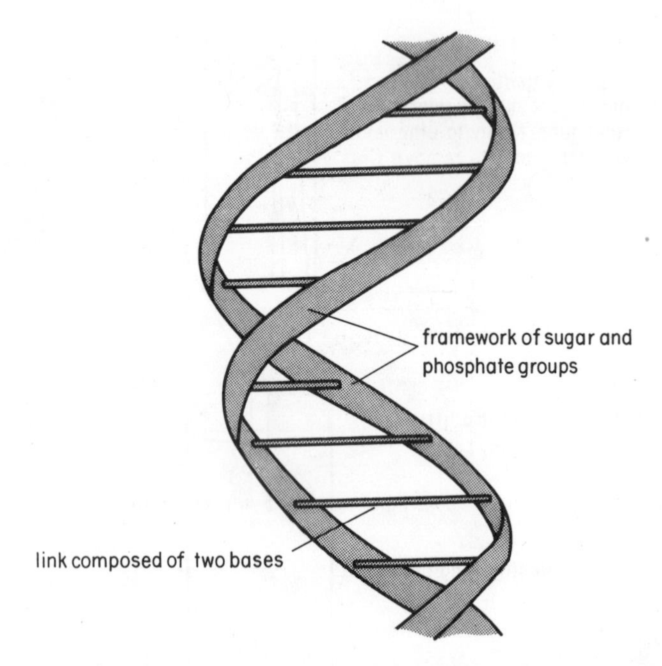

Fig. 34:13 Diagram showing part of the double helix of the DNA molecule.

worked out at Cambridge by **Francis Crick** and the American, **James Watson**, in close collaboration with **Maurice Wilkins** of King's College, London. For this achievement they were jointly awarded a Nobel prize. They showed that a DNA molecule looks rather like a spiral staircase, having a shape known as a double helix (Fig. 34:13). The framework of the staircase consists of alternating sugar and phosphate groups and the steps which join the framework together are pairs of chemical compounds called bases. These bases are of four kinds, known as A, T, C and G and they join with each other in a specific way. A will only pair with T (or T with A) and G will only pair with C (or C with G). (See Fig. 34:14). Watson and Crick suggested that such a structure would provide a mechanism whereby one DNA molecule could make an exact copy of itself. Many experiments have since been performed which confirm this view. During mitosis each chromosome forms an identical copy of itself, thus every gene is duplicated exactly. It is thought that the DNA in each chromosome is 'unzipped' down the middle by an enzyme into two complementary strands. Each strand then builds up a new complementary strand by attracting free base-sugar-phosphate groups from within the nucleus. Fig. 34:14 explains how two identical molecules of DNA can be formed.

Genes are thought to be parts of DNA molecules. Each consists of a unique sequence of bases. The chemical information is coded in the form of triplets of bases running down one side of the 'steps' within the molecule. When a gene is activated, it is thought that the code is 'read off', three bases at a time, after the DNA has been temporarily 'unzipped'. The sequence of base triplets is copied by means of a carrier molecule, RNA (ribonucleic acid). This molecule then passes out of the nucleus into the cytoplasm where the coded instructions in the form of base triplets are translated into a specific sequence of amino acids which are built up into a protein.

The genetic code enables chemical information within the genes to be translated so that the cytoplasm builds up the correct proteins within the cells of the body. These proteins, many of which are enzymes, are responsible for the metabolic processes which take place in the cells and so help to determine the particular characteristics of the individual.

The working out of the structure of DNA and the unravelling of the genetic code is one of the great scientific achievements of this century.

The genetic basis of variation

Whenever sexual reproduction takes place new combinations of genes are formed which result in a vast number of small variations in the progeny. The process has been picturesquely described as 'reshuffling the DNA pack of cards'. However, this method merely makes use of genes already present in different combinations, it does not produce entirely new ones.

Mutations

At infrequent intervals completely new genes are formed, called **mutants**. It is believed that mutation occurs as a result of some accident during the replication of the DNA molecule. If a mutation occurs in a body cell during mitosis, all the cells derived from it will have the same new characteristic, but as gametes are not involved, it will not be passed on to the progeny. This is called **somatic mutation** and produces such conditions as cells without chlorophyll in variegated plants and, in humans, the formation of a blue segment in the brown iris of an eye. If, however, the mutation occurs in a reproductive cell, it will be transmitted if the gamete takes part in fertilization, and the new individual will pass it on to future generations. Usually, such a gene mutation causes only a minor change such as red eye colour to white in *Drosophila*, but major changes occur occasionally. These major mutations are always harmful and often cause the death of the progeny at an early stage of development; they are therefore called **lethal** genes.

It is not known what causes a mutation, but under natural conditions they occur very infrequently, sometimes no more than once in a million cell divisions. The likelihood of mutation is increased by some environmental factors such as temperature and also by various forms of radiation. It is possible to produce mutations artificially by subjecting the reproductive cells of an organism to X-rays. The first person to do this was **Hermann Muller**, an American geneticist, who in 1927 experimented on *Drosophila* and produced all sorts of mutants including some with various eye colours and others with abnormally shaped wings and appendages. It is because X-rays may affect genes that it is wise not to subject the gonads to X-ray treatment. For the human race the greatest potential mutation hazard comes from atomic radiation. The higher the level of radiation in the environment the more likely it is for mutations to arise. Nuclear warfare is the ultimate hazard in this respect.

The majority of natural mutations cause very small changes; some may be beneficial, others harmful; the majority are recessive. This means that they may not become apparent for a few generations. Only when the gene has spread within the population is it likely to com-

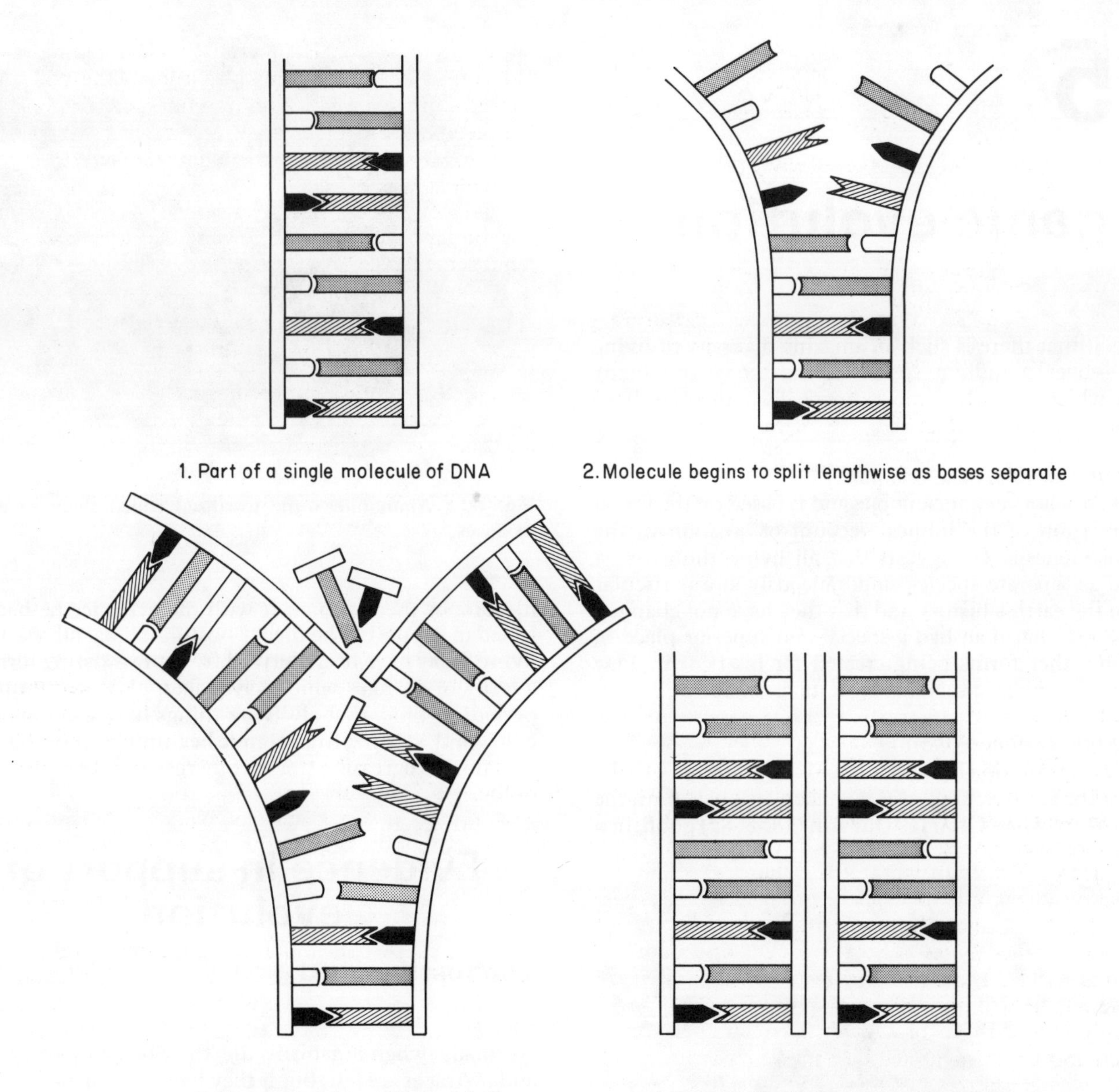

Fig. 34:14 Diagram illustrating the replication of the DNA molecule.

bine at fertilization to form a double recessive and so become visible.

Notable exceptions are mutations which affect genes carried by the X chromosome. Haemophilia, for example, is a condition where blood fails to coagulate properly and the slightest injury causes persistent haemorrhage. The mutant gene is recessive but males (XY) who carry the gene are haemophiliacs; there is no corresponding normal allele on the Y chromosome. On the other hand, females (XX) can carry the mutant gene on one X chromosome and be perfectly normal. However, half her eggs will carry the mutation. If she marries a normal man, half her sons would theoretically be haemophiliacs and half her daughters would be carriers of the mutation. A far less serious condition, red-green colour blindness, is inherited in a similar way.

Chromosome re-arrangement and multiplication

Mutations do not always result from changes in a gene. Sometimes they occur when whole chromosomes behave abnormally. Two examples will be given:

1. Occasionally during meiosis, when gametes are formed, a complete chromosome pair passes into one gamete and none into the other. If the former then fuses with a normal gamete there will be three chromosomes of one kind instead of two. This sometimes happens to chromosome 21 in man, producing in consequence a mentally-handicapped child known as a **mongol**. This abnormality becomes more probable with the increasing age of the mother; thus the risk of having a mongol baby is 1 in 2000 at the age of 20, but increases to 1 in 50 at 40.
2. Polyploidy. This is a phenomenon found in plants where the number of chromosomes is a multiple of the number normally found in the gametes, i.e. 3n, 4n, 6n, etc. Many garden varieties of flowers are polyploids and so are many modern kinds of fruits and cereals. Polyploids are often associated with an increase in size or number of parts such as petals.

35

Organic evolution

Why is it that there is such an amazing diversity of living things—over 2 million species living today and many others which have become extinct? Two theories have attempted to explain this:

1. The theory of special creation
This theory is a very ancient one and is based on the literal interpretation of the biblical account of creation in the book of Genesis. It suggests that all living things were created as separate species simultaneously at a particular time in the earth's history and that they have not changed since. Also, that man had a special and superior place in creation, other forms being created for his benefit. Few people today believe this theory in its literal form.

2. The theory of evolution
The idea of evolution is also a very old one going back to the times of the Greeks, but it was not taken seriously until the 19th century when **Charles Darwin** (1802–82) published

Fig. 35:1 Charles Darwin (1809–82).

Fig. 35:2 Ammonites—the fossilised coiled shells of extinct molluscs.

his famous theory about it with the evidence he had collected in support. The theory postulates that all organisms living today have been derived from pre-existing forms by a series of changes which have taken place over immense periods of time; that all living things have a common ancestry and, starting from simple beginnings, there has been a gradual progression towards increasingly complex forms of life.

Evidence in support of evolution

The fossil record

Fossils
Normally, when organisms die, their bodies quickly decay and no traces are left, but if they have hard structures such as shells, bones or scales they may last longer. So, for fossilisation to take place, an organism must be protected quickly from decaying action and from the effect of weathering, and this is a comparatively rare event. Usually it occurs when organisms become quickly buried in mud or sand or, less often, in volcanic laval flows or the resin which exudes from trees. Insect fossils, for example, are commonly found in amber, which is fossilised resin. Some of the most abundant fossils are the Foraminifera—those tiny Protozoa whose calcareous shells settled at the bottom of shallow seas to form an ooze which later consolidated into the chalk deposits we know today. Of the larger species, the calcareous shells of molluscs are often well preserved and may be abundant, for example, in certain limestones.

The time scale
It is estimated that the earth may be about $4\frac{1}{2}$ thousand million years old and that conditions were such that life could have come into being between 3 and $3\frac{1}{2}$ thousand million years ago. It probably took 1 or 2 thousand million years before living things became multi-cellular. The oldest fossils known are those of bacteria and blue-green

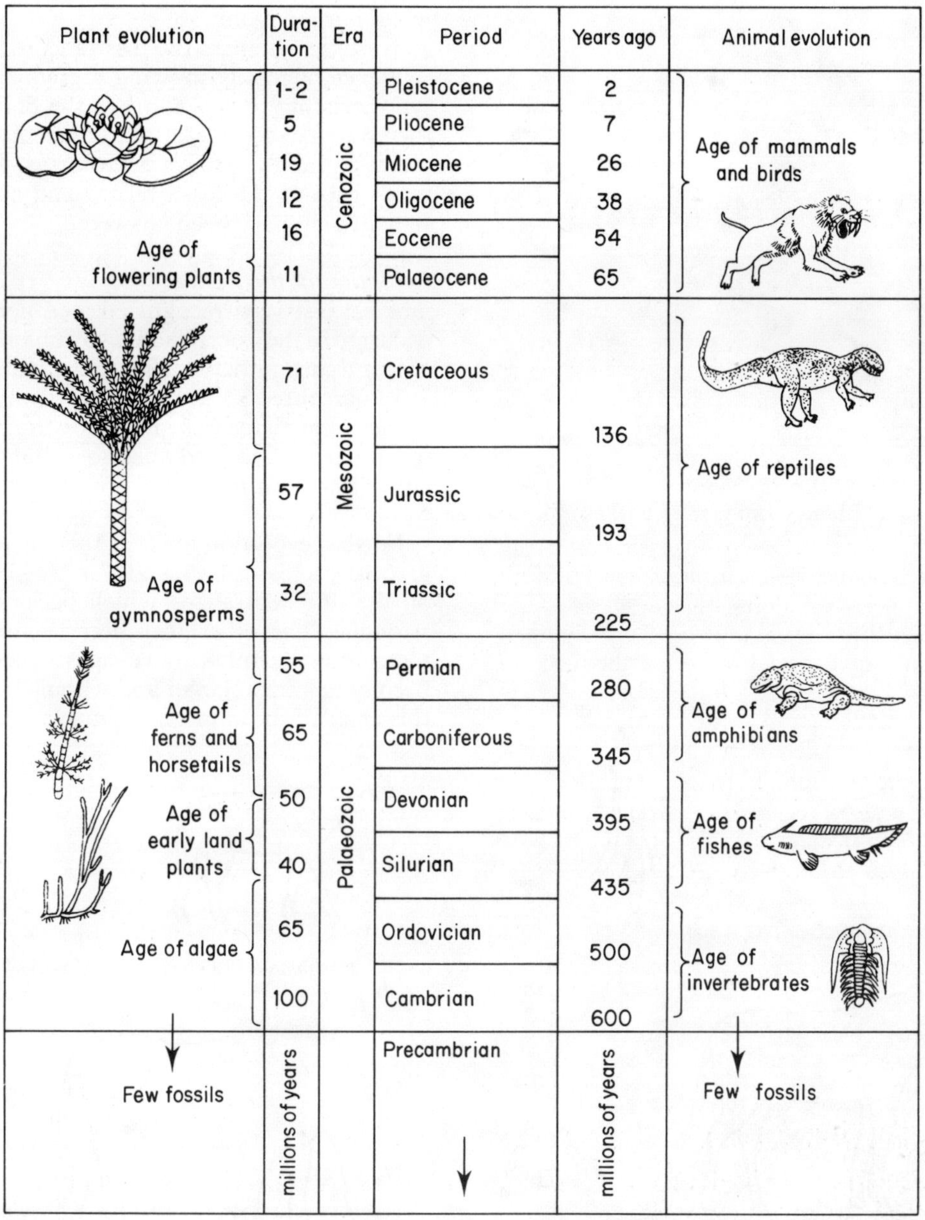

Plant evolution	Duration	Era	Period	Years ago	Animal evolution
Age of flowering plants	1-2	Cenozoic	Pleistocene	2	Age of mammals and birds
	5		Pliocene	7	
	19		Miocene	26	
	12		Oligocene	38	
	16		Eocene	54	
	11		Palaeocene	65	
	71	Mesozoic	Cretaceous	136	Age of reptiles
Age of gymnosperms	57		Jurassic	193	
	32		Triassic	225	
Age of ferns and horsetails	55	Palaeozoic	Permian	280	Age of amphibians
	65		Carboniferous	345	
Age of early land plants	50		Devonian	395	Age of fishes
	40		Silurian	435	
Age of algae	65		Ordovician	500	Age of invertebrates
	100		Cambrian	600	
Few fossils	millions of years		Precambrian	millions of years	Few fossils

Fig. 35:3 The fossil record.

algae which are estimated to be over 1600 million years old. They have been found beautifully preserved in silica.

The thrilling story of the progression of life throughout the ages is written for us in the earth's rocks. The older parts of the story are very fragmented and large parts are missing, but nevertheless in the sedimentary rocks of the past 600 million years the fossils tell a remarkable story (Fig. 35:3). These rocks were laid down as sediments—deposits of materials formed as a result of erosion by water, wind or ice. As other layers formed above them they became more and more compressed to form the sandstones, limestones and shales that we know today.

What does the fossil record tell us?

1. It shows that the diversity of life has come about very gradually over immense periods of time: that it was not created all at once.

2. By dating the rock strata (modern methods are very accurate) and considering their natural sequence in the earth's crust, it has been shown that in passing from the older rocks to the more recent, the fossils gradually increase in complexity. Complex forms are never found in the oldest rocks. It is also the case that all the earliest fossils are of aquatic species. Terrestrial types such as amphibians, insects and primitive land plants did not appear until about 400 million years ago.

3. Some fossils show characters intermediate between one group and another. For example, the earliest fossil evidence of a true bird was that of *Archaeopteryx* in the Jurassic period; it showed remarkable characters intermediate between those of a lizard and a bird. It had a bird-like beak, but rows of teeth were present as in reptiles. It had a long lizard-like tail, but with a double row of feathers attached to it. It also had feathers on its wings, but three-clawed fingers were present which it probably used for clambering. This fossil alone provides powerful evidence

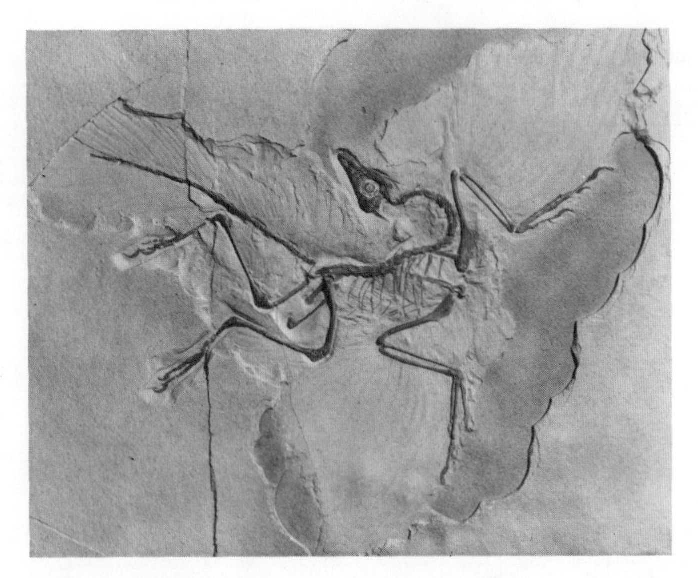

Fig. 35:4 Fossil of Archaeopteryx—the earliest known bird.

in support of evolution.

4. In some instances the fossil record shows a remarkable sequence of fossils showing a gradual evolution towards types existing today. Some of the best examples concern the ancestry of some of the larger mammals such as elephants, giraffes, camels and horses. However, there are many gaps in the fossil record and more question marks remain than have been answered.

Evolution of the horse

The earliest type recognised as a horse, *Hyracotherium*, occurred in the Eocene some 50 million years ago. It was no bigger than a rather small dog but had relatively long legs. It ran on three toes, each encased in a small hoof, but a reduced fourth toe was present on the fore limbs. No doubt it arose from five-toed ancestors, but fossils of these have either not been found, or not been recognised as related forms.

Further evolution towards the horses of today showed a gradual increase in size and a growing tendency to run on the one central toe which became much longer and stronger. The other toes became further reduced until today they are only tiny vestiges. The teeth also changed, becoming longer, larger and with flatter grinding surfaces; this enabled the animals to graze more efficiently.

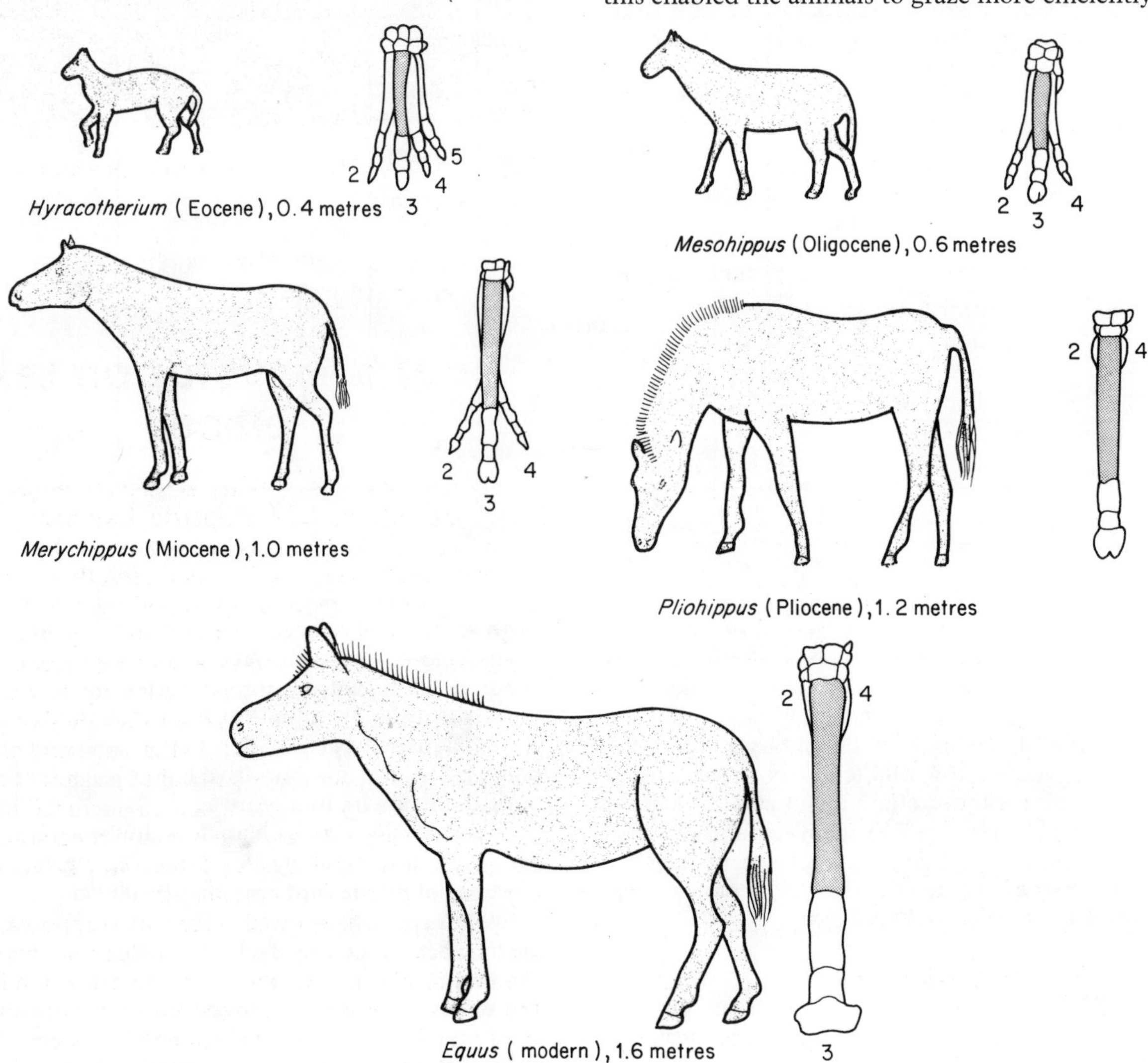

Fig. 35:5 Diagrams illustrating the evolution of horses (only 5 species shown).

Horses were very successful and the fossil record shows that many types evolved (Fig. 35:5). However, after a time, many of these became extinct. Today we only have three remaining species of the genus *Equus*—the true horses, the zebras and the asses.

Comparative anatomy

The theory of evolution explains a mass of facts which are difficult to explain in any other way. Some of these concern the structure of organisms. Consider these questions and whether evolution explains them:

1. Why is it that plants and animals can be arranged in a progression from simple forms to complex?
2. Why should a phylum or class be built on a common plan of structure, the same structures being used for very different purposes? Take, as examples, the forelimbs of vertebrates used for running, climbing, swimming and flying (p. 78) or the mouthparts of insects (p. 40).
3. Why are **vestiges** present in animals? Vestiges are structures which are useless to the possessor, but correspond to useful organs in other animals. They include structures such as the remains of hind limbs in whales and pythons, the remnants of wings in some moths and the splint bones in the limbs of horses (Fig. 35:5 and 35:6).
4. Why is it that some animals show intermediate characters between different classes of animals? Take, for example, the lung fish which has characteristics of both fish and amphibia, having both gills and lungs; and the platypus which shows some characters intermediate to those of reptiles and mammals, that is, laying eggs but having hair and suckling its young.

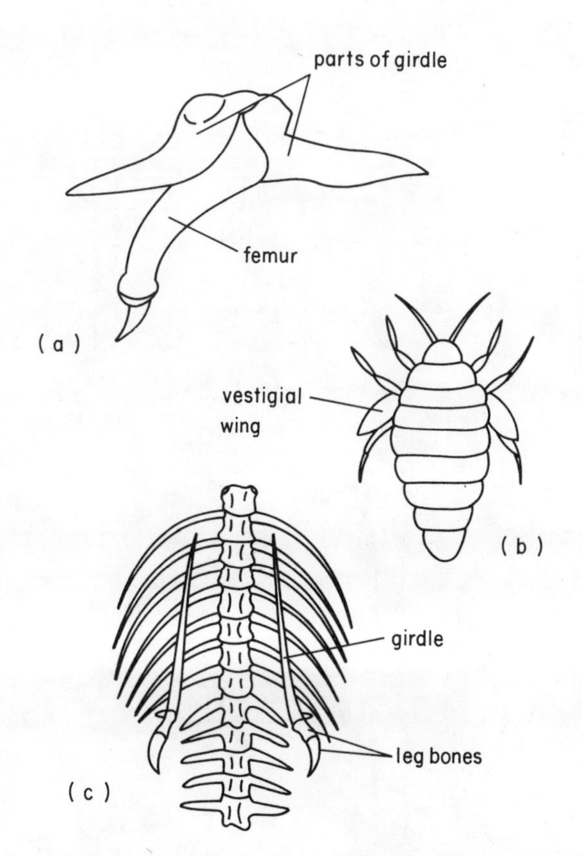

Fig. 35:6 Some examples of vestiges: a) Hind limb of a whale (buried in blubber under the skin). b) Reduced wings in a female vapourer moth. c) Hind limb of a python which projects as a claw.

Fig. 35:7 African lungfish.

Fig. 35:8 Duck-billed platypus: a mammal which lays eggs.

How has evolution taken place?

The mechanism of evolution has been the subject of much study and debate. **Jean Baptiste Lamarck**, a French scientist, published his theory in 1801. He believed in the **inheritance of acquired characters**; that is that physical characteristics acquired by organisms during their lifetime could be passed on to their progeny. For example, ancestral giraffes with short necks would acquire slightly longer ones by constantly stretching up to reach leaves almost out of reach. This characteristic would be passed on to the progeny which would do the same and after many generations produce the long-necked giraffes of today. By contrast, Lamarck thought that if an organ fell into disuse it would degenerate and over many generations would become reduced and useless. In this way he attempted to explain the presence of vestigial organs.

There is no experimental evidence to support Lamarck's theory as he envisaged it so it has fallen into disrepute.

It is **Charles Darwin** who has gone down in history as the one who made the biggest contribution towards an understanding of the evolutionary process. Many of Darwin's ideas about evolution stemmed from his five-year voyage (1832–37) round the world in the *Beagle*, a British naval ship carrying out survey work. Darwin joined the ship as naturalist to the expedition. Darwin had an enquiring mind and displayed enormous powers of critical observation; he also paid meticulous attention to detail. On his return to England he patiently and methodically assembled his evidence and continued his researches. By an extraordinary coincidence, another naturalist and explorer, **Alfred Russell Wallace** (1823–1913), when working in South-East Asia had come to similar conclusions to those of Darwin although neither was aware of the other's work. Wallace actually sent Darwin an account of his theory, asking him for his opinion about it and whether it should be published! Eventually it was decided that they should put forward their theory jointly at a meeting of the Linnaean Society in London. Darwin's book, *The Origin of Species by Means of Natural Selection* was published the year after, in 1859. This book not only provided a considerable body of evidence in support of evolution but it also went a long way to explain the means by which it could have been accomplished.

The theory of natural selection

When Darwin propounded his theory he had no knowledge of chromosomes, genes or mutations and therefore had no idea how variations were inherited; all this came later, but otherwise the modern theory is largely based on his conclusions. It may be summarised as follows:

1. In every species the individuals vary greatly among themselves; some of these variations are acquired during the life of the individual as a result of environmental factors—these are *not* inherited and hence are of *no* significance in evolution. However, many changes involving genes and chromosomes *are* inherited and these *can* lead to evolution.
2. In most species the number of progeny is far greater than is necessary to replace the parents when they die. Many in fact are so prolific that if there were no check to their numbers they would soon be unable to acquire the necessities for life. In consequence there must be competition among members of the same species for all these necessities. This is what Darwin called **'the struggle for existence'**.
3. As a result of this competition, those individuals with variations which gave some advantage in this struggle for survival would be more likely to survive and breed. Darwin described this as **'the survival of the fittest'**, and he reasoned that it was nature (meaning all the environmental factors, physical and biotic) which did the selecting of the fittest individuals; so he called the process **'natural selection'**.

The term 'fittest' does not merely mean the most healthy, but describes those with the best adaptations to survive in a particular environment *and which produce the most offspring*. It is only through successful reproduction that genes giving rise to useful variations can be passed to the next generation.

4. As a consequence of natural selection occurring generation after generation over immense periods of time, populations would gradually change and new species evolve. It

Fig. 35:9 The peppered moth (*Biston betularia*) and its black mutant: (left) both kinds at rest on a soot-covered tree trunk; (right) both kinds at rest on a lichened tree trunk in unpolluted countryside.

also follows that when climatic and other dramatic changes take place, new selective factors act upon populations causing some to die out, such as the large dinosaurs, and others to evolve further, such as the mammals which took their place. Evolution is a dynamic process which is still going on today.

Evidence in support of the theory of natural selection

An example of natural selection today

One of the best documented examples of natural selection concerns the appearance as a mutation of a black variety of the peppered moth, *Biston betularia*, in northern England towards the middle of the last century. The normal form of this moth is speckled and light in colour. The mutant resulted from a change in a single gene and rather exceptionally it happened to be a dominant. The mutant type quickly spread, and in less than a century had almost completely replaced the normal type in many parts of Britain. In the 1950s a countrywide project was mounted to work out the distribution of the two forms and it was found that the black mutant was most common in industrial regions where, because of air pollution, the bark of trees and the sides of buildings on which the moths normally rested during the day were dark with soot deposits. The light variety was more common in unpolluted areas. It seemed a likely theory that birds were selectively eliminating the light ones in the polluted areas, because they were more conspicuous against the dark background (Fig. 35:9). To test this idea, large numbers of both varieties were bred and released in both polluted and unpolluted areas and after a short period, as many as possible of those which survived the predation of birds were attracted to mercury vapour moth traps and counted. Results showed that in the polluted area a higher percentage of the black form was recaptured compared with the normal form, and the opposite occurred in the unpolluted area.

A film was also taken which showed the birds actually searching the tree trunks; this confirmed that they selected in each case more of the variety which was most easily seen.

An interesting sequel to the story is that, since the Clean Air Act of 1956, pollution has gradually become reduced and now the light varieties are once more becoming common near industrial cities.

This is an example of a mutation which provided an advantageous characteristic in a polluted area, its survival depending on the selective effect of the environment, in this case birds. But it is important to realise that the same mutation when it occurred in an unpolluted area, was disadvantageous. So the principle is that natural selection of the fittest types concerns only those individuals which survive at a definite time in a particular place. It follows therefore that when environmental conditions remain fairly constant over long periods, the organisms living there will gradually become more perfectly adapted to those conditions due to natural selection. So if mutations occur under these circumstances, the likelihood of their being advantageous is not great as the organisms are already so well adapted. However, such a mutation, if recessive, will nevertheless become part of what is called the **gene pool**, i.e. the total number of genes present in a given population. It may therefore appear from time to time as a double recessive and if by that time environmental conditions have changed, the mutation may possibly be advantageous under these new conditions and consequently it may spread through the population.

To summarise, we can say that mutation and gene shuffling during sexual reproduction produce the variations, Mendelian inheritance governs their transfer from one generation to the next, and the environment brings about the selection of individuals which, on balance, have the most advantageous combination of genes. So although mutations occur in a random manner, selection ensures that evolution is directional.

Artificial selection

Darwin pointed out that just as 'nature' could select useful variations and so change the character of a population, man had been doing it artificially for centuries. By selecting the animals and plants which had the characteristics he desired and breeding from them he has been able to produce a great variety of domestic animals and crops. Think of the many kinds of dogs, horses, cattle, cereals and fruits, such as apples, which exist today. During this century, with his increasing knowledge of genetics, man has artificially accelerated the evolutionary process (p. 119), although he has not produced a completely new species. This evidence should be considered as confirmatory rather than as proof that selection occurs in nature.

How do new species evolve?

We saw with the peppered moth how a mutant gene could bring advantage to a species and cause a change in the population, but this is a long way from becoming a new species. For this to happen some kind of **isolation** must take place. This is the separation of one population of a species from another, so that inter-breeding is avoided or becomes less likely. There are many geographical barriers both large and small which can separate populations, such as land barriers which divide aquatic populations from each other and water barriers, mountain ranges or deserts which divide terrestrial populations.

Let us assume that two populations have been isolated by some geographical barrier. We can postulate that mutations will occur in each, and natural selection will take place according to the environmental conditions prevailing in each region. As these conditions will be different in the two regions, the differences between the two populations will gradually become greater under the influence of natural selection until they are sufficiently different to prevent inter-breeding and the production of fertile off-spring. In this way a new species is evolved. That is the theory; but what evidence is there in its favour?

Islands

These are excellent examples of isolation, so it is logical to assume that if they have been islands for a very long time they will probably contain species found nowhere else. This is exactly what is found! Perhaps the most famous examples are the Galapagos islands, 1000 km off the west coast of Ecuador in the Pacific, and the Seychelles and Aldabra 1400 km from the African mainland in the Indian Ocean. All these islands have many species found nowhere else in the world.

Fig. 35:10 Giant tortoise from the island of Aldabra.

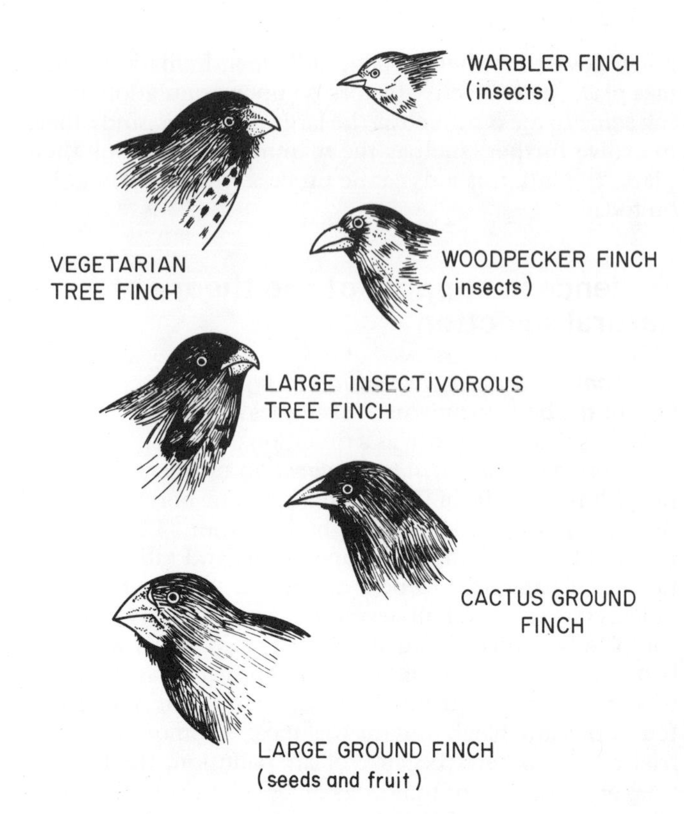

Fig. 35:11 Six of the 13 species of finch-like birds found on the Galapagos Islands. They are believed to have evolved from a single species as a result of isolation. The differences in beak size and shape are adaptations related to variation in their diets.

The Galapagos islands probably arose as a result of volcanic action and gradually became populated by plants and animals which reached them by flying, swimming and being carried by ocean currents or by the wind. In this way they became isolated from the populations on the mainland. Today many of the animals that live there still resemble the mainland forms in many ways, but they differ markedly in others. Thus there are flightless cormorants which resemble the Chilean species except for the reduction of the wings. Others, such as the giant tortoises and the iguanas, resemble species which have long since become extinct elswhere, but have survived on the islands because of lack of competition or predation.

Additional evidence for the evolution of new species occurs within the islands themselves. For example, the giant tortoises occur on several of these islands and although the populations are not separated from each other by great distances, they nevertheless show distinct differences between them. But the most spectacular example is that of the finch-like birds. There are 13 species of these which differ markedly from each other, although they have almost certainly evolved from a single species. They differ particularly in the shape of the beak, which is adapted in different species, either for feeding on various kinds of seeds, or on cacti or on insects (Fig. 35:11), but they also vary in size and behaviour. Some species are found singly on the more remote islands while others occur together on the same island, but occupy different niches in the island ecosystem. So islands provide remarkable evidence that evolution of new species has occurred.

So it is generally believed that natural selection is the best theory to explain how evolution has taken place. But like any other scientific theory it may be modified in the light of fresh evidence.

The evolution of Man

As a result of the most recent fossil finds near Lake Turkana in East Africa, it appears that Man as an upright species with a large brain probably existed at least 3 million years ago. It is likely, therefore, that his evolution from ape-like ancestors occurred several million years before.

Man owes his dominant position in the world today primarily to his large brain and his ability to hold objects and use them as tools. His opposable thumb, flexible fingers and sensitivity of touch have played a large part in making this possible. He has also evolved a long life-span and a slow rate of development which have enabled him to learn extensively and gain experience. Also, in his evolution towards a more social life he has developed aspects of parental care and co-operation far greater than in any other species. All these factors have been important in his rise to dominance.

Our most recent evolution has been cultural rather than structural. We not only inherit characteristics which are passed on by genes, but because we have learnt to speak and record, we are born into a society with a vast accumulation of knowledge and experience which we can utilise if we have the wisdom to do so. With the help of science and technology Man has built up a considerable degree of dominance over nature; he has conquered many diseases and reached a population density previously unheard of in any species of comparable size which is at the top of a food web. What then of the future?

The future lies with us, but the chief stumbling block towards our further evolution is ourselves. We have still not learnt the art of living together in harmony. Our barbarism is still uncomfortably near the surface of our natures. Our aggression which has served us so well in the past is still with us and is now a potential danger to our very existence, as its ultimate expression is war. Man has now reached a stage when he can destroy himself, and his future, more than ever before, depends upon whether he can evolve into a new unselfish type of man capable of tolerance, sympathy and generosity between individuals, factions and nations. This is the challenge for all of us.

Acknowledgements

For permission to reproduce photographic illustrations, acknowledgement is due as follows:
Bryan and Cherry Alexander: 31:3; Vivian Almy: 26:4; C. Ashall/Centre for Overseas Pest Research: 6:9, 6:10; M. Allman: 35:2; Australian Information Service, London: 35:8; Stewart Bale Ltd: 19:21; J. Beach/Wildlife Picture Agency: 31:8; Dr Alan Beaumont: 16:9; Blandford Press Ltd, reproduced from *Mammals of Britain: Their Tracks, Trails and Signs* by M. Lawrence and R. Brown, photograph by C. Atherton: 11:7; Jon Blau/Camera Press, London: 1:1; Tony Boxall/Barnaby's Picture Library: 19:9; British Museum (Natural History): 35:4; British Trust for Conservation Volunteers: 33:11; C. G. Butler: 7:17, 24:14; Canadian High Commission: 33:3; Eva Crawley: 11:6; D. N. Dalton/Natural History Photographic Agency: 30:11; S. Dalton/Natural History Photographic Agency: 10:12; John East: 29:9; *Farmer's Weekly*: 7:19, 11:15; Douglas Fisher: 30:1; Robin Fletcher: 6:19, 9:6, 9:8; Dr T. H. Flewett: 19:18; Food and Agriculture Organisation: 16:10; Forestry Commission: 21:10b; Dr. K. A. Harrow: 19:19; Philip Harris Biological Ltd: 1:3, 2:2a, 2:2b, 2:2c, 2:3, 2:6a, 2:6b, 2:9, 2:13, 2:17, 4:7, 4:19, 4:20, 4:22, 4:31, 5:2, 5:5, 5:6, 5:8, 5:9, 5:14, 10:8, 13:6, 15:8, 15:9, 15:11, 17:4, 17:9, 17:10, 17:12, 18:8, 18:9, 19:2, 19:6, 19:16, 20:1, 20:12, 23:5, 23:8, 24:15, 24:19, 25:2a, 25:2b, 25:2c, 25:2d, 25:2e, 25:9, 25:10, 25:11, 26:6, 27:1, 27:6, 27:14, 29:5, 34:9; Eric Hosking: 10:1; Imperial Chemical Industries Agricultural Division: 7:9, 33:5; Geoffrey Kinns: 16:11; M. King & M. Read: 9:9; Longman, Green & Co., reproduced from *An Atlas of Skiagrams* by Johnson and Symington: 17:7; Mansell Collection: 19:3, 19:10, 19:12, 19:14, 19:15, 35:1; M. Mockler: 10:14; H. Moussali: 19:13; A. Neal: 16:8, 33:6; E. G. Neal: 1:2, 4:25, 4:29, 4:30, 6:11, 7:18, 11:4, 19:7, 24:16, 24:17, 30:5, 30:6, 30:7, 30:8, 30:9, 30:10, 30:12, 30:13, 30:14, 30:15, 30:16, 31:4, 31:5, 31:6, 31:7, 31:14, 32:9, 33:4, 33:7, 35:10, H. E. Neal: 17:3, 35:7; K. R. C. Neal: 3:4, 3:5, 11:1, 13:8; H. C. F. Needham: 21:10a; P. O'Hanlon: 33:2; A. E. Mc. R. Pearce: 6:13, 6:15a, 6:15b; K. G. Preston-Mafham/Premaphotos Wildlife: 6:18; R. A. Preston-Mafham/Premaphotos Wildlife: 6:12; M. W. Richards/Royal Society for the Protection of Birds: 33:9; St Mary's Hospital Medical School: 4:24, 27:7, 27:8, 29:8; Shell Photographic Centre: 7:6; Sporting Pictures (UK) Ltd: 13:7; John Topham Picture Library: 33:8; M. Tweedie/Natural History Photographic Agency: 35:9a, 35:9b; C. James Webb: 7:8, 16:5, 27:9

Also to the following sources for the information listed: Edward Arnold Ltd: 30:4 from *The Study of Behaviour* by J. D. Carthy, with the kind permission of Professor N. Tinbergen; André Deutsch Ltd: quotation (p. 265) from *Only One Earth* by B. Ward and R. Dubos; Food and Agriculture Organization: 16:4 from data supplied; Heinemann Ltd: 27:16 based on *Biology by Enquiry* by Clarke et al.; Holt, Rinehart and Winston Inc.: 31:2 and 35:6c based on *Modern Biology* by Moon, Otto and Towle; Longman, Green and Co. Ltd: 27:15 and 27:17 based on *Man with Two Environments* by M. Rutherford; Longman Penguin: Tables (p. 97) base on data from *Nuffield Biology Text III*, Table 15:1 based on data from *Nuffield Biology Text IV* and *Nuffield Biology Teachers Guide IV*, 34:11b based on *Nuffield Biology Teachers Guide V* after B. J. Haller; John Murray (Publishers) Ltd: 7:13 based on *Introduction to Biology* by McKean, 28:2 based on *Basic Anatomy and Physiology* by H. G. Q. Rowett; Nelson, Thomas and Sons Ltd: 35:11 based on *An Atlas of Evolution* by De Beer; Open University Press: 31:1 from *Species and Populations*; Oxfam: 16:6 and 16:7 based on data supplied; Oxford University Press: 20:8 based on *Biology* by B. S. Beckett; Pitman Medical Publishing Co. Ltd: 14:12 from *Smoking and Health Now* issued by the Royal College of Physicians of London; Rothamsted Experimental Station: 15:15 based on data from 1968 Report, Part 2.

Index

Figures in **bold** type show that the subject is illustrated, but in addition there may be a further reference in the text on the same page.
An explanation of *biological terms* can, in most cases, be found on the page referred to first.